Hybrid Imaging in Cardiovascular Medicine

Hybrid Imaging in Cardiovascular Medicine

Edited by

Yi-Hwa Liu, PhD
Albert J. Sinusas, MD, FACC, FAHA
Yale University School of Medicine

CRC Press
Taylor & Francis Group
Boca Raton London New York

CRC Press is an imprint of the
Taylor & Francis Group, an **informa** business

CRC Press
Taylor & Francis Group
6000 Broken Sound Parkway NW, Suite 300
Boca Raton, FL 33487-2742

First issued in paperback 2020

ISBN-13: 978-1-4665-9537-8 (hbk)
ISBN-13: 978-0-367-78174-3 (pbk)

This book contains information obtained from authentic and highly regarded sources. Reasonable efforts have been made to publish reliable data and information, but the author and publisher cannot assume responsibility for the validity of all materials or the consequences of their use. The authors and publishers have attempted to trace the copyright holders of all material reproduced in this publication and apologize to copyright holders if permission to publish in this form has not been obtained. If any copyright material has not been acknowledged please write and let us know so we may rectify in any future reprint.

Visit the Taylor & Francis Web site at
http://www.taylorandfrancis.com

and the CRC Press Web site at
http://www.crcpress.com

I dedicate this book to my wife, Yi-Chen, and my children, Patricia and Bryan, for their love and inspiration.

Yi-Hwa Liu

I dedicate this book to my wife, Michele, for her love and support and most importantly her patience and understanding of my work schedule. I would also like to acknowledge the support and guidance of my colleagues at Yale University and the many research and clinical fellows that I have had the pleasure of working with and mentoring over the years.

Albert Sinusas

Contents

Series preface

Advances in the science and technology of medical imaging and radiation therapy are more profound and rapid than ever before since their inception over a century ago. Further, the disciplines are increasingly cross-linked as imaging methods become more widely used to plan, guide, monitor, and assess treatments in radiation therapy. Today, the technologies of medical imaging and radiation therapy are so complex and so computer driven that it is difficult for the persons (physicians and technologists) responsible for their clinical use to know exactly what is happening at the point of care when a patient is being examined or treated. The persons best equipped to understand the technologies and their applications are medical physicists, and these individuals are assuming greater responsibilities in the clinical arena to ensure that what is intended for the patient is actually delivered in a safe and effective manner.

The growing responsibilities of medical physicists in the clinical arenas of medical imaging and radiation therapy are not without their challenges, however. Most medical physicists are knowledgeable in either radiation therapy or medical imaging and expert in one or a small number of areas within their discipline. They sustain their expertise in these areas by reading scientific articles and attending scientific talks at meetings. In contrast, their responsibilities increasingly extend beyond their specific areas of expertise. To meet these responsibilities, medical physicists periodically must refresh their knowledge of advances in medical imaging or radiation therapy, and they must be prepared to function at the intersection of these two fields. How to accomplish these objectives is a challenge.

At the 2007 annual meeting of the American Association of Physicists in Medicine in Minneapolis, this challenge was the topic of conversation during a lunch hosted by Taylor & Francis Group and involving a group of senior medical physicists (Arthur L. Boyer, Joseph O. Deasy, C.-M. Charlie Ma, Todd A. Pawlicki, Ervin B. Podgorsak, Elke Reitzel, Anthony B. Wolbarst, and Ellen D. Yorke). The conclusion of this discussion was that a book series should be launched under the Taylor & Francis banner, with each volume in the series addressing a rapidly advancing area of medical imaging or radiation therapy of importance to medical physicists. The aim would be for each volume to provide medical physicists with the information needed to understand technologies driving a rapid advance and their applications to safe and effective delivery of patient care.

Each volume in the series is edited by one or more individuals with recognized expertise in the technological area encompassed by the book. The editors are responsible for selecting the authors of individual chapters and ensuring that the chapters are comprehensive and intelligible to someone without such expertise. The enthusiasm of volume editors and chapter authors has been gratifying and reinforces the conclusion of the Minneapolis luncheon that this series of books addresses a major need of medical physicists.

Imaging in Medical Diagnosis and Therapy would not have been possible without the encouragement and support of the series manager, Lu Han of Taylor & Francis Group. The editors and authors, and most of all I, are indebted to his steady guidance of the entire project.

William Hendee
Founding Series Editor
Rochester, Minnesota

Preface

Hybrid cardiovascular imaging holds incredible promise for preclinical research and clinical practice, providing simultaneous acquisition and coregistration of anatomical, functional, and molecular data from a target of interest and achieving extraordinary comprehensive information about the targeted object. Over the past decade, the developments of hybrid imaging technology have drawn tremendous attention from the research and clinical communities, particularly in the area of molecularly targeted imaging. With recent advancements of imaging system design and computing power, multiple imaging systems with different functionalities can be integrated into one system to simultaneously acquire the composite information about the object, from the macro level of organs (e.g., heart) to microcellular details (e.g., myocytes). The innovation of high-sensitivity detectors and fast circuitry associated with improved iterative image reconstruction algorithms further enables the acquisition and reconstruction of high-quality images with reduced acquisition and processing time. These innovative hybrid imaging technologies and reconstruction algorithms have also propelled the field of quantitative analysis of molecularly targeted imaging to the next level, increasing the reliability and reproducibility of hybrid imaging data.

Although hybrid imaging techniques have been introduced and developed over several decades with application in both the clinical or research settings, to our knowledge, a textbook encompassing a wide spectrum of hybrid imaging systems and applications is not currently available. We hope that this book will provide not only comprehensive reviews on the principles and techniques of various hybrid imaging modalities but also up-to-date applications and clinical and preclinical cases illustrations with an emphasis on cardiovascular medicine. While this book, as reflected from its title, is mainly focused on the latest multimodality imaging technology and quantification for the detection of cardiovascular diseases, applications of the hybrid imaging instrumentation and technology described herein are not limited to cardiovascular medicine per se. More specifically, other clinical and preclinical studies of hybrid imaging are also covered by this book in which image illustrations and quantitative results of preclinical and clinical studies from *in vitro* or *in vivo* studies in experimental animal models or human subjects are presented. Due to the wide range of the contents and more general applicability, it is also our expectation that this book will be beneficial to basic research scientists and engineers, as well as a large audience of medical specialists in radiology, medicine, and surgery.

This book contains a total of 20 chapters, contributed by 50 distinguished authors who are renowned experts in their respective fields. The book is divided into four parts and organized as follows: There are nine chapters in Part I dedicated to the review of the principles, instrumentation, techniques and applications of hybrid imaging with specific case illustrations, including single-photon emission computed tomography (SPECT)-computed tomography (CT) (Chapter 1), positron emission tomography (PET)-CT (Chapter 2), SPECT-magnetic resonance imaging (MRI) (Chapter 3), PET-MRI (Chapter 4), CT-MRI (Chapter 5), x-ray-optical (Chapter 6), x-ray fluoroscopy-echocardiography (Chapter 7), photoacoustic imaging (Chapter 8), and intravascular imaging (Chapter 9). Part II includes two chapters focused on multimodality probes for hybrid imaging; preclinical evaluation of multimodality probes (Chapter 10) and multimodality probes for molecular imaging (Chapter 11). The methods and illustrations for quantitative image analyses are described and presented in Part III, which contains six chapters (Chapters 12 through 17) dedicated to numerous state-of-the-art quantitative analytic methods and computer algorithms for quantification of the images acquired using the hybrid imaging systems and probes described in Parts I and II of this book. Finally, the book is concluded with Part IV, on future challenges of hybrid imaging, which elaborates on potential challenges associated with hybrid imaging (Chapter 18) and some concerns of radiation safety (Chapter 19) and suggests future directions for the developments and applications of hybrid imaging techniques (Chapter 20).

Potential readership and usage of this book may include but is not limited to (1) medical physicists, chemists, molecular biologists, and other basic scientists, (2) medical students, interns, fellows, researchers, and clinical professionals whose primary interests and practices are in cardiovascular imaging, and (3) engineering/science graduate students focused on instrumentation development and studies of medical physics and/or imaging science. Additionally, this book can be used as a textbook for a graduate-level course, potentially entitled, "New Techniques and Applications for Advanced Hybrid Medical Imaging Systems and Quantitative Analyses." For a full-year course, an instructor can make good use of all the materials covered by this book and offer the entire course in two semesters, the first focused on Parts I and II and the second on Parts III and IV. However, as an alternative, the course can also be offered in a condensed manner in one semester with a specific focus on one or two of the major sections or selected chapters from the book. The use of this book in a graduate course would provide students a detailed and up-to-date review of multimodality medical imaging techniques and new quantitative analytic methods with abundant preclinical and clinical cases and illustrations having high relevance to both basic scientists and medical specialists in training.

Yi-Hwa Liu
Albert J. Sinusas

Acknowledgments

We heartily appreciate all the contributors for their great dedications and efforts to this book. We also thank the anonymous reviewers for their helpful and invaluable suggestions and comments for this book.

Editors

Yi-Hwa Liu, PhD, is a senior research scientist in cardiovascular medicine at Yale University School of Medicine, New Haven, Connecticut; an associate professor (adjunct) of Biomedical Imaging and Radiological Sciences at National Yang-Ming University, Taipei, Taiwan; and a professor (adjunct) of biomedical engineering at Chung Yuan Christian University, Taoyuan, Taiwan. He is an elected senior member of the Institute of Electrical and Electronic Engineers and a full member of Sigma Xi of The Scientific Research Society of North America. He has served for many a years on the editorial boards of the *World Journal of Cardiology*, *Journal of Clinical and Experimental Cardiology*, *American Journal of Nuclear Medicine and Molecular Imaging*, *Current Molecular Imaging Journal*, and *American Journal of Nuclear Medicine and Molecular Imaging*. He has also served as a National Member of the American Heart Association grants review committee since 2004 and as associate editor of *Medical Physics* since 2009. Dr. Liu earned his BS degree in biomedical engineering at Chung Yuan Christian University, Taoyuan, Taiwan; MS degree in electrical and computer engineering at University of Missouri, Columbia, Missouri; and PhD degree in electrical and computer engineering at Rensselaer Polytechnic Institute, Troy, New York. He completed post-doc trainings in electrophysiology and cardiovascular physiology at Georgetown University School of Medicine and in nuclear cardiology at Yale University School of Medicine. He joined the faculty at Yale University School of Medicine as assistant professor (1998–2004) and associate professor of medicine (2004–2014). His primary research involves noncoherent image restoration, nuclear cardiac image reconstruction, and quantification. He is one of the pioneers in the fields of fluorescence microscopic image restoration and nuclear cardiac image reconstruction and quantification. He is the author of over 50 peered review publications, the leading editor of a book entitled *Cardiovascular Imaging* (CRC Press, Taylor & Francis Group, London, UK) and coinventor of the Wackers-Liu CQ SPECT Quantification Method, Food and Drug Administration-approved Commercial Software Package.

Albert J. Sinusas, MD, FACC, FAHA, is professor of medicine (Section of Cardiovascular Medicine) and radiology and biomedical imaging at Yale University School of Medicine, director of the Yale Translational Research Imaging Center, and director of Advanced Cardiovascular Imaging at Yale New Haven Hospital. He earned his BS degree from Rensselaer Polytechnic Institute and his MD degree from University of Vermont, College of Medicine, and completed training in internal medicine at the University of Oklahoma and training in cardiology and nuclear cardiology at the University of Virginia. He joined the faculty at Yale University School of Medicine in 1990, where he has remained. Dr. Sinusas has served as a standing member of the Clinical and Integrated Cardiovascular Sciences and Medical Imaging study sections of the National Institutes of Health. Dr. Sinusas has been a member of the Board of Directors of the Cardiovascular Council of the Society of Nuclear Medicine (SNM), the SNM Molecular Imaging Center of Excellence, and the American Society of Nuclear Cardiology. He was the 2008 recipient of the SNM Hermann Blumgart Award. His research is directed at development, validation, and application of noninvasive cardiovascular imaging approaches for the assessment of cardiovascular pathophysiology, including the targeted molecular assessment of myocardial ischemic injury, angiogenesis, arteriogenesis, and postinfarction atrial and ventricular remodeling. The investigation of these biological processes involves *ex vivo* and *in vivo* imaging in animal models of cardiovascular disease and humans. This translational research employs the three-dimensional modalities of x-ray computed tomography (CT) and fluoroscopy, single-photon emission CT/CT, positron emission tomography/CT, echocardiography, and magnetic resonance imaging in an animal physiology laboratory and clinical environment. Dr. Sinusas has been the principal investigator of several National Institutes of Health (NIH) grants involving multimodality cardiovascular imaging and directs an NIH-funded T32 grant providing training in multimodality molecular and translational cardiovascular imaging. He is the author of over 200 peer reviewed publications and invited reviews related to cardiovascular imaging and coedited a textbook on cardiovascular molecular imaging published in 2007.

Contributors

Frank M. Bengel
Department of Nuclear Medicine
Hannover Medical School
Hannover, Germany

James Bennett
Department of Biomedical Engineering
Rensselaer Polytechnic Institute
Troy, New York

Daniel S. Berman
Department of Medicine
Cedars-Sinai Medical Center
University of California
Los Angeles, California

Christos V. Bourantas
Erasmus Medical Center
Rotterdam, the Netherlands

Carlos A.M. Campos
Erasmus Medical Center
Rotterdam, the Netherlands

Ciprian Catana
Martinos Center for Biomedical Imaging
Massachusetts General Hospital
Boston, Massachusetts

Etienne Croteau
Université de Sherbrooke
Center for Research on Aging
Sherbrooke, Québec, Canada

Nicholas Dana
Department of Biomedical Engineering
University of Texas, Austin
Austin, Texas

Robert A. DeKemp
Department of Medicine (Cardiology)
University of Ottawa Heart Institute
Ottawa, Canada

Andrew J. Einstein
Department of Medicine
Division of Cardiology
Columbia University Medical Center
New York City, New York

Stanislav Emelianov
School and Electrical and Computer Engineering
and
Wallace H. Coulter Department of Biomedical
Engineering
Georgia Institute of Technology and Emory
University School of Medicine
Atlanta, Georgia

Javier Escaned
Erasmus Medical Center
Rotterdam, the Netherlands

Yingli Fu
Department of Radiology
Johns Hopkins University
Baltimore, Maryland

James R. Galt
Department of Radiology
Emory University
Atlanta, Georgia

Ernest V. Garcia
Department of Radiology
Emory University
Atlanta, Georgia

Hector M. Garcia-Garcia
Erasmus Medical Center
Rotterdam, the Netherlands

Guido Germano
Department of Medicine
Cedars-Sinai Medical Center
University of California
Los Angeles, California

Grant T. Gullberg
Department of Radiology and Biomedical
Imaging
University of California
San Francisco, California

R. James Housden
Division of Imaging Sciences and Biomedical
Engineering
King's College London
London, United Kingdom

James W. Hugg
Kromek/eV Products, Inc.
Saxonburg, Pennsylvania

Brian F. Hutton
Institute of Nuclear Medicine
University College of London
London, United Kingdom

Sami Kajander
Turku PET Centre
University of Turku
Turku, Finland

Andrei Karpiouk
School and Electrical
and Computer Engineering
Georgia Institute of Technology
Atlanta, Georgia

Ran Klein
Department of Nuclear Medicine
The Ottawa Hospital
Ottawa, Canada

Juhani Knuuti
Turku PET Centre
University of Turku
Turku, Finland

Dara L. Kraitchman
Department of Radiology
Johns Hopkins University
Baltimore, Maryland

Chi Liu
Departments of Radiology and Biomedical
Imaging and Biomedical Engineering
Yale University School of Medicine
New Haven, Connecticut

Michael V. McConnell
Department of Medicine
Stanford University
Palo Alto, California

Leon J. Menezes
Institute of Nuclear Medicine
University College of London
London, United Kingdom

Mathew Mercuri
Department of Medicine
Division of Cardiology
Columbia University Medical Center
New York City, New York

Kevin B. Parnham
TriFoil Imaging, Inc.
Chatsworth, California

Marina Piccinelli
Department of Radiology
Emory University
Atlanta, Georgia

Manuja Premaratne
Department of Non-invasive Imaging
Peninsula Health
Frankston, Australia

P. Hendrik Pretorius
Department of Radiology
University of Massachusetts
Worcester, Massachusetts

Jennifer M. Renaud
Department of Cardiac Imaging
University of Ottawa Heart Institute
Ottawa, Canada

Kawal S. Rhode
Division of Imaging Sciences and Biomedical
Engineering
King's College London
London, United Kingdom

Andrew Rittenbach
Department of Radiology
Johns Hopkins University
Baltimore, Maryland

Antti Saraste
Turku PET Centre
University of Turku
Turku, Finland

Patrick W. Serruys
Erasmus Medical Center
Rotterdam, the Netherlands

Albert J. Sinusas
Departments of Medicine Radiology and
Biomedical Imaging
Yale University School of Medicine
New Haven, Connecticut

Piotr J. Slomka
Department of Medicine
Cedars-Sinai Medical Center
University of California
Los Angeles, California

David E. Sosnovik
Martinos Center for Biomedical Imaging
Massachusetts General Hospital
Boston, Massachusetts

James T. Thackeray
Department of Nuclear Medicine
Hannover Medical School
Hannover, Germany

Benjamin M.W. Tsui
Department of Radiology
Johns Hopkins University
Baltimore, Maryland

Ge Wang
Department of Biomedical Engineering
Rensselaer Polytechnic Institute
Troy, New York

R. Glenn Wells
Division of Cardiology
University of Ottawa Heart Institute
Ottawa, Canada

Eleanor C. Wicks
Institute of Nuclear Medicine
University College of London
London, United Kingdom

Lei Xing
Department of Radiation Oncology
Stanford University
Palo Alto, California

Jingyan Xu
Department of Radiology
Johns Hopkins University
Baltimore, Maryland

Doug Yeager
Department of Biomedical Engineering
University of Texas, Austin
Austin, Texas

Raiyan T. Zaman
Department of Medicine
Stanford University
Palo Alto, California

PRINCIPLES, INSTRUMENTATION, TECHNIQUES, APPLICATIONS, AND CASE ILLUSTRATIONS OF HYBRID IMAGING

Principles and instrumentation of SPECT/CT

R. GLENN WELLS

1.1 INTRODUCTION

Single-photon emission computed tomography (SPECT) is technology for creating three-dimensional (3-D) images of the distribution of radioactively labeled substances within a subject. The energy of the radiation emitted is high enough to penetrate the patient tissues, allowing visualization of structures at all depths inside the patient. The energy is too high to be seen directly with the human eye and so a specially designed high-density detector is used to measure the emitted signals. The detector provides a 2-D picture of the radiation. By rotating the detector around the patients, a collection of pictures is obtained that can be converted into a 3-D image of the radioactivity distribution. Because the radioactive label is attached to a substance, the images track where that substance goes after being injected into the body. Thus, images of the radioactivity distribution can provide information on the function of different organs and physiologic systems with respect to the injected substance. For example, images of the distribution of ^{99m}Tc-tetrofosmin indicate how well blood is flowing to the myocardial tissues. The information in the images is degraded, however, by interactions of the emitted radiation with the surrounding tissues in the patient. Computed tomography (CT) can provide an accurate picture of the patient's anatomy, which can be used to significantly enhance the quality of the SPECT information. The combination of these two modalities thus provides a powerful tool for evaluating the heart. This chapter will describe the principles and instrumentation behind hybrid imaging with SPECT/CT.

1.2 RADIOISOTOPES USED IN SPECT

The radioisotopes used in SPECT imaging decay through either the direct emission of gamma rays from a meta-stable state (isomeric transition [IT]); the emission of electrons from the nucleus (β– particles) followed promptly by gamma-ray emission (β–,γ); internal conversion (IC), whereby the excess energy of the nucleus is transferred to an inner-shell electron, which is subsequently ionized; or by electron capture (EC) wherein an inner-shell electron is absorbed by the nucleus. Internal conversion and EC result in characteristic x-ray production as an outer-shell electron fills the vacancy left by the ionized or absorbed inner-shell electron. The energies of the gamma-rays or characteristic x-rays that are of use in nuclear medicine are between 69 keV and 364 keV. These energies are high enough that the photons have a reasonable probability of exiting the patient without interacting with the patient tissues and yet are low enough that they are efficiently detected by the gamma camera. The most common isotope used in gamma-camera imaging is ^{99m}Tc. It is used in approximately 85% of nuclear medicine tests (Eckelman 2009). Some of the more common isotopes used in SPECT cardiac imaging are given in Table 1.1. The half-life of an isotope is the time required for the activity

Table 1.1 Properties of common radionuclides used in SPECT imaging

Isotope	Half-life (hour)	Emission energies (keV)	Decay mode	Production	Typical uses
^{99m}Tc	6.02	140.5 (89%)	IT	Generator	Sestamibi (perfusion) Tetrofosmin (perfusion) Red-blood cells (ventricular function)
^{201}Tl	72.9	γ-rays: 167 (10%), 135 (3%) X-rays: 69–70 (73%), 80–82 (20%)	EC	Cyclotron	Tl-chloride (perfusion)
^{123}I	13.3	159 (83%), 529 (1.3%)	EC	Cyclotron	MIBG (heart failure)
^{131}I	192.5	364 (82%), 637 (7%), 284 (6%)	(β–,γ)	Nuclear Reactor	Alternative to 123-I
^{111}In	67.3	171 (90%), 245 (94%)	EC	Cyclotron	Oxine (cell-labeling)

Note: MIBG, metaiodobenzylguanidine.

of an isotope to decay to half of its original value. Half-lives for SPECT isotopes range from hours to days, which facilitates distribution of the isotopes and avoids the need for on-site production facilities. Isotopes like ^{201}Tl and ^{123}I are produced in high-energy cyclotrons, but ^{99m}Tc is usually obtained from a generator system. The parent isotope in the ^{99m}Tc generator is ^{99}Mo, which is typically produced in a nuclear reactor. The ^{99}Mo (66-hour half-life) is bound to an alumina column. Once it decays, it no longer binds and ^{99m}Tc can be eluted from the column with a simple saline rinse. The elute, ^{99m}Tc-pertechnetate, is then bound to a pharmaceutical using preformulated chemistry kits that only require the pertechnetate to be injected into the precursor vial, mixed, and possibly heated for a short period of time. The effective patient dose from the tracers used in cardiac SPECT usually range from 6 mSv (for ^{99m}Tc-labeled red blood cells) to 11 mSv for (^{99m}Tc-sestamibi) but can be as high as 32 mSv for ^{201}Tl stress-reinjection perfusion protocols (Einstein et al. 2007).

1.3 THE GAMMA CAMERA

The camera used in SPECT imaging is the gamma camera. The gamma camera was invented by Hal Anger in 1958 (Anger 1958) and is often referred to as the Anger camera. The camera can be divided into four primary components: the scintillation detector, an array of photomultiplier tubes (PMTs), processing electronics, and a collimator (Figure 1.1).

1.3.1 NaI(Tl) SCINTILLATION DETECTOR

As aforementioned, the typical energies emitted by isotopes used in nuclear medicine are between 69 and 364 keV. Detecting photons in this energy range requires a thick, dense material. The most commonly used detector for nuclear medicine is the thallium-doped sodium iodide (NaI(Tl)) scintillation crystal. Scintillation crystals convert the high-energy gamma ray into a shower of lower-energy light photons that can in turn be converted into an electrical signal, which can be measured and recorded. The incoming gamma ray transfers its energy by raising electrons in the crystal from their ground state up into an excited state. The difference in energy between these two states is only a few eV, and so it takes many interactions before the gamma ray is fully absorbed or stopped by the detector. For NaI crystals, the energy required to excite an electron is ~20 eV. The energy states of a pure crystal are discrete, and when an excited electron drops back down to the ground state, it would release all of its energy, creating a photon that is typically beyond the visible range. To improve the efficiency of de-excitation and to create visible light photons, impurities, called activators, are introduced

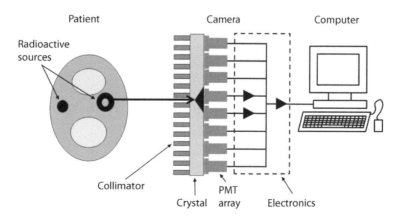

Figure 1.1 The gamma camera. The camera has four components. The collimator acts as the lens of the camera and fixes the direction of the detected gamma rays. The scintillation crystal converts the gamma ray into a shower of light photons. The PMTs convert the light signal into an amplified electrical signal, which is then processed by the electronics to determine the position that the gamma ray was detected at. Electronic data are then stored in a computer for further processing and display.

into the crystal. This creates energy states between the ground and excited states and facilitates de-excitation. The energy difference between the activator states and ground state is much less, ~3 eV for NaI(Tl), and so leads to the generation of blue light photons at 410 nm (Melcher 2000). NaI is a popular crystal because it is bright, generating 38 photons/keV of gamma-ray energy absorbed. Also, the strength of the signal from the crystal, the number of photons, is proportional to the energy of the gamma ray absorbed by the crystal. Finally, NaI is dense (3.67 g/mL), giving it a good stopping power, and it is relatively inexpensive to manufacture. Some disadvantages of NaI are that it is fragile and hydroscopic, meaning that it reacts easily with water. For this reason, the crystal must be sealed to prevent exposure to the air. Contact with moisture renders the crystal opaque and degrades its performance.

1.3.2 PHOTOMULTIPLIER TUBES

Though NaI(Tl) is considered a bright crystal, it still only emits a very dim glow when it is struck by gamma rays. Manipulating this light signal requires both conversion into an electrical signal and amplification. The photomultiplier tube (PMT) performs this function. The PMT is a vacuum tube that uses a series of large potential differences to amplify an electrical signal (Figure 1.2). The light photons are absorbed by the photocathode and their energy causes an electron to be ionized. The light conversion process is imperfect and only one to three electrons are generated for every 10 light photons incident on the photocathode. The ionized electrons are then accelerated by an electric field created with a stepped high-voltage applied to a series of dynodes. Dynodes act as anodes in one direction and cathodes in the other. The electrons ionized at the photocathode are accelerated toward the first dynode. When they collide with the dynode, their kinetic energy is transferred to other dynode electrons, ionizing them in turn. About six electrons are ionized for every one that strikes the dynode. This group of electrons is then accelerated by a potential difference toward the second dynode and the process repeats itself down the chain. At the end of the dynode chain is an anode from which the output signal from the PMT is read. A typical PMT has 10–12 dynodes and so the net amplification of the PMT is on the order of 10^7 or 10 million times.

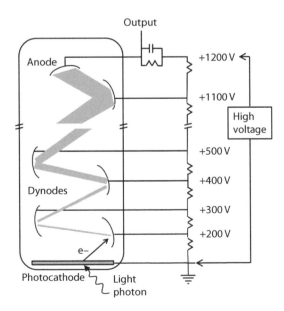

Figure 1.2 The PMT. Incident light photons are absorbed by the photocathode and their energy is used to ionize electrons. The electrons are accelerated across an electrical potential to the first dynode, where their kinetic energy ionizes approximately six electrons. These are accelerated to the next dynode, where the process repeats. The signal is read out from the anode at the end of the dynode chain.

1.3.3 Positioning electronics

The key idea that Anger had that allowed him to create a 2-D detector was what is now known as Anger logic (Figure 1.3). He realized that the light shower created by the interaction of the gamma ray in the scintillation crystal spread out as it travelled to the PMT. As such, if one placed an array of PMTs on the back of a crystal, the signal from a gamma-ray interaction would be distributed over the set of PMTs. The closer the event was to the center of a given PMT, the larger the fraction of the light shower it would capture, and the stronger would be the signal generated. Thus, by taking a weighted combination of the PMT signals, one could estimate the position of the event in the crystal. Consider the 1-D case illustrated in Figure 1.3. The sum of the signals is weighted according to how close it is to the left (W^-) or right (W^+) side of the array, generating a left signal (X^-) and a right signal (X^+). The difference in these two signals, normalized by the total signal strength, gives the fractional distance from the center to the edge of the crystal. Recall that the total signal strength is proportional to the gamma-ray energy absorbed by the crystal, and so, without normalization, the position would depend on the energy of the source radiation. The signal is separately measured in two orthogonal directions, giving its coordinates on the 2-D surface of the crystal. This positioning logic was originally implemented with analogue circuitry, but modern camera now tends to convert the analogue PMT signal into a digital one and then use maximum likelihood algorithms to estimate the position of the event. Though the implementation is more sophisticated, the principle remains the same; the 2-D position of the event is estimated by comparing the relative strengths of the signals from an array of PMTs.

1.3.4 Energy and spatial resolution

The uncertainty in the measured signal strength determines the energy resolution and intrinsic spatial resolution of the camera. There are statistical fluctuations in the different steps of the signal detection process. There is variation in the number of photons produced per keV of the gamma ray, the number of electrons generated per light photon at the photocathode, and the number of secondary electrons ionized per primary electron striking the dynode of the PMT. There is also variation in sensitivity over the surface of the photocathode and variation in sensitivity of the PMT array to light emitted from different locations in the crystal. Finally, there are variations in the high voltages applied to the dynodes and electronic noise in the PMT amplifier. All of these factors introduce uncertainty into the signal measured by each PMT, which degrades the accuracy of spatial positioning, and the total signal put out by the PMT array, which degrades energy resolution.

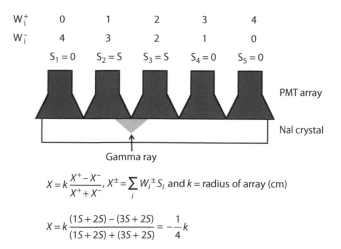

Figure 1.3 Anger logic. The position where the incident gamma ray interacts with the NaI scintillation crystal is determined using the weighted sum of the signals from the PMT array. An X^+ and X^- signal is generated by weighting the signal according to how close the PMT is to the right (+) or left (−) side of the array. The difference in these values, normalized by the sum, gives the position on the crystal. A 1-D example is illustrated.

The typical spatial positioning accuracy (R_{int}) for 140 keV photons with a NaI detector is 3–4 mm, while the typical energy resolution is 10%. The total number of light photons is proportional to the energy of the gamma ray, so statistical uncertainty decreases with increasing gamma-ray energy. If uncertainty was due entirely to the number of light photons produced in the scintillation crystal, the resolution would change as $E^{-1/2}$, but other contributing factors cause the resolution to vary slightly from this relationship (Cherry et al. 2003).

1.3.5 COLLIMATORS

Detecting the position of an event on the face of the detector crystal is one component of making a picture. Another is focusing on the object of interest. For optical imaging, the object is brought into focus with lenses. For the gamma camera, the lens is the collimator. The collimator defines the direction that the detected photons are travelling in. It is a large block of highly absorbing material, typically lead or tungsten, with well-defined holes passing through it. The collimator specifies photon direction by blocking or absorbing any photons that are not travelling in the specified direction. Collimators are classified according to their focal length, the photon energy they're designed for, and their sensitivity/resolution trade-off. The most common collimator used for cardiac imaging is the low-energy parallel-hole collimator.

1.3.5.1 PARALLEL-HOLE COLLIMATORS

As its name implies, the holes in a parallel-hole collimator are all aligned parallel to one another and typically run perpendicular to the surface of the scintillation crystal. The important parameters of this collimator are the length (L) and diameter (d) of the holes and the thickness (T) of the walls between them—the septa (Figure 1.4). The spatial resolution of the collimator is determined by the hole length and diameter. Any photon travelling at an angle less than $\tan^{-1}(d/L)$, the acceptance angle, will pass through the collimator hole and reach the crystal. Thus, the full-width half-maximum (FWHM) of the point-spread function (PSF) of a collimator is given by $R_{par} = d\,(L_{eff} + b)/L_{eff}$, where b is the perpendicular distance of the source from the collimator surface and L_{eff} is the "effective" hole length (Cherry et al. 2003). L_{eff} takes into account the probability of photons penetrating through the corner of the collimator at either end of the hole: $L_{eff} = L - 2/\mu$, where μ is the linear attenuation coefficient for the collimator material. The collimator resolution thus falls off linearly with distance from the collimator. The spatial resolution of the camera (R_{sys}) is a combination of the collimator and intrinsic resolutions. As these are independent uncertainties, $R_{sys}^2 = R_{par}^2 + R_{int}^2$. The spatial resolution of the camera is dominated by the collimator resolution at distances typical of clinical

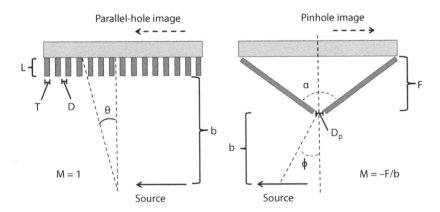

Figure 1.4 The parallel-hole and pinhole collimators. The characteristics of the parallel-hole collimator are determined by the hole length (L), the septal thickness (T), and the hole diameter (D). The acceptance angle (θ) is given by $\tan^{-1}(D/L)$. The magnification (M) of the collimator is one. The performance of the pinhole collimator is determined by the pinhole diameter (D_p), the opening angle (α), and the focal length of the collimator (F). Magnification (M) depends on the source distance (b) and sensitivity depends on b and the angle φ of the source from the central axis of the pinhole (see text).

imaging and a low-energy high-resolution collimator usually has a FWHM of 7 mm at 10-cm distance from the collimator.

The thickness of the septa is designed to minimize the septal penetration of the gamma ray and thus depends on the gamma-ray energy. The purpose of the collimator is to ensure that the photons that reach the crystal must have passed through a hole. If a photon passes through a septum, then the accuracy of the directional information is reduced. Attenuation of photons is exponential, and therefore, it is never possible to completely stop all penetration and instead the collimator is designed to have a maximum of 5% septal penetration. The geometry of the collimator then requires that $T > 6d/(\mu L - 3)$ (Cherry et al. 2003).

The sensitivity of the camera depends on the thickness of the detector crystal and the collimator geometry. A 9-mm-thick NaI crystal stops more than 90% of 140 keV photons. The geometric sensitivity (G_{par}) of the collimator is the fraction of photons exiting the source that pass through the collimator to the detector crystal. The number of photons per unit area falls off as $1/r^2$, but the area of the crystal illuminated by the source (the PSF) goes as $r^2(D/L_{eff})^2$. The area of the collimator that photons can pass through is reduced by the thickness of the septa. Therefore, $G_{par} = k(d/L_{eff})^2 (d/[d + T])^2$, where the proportionality constant k depends on the hole shape and is 0.0676 for hexagons (Cherry et al. 2003). For dimensions of d = 1 mm, L_{eff} = 2.06 cm, and T = 0.15 mm, this gives a camera efficiency of 1.2E-4, or about 0.02% for a dual-head gamma camera. Of note is that the geometric efficiency of the parallel-hole collimator does not depend on source distance, whereas the resolution degrades with distance, suggesting that the best image quality will be obtained with the camera placed as close to the patient as possible. Also, note that G_{par} is proportional to R_{par}^2, and so as collimator sensitivity goes up, collimator resolution degrades. Higher-resolution collimators have worse sensitivity and vice versa.

1.3.5.2 PINHOLE COLLIMATORS

The original design proposed by Hal Anger in 1958 was based on pinhole collimators; it was not until 1964 (Anger 1964) that he proposed the use of parallel-hole collimators. Though the parallel-hole collimator is used almost exclusively for SPECT imaging, recent development of dedicated cardiac cameras (Slomka et al. 2009b) has brought back a number of different collimator designs, including the pinhole collimator. Multipinhole imaging has been used for small-animal imaging to capitalize on the magnification properties of the collimator (Franc et al. 2008), but human pinhole cameras use the collimator to shrink the image to fit on a smaller detector.

A pinhole collimator has a single hole through which all of the detected gamma rays must pass. The defining parameters for a pinhole collimator (Figure 1.4) are the diameter of the pinhole (d_p) and the focal length (F). Because all of the detected-photon paths pass through the pinhole, this collimator inverts the image of the source in both the axial and transverse directions and changes the size of the image depending on the distance of the source from the pinhole. The magnification (M) is given by M = $-b/F$, where b is the perpendicular distance from the collimator. The resolution of the collimator (R_{pin}) is given by the FWHM of the image of a point source, corrected for the magnification of the collimator. $R_{pin} = [d_{eff} (b + F)/b] \times M = d_{eff} (b + F)/F$. d_{eff} is the effective pinhole diameter, taking into account penetration of photons at the edge of the pinhole. $d_{eff} = d_p + \ln(2)\tan(\alpha/2)/\mu$, where α is the opening angle of the pinhole (Accorsi and Metzler 2004). The geometric sensitivity of the pinhole collimator is $G_{pin} = d_{eff,G}^2 \, cos^3 (\phi)/(16b^2)$, where ϕ is the angle of the source away from the pinhole axis and $d_{eff,G}$ is the effective diameter for geometric sensitivity. $d_{eff,G}^2 = d[d + (2/\mu)\tan(\alpha/2)] + [(2/\mu^2)\tan^2(\alpha/2)]$ (Smith and Jaszczak 1997). Unlike the parallel-hole collimator, the geometric sensitivity is not constant with source distance but rather falls off as the square of the distance from the pinhole (Figure 1.5). For an effective hole diameter of 2.5 mm and a focal length of 10 cm, the sensitivity on axis at 6.2-cm distance is 0.012%. Close to the pinhole aperture, the sensitivity can exceed that of a parallel-hole collimator with similar resolution, but it has lower sensitivity at typical organ-camera distances for cardiac imaging. Like the parallel-hole collimator, the pinhole collimator still has a tradeoff between resolution and sensitivity with respect to the pinhole diameter.

For cardiac imaging, the field of view (FOV) of interest is typically less than 20 cm across. It is possible, therefore, to fit multiple FOVs onto a modern camera detector surface of 40 × 50 cm. The magnification properties of the pinhole collimator can be used to shrink images and further increase the number of

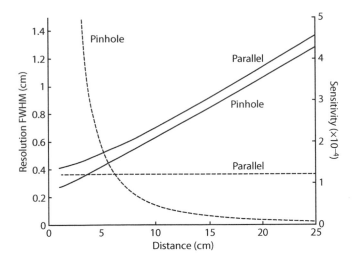

Figure 1.5 Collimator resolution and sensitivity. Spatial resolutions (FWHM) of the parallel-hole and pinhole collimators (solid lines) fall off linearly at typical imaging distances. The sensitivity (dashed lines) of the parallel-hole collimator is constant, whereas that of the pinhole collimator falls off as the square of the source distance.

FOVs. The use of multiple pinholes increases the overall sensitivity of the system by roughly the number of pinholes used and is a technique used to improve sensitivity in small-animal imaging (Franc et al. 2008). In 2006, Funk et al. showed that one could use a nine-pinhole collimator to image the human heart and obtain a sensitivity approximately 5× that of a parallel-hole collimator of the same resolution (Funk et al. 2006). The multipinhole approach was adopted by GE Healthcare in the design of their dedicated cardiac camera.

1.3.6 CADMIUM-ZINC-TELLURIDE DETECTORS

Cadmium-zinc-telluride (CZT) detectors have been introduced as an alternative detector for nuclear medicine applications. CZT is a semiconductor that can operate at room temperatures and directly converts the gamma ray into electron-hole pairs that can be read out by dedicated electronics. Thus, the CZT detector replaces both the NaI scintillation crystal and the PMT of the standard gamma camera. The advantage of CZT is that the energy required to create an electronic-hole pair is only ~4.5 eV, compared to ~20 eV for NaI. In addition, only one in five light photons is converted back into an electron by the photocathode. So, the initial charge available for amplification is >20× higher for CZT than for the NaI-PMT combination. The stronger signal translates into an improved energy resolution for the CZT-based camera (6%) compared to the standard NaI-based system (10%). The density of the CZT material is higher (5.81 g/cm³) than NaI (3.67 g/cm³) giving it a higher stopping power per unit length, but the detector thickness used for nuclear medicine applications is 5 mm, giving it an intrinsic efficiency for 140 keV ^{99m}Tc photons of 80%, which is very similar to the 85% obtained with a 9-mm-thick NaI crystal. The CZT detectors are manufactured as a pixilated array with a 2.5-mm pitch and the intrinsic spatial resolution is equal to the pixel size (Erlandsson et al. 2009; Bocher et al. 2010). One disadvantage of the CZT material is that partial charge collection at the edge of pixels, electron-hole trapping, and other effects lead to a low-energy tail (Blevis et al. 2004; Wagenaar 2004). The energy spectrum of a source is not a symmetric Gaussian, but instead has an increased number of photons detected with lower energies. This leads to a small decrease in the peak efficiency, the fraction of unscattered photons detected in the photopeak energy window. A large advantage of this detector is its compact size, allowing cameras to be designed that have many detectors mounted onto a single gantry (Erlandsson et al. 2009; Bocher et al. 2010). A large number of detectors means that these cameras are able to sample enough angles simultaneously to perform 3-D image reconstruction with little or no gantry rotation.

1.4 3-D IMAGE RECONSTRUCTION

1.4.1 SAMPLING REQUIREMENTS

The gamma camera creates a 2-D view of the tracer distribution inside the patient. By acquiring many such 2-D views, called projections, from different angles around the patient, it is possible to reconstruct a 3-D SPECT representation of the tracer distribution. Accurately describing a point in the image requires views from "all" angles around that point. For parallel-hole collimated imaging, this requires rotating the camera over a range of 180° around the patient (Defrise et al. 1995). A rotation of less than this risks introducing artifacts into the image reconstructed because of structures that are not adequately resolved by any of the acquired views. The resolution of the image is dictated by the density of samples taken, in the angular as well as in the axial and transverse directions. Accurately representing a resolution requires approximately three samples per FWHM, that is, a projection pixel size of FWHM/3. The angular sampling density follows a similar rule-of-thumb. The distance between samples on the circumference of the desired FOV should be FWHM/3 (Cherry et al. 2003). Thus, for a FOV diameter of d_{FOV}, the number of projections (N) required for a resolution of FWHM is $N = 3\pi d_{FOV}/(2\ FWHM)$. For example, supporting a system resolution of 1 cm FWHM over an FOV of 20-cm diameter requires projection pixels that are 3.3 mm and 94 projections over 180°.

1.4.2 FILTERED BACKPROJECTION

For many years, the most common way to reconstruct 3-D SPECT images was to perform filtered backprojection (FBP). FBP makes use of the Central Section Theorem (Keinert 1989) which states that the Fourier transform of a projection through an object at an angle θ corresponds to the values along a line through the Fourier transform of the object at angle θ (Figure 1.6). With a parallel-hole collimator, as the camera rotates around the source object, a radial sampling of Fourier space is obtained like spokes on a wheel. For an ideal parallel-hole collimator, each axial row of the projection data set can be considered independent, and thus, the 2-D image for each row can be reconstructed separately and then stacked back together to form a 3-D volume. With this assumption, the reconstruction problem simplifies to the reconstruction of a 2-D image

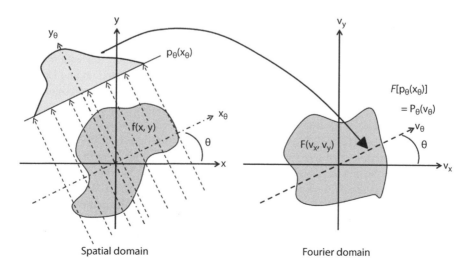

Figure 1.6 Central section theorem. If the Fourier transform of an activity distribution f(x,y) is $F(v_x,v_y)$ and the projection, $p_\theta(x_\theta)$, is the sum of f(x,y) along the direction $y_\theta = y cos(\theta) - x sin(\theta)$, then the Fourier transform of $F[p_\theta(x_\theta)] = P_\theta(v_\theta)$ is the line through $F(v_x,v_y)$ at angle θ.

from a set of 1-D projections. So, if the source distribution is described in 2-D by the function f(x,y), then a projection at angle θ is $p_\theta(x_\theta)$.

$$p_\theta(x_\theta) = \int_{-\infty}^{\infty}\int_{-\infty}^{\infty} f(x,y)\delta(x\cos(\theta) + y\sin(\theta) - x_\theta)dx\,dy$$

$$= \int_{-\infty}^{\infty} f(x,y)dy_\theta \tag{1.1}$$

where δ is the Dirac delta function and (x_θ, y_θ) refers to the Cartesian coordinate system rotated by θ with respect to the original coordinate system (x, y), e.g., $x_\theta = x\cos(\theta) + y\sin(\theta)$. The full set of projections acquired is then $p(x_\theta, \theta)$, where $p(x_\theta, \theta)|_{\theta=\phi} = p_\phi(x_\phi)$. The Central Section Theorem says that a line through the Fourier transform of the image $F(v_x, v_y) = F(f(x,y))$ at the angle θ given by $P_\theta(v_\theta)$ is

$$P_\theta(v_\theta) = \int_{-\infty}^{\infty} F(v_\theta, v_\theta)\,\delta(v_x\cos(\theta) + v_y\sin(\theta) - v_\theta)\,dv_y = F(p_\theta(x_\theta)) \tag{1.2}$$

The full set of projections in the Fourier domain is then denoted $P(v_\theta, \theta)$ and samples Fourier space radially, which produces a variable sampling density. Because the number of "spokes" (projections) is constant, as the radius of the wheel (spatial frequency) increases, the sampling density falls as $1/|v|$, where $|v| = \sqrt{(v_x^2 + v_y^2)}$. To correct for this, the Fourier domain image is filtered by multiplying it with the function $h(v_x, v_y) = \sqrt{(v_x^2 + v_y^2)} = |v|$. This filter is referred to as the ramp filter because of its simple shape in the Fourier domain. The inverse Fourier transform of the filtered image gives us an image of the tracer distribution.

$$f(x,y) = F^{-1}\left[|v|P(v_\theta, \theta)\right] = F^{-1}\left[|v| F\left(p(x_\theta, \theta)\right)\right] \tag{1.3}$$

Although this discussion has been done in the context of 2-D reconstruction from parallel-hole collimators, the theory can be adapted to converging collimators like the pinhole collimator as well (Datz et al. 1994).

1.4.3 ITERATIVE RECONSTRUCTION

The FBP algorithm inherently assumes that the projection data are consistent. In other words, FBP assumes that projection at each different angle is a sum of the same object. However, this assumption is violated by noise. Each projection is a sum of the object plus noise and the noise is random and hence different in each projection. Other factors like attenuation, scatter, distance-dependent collimator resolution, and patient motion (Section 1.5) introduce further inconsistencies between the projections. This has led to the development of alternative methods of reconstruction. Because the data are inconsistent, the solution is often pursued using iterative methods (Figure 1.7) that try to find the solution that is the best compromise between the measured data, something that minimizes the errors. The most popular algorithm used clinically is the ordered subset expectation maximization (OSEM) algorithm (Hudson and Larkin 1994) that is a variant on the maximum likelihood expectation maximization algorithm that was introduced into nuclear medicine by Shepp and Vardi (1982). This algorithm is built on the assumption that the noise in the data is Poisson distributed. The idea of iterative reconstruction is illustrated in Figure 1.7. An estimate is made of the distribution of activity. Having no other information available, it is usually initially assumed that the activity is uniformly distributed throughout the FOV, though any positive, nonzero distribution is a valid starting point. Using an understanding of the geometry of the camera and how photons travel from points within the FOV to the detector, a set of projections are computed that show what the camera would have measured given the current estimate of the activity distribution (the forward projection). Then the calculated projections are compared

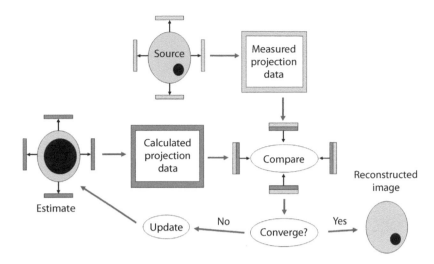

Figure 1.7 Iterative reconstruction. An estimate of the activity distribution is used to calculate its correspond-ing projection data. These calculated projections are compared to those measured from the true source dis-tribution in the camera. If the projections are not the same, then the differences or ratios are used to update the estimate and a new set of calculated projections are generated. Once the projections are the same, then the estimate has converged and is the reconstructed image of the source.

to the ones measured with the camera. Where the calculated values are higher than the measured ones, we reduce the activity in the object that contributes to that part of the projections. Where the calculated values are lower, we increase our estimate. This is done by backprojecting the ratio of measured over calculated pro-jections. The backprojected ratios are used to rescale the original estimate and generate a better one. We then repeat or iterate the process until the estimate stops changing, giving us our reconstructed image.

Mathematically, if $f_i^{(k)}$ denotes the value of the image at voxel i and iteration k, then the OSEM algorithm is written as

$$f_i^{(k+1)} = f_i^{(k)} \frac{\sum_{j \in J} A_{ij} \left(\frac{p_j}{g_j} \right)}{\sum_{j \in J} A_{ij}} \tag{1.4}$$

$$g_j = \sum_i A_{ij} f_i^{(k)} \tag{1.5}$$

where p_j is the measured value of projection pixel j, A_{ij} is the projector/backprojector function that defines the probability that a photon emitted from image voxel i will be detected in projection pixel j, and g_j is the calcu-lated value of projection pixel j given the current image estimate $f^{(k)}$. At each iteration k, only the projection pixels j from subset J of the full projection data set are projected and backprojected to update $f^{(k)}$ to $f^{(k+1)}$. The algorithm then cycles through all of the subsets as it iterates.

Because the algorithm is derived with the assumption that the projection data are Poisson distributed, OSEM weights the projection values appropriately. This changes the noise characteristics of the image. Whereas FBP produces noise levels that are similar over the entire image, OSEM produces an image with noise that is correlated with the amplitude of the underlying signal (Barrett et al. 1994). Thus, in OSEM images, the noise in low-count regions is lower than the noise in high-count regions. Another feature of the OSEM reconstruction is positivity. Because the algorithm always scales the estimate by a positive number (the ratio of measured and calculated projections is always greater than or equal to zero), if the initial estimate of activity is nonnegative, it will always remain so. In contrast, the ramp filtering of the FBP algorithm can lead to areas of "negative activity" in the image, particularly close to the very "hot" structures that contain high concentrations of radioactivity. This is the source of the so-called ramp artifact, which can suppress activity in the inferior wall of the heart when a liver with very high activity levels is nearby (Burrell and MacDonald 2006).

The greatest advantage of the iterative algorithms, though, is that they allow accurate correction of various significant image degrading factors like attenuation, scatter, and resolution loss due to the geometry of the collimator. By including these effects into the forward projector (A_{ij}), the acquisition process of the camera is more accurately modeled, and thus, the underlying estimate is adjusted to compensate for these effects. Because the distributions of the tracer activity and of the patient tissues are estimated, the variable effects of attenuation and scatter and the distance-dependent nature of the collimator resolution can be applied accurately.

1.5 FACTORS THAT INFLUENCE SPECT IMAGE QUALITY

1.5.1 ATTENUATION

There are many factors that influence the appearance of the image acquired by the gamma camera. The most significant is photon attenuation. As the gamma rays emitted by the tracer transit the patient en route to the detector, they can interact with the patient tissues. The two primary types of interaction of interest in nuclear medicine are photoelectric absorption and Compton scattering. Photoelectric absorption occurs when a gamma ray is completely absorbed by an atom and its energy transferred to an electron which is then ejected from the atom. In Compton scattering, the gamma ray transfers only some of its energy to the electron. The electron is again ejected from the atom, but the photon is not completely absorbed and instead travels on in a new direction, with reduced energy. Both types of interaction combine to produce attenuation (loss of photons with the original energy of emission, e.g., for ^{99m}Tc, loss of 140 keV photons). The probability of attenuation is given by the linear attenuation coefficient (μ), which depends on both the energy of the gamma ray (E) and the atomic number of the attenuating material (Z) and has units of 1/distance. The fraction of gamma rays that pass through a material unattenuated (I/I_0) is given by $I/I_0 = exp(-\mu(E,Z) t)$, where t is the thickness of the material.

The loss of signal is greater for structures that are deeper within the patient and so the effects of attenuation produce a depression in the apparent activity concentration in the center of the patient. The presence of a large amount lower-abdominal tissue can cause greater attenuation of signals from the inferior wall of the heart and thereby generate an apparent reduction or attenuation artifact (more commonly seen in males). In contrast, the presence of breast tissue that shadows the top half of the heart can increase the attenuation of the anterior wall, producing a breast attenuation artifact (more commonly seen in females) (Figure 1.8). Both

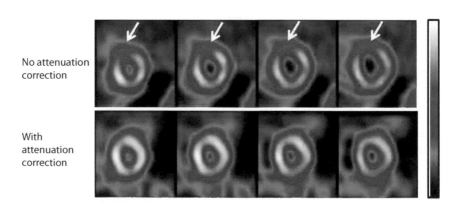

Figure 1.8 Breast attenuation artifact. The shadow of the breast can cause an apparent reduction in the anterior wall of the heart (arrow), which is not present with attenuation correction.

of these artifacts make it more difficult to interpret the images, reducing reader confidence. Two approaches are used to compensate for attenuation: prone imaging (Germano et al. 2007) and transmission-based attenuation correction (AC) (Garcia 2007). In prone imaging, the patient is imaged in both the supine and prone positions. An attenuation artifact will tend to reduce because the change in orientation will shift the attenuating structures with respect to the heart. A true area of reduced myocardial tracer-uptake will not move, allowing the reader to distinguish between the two. Transmission-based AC removes the effects of attenuation by modeling it in the reconstruction. This requires a patient-specific transmission measurement of the attenuating structures, which is now most commonly acquired from an x-ray CT image.

1.5.2 SCATTER

Gamma rays with Compton scatter in the patient may still be detected by the camera. Presence of scattered photons in the image degrades the accuracy of the information as the direction that the photon is detected as coming from does not correspond to the direction of the source that originally emitted the photon. This leads to reduced contrast in the image and an overcorrection for attenuation when linear attenuation coefficients are used for AC. More specifically, the linear attenuation coefficient assumes that a scattered photon is lost—attenuated—and not detected by the camera. The magnitude of scatter is less than that of attenuation, but when AC is applied without scatter correction (SC), scatter artifacts can become apparent. The most common of these artifacts in nuclear cardiac imaging is an apparent brightening of the inferior wall (Figure 1.9) caused by photons from the liver, which pass through the lungs and then scatter into the inferior wall of the heart toward the detector (King et al. 1995).

Energy discrimination is the primary method of reducing the number of scattered photons detected. However, due to the 10% energy resolution of NaI detectors, rejection of scattered photons is imperfect. To avoid loss of primary (unscattered) photons, an energy window of ±10% (or ±7.5%) is chosen, accepting 98% (or 92%) of the expected distribution of photons. However, a ±10% window means that photons with a 10% true loss in energy still have a 50% chance of being accepted. For ^{99m}Tc photons, 126 keV (140 keV − 14 keV) is the energy of a photon that has Compton scattered through 53° (Cherry et al. 2003). Improving the energy resolution of the camera can be beneficial because it allows reduction of the acceptance window without loss in the number of primary photons detected. Narrowing the acceptance window will reject a larger number of scattered photons, and those scattered photons that are accepted within the window will have on average a smaller scattering angle and thus less error in their position information.

Removal of scatter artifacts and accurate quantification of the activity concentration measured in images requires correction for the residual scatter not rejected by the primary (photopeak) energy window. Many approaches to estimating the scatter in an image that have been proposed include energy-window-based, modeling, and other scenarios (Hutton et al. 2011). The energy-window methods use the distribution of

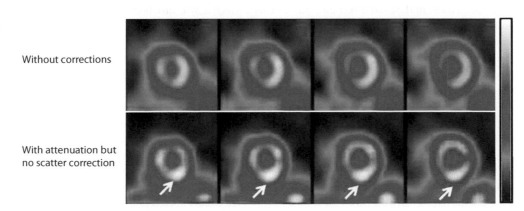

Without corrections

With attenuation but no scatter correction

Figure 1.9 Inferior wall scatter artifact. When AC is applied with no SC, scatter from sub-diaphragmatic sources can cause an apparent increase in counts in the inferior wall (arrows).

photons detected at energies outside the photopeak to estimate the scatter detected within the photopeak. The modeling approaches use an estimate of the activity distribution and knowledge of the scattering medium to calculate the scatter distribution. The modeling methods tend to be the most accurate but are computationally intensive and take longer to generate a scatter estimate. Once scatter has been estimated, the two most common approaches to correcting it are to either subtract it from the projection data prior to reconstruction or include it within the projectors of the reconstruction algorithm. The latter approach has been shown to provide a more accurate and precise image (Beekman et al. 1997) but also can significantly increase the time required for reconstruction.

1.5.3 DISTANCE-DEPENDENT COLLIMATOR RESOLUTION

The resolution of the gamma camera is described by the system PSF, which is simply the image created by the gamma camera from a point source of activity. The system PSF is dominated by the collimator resolution, which depends on the distance from the source to the collimator (Figure 1.5). The gamma camera is a 2-D imaging system and so the PSF introduces blurring in both the transverse and the axial directions. An important consequence of this is that the activity in one transverse plane contributes to the signal detected at other transverse planes on the detector. The signals from the different planes are intermingled and accurate reconstruction must take this into consideration: SPECT reconstruction is a 3-D problem, not an independent set of 2-D problems. Compensation for the system PSF has the potential to recover lost resolution in the images and so is often referred to as resolution recovery. While efforts have been made to include compensation for distance-dependent resolution in FBP reconstruction (Glick et al. 1994), accurate correction is usually done within an iterative reconstruction algorithm. The inclusion of resolution compensation (or system PSF modeling) within the reconstruction improves the accuracy of the camera model used in the projector and leads to better image quality (Narayanan et al. 2003). Resolution recovery also associates each voxel in the image to a larger number of pixels in the projection data, making the image less sensitive to noise. This aspect has been capitalized on to reduce the time required for imaging (Ali et al. 2009; Bateman et al. 2009; DePuey et al. 2011) or equivalently reduce the amount of activity injected into the patient and hence the patient effective dose. Dose reductions of 50% or more are possible with currently available clinical software (DePuey et al. 2011; Slomka et al. 2012).

1.5.4 PATIENT MOTION

Movement of the patient during imaging introduces inconsistencies into the projection data as projections are acquired sequentially while the camera rotates around the patient. These inconsistencies can lead to artifacts in the reconstructed images (Botvinick et al. 1993). Motion of this sort can be seen and often corrected for by examining the projection data as a function of projection angle (O'Connor et al. 1998; Matsumoto et al. 2001). Movement in the axial direction is most easily corrected by translation of the affected projections to realign the activity sources. However, movement in the transverse direction and rotation about the axis of the camera are more difficult to accurately identify and correct.

Movement during acquisition of the projection, whether voluntary motion or involuntary motions such as cardiac contraction and respiration, are more difficult to detect and also cause image blur. The effects of periodic movement like cardiac contraction and respiration can be minimized by gating. Gating uses a signal such as the electrocardiogram (ECG) to generate a repeating trigger signal, corresponding to a particular point within the motion cycle, and thereby allow the camera to sort the data acquisition based on the time from the most recent trigger event. Dividing the data into separate gates in this fashion minimizes the movement within each gate and can provide valuable information about the motion. For example, cardiac ECG-gating of SPECT perfusion and blood-pool images allows measurement of wall motion and ejection fraction and has proven to be incrementally beneficial over ungated imaging alone (Mansoor and Heller 1999). However, subdividing the data set also decreases the number of detected counts in each image and hence increases the noise. The increase in noise can be eliminated by measuring the motion between gates and then coregistering the gates to a fixed reference frame or incorporating the motion within the reconstruction algorithm to create a single motion-compensated image (Gravier et al. 2006; Gilland et al. 2008; Slomka et al. 2009a). Methods are similarly being developed to compensate for respiratory motion (Kovalski et al. 2007; McNamara et al. 2009). Further discussions on motion corrections are described in Chapters 14 and 16 of this book.

1.6 COMPUTED TOMOGRAPHY

Correcting for attenuation and using model-based methods of scatter compensation require knowledge of the distribution of patient tissues. This is typically acquired through a transmission scan. Though many approaches were suggested that used radioisotope transmission sources (King et al. 1995), the most common approach being employed today is to use an x-ray CT scan of the patient.

1.6.1 BASICS OF CT

The CT scan uses a beam of x-rays as a source for transmission imaging. The x-ray beam is created with an x-ray tube (Figure 1.10). A current is run through a wire filament to heat it and generate a cloud of loosely bound electrons. Applying an electric potential between the filament and a target strips the electrons off of the filament and accelerates them toward the target. When the accelerated electrons strike the target, most of their energy (99%) is converted to heat in the target material, but only a small amount of the energy produces x-rays, either by characteristic x-rays or bremsstrahlung radiation.

Characteristic x-rays are produced when energy is transferred to an inner-shell electron of the target atom, in this case by collision with incoming electron. The collision does not involve direct physical contact, but rather a collision of their respective electric fields: the two electrons have the same charge and so there is an electro-static repulsive force between them. The inner-shell electron is ionized and so escapes from the atom, leaving behind a hole in the electron shell. An electron from a higher-energy shell drops down to fill this hole, releasing a photon with energy equivalent to the difference in energy between the two shells. This

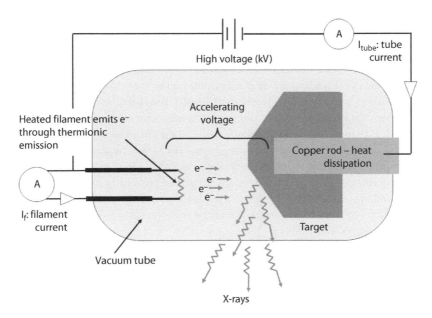

Figure 1.10 The x-ray tube. A current I_f is run through a filament to generate a source of electrons. The accelerating potential pulls the electrons from the filament and accelerates them toward the target. When the electrons strike the target, x-rays are generated in the form of characteristic x-rays and bremsstrahlung radiation. The tube current, I_{tube}, is a measure of the flow of electrons from the filament to the target and is proportional to the intensity of the x-ray beam.

photon is called a characteristic x-ray as the energy-level structure, and hence, the x-ray energies are specific to the target atom.

Bremsstrahlung or "braking" radiation is emitted when the incoming electron interacts with an electric field of the positively charged nucleus of a target atom and is decelerated and changes direction. The energy lost in this process is emitted as x-rays. Because the amount of deceleration depends continuously on how close the electron happens to pass to the atom, the bremsstrahlung radiation has a continuous spectrum of energies up to the maximum energy of the incoming electron and peaks at roughly one third of the maximum energy. Because the very low-energy x-rays are less able to penetrate the body, they contribute significantly to patient dose. To reduce dose and increase the mean energy of the x-ray beam, the low-energy x-rays are preferentially attenuated by filtering the beam with a thin layer (few mm) of aluminum (Al) or similar material. The preferential attenuation of lower-energy x-rays is called beam hardening, and prefiltering of the beam serves to reduce the effects of beam-hardening in the patient. Beam hardening can lead to depression of the CT signal in the middle of the patient, similar to the attenuation artifact seen in SPECT imaging. The intensity of the x-ray beam is proportional to the tube current, the number of electrons flowing from the filament to the target, and depends on the filament current (how tightly bound electrons are to the filament) and the accelerating potential (how strongly electrons are being pulled toward the target).

The CT scanner is configured with the x-ray tube opposite a detector array. The x-ray beam is collimated to produce a fan of x-rays that matches the width of the detector. The relative positions of the x-ray source and detector are fixed, but the whole unit rotates rapidly around the patient. Modern CT scanners rotate in times as low as 0.3 s. The intensity of the x-ray beam with a patient in the scanner is compared to that acquired without the patient. The log of that ratio ($\ln(I/I_o)$) equals the line integral of the linear attenuation through the patient. As the scanner rotates around the patient, projections are acquired at many different angles exactly analogous to the SPECT data acquisition process described previously (Section 1.4). The projection data are then reconstructed to create a tomographic image of attenuation. The CT numbers correspond to the linear attenuation in each pixel of the image and are often given in terms of Hounsfield units (HU; after Godfrey Hounsfield, the inventor of CT), which relates the attenuation back to attenuation of water.

$$HU = 1000 \times (\mu_{material} - \mu_{water})/\mu_{water} \tag{1.6}$$

In HU, water is always 0 and vacuum (or air) is −1000.

Image reconstruction is frequently done with FBP. The assumptions of the algorithm are better satisfied by CT data than they are by SPECT data. Rather than counting photons, CT detectors measure a photon current and the count rate in CT is many orders of magnitude higher than that in SPECT. Thus, the statistical noise in the CT data sets is correspondingly less. With SPECT, the signal being measured is the number of photons emitted from the radioactive tracer and so photon attenuation introduces inconsistencies into the projections. With CT, attenuation is not a problem for the FBP algorithm because the incident X-ray intensity is known and the signal being measured is the amount that the X-rays are attenuated. Also, the fraction of scattered photons detected tends to be less as the fan-beam design of CT allows use of scatter rejection grids at the detector. Nevertheless, iterative reconstruction has gained popularity for CT reconstruction as computers have increased in speed, making CT iterative reconstruction times practical. Iterative reconstruction has improved the signal-to-noise in CT images, allowing a corresponding reduction in the patient exposure delivered during imaging (Kordolaimi et al. 2013).

Images with CT are usually obtained with a helical or spiral acquisition. In this mode, the patient bed slides continuously through the scanner while the detector is rotating around the patient. Thus, the detector describes a helical path around the patient. The advantage of this acquisition mode is that it greatly decreases the scanning time. The alternative approach of step-and-shoot accelerates the patient to move them between scan positions and then halts their motion for each acquisition. Inertia causes the patient tissues to oscillate, blurring the image unless the acquisition is delayed until the oscillations damp out. The delay at each step can significantly increase the total scanning time. The disadvantage of helical scanning is that it introduces inconsistencies between the acquired projections. Because the patient is continuously moving, each projection looks like a projection from a "different patient" rather than a projection acquired later in the rotation.

The inconsistency is corrected by interpolating between successive rotations to generate a full set of consistent projections at a specified axial location on the patient. The interpolation approach also allows image planes to be chosen at arbitrary intervals, which facilitates coregistration with nuclear medicine images.

Gating is also used in CT acquisitions to reduce patient motion or to examine how the CT image of the heart changes over the course of the cardiac or respiratory cycle. As with gating in SPECT, the trigger signal is generated based on the ECG and an image or a set of images is created of a specific time interval following the ECG-triggered signal. Thus, an image can be acquired, for example, of a particular phase of the cardiac cycle. It is possible to gate in a fashion similar to SPECT, acquiring data over the full duration of the cycle, sorting the data according to the time since the trigger signal, and then reconstructing images for all of the phases throughout the cardiac cycle (retrospective gating). However, unlike SPECT, where the radiation exposure to the patient is fixed regardless of how many images are acquired, with CT, radiation exposure increases when images from more phases are taken. An alternative approach is to turn down the tube current, except during the phase of interest (prospective gating), and thereby obtain a high-quality image at one predetermined cardiac phase while minimizing patient exposure over the rest of the cardiac cycle.

A significant advance in CT technology has been the introduction of multislice detectors. Multiple rows of detectors in the axial direction has increased axial coverage and thus decreased the scanning times as more of the patient is imaged with each rotation of the scanner. The increase in axial detector size has changed the CT scanner geometry from a 2-D fan-beam to a 3-D cone-beam geometry, which has complicated the reconstruction problem. However, algorithms have been developed to compensate for this, providing true 3-D reconstruction (Defrise et al. 1995; Kachelrieß et al. 2004). Vendors now provide scanners with up to 320 detector rows, providing greatly expanded coverage for cardiac imaging. High-speed rotation (up to 0.3 s/rotation) and the introduction of dual-source CT systems further reduce the effective scanning time, making it possible to accurately track a contrast bolus through the coronary arteries and perform CT-based coronary angiography (CTA) (Flohr et al. 2009).

1.6.2 CT-BASED CORRECTION OF NUCLEAR MEDICINE IMAGES

One of the primary uses of CT for cardiac SPECT imaging has been to provide a patient-specific attenuation map for attenuation correction (CT-AC). The CT scan directly measures photon attenuation properties of the tissue; however, a few processing steps are required before this information can be used to apply attenuation correction in a SPECT reconstruction. The CT scan is based upon a polyenergetic x-ray beam. As the linear attenuation coefficients are dependent on both the energy of the photon and attenuating material, the CT numbers need to be converted into the attenuation coefficients appropriate to the energy of the gamma ray(s) emitted by the SPECT isotope. Unfortunately, due to the different effective atomic numbers of different tissues in the body, a single conversion factor is insufficient. In particular, although most soft tissues scale in a fashion very similar to water, bone does not; the effective atomic number of bone is 12.3 compared to water at 7.5. A common approach to solving this problem is to reconstruct the CT image, separate each pixel into fractions of water and bone based on the CT number, scale the "water" and bone images independently, then recombine them (Seo et al. 2008). In practice, this can also be easily implemented using a look-up table of conversion values. An additional complication is the presence of contrast. Iodinated contrast has a very high atomic number (Z_{iodine} = 53) and so greatly attenuates the x-ray beam. In a simple segmentation scheme, it can be included in the bone image, but its attenuation is proportionately much different at the energies used in nuclear medicine. Thus, care needs to be taken if a contrast-CT image is to be used to generate an attenuation map.

The CT image has much higher spatial resolution than the nuclear medicine image, which is very helpful when locating the site of nuclear medicine tracer accumulation (as is done in cancer imaging), but can also introduce AC artifacts if it is not matched to the SPECT resolution (Meikle et al. 1993). The poor resolution of the SPECT image causes the apparent source of activity to blur from one region into the next (for example, from the heart to the lung). If the source appears to come from a region of lower attenuation, it will be undercorrected, and if it appears to be from a region of higher attenuation, it will be overcorrected. This is easily avoided by smoothing the CT to match the SPECT image resolution.

Finally, the CT image needs to be registered to the emission data set. Like the problem of mismatched resolution, if the emission data are registered to the wrong region, it can lead to overcorrection or undercorrection. In cardiac imaging, the most problematic error occurs if the emission image of the myocardial wall extends into the lung. The reduced attenuation of the lung can suppress the apparent activity in the myocardium, causing the artifactual appearance of reduced tracer uptake. Hybrid SPECT/CT imaging systems where the patient is on the same imaging pallet and CT and SPECT imaging are performed in rapid succession reduce the chance of patient movement between scans and facilitates the SPECT/CT coregistration. However, even with the hybrid systems, the coregistration needs to be confirmed prior to attenuation correction as movement, for example, respiratory motion, can occur. Part of the problem in registering CT images to SPECT images is that the CT FOV may be smaller than the SPECT FOV and truncation can occur. Truncation can lead to artifacts in the CT image, which may propagate into the corrected SPECT image (Lalush and Tsui 2000; Chen et al. 2005). Algorithms are available with many hybrid systems to extend the FOV of the CT reconstruction and minimize the CT artifacts (Beyer et al. 2006), but effort should be made to minimize truncation if possible and care should be taken in evaluating images near the edge of the true FOV.

1.6.3 HYBRID SPECT/CT CAMERA DESIGNS

Software coregistration of a CT scan acquired on a separate scanner can be used to perform CT-AC for SPECT (Wells et al. 2012). However, combining a CT and a SPECT camera to create a hybrid imaging system greatly increases the ease with which the CT can be applied to the SPECT reconstruction. Hybrid imaging protocols reduce the time between scans and allow the patient to be imaged on the same bed in exactly the same position, which minimizes changes in the patient orientation between the CT and SPECT images. There are a wide variety of hybrid systems available for SPECT/CT imaging. The CT component divides roughly into two categories: slow-rotation (nondiagnostic) and fast-rotation (diagnostic) scanners as described in the next sections.

1.6.3.1 SLOW-ROTATION CT

The slow-rotation systems mount the CT x-ray tube and detector on the same gantry as the SPECT detectors. The rotation speed of these systems is therefore much slower than that of a dedicated CT gantry. Typical rotation speed is 2–5 rotations/min and acquisition time for a cardiac FOV are as long as 2–3 min. The x-ray tube current is reduced to keep patient exposure low despite the long acquisition times, and the effective dose for these studies is similar to that from a diagnostic scanner attenuation map acquisition (~0.4 mSv). Two examples of this type of approach are the Hawkeye system (GE Healthcare) that uses a four-slice CT detector with a 2-cm axial FOV and the XCT system (Philips Medical Systems) which uses a 30 × 40-cm flat-panel CT detector with a 14-cm axial FOV. With a slow-rotation CT, image quality for a static phantom is very good, but respiratory motion can introduce significant CT artifacts into images of the thorax. Nevertheless, these CT images are sufficient for AC of SPECT data and no significant clinical difference was observed between images corrected with a slow-rotation CT attenuation map and those corrected by a map based on a diagnostic-quality CT scan (Wells et al. 2012).

1.6.3.2 FAST-ROTATION CT

The other approach to hybrid SPECT/CT systems is to use separate gantries for SPECT and CT. The two scanners are bolted back to back and a single bed moves the patient from one scanner to the next. Decoupling the gantries allows the technology developed for diagnostic CT to be applied and rapid-rotation multislice CT scanners are available. Scanners in SPECT/CT hybrid systems have as many as 64-slice detectors, and subsecond rotation speeds are typical. The use of high-speed, multislice detectors means that acquisitions of only a few seconds are required for a full cardiac FOV. This greatly reduces respiratory-motion artifacts in the CT images. Although the cost of these systems tends to be higher than that of the slow-rotation scanners, the image quality is better. The higher-end fast-rotation systems are equivalent to diagnostic CT scanners and thus capable of diagnostic CT protocols like calcium scoring and angiography (CTA). This allows both nuclear and CT diagnostic evaluation of a patient in a single imaging session, and acquisition of CTA on the same camera as the SPECT may facilitate coregistration of the resulting datasets. Examples of this type of system include the Discovery 670 (GE Healthcare) and the 64-slice Lightspeed CT available with the Ventri and Discovery NM 570c cameras (GE Healthcare); the Symbia series of SPECT/CT systems (Siemens Medical), which offers from 2- to 16-slice CT scanners; and the Precedence system (Philips Medical Systems).

1.6.4 Advantages and disadvantages of SPECT/CT

Compared to radioisotope transmission systems, the advantages of using a CT scan to provide the attenuation map for SPECT AC are speed of acquisition, low noise, lack of interference with the emission data, and no decay of the radioactive sources. Even the slow-rotation CT system produces images in a few minutes, which is still faster than the 5–10 min typically needed for transmission scanning. The high-speed CT scanners can acquire images in a few seconds, which makes the time for transmission imaging negligible next to the time required for the SPECT study. CT images also have considerably lower image noise than radioisotope transmission studies. However, the calculation of attenuation factors is based on a line integral through the attenuation map, so noise in the attenuation map is averaged, and at typical count levels, the noise in the attenuation-corrected image remains dominated by emission noise even with radioisotope transmission scans (Tung and Gullberg 1994). Nevertheless, as transmission sources decay, the noise levels increase and they must be replaced periodically to ensure the noise does not propagate into the corrected images. Noise in the CT images is much less than that in the SPECT images, so its contribution to the corrected emission image is negligible. Because the CT x-ray beam intensity is orders of magnitude higher than the SPECT emissions, the interference of the radioisotope emission signal into the CT signal is also negligible. The CT is turned off when the SPECT image is acquired, so there is no interference from the CT into the emission image either. Finally, radioisotope transmission sources require periodic evaluation for radioisotope leakage and damage and require replacement as they decay, which increases exposure to staff and requires constant adjustments to correct for changing source activity levels.

An advantage of the hybrid camera over CT images acquired on a separate camera, as mentioned previously, is that coregistration is much easier and more accurate. Patient movement between imaging sessions is reduced and hardware coregistration accurately aligns the image coordinate systems of the two cameras.

A disadvantage of the CT is that the faster speed compared to the SPECT acquisition can introduce additional artifacts due to respiratory motion mismatch. In addition, artifacts in the CT image, such as those from metal implants, contrast, or truncation, can propagate into the attenuation corrected image. The addition of a second imaging modality to the camera increases the complexity of the system, which increases the maintenance required, the quality assurance needed, and the likelihood of downtime simply because there are more things that can go wrong. The higher image quality of the CT scan comes with greater patient radiation exposure. Though the effective dose from attenuation-map CT scans is relatively low (~0.4 mSv), the dose is still more than 10× higher than that from radioisotope transmission imaging. However, compared with the effective dose from a standard cardiac perfusion SPECT rest/stress study, the dose remains quite small.

A disadvantage of the hybrid camera compared to using a separate system to acquire the CT scan is that only one modality is being used at once. On a SPECT/CT camera, while the SPECT acquisition is taking place, the CT system sits idle. With a separate CT system, scheduling of CT-only studies could fill in these idle periods, making the use of the CT scanner much more efficient and hence cost-effective.

1.6.5 Synergy of SPECT and CT

Most of the discussion in this chapter has focused on the use of CT to compensate for degrading aspects of SPECT imaging, primarily for attenuation correction but also for model-based approaches to scatter compensation. There is potentially a much greater synergy that could be achieved with the two systems. Though not as important in cardiac SPECT as it is in oncology, localization of tracer uptake is helpful in assessing biodistribution during the development of new tracers with, for example, small-animal SPECT/CT. There, the use of attenuation correction has less impact due to the smaller subject size, but tracer distributions may not be well identified, and thus, the CT can provide a useful reference. More importantly, high-quality CT provides a great deal of valuable information on its own, including coronary calcium scores and CTA. Integration of this information may improve risk stratification of patients undergoing stress perfusion imaging (Slomka et al. 2009a; Ghadri et al. 2012; Kirisli et al. 2014; Mouden et al. 2014). The fusion of anatomical information from CT and physiologic information from SPECT can also help identify stenoses that are hemodynamically significant. Physiologically guided intervention has been shown to be beneficial (Pijls et al. 2010; De Bruyne et al. 2012) and nuclear medicine techniques may be able to noninvasively provide that information (Phillips et al. 2013).

1.7 CONCLUSION

SPECT is a tracer-based imaging technology that provides valuable information on cardiac function. The accuracy of its images is degraded by photon interactions within the patient tissues, the distance-dependent collimator resolution, and patient motion. Transmission-based methods can be incorporated into iterative reconstruction algorithms to accurately compensate for many of these effects. The CT component of hybrid SPECT/CT systems is most commonly used to provide the required transmission maps. CT scanning is less noisy, faster, and avoids the interference of radioisotope transmission approaches, which has led to increased interest in SPECT/CT hybrid systems. Integration of diagnostic quality CT scanners also offers future opportunities to synergistically combine SPECT perfusion with CT-based calcium scoring and angiography. The combination of high-quality anatomical detail with cardiac physiologic imaging makes hybrid SPECT/CT a versatile and powerful diagnostic tool.

REFERENCES

Accorsi R, Metzler SD. 2004. Analytic determination of the resolution-equivalent effective diameter of a pinhole collimator. *IEEE Trans Med Imaging*; 23:750–63.

Ali I, Ruddy TD, Almgrahi A, Anstett FG, Wells RG. 2009. Half-time SPECT myocardial perfusion imaging with attenuation correction. *J Nucl Med*; 50:554–62.

Anger HO. 1958. Scintillation camera. *Rev Sci Instr*; 29:27–33.

Anger HO. 1964. Scintillation camera with multichannel collimators. *J Nucl Med*; 5:515–31.

Barrett HH, Wilson DW, Tsui BM. 1994. Noise properties of the EM algorithm: I. Theory. *Phys Med Biol*; 39:833–46.

Bateman TM, Heller GV, McGhie AI, Courter SA, Golub RA, Case JA, Cullom SJ. 2009. Multicenter investigation comparing a highly efficient half-time stress-only attenuation correction approach against standard rest-stress Tc-99m SPECT imaging. *J Nucl Cardiol*; 16:726–35.

Beekman FJ, Kamphuis C, Frey EC. 1997. Scatter compensation methods in 3D iterative SPECT reconstruction: A simulation study. *Phys Med Biol*; 42:1619–32.

Beyer T, Bockisch A, Kühl H, Martinez MJ. 2006. Whole-body 18F-FDG PET/CT in the presence of truncation artifacts. *J Nucl Med*; 47:91–9.

Blevis IM, Tsigelman A, Pansky A, Altman H, Hugg JW. 2004. Investigation of spectral response of monolithic CZT for imaging. *IEEE Nucl Sci Symp/Med Imaging Conf Rec*; 7:4556–9.

Bocher M, Blevis IM, Tsukerman L, Shrem Y, Kovalski G, Volokh L. 2010. A fast cardiac gamma camera with dynamic SPECT capabilities: Design, system validation and future potential. *Eur J Nucl Med Mol Imaging*; 37:1887–902.

Botvinick EH, Zhu YY, O'Connell WJ, Dae MW. 1993. A quantitative assessment of patient motion and its effect on myocardial perfusion SPECT images. *J Nucl Med*; 34:303–10.

Burrell S, MacDonald A. 2006. Artifacts and pitfalls in myocardial perfusion imaging. *J Nucl Med Technol*; 34:193–211.

Chen J, Galt JR, Case JA, Ye J, Cullom SJ, Durbin MK, Shao L, Garcia EV. 2005. Transmission scan truncation with small-field-of-view dedicated cardiac SPECT systems: Impact and automated quality control. *J Nucl Cardiol*; 12:567–73.

Cherry SR, Sorenson JA, Phelps ME. 2003. *Physics in Nuclear Medicine* (3rd edition). Saunders (Philadelphia, PA, USA).

Datz FL, Gullberg GT, Zeng GL, Tung CH, Christian PE, Welch A, Clack R. 1994. Application of convergent-beam collimation and simultaneous transmission emission tomography to cardiac single-photon emission computed tomography. *Semin Nucl Med*; 24:17–37.

De Bruyne B, Pijls NH, Kalesan B, Barbato E, Tonino PA, Piroth Z, Jagic N, Möbius-Winkler S, Rioufol G, Witt N, Kala P, MacCarthy P, Engström T, Oldroyd KG, Mavromatis K, Manoharan G, Verlee P, Frobert O, Curzen N, Johnson JB, Jüni P, Fearon WF; FAME 2 Trial Investigators. 2012. Fractional flow reserve-guided PCI versus medical therapy in stable coronary disease. *N Engl J Med*; 367:991–1001.

Defrise M, Clack R, Townsend DW. 1995. Image reconstruction from truncated, two-dimensional, parallel projections. *Inv Prob*; 11: 287–313.

DePuey EG, Bommireddipalli S, Clark J, Leykekhman A, Thompson LB, Friedman M. 2011. A comparison of the image quality of full-time myocardial perfusion SPECT vs wide beam reconstruction half-time and half-dose SPECT. *J Nucl Cardiol*; 18:273–80.

Eckelman WC. 2009. Unparalleled contribution of technetium-99m to medicine over 5 decades. *JACC Cardiovasc Imaging*; 2:364–8.

Einstein AJ, Moser KW, Thompson RC, Cerqueira MD, Henzlova MJ. 2007. Radiation dose to patients from cardiac diagnostic imaging. *Circulation*; 116:1290–305.

Erlandsson K, Kacperski K, van Gramberg D, Hutton BF. 2009. Performance evaluation of D-SPECT: A novel SPECT system for nuclear cardiology. *Phys Med Biol*; 54:2635–49.

Flohr TG, Raupach R, Bruder H. 2009. Cardiac CT: How much can temporal resolution, spatial resolution, and volume coverage be improved? *J Cardiovasc Comput Tomogr*; 3:143–52.

Franc BL, Acton PD, Mari C, Hasegawa BH. 2008. Small-animal SPECT and SPECT/CT: Important tools for preclinical investigation. *J Nucl Med*; 49:1651–63.

Funk T, Kirch DL, Koss JE, Botvinick E, Hasegawa BH. 2006. A novel approach to multipinhole SPECT for myocardial perfusion imaging. *J Nucl Med*; 47:595–602.

Garcia EV. 2007. SPECT attenuation correction: An essential tool to realize nuclear cardiology's manifest destiny. *J Nucl Cardiol*; 14:16–24.

Germano G, Slomka PJ, Berman DS. 2007. Attenuation correction in cardiac SPECT: The boy who cried wolf? *J Nucl Cardiol*; 14:25–35.

Ghadri JR, Fiechter M, Veraguth K, Gebhard C, Pazhenkottil AP, Fuchs TA, Templin C, Gaemperli O, Kaufmann PA. 2012. Coronary calcium score as an adjunct to nuclear myocardial perfusion imaging for risk stratification before noncardiac surgery. *J Nucl Med*; 53:1081–6.

Gilland DR, Mair BA, Parker JG. 2008. Motion estimation for cardiac emission tomography by optical flow methods. *Phys Med Biol*; 53:2991–3006.

Glick SJ, Penney BC, King MA, Byrne CL. 1994. Noniterative compensation for the distance-dependent detector response and photon attenuation in SPECT imaging. *IEEE Trans Med Imaging*; 13:363–74.

Gravier E, Yang Y, King MA, Jin M. 2006. Fully 4D motion-compensated reconstruction of cardiac SPECT images. *Phys Med Biol*; 51:4603–19.

Hudson HM, Larkin RS. 1994. Accelerated image reconstruction using ordered subsets of projection data. *IEEE Trans Med Imaging*; 13:601–9.

Hutton BF, Buvat I, Beekman FJ. 2011. Review and current status of SPECT scatter correction. *Phys Med Biol*; 56:R85–112.

Kachelrieß M, Knaup M, Kalender WA. 2004. Extended parallel backprojection for standard three-dimensional and phase-correlated four-dimensional axial and spiral cone-beam CT with arbitrary pitch, arbitrary cone-angle, and 100% dose usage. *Med Phys*; 31:1623–41.

Keinert F. 1989. Inversion of k-plane transforms and applications in computer tomography. *SIAM Rev*; 31:273–98.

King MA, Tsui BM, Pan TS. 1995. Attenuation compensation for cardiac single-photon emission computed tomographic imaging: Part 1. Impact of attenuation and methods of estimating attenuation maps. *J Nucl Cardiol*; 2:513–24.

Kirisli HA, Gupta V, Shahzad R, Al Younis I, Dharampal A, Geuns RJ, Scholte AJ, de Graaf MA, Joemai RM, Nieman K, van Vliet L, van Walsum T, Lelieveldt B, Niessen WJ. 2014. Additional diagnostic value of integrated analysis of cardiac CTA and SPECT MPI using the SMARTVis system in patients with suspected coronary artery disease. *J Nucl Med*; 55:50–7.

Kordolaimi SD, Argentos S, Pantos I, Kelekis NL, Efstathopoulos EP. 2013. A new era in computed tomographic dose optimization: The impact of iterative reconstruction on image quality and radiation dose. *J Comput Assist Tomogr*; 37:924–31.

Kovalski G, Israel O, Keidar Z, Frenkel A, Sachs J, Azhari H. 2007. Correction of heart motion due to respiration in clinical myocardial perfusion SPECT scans using respiratory gating. *J Nucl Med*; 48:630–6.

Lalush DS, Tsui BM. 2000. Performance of ordered-subset reconstruction algorithms under conditions of extreme attenuation and truncation in myocardial SPECT. *J Nucl Med*; 41:737–44.

Mansoor MR, Heller GV. 1999. Gated SPECT imaging. *Semin Nucl Med*; 29:271–8.

Matsumoto N, Berman DS, Kavanagh PB, Gerlach J, Hayes SW, Lewin HC, Friedman JD, Germano G. 2001. Quantitative assessment of motion artifacts and validation of a new motion-correction program for myocardial perfusion SPECT. *J Nucl Med*; 42:687–94.

McNamara JE, Pretorius PH, Johnson K, Mukherjee JM, Dey J, Gennert MA, King MA. 2009. A flexible multicamera visual-tracking system for detecting and correcting motion-induced artifacts in cardiac SPECT slices. *Med Phys*; 36:1913–23.

Meikle SR, Dahlbom M, Cherry SR. 1993. Attenuation correction using count-limited transmission data in positron emission tomography. *J Nucl Med*; 34:143–50.

Melcher CL. 2000. Scintillation crystals for PET. *J Nucl Med*; 41:1051–5.

Mouden M, Ottervanger JP, Timmer JR, Reiffers S, Oostdijk AH, Knollema S, Jager PL. 2014. The influence of coronary calcium score on the interpretation of myocardial perfusion imaging. *J Nucl Cardiol*; 21(2):368–74.

Narayanan MV, King MA, Pretorius PH, Dahlberg ST, Spencer F, Simon E, Ewald E, Healy E, MacNaught K, Leppo JA. 2003. Human-observer receiver-operating-characteristic evaluation of attenuation, scatter, and resolution compensation strategies for (99m)Tc myocardial perfusion imaging. *J Nucl Med*; 44:1725–34.

O'Connor MK, Kanal KM, Gebhard MW, Rossman PJ. 1998. Comparison of four motion correction techniques in SPECT imaging of the heart: A cardiac phantom study. *J Nucl Med*; 39:2027–34.

Phillips LM, Hachamovitch R, Berman DS, Iskandrian AE, Min JK, Picard MH, Kwong RY, Friedrich MG, Scherrer-Crosbie M, Hayes SW, Sharir T, Gosselin G, Mazzanti M, Senior R, Beanlands R, Smanio P, Goyal A, Al-Mallah M, Reynolds H, Stone GW, Maron DJ, Shaw LJ. 2013. Lessons learned from MPI and physiologic testing in randomized trials of stable ischemic heart disease: COURAGE, BARI 2D, FAME, and ISCHEMIA. *J Nucl Cardiol*; 20:969–75.

Pijls NH, Fearon WF, Tonino PA, Siebert U, Ikeno F, Bornschein B, van't Veer M, Klauss V, Manoharan G, Engstrm T, Oldroyd KG, Ver Lee PN, MacCarthy PA, De Bruyne B, and FAME Study Investigators. 2010. Fractional flow reserve versus angiography for guiding percutaneous coronary intervention in patients with multivessel coronary artery disease: 2-year follow-up of the FAME (fractional flow reserve versus angiography for multivessel evaluation) study. *J Am Coll Cardiol*; 56:177–84.

Seo Y, Mari C, Hasegawa BH. 2008. Technological development and advances in single-photon emission computed tomography/computed tomography. *Semin Nucl Med*; 38:177–98.

Shepp LA, Vardi Y. 1982. Maximum likelihood reconstruction for emission tomography. *IEEE Trans Med Imaging*; 1:113–22.

Slomka PJ, Cheng VY, Dey D, Woo J, Ramesh A, Van Kriekinge S, Suzuki Y, Elad Y, Karlsberg R, Berman DS, Germano G. 2009a. Quantitative analysis of myocardial perfusion SPECT anatomically guided by coregistered 64-slice coronary CT angiography. *J Nucl Med*; 50:1621–30.

Slomka PJ, Patton JA, Berman DS, Germano G. 2009b. Advances in technical aspects of myocardial perfusion SPECT imaging. *J Nucl Cardiol*; 16:255–76.

Slomka PJ, Dey D, Duvall WL, Henzlova MJ, Berman DS, Germano G. 2012. Advances in nuclear cardiac instrumentation with a view towards reduced radiation exposure. *Curr Cardiol Rep*; 14:208–16.

Smith MF, Jaszczak RJ. 1997. The effect of gamma ray penetration on angle-dependent sensitivity for pinhole collimation in nuclear medicine. *Med Phys*; 24:1701–9.

Tung CH, Gullberg GT. 1994. A simulation of emission and transmission noise propagation in cardiac SPECT imaging with nonuniform attenuation correction. *Med Phys*; 21:1565–76.

Wagenaar, DJ. 2004. CdTe and CdZnTe semiconductor detectors for nuclear medicine imaging in *Emission Tomography* (Wernick and Aarsvold, editors, Elsevier Academic Press, London); 269–91.

Wells RG, Soueidan K, Vanderwerf K, Ruddy TD. 2012. Comparing slow- versus high-speed CT for attenuation correction of cardiac SPECT perfusion studies. *J Nucl Cardiol*; 19:719–26.

Cardiovascular PET-CT

ETIENNE CROTEAU, RAN KLEIN, JENNIFER M. RENAUD, MANUJA PREMARATNE, AND ROBERT A. DEKEMP

2.1 POSITRON EMISSION TOMOGRAPHY PRINCIPLES

2.1.1 RADIOTRACERS

Positron emission tomography (PET) is the leading tool in nuclear cardiology for noninvasive assessment of molecular function. Images are obtained via detection of positrons emitted from the decay of an injected radiotracer. Radiotracers are either short-lived isotopes themselves, such as ^{82}Rb, or isotopes that have been incorporated into biological or drug compounds, such as ^{18}F-fluoro-deoxyglucose or ^{11}C-methyl-losartan, respectively. The amount of tracer injected is low enough such that it does not affect the physiological process being imaged. Most isotopes are produced in a cyclotron and undergo radiochemical synthesis to be incorporated into a tracer molecule, requiring an onsite or local cyclotron due to the short half-lives. There has been a shift toward simpler and more cost-effective onsite alternatives, such as the generator-produced tracer ^{82}Rb. Some of the most common clinical and research-based cardiac PET tracers, their characteristics, applications, and the associated imaging protocols are listed in Table 2.1 (Zober et al. 2006; Thackeray and Bengel 2013; Danad, Raijimakers, and Knaapen 2013).

2.1.2 COINCIDENCE DETECTION

PET is built on the technique of coincidence photon detection. When a positron is emitted as the radioisotope nucleus decays, it travels a short distance in the surrounding tissue, referred to as the positron range (Table 2.1), until it annihilates with an electron, resulting in the production of two 511 keV photons propagating in nearly opposite directions (Figure 2.1). Longer positron ranges result in lower reconstructed image spatial resolution. Upon leaving the body, if the photons are detected within the short interval of the coincidence timing window (3 to 12 ns), a prompt coincidence is recorded. Due to the finite width of the timing window, four types of coincidences can be detected: true (unscattered), Compton scattered, random (accidental), and cascade (prompt-gamma). True coincidences are recordings of photon pairs that originated from a single decay event and did not undergo any scatter or attenuation by molecules in the body. Scattered events are recorded when one or both of the photons from a single decay event are Compton scattered within the tissue and subsequently misaligned. Random coincidences are the result of two photons from separate decay events simultaneously striking opposing detectors (Figure 2.1) by chance. Cascade prompt-gamma coincidences occur when an annihilation photon is detected together with a scattered cascade gamma photon. Prompt-gamma, random, and scattered coincidences increase image noise and decrease contrast, depending on the amount of

Table 2.1 Commonly used cardiovascular PET tracers

Radiotracer	Half-life	Positron range (RMS)	Physiologic application	Imaging protocol
Clinical				
⁸²Rb-rubidium	1.27 min	2.6 mm	Perfusion	Rest/stress and gated
¹³N-ammonia	9.97 min	0.57 mm	Perfusion	Rest/stress and gated
¹⁸F-FDG	110 min	0.23 mm	Glucose metabolism	Rest/gated
Research				
¹⁵O-water	2.07 min	1.0 mm	Perfusion	Rest/stress
¹⁵O-carbon monoxide	2.07 min		Blood volume/contractile function	Rest/gated
¹⁸F-flurpiridaz	110 min	0.23 mm	Perfusion	Rest/stress and gated
¹⁸F-FTHA	110 min		Fatty acid metabolism	Rest and gated
¹¹C-palmitate	20.4 min	0.39 mm	Fatty acid metabolism	Rest
¹¹C-acetate	20.4 min		Oxidative metabolism	Rest/stress and gated
¹¹C-KR31173	20.4 min		Angiotensin II type I receptor	Rest
¹¹C-hydroxyephedrine	20.4 min		Adrenergic receptor	Rest

Source: Zober TG, Mathews WB, Seckin E et al. *Nucl Med Biol*, 33, 5–13, 2006; Thackeray JT, Bengel FM. *J Nucl Cardiol*, 20, 150–165, 2013; Danad I, Raijmakers PG, Knaapen P. *J Nucl Cardiol*, 20, 874–890, 2013.

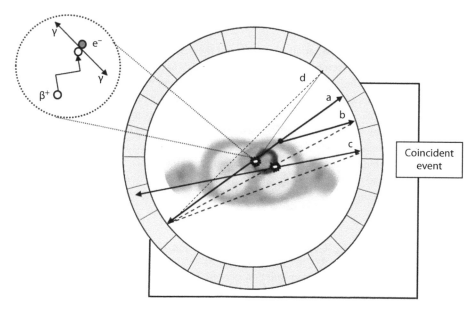

Figure 2.1 Upon radioisotope decay (inset), a positron (β^+) is emitted from the atomic nucleus, which travels through the tissue until it annihilates with an electron (e^-). The resultant collinear annihilation photons (solid black lines) escape from the body and are detected as a coincidence event, indicating that the decay occurred along the LOR (dashed lines) connecting the two detectors. Four types of coincidences can be registered: (a) true (unscattered), (b) Compton scattered, (c) random, and (d) prompt-gamma (dotted line).

radioactivity present in the scanner field of view (FOV), as well as the mass, density, and extent of the tissue traveled through. Corrections for these physical effects are required to obtain an accurate representation of the radiotracer distribution in the subject.

To detect many annihilation photons, a PET scanner is typically constructed of multiple circular rings of scintillation crystal detector blocks stacked together, allowing line of response (LOR) projections through the

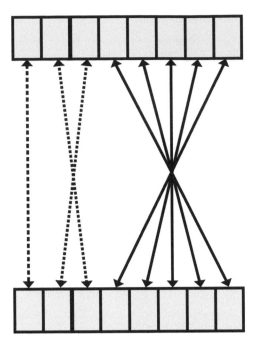

Figure 2.2 Axial cross-section of a multiring PET system illustrating subsets of direct (dashed line), cross-plane (dotted lines), and oblique (solid lines) LORs.

subject to be obtained simultaneously at all angles around the object. The crystals are most often coupled to photomultiplier tubes that convert incoming scintillation photons to photoelectrons, which can then be processed by the coincidence timing circuitry to determine if a valid event occurred, that satisfies the minimum energy and timing requirements. In the two-dimensional (2D) imaging mode used on previous generations of PET scanners, thin septa were placed between adjacent crystal rings to attenuate scattered photons, so that coincidences were only recorded using detectors within the same ring (direct-plane LORs) or in closely neighboring rings (cross-plane LORs). On current 3D PET systems, the septa have been removed, allowing for coincidence events along all oblique LORs to be recorded (Figure 2.2). While this mode increases the sensitivity of the system to true coincidences allowing for higher image quality, the sensitivity to scattered and random events is also increased so accurate corrections are even more essential compared to the older 2D imaging mode.

2.1.3 Data storage, correction, and image reconstruction

A coincidence event indicates that an annihilation occurred somewhere along the line between two PET detectors, implying that the radioactive atom that decayed should be located along that same LOR. Using this method, coincidence events are recorded in projection LORs, defined by a particular radial and angular coordinate corresponding to each possible detector pair. Data from the LORs can then be organized into a histogram, where each bin corresponds to a unique radial and angular location in the FOV. A common histogram format is referred to as a sinogram, indicating that a point source of activity traces a sinusoidal shape when stored in this matrix form. Coincidence events can alternatively be stored in list-mode format, where each detector pair defining the LOR, along with the corresponding time of arrival, are stored as separate entries. Multiple sinograms can then be created retrospectively as desired following data acquisition.

To correct for random, scattered, and prompt-gamma coincidences, subtraction of their measured or modeled event rates from the total measured prompts is performed prior to image reconstruction, or addition of their event rates during reconstruction. Another physical effect that must be corrected is photon attenuation.

Annihilation photons can be scattered out of the scanner FOV or absorbed (attenuated) by the tissue, decreasing the probability that they will exit the body and be detected. To correct for this effect, acquired data are scaled using a map of the tissue attenuation coefficients in the FOV, typically obtained via a fast, low-dose coregistered computed tomography (CT) scan. Additionally, data must be corrected for the decrease in activity over time due to the effect of radioisotope decay.

Further corrections are required to account for the detector-related effects of nonuniform efficiency and dead-time losses. To ensure optimal detection sensitivity, daily quality assurance is required to check that the detector response has not drifted in comparison to an established reference scan. Dead time is dependent on the type of scintillation material used and is defined as the period during which the detector (and associated electronics) is processing an event, rendering it unable to detect any additional events, resulting in an underestimation of tracer activity. The lost counts can be compensated using a model of the scanner's response to a range of measured count rates acquired during a decaying-isotope calibration scan. All of these previous corrections must be performed as part of the tomographic reconstruction process to obtain accurate images of isotope activity concentration (Bq/cc).

To obtain an image of the tracer distribution from a sinogram, 2D image reconstruction can be performed. In this approach, each tomographic plane of data is treated independently and a single transaxial plane is reconstructed at a time. While analytic filtered back projection (FBP) has been the most common reconstruction method, it has been largely replaced by iterative techniques including maximum likelihood expectation maximization and ordered subset expectation maximization, due to the availability of more powerful computers and the shift to 3D imaging technology.

Iterative reconstruction techniques provide images with decreased noise in comparison to FBP by modeling the Poisson noise from the raw coincidence counts, particularly within areas of high tracer uptake. With 3D reconstruction, additional information about the detector geometry and timing of the coincidence events can also be incorporated to yield more accurate images. When a photon near the edge of the transaxial FOV strikes a detector, it does so at an angle and may first travel through (or be scattered in) an adjacent crystal, causing a distortion in the placement of the LOR. To compensate for this effect, the detector point spread function (PSF) can be incorporated into the reconstruction algorithm. The PSF describes the response of each detector to a point source, and it can be obtained analytically, through simulations or experimental measurements. Once incorporated into the reconstruction algorithm, the recorded counts can be more accurately positioned in their original spatial locations, thereby improving spatial resolution. Furthermore, fast detectors with high light output can provide picosecond timing resolution, allowing the approximate location of the coincidence event along the LOR to be determined using the difference in arrival times of the two annihilation photons. This technique is referred to as time-of-flight (ToF), and it allows the spatial locations of individual events to be constrained to a shorter distance along the LOR during reconstruction, resulting in a higher image signal-to-noise ratio for the same number of recorded counts. Both ToF and PSF modeling results in better quality images and provide higher accuracy for estimates of physiologic quantities, including myocardial blood flow (MBF), metabolism, and cell signaling (Akamatsu et al. 2012; Armstrong, Tonge, and Arumugam 2014).

2.1.4 QUANTIFICATION

From the reconstructed tomographic images, the spatial and temporal distributions of the radiotracer in the organ of interest are obtained. With the appropriate corrections, calibration, and knowledge of the specific activity (GBq/mmol) of the administered tracer, the isotope activity or tracer molar concentration per unit of tissue volume (Bq/cc or mmol/L) can be assessed quantitatively. Semiquantitative biodistribution analysis can also be performed, where the activity concentration is normalized by the total amount of activity injected and the subject body weight to determine the standard uptake value in g/cc or g/mL. Additionally, dynamic PET imaging can be used to assess in vivo physiology, biochemistry, or receptor binding of a tracer by quantifying properties such as organ perfusion (mL/min/g), substrate metabolism (mol/min/g), receptor density (pmol/cc), or receptor-ligand binding potential (B_{max}/K_d) (Beauregard et al. 2007).

2.2 X-RAY CT PRINCIPLES

2.2.1 HISTORY

First invented by Godfrey Hounsfield in the 1960s, the CT scanner produces tomographic slices of the object in the FOV. The CT consists of a rotating gantry with an x-ray source that "illuminates" the subject and an array of x-ray detectors that measure the x-ray intensity that penetrates the subject. The detected signal forms a projection (shadow) that corresponds to the integral attenuation by the subject along lines between the corresponding sources and detector, as in radiography. The rotating gantry, depicted in Figure 2.3a, enables imaging from all projection angles, which are then used to reconstruct transaxial images (tomograms) of the relative attenuation distribution within the FOV. Using a combination of multislice detectors and a moving patient bed, many tomograms can be stacked to generate a volume image of the subject anatomy (Kalender 2005).

2.2.2 X-RAY SOURCES

X-ray (initially called Röentgen) radiation is created by bombarding a target material (e.g., tungsten) with electrons that have been accelerated across an electrically charged gap. Most commonly, an x-ray tube such as depicted in Figure 2.3b is employed. A filament coil is heated using an electric current generating a cloud of excited electrons. A strong electric field is induced between the cathode (negatively charged) filament and the anode (positively charged) target, which accelerates the electrons toward the target. The target is typically devised as a rapidly rotating disk coupled to a heat-dissipating material (e.g., copper) and liquid cooled.

As the electrons bombard the target, they undergo physical interactions, resulting in the production of x-radiation emission. A fraction of the electrons will "knock out" orbital electrons in the target material, and as these vacancies become filled by free electrons, discrete, characteristic energy photons are emitted. The majority of the photons are produced by scattering of the electrons within the target material, producing a broad spectrum of photon energies called Bremsstrahlung radiation. The maximum energy photons (and the photon energy spectrum) are determined by the *tube voltage* measured in kV, and the quantity (flux) of photon is determined by the electric *tube current* measured in mA; these two parameters are important controls of the x-ray radiation used during imaging. Higher kV and mA values result in higher-quality images, but also increased radiation dose to the patient.

Since low-energy photons are readily absorbed by the body, they do not contribute to the imaging process but greatly increase the patient radiation dose; a hardening filter (typically made of aluminum) is used to

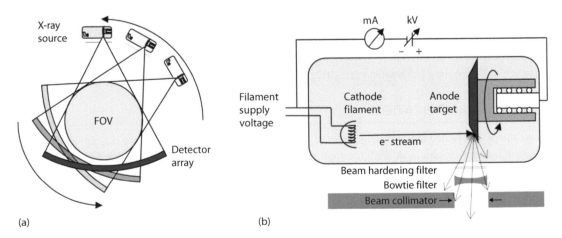

(a)   (b)

Figure 2.3 (a) Typical CT scanner configuration consisting of a rotating gantry with x-ray source and curved detector array defining the transaxial FOV. (b) X-ray source diagram consisting of the x-ray tube, filters, and collimator. The tube voltage (kV) and current (mA) are important controls in a CT scanner.

attenuate these low-energy photons, while transmitting most of the high-energy photons. A bow-tie filter may also be employed to reduce beam intensities at the edge of the FOV, so as to reduce radiation exposure to the extremities, which are thinner than the torso. Finally, a beam collimator is used to absorb all photons that are outside of the detector coverage.

2.2.3 X-RAY DETECTORS

Current CT detectors are typically composed of an array of fast scintillating materials coupled to semiconductor detectors (Kalender 2005). Most commonly, the detectors are arranged in a long row (arc with ~1 m radius centered about the x-ray source) consisting of hundreds of individual detector elements, enabling to capture projections of a single slice through the patient. The detectors are coupled to an antiscatter collimator, which is designed to reject off-axis photons that have scattered in the patient and are a source of image noise. Variability in the sensitivity of each detector element is expected due to inconsistent manufacturing and geometric efficiency. A periodic air calibration scan, without anything in the FOV, is conducted as part of a quality assurance system to account for the relative sensitivity of each detector element (Goldman 2007).

2.2.4 2D RECONSTRUCTION (FULL-SCAN AND HALF-SCAN)

CT instrumentation measures the integral x-ray attenuation along finely spaced line integrals or projections between the x-ray source and the detector elements. Projections from all (360°) angles are fed into a computer algorithm to reconstruct images of the FOV. The realization that opposing line integrals offer redundant information has resulted in the use of half-scan (180° + ½-scan-angle) over full-scan (360°) image reconstruction (Kalender 2005). Half-scan imaging can be used to increase axial sampling (with x-ray source focal spot axial shifting) and improve temporal resolution.

Image reconstruction is developed as an integral part of any CT scanner and must account for the specific scanner geometries. FBP has been used in CT reconstruction since its conception (Webb 1992). Prior to "smearing" the projection signal back onto the image space, it is ramp-filtered to correct for the higher sampling density at the center of FOV (rotation) compared to the radial edges. Additional (low-pass) filtering may be applied to reduce image noise. With modern parallel computers, FBP tomographic image slices are reconstructed in a fraction of a second.

2.2.5 RADIODENSITY QUANTIFICATION (HU)

CT images represent the relative attenuation, or radiodensity, of the materials in the FOV. Image intensity is commonly scaled to Hounsfield units (HU), which is defined to be 0 for distilled water at standard room temperature and pressure and −1000 HU for air, thus giving consistent numbers for soft tissues as demonstrated in Figure 2.4 (Mark et al. 2010). Accurate conversion between HU and linear attenuation coefficients is typically not possible due to the polyenergetic spectrum of x-ray sources (as illustrated for aluminum). Nonlinear conversion to electron-density units (e.g., for the purpose of proton therapy planning), however, has been demonstrated (Matsufuji et al. 1998). CT images are usually represented on a 12-bit scale of −1000 (air) to 3091 (dense bone) HU; however, depending on the FOV and organ of interest, the image intensity can be "windowed" to increase image contrast. Contrast agents (e.g., iodine-based to highlight blood) may be administered to enhance contrast between tissues or to evaluate physiologic function (e.g., ventilated lung volume with xenon gas, perfusion with iodinated blood contrast) (Lusic and Grinstaff 2013). The contrast-enhanced HU values are dependent on the in vivo contrast concentration.

2.2.6 DATA ACQUISITION

Data acquisition commences with positioning of the patient on the scanner bed and determining the axial range for imaging. Based on multiple considerations such as organ of interest, camera type, patient size, radiation dose,

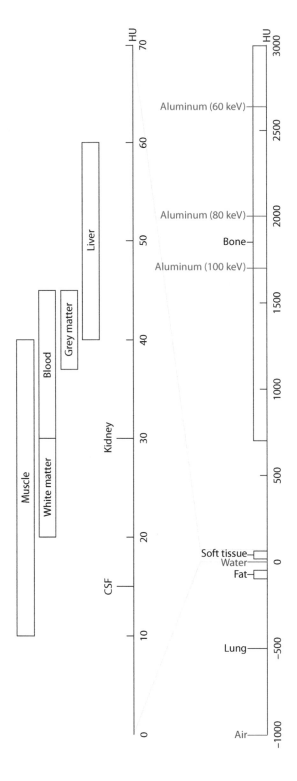

Figure 2.4 HU of human tissues. The HU scale is defined as 0 for distilled water, −1000 for air, with other CT numbers interpolated/extrapolated accordingly, but variations can occur depending on tube voltage and beam hardening, as demonstrated for aluminum.

Figure 2.5 CT acquisition modes with rings representing scanner projection angles over the scan time (ranging from dark to light). (a) Step-and-shoot, (b) spiral-helical, and (c) dynamic-cine.

image quality, and temporal resolution, exposure parameters must be preconfigured prior to data acquisition. For all imaging modes, the tube voltage (kV), tube current (mA), and gantry rotation time (ms) are required.

In order to acquire images with large axial range (e.g., whole body), the step-and-shoot acquisition mode was first introduced, in which the bed is shifted at fixed intervals and remains stationary during acquisition, as illustrated in Figure 2.5a. To shorten scans with long axial extents and to increase spatial resolution, spiral (or helical) acquisition was later developed. The bed moves at a constant velocity through the scanner, as the gantry is rotated continuously, resulting in spiral coverage of the patient, as illustrated in Figure 2.5b. The rate at which the bed moves compared to the gantry rotation time is referred to as the pitch. Large pitches result in shorter scan times at the expense of poorer spatial resolution and possible gaps in the coverage.

Dynamic-cine acquisitions (Figure 2.5c) can be used to monitor changes in the FOV over time, especially as a contrast agent is administered and distributed. The bed remains stationary throughout the imaging session and a series of images is acquired at different times, generating a cine-graph. Variations of the dynamic acquisition exist in which the bed is "rocked" between two positions, increasing the FOV at the expense of temporal resolution.

Cardiac CT has seen tremendous growth in recent years and requires special considerations due to cardiac motion (Mark et al. 2010). Gantry rotation speeds (3 to 0.5 rotations per second) are on the order of a resting heart rate (HR) (1 beat per second). Thus, image reconstruction using a single continuous acquisition results in imaging artifacts associated with projections being acquired at different phases of the cardiac cycle (Mark et al. 2010). In order to synchronize all projections to the same heart phase, electrocardiography (ECG) gating is recorded during data acquisition. ECG gating can be used either prospectively or retrospectively (Moorees and Bezak 2012).

In *prospectively gated* acquisitions, the x-ray beam is modulated based on the ECG to turn on only during predefined phases of the cardiac cycle (e.g., 70%–80% of the R–R interval associated with end-diastole, in which dilation is maximal and motion is minimal), resulting in reduced radiation dose to the patient. Since several heart cycles are required for a single acquisition of any slice, prospective acquisitions are used in the step-and-shoot mode.

Retrospectively gated acquisitions consist of continuous x-ray acquisition (and beaming) with simultaneous ECG signal recording. After acquisition is complete, any cardiac phase may be reconstructed, affording the ability to monitor heart wall motion, but at the cost of greater radiation dose to the patient. Retrospective gating can be performed with spiral CT and a low pitch, to avoid large gaps in the projection data of each cardiac phase (Mark et al. 2010).

To reduce patient radiation doses, the tube current can be modulated to the parts of the image acquisition that are of interest and to achieve a target signal to noise ratio. In prospective gating, the x-ray beam is switched on only for portions of the cardiac cycle that are on interest (e.g., 70%–75% of the R–R interval) (Goldman 2007; Mark et al. 2010). Current modulation is further applied based on gantry rotation angle so that higher beam intensities are used to image thick cross-sections and lower intensities are used for thin cross-sections (e.g., sagittal vs. coronal views of the thorax), thus producing a relatively constant detected signal intensity for all projection angles.

2.2.7 MULTISLICE DETECTION GEOMETRY

Modern clinical scanners employ multirow detectors enabling to reconstruct multiple slices from a single bed position. While the number of detector rows is an important property of a CT scanner, vendors often report the number of reconstructed slices, which can be doubled by longitudinally shifting the z-ray source slightly throughout the gantry rotation. The total detector width determines the volume of coverage of the scanner, and the number of rows (and their corresponding width) is an important factor of the axial spatial resolution. Typically, detector rows are (0.5–1.5 mm) wide and may vary from thin (center) to wide (axial extents) in order to account for varying angles of intersection of the x-ray beam with the detector (Kalender 2005). On single-slice scanners, axial coverage is therefore as small as 5 mm; however, with multislice (as many at 320) scanners, axial coverage as large as 12 cm is now available, enabling to image the entire heart within a single bed position—which is particularly advantageous for contrast imaging of rapid concentration changes.

2.2.8 Dual-source and dual-energy systems

CT scanners with two sets of sources and detectors have seen increased interest in recent years due to their improved temporal resolution and ability to image at two tube voltages (energies) simultaneously. The improved temporal resolution of dual-source systems is especially of interest in cardiac applications, enabling motion artifact free imaging in patients with elevated HRs due to disease or stress conditions. While numerous applications of dual-energy CT imaging have been demonstrated (e.g., reducing beam-hardening artifacts, tissue characterization, contrast enhancement, and virtual enhancement) and are possible with sequential imaging on a single-source scanner, dual-source scanners can simultaneously acquire both energy projections, improving data coregistration, especially in the presence of motion (cardiac, respiratory, patient, and contrast distribution) (Silva et al. 2011).

2.2.9 Cone-beam reconstruction

Fan-beam image reconstructions has been successfully employed for decades on single-slice CT and scanners with few slices, where the x-ray beam can be assumed to be thin in the axial direction. However, with multislice CT (>four slices) with large axial coverage, the x-ray beam covers a conical volume in order to illuminate all the detectors. Off-center objects are projected onto different rings as the gantry is rotated, and therefore, standard 2D reconstruction produces artifacts (Kalender 2005). Specialized cone-beam reconstruction is required for cone-beam geometries; typically, these algorithms "rebin" the projection data into parallel off-axis slices before employing fast 2D reconstruction algorithms to generate slice images.

2.2.10 Iterative and model-based reconstruction

More advanced iterative reconstruction methods have been developed that model the imaging physics (e.g., beam hardening and photon scatter) more accurately, producing images with improved spatial resolution and reduces noise (Pan, Sidky, and Vannier 2009). Due to significantly more complex computations, specialized computing hardware is required to iteratively reconstruct images within a reasonable time that would not adversely impact clinical workflow. The higher quality of iteratively reconstructed images compared to FBP can be traded off to reduce patient radiation exposure while maintaining desired image quality. Iterative image reconstruction is typically offered as an optional add-on and therefore is not available in all installations (Pan, Sidky, and Vannier 2009).

2.3 PET IMAGING TECHNIQUES

Several imaging modes are used commonly in clinical practice and research studies to measure myocardial perfusion, metabolism, and cell signaling:

1. Whole-body/static imaging of relative organ distribution after blood clearance
2. ECG-gated cine-imaging of contractile function after blood clearance
3. Respiratory-gated cine-imaging of respiratory motion after blood clearance
4. Dynamic imaging of temporal kinetics during blood clearance and organ distribution

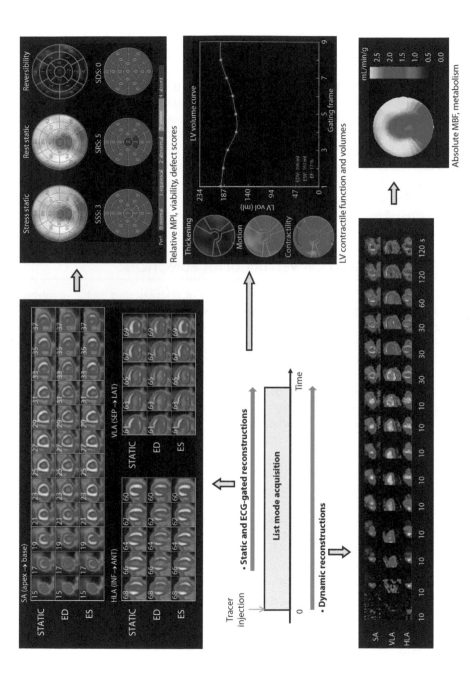

Figure 2.6 List-mode acquisition allows reconstruction of ECG-gated, static (summed), and dynamic images from a single injection. During the uptake phase of the scan, starting ~2 minutes after tracer injection, a static uptake image is reconstructed for myocardial perfusion interpretation. Simultaneous gating of the image with an ECG trigger is used to reconstruct gated images of the beating heart over a similar time period, to interpret contractile function including wall motion, thickening, contractility, and ejection fraction. Finally, a series of dynamic images is reconstructed from the start of tracer injection, to quantify myocardial blood flow. (With kind permission from Springer Science + Business Media: *Handbook of Nuclear Cardiology: Cardiac SPECT and Cardiac PET*. Heller GV, Hendel RC, (Eds), PET instrumentation, Chapter 14, 2012, Renaud J, Beanlands RSB, deKemp RA.)

List-mode imaging is standard on current PET-CT systems, recording individual coincidence counts in a file that can be "replayed" to reconstruct images in any/all of the modes listed previously, from a single injection (Figure 2.6). This allows tremendous flexibility for clinical applications and research.

2.3.1 WHOLE-BODY IMAGING OF TRACER BIODISTRIBUTION

High-throughput PET imaging studies are typically performed in static imaging mode, after clearance from the blood and following stable uptake into the tissues/organs of interest. Routine clinical perfusion/viability studies are a typical example, as shown in Figure 2.7. The tracer (e.g., fluorodeoxyglucose [FDG] for viability) can be administered to the patient outside the scanner, and after some time is allowed for tissue/organ uptake (40–60 min delay), the patient is positioned on the scanner bed for imaging (2–20 min scan). If the heart is the only organ of interest, a single bed position is sufficient to complete the scan. However, if other organs are targeted, then multiple (overlapping) bed positions are scanned, reconstructed, and assembled into a "whole-body" image with the prescribed axial coverage (Figure 2.8a). The scan time depends generally on the injected activity and desired spatial resolution; therefore, the radiation dose to the patient can be minimized in some cases by injecting less activity, at the expense of longer scan times and/or lower resolution images reconstructed with more spatial filtering. For some short half-life perfusion tracers such as ^{82}Rb and ^{13}N-ammonia, the uptake time is very short (1–2 min) so the injection is typically performed with the patient on the scanner bed for added convenience.

2.3.2 ECG-GATED IMAGING OF VENTRICULAR FUNCTION

If cardiac triggers are recorded in the list-file from the R-wave of the ECG signal, then each R–R interval can be divided into a number of equally spaced phases or "gates." When scanned over many heartbeats, the counts in each cardiac phase are used to reconstruct cine-loop images of the beating heart, in a single-bed-position scan. The number of gates depends on the desired image quality, again determined by the injected activity and scan time. For perfusion imaging, 8–16 gates/cycle are generally sufficient for analysis of left and right ventricle ejection fractions, regional wall-motion, and wall thickening. For blood-pool imaging, 24–32 gates/cycle have been used to also evaluate second-order contractile effects such as atrial-kick. ECG-gated perfusion imaging with ^{82}Rb or ^{13}N-ammonia is used commonly to measure the stress-rest left ventricle (LV) ejection fraction reserve, reflecting the immediate contractile effects of ischemia at the exact time of peak-stress.

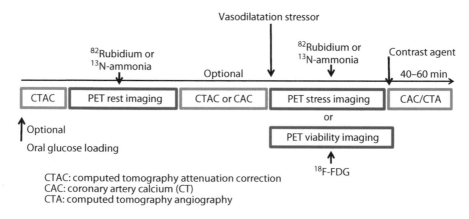

Figure 2.7 Integrated cardiac PET/CT protocol for myocardial perfusion/viability imaging.

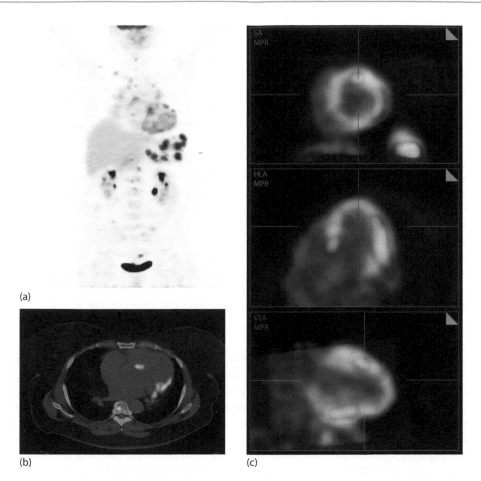

(a)

(b)

(c)

Figure 2.8 FDG PET-CT images of cardiac sarcoid in a 56-year-old woman with newly diagnosed third-degree atrioventricular block due to cardiac sarcoidosis. (a) whole-body MIP image shows multi-focal FDG uptake in the mediastinal and hilar lymph nodes, heart, and spleen. (b) Transaxial slice of fusion PET-CT shows focal myocardial FDG uptake in the basal interventricular septum and lateral wall. (c) Cardiac fusion FDG + perfusion images demonstrate focal FDG uptake in the septal, anterior, inferior, and lateral walls (color), while Rb-82 perfusion is normal (grey).

2.3.3 RESPIRATORY-GATED IMAGING

Analogous to cardiac-gating, a breathing monitor such as pressure-belt or fiducial-marker tracking system can be used to record respiratory-gated triggers into the list-data and reconstruct cine-loop images showing heart displacement over the respiratory cycle. This is extremely useful to detect inferior-wall artifacts that may result from excessive breathing motion such as upward-creep following exercise stress, or in patients with chronic obstructive pulmonary disease or large breathing response to pharmacologic stress. The normal incidence of respiratory-motion artifacts appears to be low (~1%–2%) for routine clinical rest-stress perfusion imaging. Dual cardiac- and respiratory-gating has the potential to improve PET image resolution using "motion-frozen" techniques developed originally for single photon emission computed tomography (SPECT).

CT attenuation correction (CTAC) is an integral component required for accurate PET-CT imaging. For most cardiac applications, a fast (1–2 s) normal-end-expiration CT scan is adequate in patients with quiet shallow breathing because more time is spent in the expiratory- vs. inspiratory-phase during the PET acquisition. In some cases, end-expiration PET or respiratory phase-matched PET-CTAC has been suggested to improve accuracy, for example, following treadmill exercise or hyper-capnea stress. Phase-matched attenuation correction also requires respiratory-gated 4D-cine-CT, which can increase the radiation dose to the patient, unless additional CT dose-reduction strategies are employed. For vascular imaging applications (e.g., aortic plaque), small offsets in the PET-CTAC misalignment outside the heart can produce apparent uptake in the vessel wall that may be misinterpreted as physiologically relevant.

2.3.4 DYNAMIC IMAGING FOR QUANTIFICATION OF PHYSIOLOGIC AND MOLECULAR FUNCTION

An important PET advance on static imaging is to characterize the behavior of the radiotracer with dynamic imaging starting immediately from the time of injection to assess quantitative kinetic physiologic rate. The dynamic PET molecular imaging can characterize rate of perfusion, substrate metabolism, and receptor binding. To obtain the tissue response, the PET kinetic analysis needs usually the assessment of the radioactivity in arterial blood vs. time, the input function, and a physiological model describing the behavior in the myocardial tissue, the kinetic model. Compartment models convolving the input function with a tissue response equation, Patlak graphical analysis for an irreversible radiotracer like ^{18}F-FDG phosphorylated and trapped within the cell (Logan 2000), and Logan graphical analysis for a reversible binding radiotracer like ^{11}C-epinephrine to assess sympathetic neuronal activity (Bengel 2011) are the main PET cardiac quantitative kinetic analysis.

2.3.4.1 PERFUSION

Myocardial perfusion imaging (MPI) is the most common test used to diagnose coronary artery disease (CAD) using relative tracer uptake (0%–100% of maximum) (Figure 2.9a). Absolute MBF uses tracer kinetic modeling to assess quantitative perfusion (mL/min/g) as illustrated in Figure 2.9b. This approach is helpful to characterize the difference between normal and abnormal perfusion or specific diseases where relative imaging cannot distinguish between normal and globally reduced perfusion, such as balanced triple-vessel disease or microvascular disease. There is a distinction in the imaging protocol used to perform MPI or MBF imaging; measurement of the early phase (first-pass) of the radiotracer transit, when the maximum activity is in the blood, is critical for accurate assessment of MBF using dynamic imaging. The ideal characteristics of a radiotracer to assess MBF are high first-pass extraction and subsequent retention in the myocardial tissue, combined with a fast clearance from the blood. ^{15}O-water, ^{13}N-ammonia, and ^{82}Rb-chloride can reliably assess quantitative MBF. The first-pass extraction and net retention fractions are well known (Figure 2.10) in comparison to ^{18}F-flurpiridaz, which appears to be a promising cardiac perfusion radiotracer (Danad, Raijmakers, and Knaapen 2013).

Coronary atherosclerosis limits the vasodilatation capacity of the arterial vessels. Performing PET perfusion scans at rest and stress combined with a CT scan for attenuation correction (and coronary artery calcium [CAC] or angiography [CTA]) is a highly accurate technique to assess CAD. Pharmacological stressors for coronary vasodilation of the epicardial vessels and the microcirculation are typically used rather than exercise due to the short half-lives of PET perfusion tracers. The common stressor drugs are adenosine, adenosine triphosphate (ATP), regadenason, dipyridamole, and dobutamine (Table 2.2).

Adenosine and regadenason primarily induce vasodilatation of the vascular smooth muscle in the epicardial and microcirculation; indirectly, this also causes vascular endothelium-mediated vasodilatation in response to shear stress to obtain maximal MBF response (Lupi et al. 1997). Dipyridamole acts indirectly by inhibiting adenosine reuptake, and this increased intramyocardial adenosine concentration activates the coronary A_{2A} adenosine receptors (Kjaer et al. 2003). With a different mechanism, dobutamine acts through the adrenergic receptor system, as a strong β_1, moderate β_2, and mild α_1 agonist to increase myocardial work and oxygen demand (Miyashiro and Feigl 1993; Craig, Haskins, and Hildebrand 2007). Dobutamine increases the HR and contractility, in turn increasing the oxygen demand, and thus induces coronary vasodilatation to supply the required blood and nutrients. The maximal MBF evaluation measured using these pharmacological stressors gives a measure of the absolute perfusion reserve (stress/rest ratio), which is an indicator of the health of the coronary vasodilator function or ability to provide blood according to the cardiac demand. The cutoff used to describe an abnormal flow reserve is typically a value

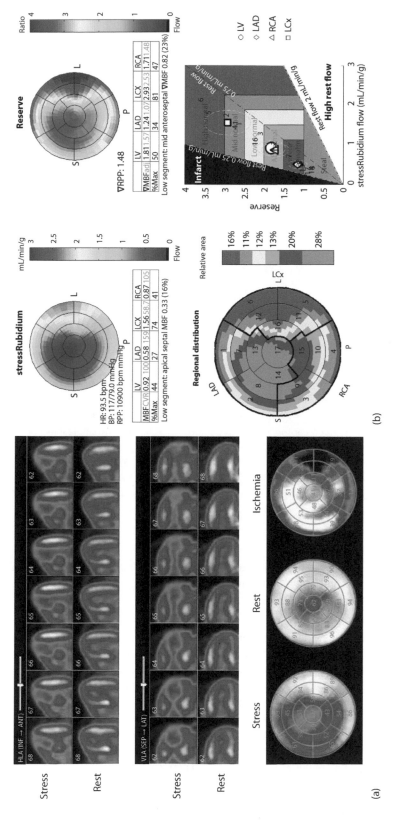

Figure 2.9 (a) Rubidium PET example of severe reversible ischemia in the territory of the left anterior descending (LAD) artery. (b) MBF quantification demonstrates severely impaired flow reserve with coronary steal (stress < rest MBF) in most of the LAD territory.

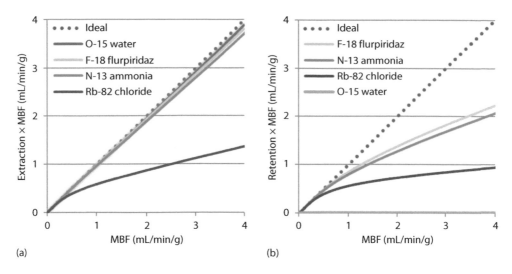

Figure 2.10 (a) PET radiotracer uptake rate (MBF × extraction-fraction = K1) and (b) retention rate (MBF × retention-fraction) as a function of MBF.

Table 2.2 Pharmacologic stress agents for PET MPI

Stressor	Target	Mechanism	Dose	Half-life
Adenosine	Purinergic (P1) receptor agonist	Systemic smooth muscle vasodilatation	0.14 mg/kg/min for 4–6 min	~30 s
Dipyridamole	Adenosine reuptake antagonist	Systemic smooth muscle vasodilatation	0.14 mg/kg/min for 4–5 min	3–12 min
Regadenason	A_{2A} adenosine receptor agonist	Coronary smooth muscle vasodilatation	0.4 mg single injection	2–3 min
Dobutamine	Strong β_1 agonist, moderate β_2, and mild α_1 adrenergic receptor agonist	Coronary neuronal and metabolic mediated vasodilatation	10, 20, and 40 µg/kg/min incremental doses ± atropine	2 min

lower than 2.0 mL/min/g, but there are many factors such as resting hemodynamics and age that should also be taken into consideration when interpreting myocardial flow reserve.

2.3.4.2 METABOLISM

The mechanism of cardiac metabolism is important for understanding how to assess normal and disease adaptations of the heart. Energy metabolism and control are complex and good knowledge of those pathways and the cardiac remodeling pattern can improve the cardiovascular imaging investigation. ATP reserve in the myocardial cells supports only a few heartbeats before failure, and this could explain why the heart easily adapts itself to be able to burn a broad range of energy substrates (Stanley, Recchia, and Lopaschuk 2005). Oxidation of the fatty acids and carbohydrates are the most predominant energy sources for the heart and to a lesser degree the ketones, lactate, pyruvate, and amino acids. The strength of cardiovascular PET-CT imaging is the ability to evaluate MBF through the coronary arteries and microcirculation and the assessment of heart metabolism. Several PET radiotracers are available to perform those studies (Figure 2.10 and Table 2.1). ^{13}N-ammonia, [82]Rb-rubidium, and ^{18}F-FDG are the most common and clinically accepted radiotracers used to diagnose CAD and myocardial viability. For cardiac research, the same radiotracers can be used; others are mentioned in the Table 2.1 and many more are available to investigate more specific receptor targets, pathology pathways or metabolism to increase our knowledge about cardiac disease, therapy, and novel treatment.

Normal heart oxidative metabolism is usually one-third glucose oxidation and two-thirds fatty acid beta-oxidation. Glucose is converted to pyruvate and the fatty acids to acyl-Coenzyme A (CoA), and both of them turn to acetyl-CoA to be metabolized in the Krebs cycle to generate energy (ATP) through oxidative phosphorylation (Lopaschuk et al. 2010). Under ischemic conditions, when there is reduced blood supply, lactate production and anaerobic glucose consumption (glycolysis) are promoted (Yoshinaga and Tamaki 2007). Repetitive ischemic episodes lead to an adaptation of the myocardial cell to down-regulate its contractile function and metabolism to preserve viability. ^{18}F-FDG uptake reflects only the initial steps of exogenous glucose metabolism (transport and phosphorylation) and stops before the process of oxidation or glycogen formation. PET-CT viability imaging assesses the regions of the heart where perfusion is

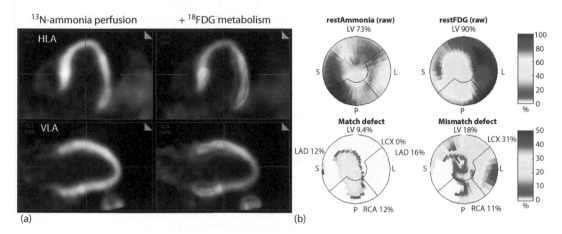

Figure 2.11 ^{13}N-ammonia (grey) and ^{18}FDG (color) perfusion + metabolism (or viability) images (a) and LV polar-maps (b) demonstrating substantial hibernating myocardium (metabolism–perfusion "mismatch") in the basal anterior and lateral walls (red), with nontransmural scar (perfusion + metabolism "match" defects) and minimal hibernating myocardium at the apex and distal inferior walls (yellow and blue). FDG uptake is assumed to be normal (viable myocardium) in the basal inferoseptal wall, which is the region of maximal perfusion.

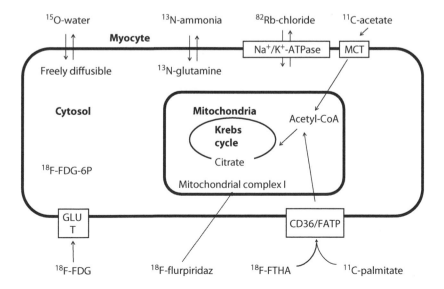

Figure 2.12 Cardiomyocyte mechanisms of uptake and clearance for the principal cardiac PET radiotracers used for perfusion and metabolism imaging.

decreased at rest, compared to uptake of ^{18}F-FDG, indicating the difference between cell survival (viability) or fibrotic tissue (scar), to determine recommendations for the benefit of revascularization (Figure 2.11). The ideal radiotracer for myocardial viability imaging should have the following characteristics: trapping inside the myocyte, inert after one metabolic step, high extraction from the blood, and fast blood clearance. If dynamic scanning is performed starting from the time of injection, then techniques of compartmental modeling can be used to assess the metabolic rate of radiotracer uptake used to generate energy to fuel cardiac contractile work.

In cardiac research, several radiotracers are used to assess different pathways of energy metabolism (Figure 2.12). For instance, ^{11}C-acetate is used to assess total oxidative metabolism (clearance rate) and also MBF (uptake rate). ^{18}F-FDG is used to measure rate of exogenous glucose uptake (μmol/min/g). Fatty acid uptake or beta-oxidation rates can be evaluated with ^{18}F-FTHA or ^{11}C-palmitate, respectively. These and other tracers are used with PET-CT imaging, which is a powerful tool for assessing the improvement, localization, survival, and metabolic adaptation in regions of myocardial infarction (MI) before/after therapy.

2.3.4.3 CELL SIGNALING

Beyond MBF and metabolism, the sympathetic nervous system, a component of the autonomic nervous system, can also be studied using PET-CT. Heart adrenergic receptor imaging of the sympathetic nerves has been used in diabetic, ischemic, and cardiac transplant states. Receptor density and function assessment may help improve the diagnosis and management of patients with heart failure following MI. Using the false neurotransmitter ^{11}C-hydroxyephedrine (^{11}C-HED), PET-CT dynamic imaging has shown that reduced retention (regional denervation) may be an early marker of CAD without evidence of MI (Thackeray and Bengel 2013) and i s associated with poor prognosis in patients receiving implantable cardioverter defibrillator (ICD) for primary prevention of ventricular fibrillation. As a second example, the renin-angiotensin system (RAS) is implicated in heart failure progression and cardiac remodeling following MI. Interest is increasing to develop strategies to mitigate the maladaptive mechanisms following myocardial injury. Medical therapies targeted at the RAS have demonstrated an outcome benefit after MI, whereas overexpression of angiotensin II type I (AT1R) has been associated with hypertrophy and fibrosis observed in the heart. ^{11}C-KR31173 imaging of AT1R receptor density has been shown to predict the risk of adverse cardiac ventricular remodeling (Zober et al. 2006).

2.4 CT IMAGING TECHNIQUES

2.4.1 CORONARY CT ANGIOGRAPHY

Coronary CT angiography (CCTA) can be performed with prospective (single-phase) or retrospective (multiphase) ECG-gating (Abbara et al. 2009). A critical difference between these two techniques is the radiation exposure, with a reduction of up to 90% using prospective gating. During retrospective gating, the x-ray source is on continuously, and only data with the least cardiac motion are selected for reconstruction and/or interpretation. With prospective gating, the x-ray source is only active during a prespecified cardiac phase, typically during diastole, in which there is less motion (Dilsizian, Pohost, and Nieman 2010). The introduction of prospective gating has seen a significant decrease in the radiation dose and hence has made CCTA more readily considered as an initial diagnostic test. The disadvantage of prospective gating is that image reconstruction is possible only during the limited prespecified phases. No assessment of ventricular contractile function can be made due to the fact that images are unavailable from the entire cardiac cycle. Retrospective gating is now typically reserved for patients who need assessment of ventricular function and who might be at risk of ventricular arrhythmias, where beat-by-beat ECG-editing can be used to select the most appropriate phase of the cardiac cycle to reconstruct for optimal visualization of the coronary arteries.

The importance of HR is shown by the improved accuracy seen with HRs <60 bpm on some cardiac CT scanners (Abbara et al. 2009). Beta-blockers have superseded calcium channel blockers as the drug of choice to lower the HR, except in patients with specific contraindications. Metoprolol is the most widely used beta-blocker due to its safety profile in congestive heart failure patients and chronic obstructive pulmonary disease patients. Atenolol may be used for patients with hepatic dysfunction. Esmolol is advantageous with its short half-life (9 min). Lanodiolol has an even shorter half-life and offers better potency compared to metoprolol. Metoprolol can be administered in one of two ways: 100 mg given 1 hour before the scan or 50 mg given 12 hours prior to the scan and another 50 mg 1 hour before the scan. Intravenous (IV) metoprolol can be used to supplement oral metoprolol or as the sole agent. Increments of 5 mg IV metoprolol can be given until achievement of the target HR. As a general rule, active bronchospastic diseases are a contraindication to beta-blockers. Administration should also be avoided in patients with advanced heart block (especially known or suspected sick sinus syndrome), unexplained presyncope or collapse, and concurrent use of other antiarrhythmics (especially calcium channel blockers, digoxin, or amiodarone).

Nitrates are typically administered immediately prior to CTA for coronary vasodilatation and enhancement of image quality; 400–800 µg of sublingual nitrates is given a few minutes prior to the scan (Abbara et al. 2009). Hypotension is to be expected, but as the procedure is performed in a supine position, this is generally safe. Contraindications to nitrates include erectile dysfunction medications: sildenafil, vardenafil, or tadalafil, or sildenafil taken for pulmonary hypertension. Other contraindications relate to deleterious consequences of systemic vasodilatation. These are inferior wall MI with right ventricle involvement, pronounced hypovolemia, raised intracranial pressure, cardiac tamponade, constrictive pericarditis, severe aortic stenosis, and hypertrophic obstructive cardiomyopathy. Breath-holding at end-inspiration is important with regard to minimizing respiratory motion; explicit instructions with a practice scan are recommended.

Acquisition data padding is defined as the time the x-ray beam is on, beyond the minimum time required to acquire an image of the heart at a single point in the middle of diastole (Labounty et al. 2010). While it would be logical to expect a higher duration of exposure to result in better interpretation due to the acquisition of more cardiac phases, this is not necessarily the case. LaBounty et al. (2010) showed in their study that there was no statistically significant improvement in image interpretability associated with the increase in radiation dose due to padding duration. Critical to the no-padding approach that was reinforced in this study was the importance of adequate HR control; all the patients had a HR < 65 bpm.

In the current era of "imaging gently," radiation exposure has come under increased clinical focus and also public awareness, and strategies to reduce the radiation dose are important to maximize the benefit-to-risk ratio. In addition to prospective gating and zero padding, reducing the energy of the scans has been explored. Decreasing the maximum energy of the x-ray beam from 120 kVp to 100 or 80 kVp has also seen significant reductions in radiation dose (Abbara et al. 2009), proportional to the square of the tube voltage change. It may be appropriate in rare circumstances to use 140 kVp in extremely obese patients, but reducing to 100 or 80 kVp sees a 30%–50% reduction in dose in small patients, with preservation of adequate contrast-to-noise ratio. Increase in tube current is proportional to radiation dose. Larger patients will need higher tube current to reduce image noise and overcome higher tissue attenuation. This should be balanced carefully according to the patient body habitus and minimal diagnostic image quality.

Intravascular iodine contrast agents are often used to improve vessel opacification and improve contrast to noise ratios (Abbara et al. 2009). Optimal images require intra-arterial opacification with target values >250 HU. Lower contrast viscosity and higher injection rates at lower injection pressures are possible with warming of the contrast agent. Injection duration should be as long or slightly longer than the scan duration. In patients with higher cardiac output, the injection rate should be increased to maintain optimal timing of arterial opacification. The typical contrast volume load ranges from 50 to 120 mL. Dual-head power injectors are preferred to single-head pumps, allowing contrast to be followed by a saline-push injection or followed by a mixture of contrast with saline. Biphasic or triphasic protocols are recommended in different clinical settings. Biphasic protocols have two stages, with the initial phase of 4–7 mL contrast (dependent on scan length) and a second injection of 40–50 mL saline push. This protocol is optimal for left ventricle cavity and coronary artery imaging, but the right heart cavities will not be enhanced. If the opacification of the right heart is desired, the saline flush may be replaced by a mixture of contrast with saline. Triphasic protocol consists of an initial high flow rate contrast injection (4–7 mL/sec), with a second injection of either contrast with saline mixture, or contrast injection at a slower flow rate, followed by a third injection of a lower volume saline push.

Test bolus vs. bolus tracking: Accurate timing of the scan to the arrival of the IV contrast to the structures of interest is crucial. Vascular enhancement has to be maintained for the duration of the scan. Due to the short duration of the CCTA scans, timing errors of 5–10 s can make a crucial difference in image quality. As a general rule, the scan delay should equal the contrast travel time from the injection vein to the ascending aorta, plus the addition of 2–3 s. Three strategies can be utilized to determine the vein-to-aorta time (called the delay time). (1) Easiest in implementation but with the least reliability is the 22–25 second best-guess rule. (2) Bolus tracking is preferred where there is automatic scan triggering; a region of interest (typically the ascending or descending aorta) is sampled every 2 s after the initiation of the contrast bolus infusion. Exceeding a preset imaging value (typically 100 HU) will start the diagnostic scan. (3) Test bolus is where a small bolus injection (typically 10–20 mL of contrast followed by saline push of 50 mL, both at a rate of 4–7 mL/s) is given with end-inspiration breath-holding; sampling at the level of the ascending aorta every 1–2 s, the delay time can be accurately determined. The benefits of this strategy include lower risk of false-starts or delays, identifying contrast dilution, IV line patency, and assessment of patient breath-holding ability prior to the actual diagnostic scan.

2.4.2 CORONARY ARTERY CALCIUM

Coronary artery calcification (CAC) was initially measured with electron beam CT but has now been replaced with fast-rotation multislice cardiac CT. The relationship between CAC and coronary artery stenosis is direct and strong; higher calcium scores are correlated with the likelihood of obstructive CAD and major adverse cardiovascular events. Measurement of CAC can be by the mass score, calcium volume method, and the Agatston score, which is most commonly used. CAC is evaluated with a noncontrast CT scan, quantifying both the spatial extent and density of calcium in several ranges (140,200,400,800 HU). Prospective triggering is used to minimize the radiation dose, with the diastolic cardiac phase typically acquired between 65% and 80% of the R–R interval.

2.4.3 CT MYOCARDIAL PERFUSION IMAGING (CTP)

2.4.3.1 PROTOCOL

An attractive property of iodinated contrast as a perfusion tracer is the property that it remains predominantly intravascular during first-pass circulation and that the iodine concentration is directly proportional to the attenuation (HU values) produced (Techasith and Cury 2011). With dynamic imaging, rates of myocardial enhancement at rest and stress are collected for the left ventricle myocardium and aorta and then MBF (perfusion) is derived by mathematical modeling. This is the basis of quantitative assessment of myocardial perfusion. Delayed myocardial contrast enhancement can also be captured by the late-time pharmacokinetics of iodine, which is relatively excluded from the intracellular spaces of intact myocardium but accumulated into the extracellular space within recently infarcted myocardium. Similarly to SPECT studies, patients are advised to abstain from caffeine, a nonselective competitive inhibitor of the adenosine receptor. Antecubital veins are typically used for high-volume and flow-rate IV administration of contrast and stress agents (adenosine, dipyradimole, or regadenoson). Beta-blockers are administered using oral and/or IV metoprolol to achieve the desired HR of <60 bpm to minimize cardiac motion artifacts (Ko et al. 2011). While there is a potential concern that beta-blockers may mask the true extent of ischemia in exercise perfusion protocols, recent studies have shown that no such effects are realized in vasodilator stress SPECT MPI or CTP studies.

Similar to nuclear MPI studies, CTP studies consist of a rest and stress phase. If imaging of myocardial scar is desired, then a delayed phase acquisition can be performed looking for late contrast enhancement. A disadvantage of a short time interval between the two studies is that contrast from the first acquisition may still be present in the myocardium when the second phase is performed. This could result in the decrease of the sensitivity for the detection of infarcted and/or ischemic myocardium.

2.4.3.2 REST VS. STRESS OR STRESS VS. REST

Rest-first CTP studies are advantageous in that they parallel real life, where the patient will only proceed to have the stress portion if coronary stenoses of at least moderate severity are identified. Stress-first allows for the optimization of ischemia detection by paralleling nuclear MPI protocols with stress and rest imaging. The administration of nitrates is recommended only with rest-first imaging. An interval of at least 20 min between scans allows the washout of contrast from the myocardium. It has also been advocated that low-risk patients should undergo a rest-first study while intermediate to high-risk patients should undergo stress-first studies.

In both cases, planar scout images are obtained to enable accurate positioning and definition of the axial scan FOV. The rest phase consists of 60–70 mL of contrast agent delivered at 5 mL/s and prospective gating at 70%–80% of the R–R interval, with tube current and voltage prescribed according to body mass index (BMI). The stress phase is performed with IV adenosine at 140 μg/kg/min for a period of up to 6 min (or 0.140 mg/kg/min of dipyradimole infused over 4–6 min), and then the contrast agent is administered as for rest. Acquisition can be triggered by either bolus tracking or test bolus timing. Stress phase acquisition may use retrospective ECG-gated scanning with tube current modulation. BMI is the main determinant of tube voltage: 100 kVp for patients with BMI <30 kg/m^2 and 120 kV for those with BMI >30 kg/m^2. For stress-first protocols, there should be a period of 20 min of delay for the resolution of the pharmacologic effects of the stress phase before rest imaging. Return of the HR to baseline and symptom resolution can be used as surrogate markers to confirm the end of the stress phase. Rest acquisition again sees a similar volume of contrast at a similar rate, with the main difference being prospective gating but with unchanged tube voltage and current settings. Assessment of coronary arteries and rest perfusion is carried out. Assessment for delayed contrast enhancement is carried out 5–10 min after the second phase, which can be either rest or stress. No additional contrast is required as the dose used for the second phase is allowed to redistribute into the extracellular space. Similar to the rest scan, it is prospectively triggered, using 100 kVp and unchanged current settings.

2.4.4 CT-FRACTIONAL FLOW RESERVE

The use of computational fluid dynamics and image-based modeling is the basis of fractional flow reserve (FFR) CT (Nam et al. 2011). CCTA images are acquired as per normal with at least 64-slice CT (Min et al. 2012) and then sent to a computational core lab to estimate discrete lesion FFR. The main disadvantage is that it takes 6 hours minimum for the processing in the core lab and that the software is currently proprietary and relatively expensive. This computational method models only the flow-limiting effects of epicardial disease, not accounting for the autoregulatory effects of microvascular vasodilatation or dysfunction. Early studies have been encouraging, with the results of the diagnosis of ischemia-causing stenoses obtained via noninvasive fractional flow reserve (DISCOVER-FLOW) trial showing that there was incremental benefit to FFR-CT compared to CT alone.

2.5 CLINICAL APPLICATIONS OF HYBRID PET/CT

At present, all major vendors manufacture PET systems in conjunction with a CT scanner, and therefore, hybrid imaging has become the convention for cardiac PET. There are multiple applications of this combined imaging modality from a cardiovascular perspective, although in the majority of scenarios, the primary application for CT is accurate attenuation correction of the PET images. However, with the use of higher-resolution CT scanners (64-slice and above) and IV iodinated contrast agents, hybrid imaging with PET and CT can be used to provide a detailed anatomical assessment of the cardiovascular system complimented by either functional or metabolic information provided by PET.

2.5.1 CORONARY ARTERY DISEASE

As described previously, there are several PET radiotracers that can be used for MPI, namely, rubidium-82-chloride (^{82}Rb), nitrogen-13-ammonia (^{13}NH$_3$), and oxygen-15-water $\left(\text{H}_2^{15}\text{O} \right)$ and each has distinct advantages and disadvantages (Table 2.1). There is now over three decades of experience using PET MPI, and there is a wealth of published data demonstrating its diagnostic accuracy for obstructive CAD (Parker et al. 2012; Mc Ardle et al. 2012) as well as its incremental prognostic value for risk stratification (Danad et al. 2013; Yoshinaga et al. 2006; Herzog et al. 2009; Fiechter et al. 2013).

Despite the higher diagnostic accuracy of PET MPI when compared to other noninvasive modalities such as SPECT or echocardiography (Bateman et al. 2006), the lack of anatomical delineation of the coronary anatomy limits the comparison of PET with invasive coronary angiography (ICA), which remains the conventional gold standard for diagnosis of obstructive CAD.

As described previously, CCTA can provide high-quality images of the epicardial coronary vessels and provides information not only on the degree of stenosis within the vessel but also regarding atherosclerotic plaque extent and morphology. While it has excellent negative predictive value for ruling out obstructive disease (Health Quality Ontario 2010), CTA tends to overestimate the severity of luminal stenosis, rendering it difficult to distinguish a truly flow-limiting lesion from moderate CAD. Based on the results of the fractional flow reserve versus angiography for multi-vessel evaluation (FAME)-1 and FAME-2 trials, it is now accepted that the outcome benefit following revascularization, particularly with percutaneous coronary intervention (PCI), is best predicted by physiologic as opposed to anatomical disease information (De Bruyne et al. 2012; Tonino et al. 2009).

Therefore, a combination of anatomical delineation with CTA and the physiologic information derived from PET MPI can offset the limitations of both modalities to a certain degree and provide a more complete assessment of CAD. The accuracy of this combined assessment has been compared to stand-alone PET and stand-alone CTA in several observational studies (Danad et al. 2013; Kajander et al. 2010; Groves et al. 2009) and is highly accurate for diagnosis of obstructive CAD using either visual analysis of ICA (stenosis >50% diameter) or invasive measurement of trans-stenotic pressures (FFR <0.8) as a gold standard. This improved accuracy is principally a result of improved specificity, whereby either a stenosis deemed to be obstructive on CTA subtends an area of normal perfusion or an area of apparent decreased perfusion is not associated with a stenosis of the artery supplying that territory, inferring the presence of microvascular disease. In addition, there is evidence that the results of hybrid PET-CTA imaging can impact on subsequent referral for ICA, particularly in the presence of equivocal CTA findings, where the MPI results may act as a gatekeeper to downstream referral for ICA targeted at revascularization instead of diagnosis (Danad et al. 2014).

However, while evaluating both anatomy and function in the same imaging session increases overall accuracy, PET-CTA is associated with both greater cost and increased radiation exposure to the patient. This concept is discussed in detail in later chapters, but it is noteworthy that recent advances in both PET (3D imaging and iterative reconstruction) and CT (128-slice hybrid PET-CT systems, prospective ECG gating, tube voltage, and current modulation) have enabled significant reductions in radiation dose from approximately 20 mSv to 5 mSv or less (Herzog et al. 2011). Radiation dose can be expected to decrease further with improved hybrid systems that incorporate the latest CT technology (320-slice high-pitch multi-detector CT [MDCT]) and potential for stress-only or stress-first PET imaging.

2.5.2 DETECTION OF SUBCLINICAL CORONARY DISEASE

Our understanding of the subclinical course of CAD has increased greatly over the past two decades and has been enhanced by the use of noninvasive imaging modalities, including PET-CT. As described previously in this chapter, dynamic imaging with PET perfusion tracers allows quantification of MBF in absolute terms, and alterations in MBF have been shown to occur in asymptomatic patients with risk factors for CAD such as smoking (Kaufmann et al. 2000), dyslipidemia (Alexanderson et al. 2010), and hypertension (Brush et al. 1988) and is reflective of endothelial dysfunction that is recognized as a precursor to the development of symptomatic overt CAD.

Moreover, the noncontrast CT scan for attenuation correction of the PET images may be used to visually assess coronary calcification and also to estimate the coronary artery calcium (CAC) score (Einstein et al. 2010; Mylonas et al. 2012). This can provide additional information, particularly in the setting of normal PET perfusion, whereby calcification identifies the presence of nonobstructive disease, which is of prognostic significance (Schenker et al. 2008). The interplay between changes in MBF and CAC has been evaluated with studies showing rather poor correlation between the two parameters (Curillova et al. 2009). In symptomatic patients, MBF has been shown to be a better predictor of subsequent cardiac risk than CAC (Naya et al. 2013).

2.5.3 IMAGING ATHEROSCLEROTIC PLAQUE

Identification of plaque "vulnerability," i.e., an atherosclerotic lesion with a thin fibrous cap at high risk of rupture that will result in a clinical event such as an acute coronary syndrome or a cerbrovascular accident, remains an elusive target, but the role of PET-CT in this area has garnered significant interest in recent years. Most experience has been with ^{18}F-FDG to image plaque inflammation secondary to macrophage infiltration, which portends an elevated risk of rupture (Naghavi et al. 2003). FDG images are coregistered with CT images, preferably following IV contrast, for accurate anatomic localization of uptake. In the carotid arteries, the degree of FDG uptake within areas of plaque, expressed as a ratio to the blood pool uptake, has been shown to correlate with macrophage infiltration based on CD-68 staining of specimens following subsequent endarterectomy (Rudd et al. 2007). In addition, increased FDG uptake within the aorta has been shown to correlate with an increase in adverse cardiovascular events in asymptomatic patients (Figueroa et al. 2013).

The ability of FDG to image plaque within the coronary arteries, however, is limited, both by the spatial resolution of PET as well as surrounding myocardial uptake despite suppression measures (low-carbohydrate, high-protein diet, prolonged fasting, and heparin) (Wykrzykowska et al. 2009). A recent observational study performed PET-CT imaging of the coronary arteries using both FDG and ^{18}F-sodium fluoride (NaF), a marker of active calcification in patients with a history of recent MI or stable angina, and they found localization of maximal NaF uptake to the culprit plaques in 93% of acute MI patients. Furthermore, they found significant correlation between NaF uptake and high-risk plaque morphology on intravascular ultrasound. FDG uptake was most commonly not discernible in the coronary arteries due to high background activity (Joshi et al. 2014). However, it should be noted that while these findings are of significant interest, their potential impact on clinical practice remains to be established.

2.5.4 METABOLIC AND VIABILITY IMAGING OF THE MYOCARDIUM

The role of revascularization with either coronary artery bypass surgery (CABG) or PCI in patients with ischemic cardiomyopathy has been an area of controversy in recent years following the publication of the surgical treatment for ischemic heart failure (STICH) trial (Velasquez et al. 2011), where the primary analysis failed to show a significant benefit for CABG over optimal medical therapy. The viability substudy also failed to show an association between the degree of viable myocardium and subsequent outcomes following revascularization (Bonow et al. 2011). These results were in stark contrast to a wealth of observational evidence (Nam et al. 2011), and it is noteworthy that FDG PET-CT was not used for viability testing in the STICH trial (Allman et al. 2002).

PET-CT viability imaging is performed with a combination of a perfusion radionuclide such as Rb-82 or [13]NH$_3$ to identify resting perfusion defects, compared with FDG, to identify areas of "viable" hibernating myocardium within the areas of decreased perfusion. A CT scan is performed for accurate attenuation correction. Hibernation refers to a phenomenon of altered metabolism within chronically ischemic myocardium with contractile dysfunction, where there is an increase in anaerobic glycolysis. This occurs in response to repeated ischemic insults and is of clinical significance, as normal contractile function may be restored by revascularization of these areas, and the degree of hibernating myocardium has been shown to predict improved outcomes following successful revascularization (D'Egidio et al. 2009) and may be a better predictor than inducible ischemia in these patients (Ling et al. 2013).

2.5.5 PULMONARY ARTERIAL HYPERTENSION

Pulmonary arterial hypertension (PAH) is a progressive condition characterized by obliteration of the pulmonary vasculature as a result of vasoconstriction and vascular remodeling. This results in an increase in pulmonary pressures, which culminates in hypertrophy and then dysfunction of the right ventricle. Survival in PAH patients is variable but is greatly reduced in those with signs of right ventricle (RV) dysfunction. Recently, there has been increasing interest in the potential role of metabolic alterations in the development of RV dysfunction (Archer et al. 2013). Under normal conditions, the predominant substrate for myocardial metabolism is fatty acids. However, in the setting of RV hypertrophy, there is a shift toward anaerobic glycolysis, a less energy-efficient form of metabolism that is likely to be maladaptive. Using FDG PET-CT, several observational studies have shown a relationship between increased RV FDG uptake and pulmonary artery pressures, as well as echocardiography (ECHO) markers of RV dysfunction (Bokhari et al. 2011; Can et al. 2011); a further study showed that a decrease in RV uptake following treatment with epoprostenol also correlated with improvements in RV strain patterns on ECHO (Oikawa et al. 2005).

2.5.6 HEART FAILURE

Beyond CAD imaging, there are several emerging applications for PET-CT in patients with heart failure that may influence patient management.

2.5.6.1 SYMPATHETIC INNERVATION

The autonomic nervous system plays an integral role in mediating cardiac function, and several imaging techniques have been developed to evaluate it. Most experience has been reported with the use of iodine-123 meta-iodo-benzylguanidine, a radionuclide used for planar SPECT imaging that reflects sympathetic innervation of the myocardium; reductions in uptake have been prospectively shown to be predictive of ICD-discharge risk in patients with implantable defibrillators (Boogers et al. 2010). More recently, the false neurotransmitter PET radionuclide ^{11}C-HED was prospectively evaluated in conjunction with resting perfusion using ^{13}NH$_3$ and viability with FDG in 204 patients with heart failure (Fallavollita et al. 2014). They demonstrated that the extent of the reduction in HED uptake was independently predictive of sudden cardiac death or ICD discharge. Furthermore, using HED measurements, the authors were able to identify a group of patients in whom the incidence of lethal arrhythmia was extremely low. While further study is required in this area, the ability to identify patients with a very low ejection fraction who remain at low risk for arrhythmia would be of significant clinical value for decision making surrounding the use of implantable defibrillators.

2.5.6.2 RENIN-ANGIOTENSIN SYSTEM

The RAS plays a key role in the pathogenesis of heart failure, mediating interstitial fibrosis and myocardial apoptosis. A variety of PET-CT radionuclides have been developed to image the Angiotensin-1 (AT1) receptor. One study using a ^{11}C-labeled tracer demonstrated up-regulation of AT1 receptors in the area of MI in pigs. They also demonstrated uptake of the radionuclide in the myocardium of healthy human volunteers (Fukishima et al. 2012). Imaging of the AT1 receptor has the potential to improve our understanding of the pathophysiology of adverse cardiac remodeling following MI and may represent a novel target to assess response to therapies in the near future.

2.5.6.3 CARDIAC STEM-CELL TRACKING

Development of viable cardiac stem cells remains an area of intense research, and methods to ascertain adequate delivery and retention of these cells is of significant value in this process. PET-CT imaging of cells labeled with a variety of PET radionuclides has been attempted primarily in animal models with varying success. FDG has been employed most frequently (Zhang et al. 2008), although its short half-life of 110 min limits the ability to ascertain cell retention over long periods. Several copper- and gallium-based tracers with longer half-lives have been evaluated with some success in animal models (Wu et al. 2013).

2.5.7 INFLAMMATION

2.5.7.1 CARDIAC SARCOIDOSIS

Cardiac sarcoidosis (CS) is an inflammatory condition characterized by the development of noncaseating granulomatous lesions within the affected organs. While lung involvement is the most frequent finding, cardiac involvement is increasingly being recognized and is associated with significant morbidity and mortality. Advanced imaging modalities including magnetic resonance imaging (MRI) as well as PET-CT have a significant role to play in the increasing recognition of CS, as it is a difficult diagnosis to make due to heterogeneity of patient presentation as well as the limited sensitivity of myocardial biopsy due to the patchy nature of the disease.

Most imaging protocols include both a whole-body FDG PET-CT scan in conjunction with a dedicated cardiac scan. A resting perfusion scan is usually performed during the same imaging session to assess viability (Mc Ardle et al. 2013). Focal FDG uptake within the myocardium in a noncoronary distribution, with or without areas of decreased resting perfusion, is the PET-CT finding most commonly suggestive of active CS, and has been shown to be a sensitive method for accurate diagnosis (Youssef et al. 2012). More recently, various quantitative measures of FDG activity have been shown to provide an estimate of response to immune-suppressive therapy (Tahara et al. 2010), correlating also with patient presentation, and therefore may be a marker of risk of significant arrhythmia (Mc Ardle et al. 2013). At present, work is underway to develop criteria for reproducible diagnosis of CS that incorporate the results of advanced imaging with PET-CT and MRI together with clinical criteria.

2.5.7.2 AORTITIS

PET-CT with FDG has been used for diagnosis and surveillance in patients with large vessel vasculitides such as giant-cell arteritis (GCA) as well as Takayasu's arteritis. Hybrid imaging with a noncontrast whole-body CT scan coregistered with FDG PET can be used to localize focal FDG uptake within the ascending aorta and aortic arch as well as its large branches that is indicative of active inflammation in these areas. In GCA, the use of FDG PET-CT can aid in the diagnosis, and in atypical cases that do not involve the temporal arteries (i.e., cases where temporal artery biopsy will be negative), PET-CT is the diagnostic modality of choice (Janssen et al. 2008). In addition, the degree of aortic FDG uptake has been shown to correlate with aortic dimension on follow-up CT scanning after a mean of 2 years (Blockmans et al. 2008).

In Takayasu's arteritis the two largest observational studies to-date yielded conflicting results (Arnaud et al. 2009; Lee et al. 2009), with one study showing a correlation between FDG uptake and serum levels of acute phase reactants (Arnaud et al. 2009) and the other showing no relationship with either serum markers or MRI findings (Blockmans et al. 2008). At present, there is a further ongoing prospective study evaluating the use of PET-CT and MRI for this indication that was due to be completed in 2014.

2.5.7.3 ENDOCARDITIS AND DEVICE INFECTION

Given the ability of FDG PET-CT to localize areas of inflammation, it is therefore of potential value to identify areas of infection. Infective endocarditis is frequently a difficult diagnosis to confirm, particularly in cases of prosthetic valve infection, where the sensitivity of the Duke Criteria (the current clinical gold-standard) is low due to limitations of ECHO in imaging prosthetic valves and the low sensitivity of blood cultures. In addition, the mortality from prosthetic valve endocarditis is high, meaning that early diagnosis is essential. A recent study evaluated the use of FDG PET-CT in 72 patients with suspected prosthetic valve endocarditis, with the final diagnosis being based on the Duke criteria after 3 months of follow-up (Saby et al. 2013). They found that adding abnormal FDG uptake around the prosthetic valve as a new major criterion increased the sensitivity of the modified Duke criteria at admission from 70% (52% to 83%) to 97% (83% to 99%) ($p = 0.008$).

For cases of possible implanted device infection, FDG PET-CT has potential value for localization of the site and extent of infection. While superficial skin erythema and purulent discharge are indicative of infection, these signs do not distinguish between superficial and deep pocket infection, which is clinically important as the latter requires full device system extraction (Chen et al. 2014). Several observational studies have shown high sensitivity for detection of pocket infections (Bensimhon et al. 2011; Cautela et al. 2013; Ploux et al. 2011). However, there are conflicting results for detection of lead infection where some studies show decreased sensitivity (Bensimhon et al. 2011; Cautela et al. 2013) and others show high negative predictive value (Ploux et al. 2011).

REFERENCES

Abbara S, Arbab-Zadeh A, Callister TQ et al. 2009. SCCT guidelines for performance of coronary computed tomographic angiography: A report of the society of cardiovascular computed tomography guidelines committee. *J Cardiovasc Comput Tomogr.* 3:190–204.

Akamatsu G, Ishikawa K, Mitsumoto K et al. 2012. Improvement in PET/CT image quality with a combination of point-spread function and time-of-flight in relation to reconstruction parameters. *J Nucl Med.* 53:1716–1722.

Alexanderson E, Garcia-Rojas L, Jimenez M et al. 2010. Effect of ezetimibe-simvastatine over endothelial dysfunction in dyslipidemic patients: Assessment by 13N-ammonia positron emission tomography. *J Nucl Cardiol.* 17:1015–1022.

Allman KC, Shaw LJ, Hachamovitch R, Udelson JE. 2002. Myocardial viability testing and impact of revascularization on prognosis in patients with coronary artery disease and left ventricular dysfunction: A meta-analysis. *J Am Coll Cardiol.* 39:1151–1158.

Archer SL, Fang YH, Ryan JJ, Piao L. 2013. Metabolism and bioenergetics in the right ventricle and pulmonary vasculature in pulmonary hypertension. *Pulm Circ.* 3:144–152.

Armstrong IS, Tonge CM, Arumugam P. 2014. Impact of point spread function modeling and time-offlight on myocardial blood flow and myocardial flow reserve measurements for rubidium-82 cardiac PET. *J Nucl Cardiol.* 21(3):467–474.

Arnaud L, Haroche J, Malek Z et al. 2009. Is 18F-fluorodeoxyglucose positron emission tomography scanning a reliable way to assess disease activity in Takayasu arteritis? *Arthritis Rheum.* 60:1193–1200.

Bateman TM, Heller GV, McGhie AI et al. 2006. Diagnostic accuracy of rest/stress ECG-gated rb-82 myocardial perfusion PET: Comparison with ECG-gated tc-99m sestamibi SPECT. *J Nucl Cardiol.* 13:24–33.

Beauregard JM, Croteau E, Ahmed N, Ouellette R, van Lier JE, Benard F. 2007. Effective specific activities determined by scintillation proximity counting for production runs of [(18)F]FES and 4F-M[(18)F] FES. *Nucl Med Biol.* 34:325–329.

Bengel FM. 2011. Imaging targets of the sympathetic nervous system of the heart: Translational considerations. *J Nucl Med.* 52(8):1167–1170.

Bensimhon L, Lavergne T, Hugonnet F et al. 2011. Whole body [(18)F]fluorodeoxyglucose positron emission tomography imaging for the diagnosis of pacemaker or implantable cardioverter defibrillator infection: A preliminary prospective study. *Clin Microbiol Infect.* 17:836–844.

Blockmans D, Coudyzer W, Vanderschueren S et al. 2008. Relationship between fluorodeoxyglucose uptake in the large vessels and late aortic diameter in giant cell arteritis. *Rheumatology.* 47:1179–1184.

Bokhari S, Raina A, Rosenweig EB et al. 2011. PET imaging may provide a novel biomarker and understanding of right ventricular dysfunction in patients with idiopathic pulmonary arterial hypertension. *Circulation Cardiovasc Imaging.* 4:641–647.

Bonow RO, Maurer G, Lee KL et al. 2011. Myocardial viability and survival in ischemic left ventricular dysfunction. *N Engl J Med.* 364:1617–1625.

Boogers MJ, Borleffs CJ, Henneman MM et al. 2010. Cardiac sympathetic denervation assessed with 123-iodine metaiodobenzylguanidine imaging predicts ventricular arrhythmias in implantable cardioverter-defibrillator patients. *J Am Coll Cardiol.* 55:2769–2777.

Brush JE, Jr, Cannon RO, 3rd, Schenke WH et al. 1988. Angina due to coronary microvascular disease in hypertensive patients without left ventricular hypertrophy. *N Engl J Med.* 319:1302–1307.

Can MM, Kaymaz C, Tanboga IH et al. 2011. Increased right ventricular glucose metabolism in patients with pulmonary arterial hypertension. *Clin Nucl Med.* 36:743–748.

Cautela J, Alessandrini S, Cammilleri S et al. 2013. Diagnostic yield of FDG positron-emission tomography/computed tomography in patients with CEID infection: A pilot study. *Europace.* 15:252–257.

Chen W, Kim J, Molchanova-Cook OP, Dilsizian V. 2014. The potential of FDG PET/CT for early diagnosis of cardiac device and prosthetic valve infection before morphologic damages ensue. *Curr Cardiol Rep.* 16:459.

Craig CA, Haskins SC, Hildebrand SV. 2007. The cardiopulmonary effects of dobutamine and norepinephrine in isoflurane-anesthetized foals. *Vet Anaesth Analg*. 34:377–387.

Curillova Z, Yaman BF, Dorbala S et al. 2009. Quantitative relationship between coronary calcium content and coronary flow reserve as assessed by integrated PET/CT imaging. *Eur J Nucl Med Mol Imaging*. 36:1603–1610.

D'Egidio G, Nichol G, Williams KA et al. 2009. Increasing benefit from revascularization is associated with increasing amounts of myocardial hibernation: A substudy of the PARR-2 trial. *JACC Cardiovasc Imaging*. 2:1060–1068.

Danad I, Raijmakers PG, Appelman YE et al. 2013. Hybrid imaging using quantitative H215O PET and CT-based coronary angiography for the detection of coronary artery disease. *J Nucl Med*. 54:55–63.

Danad I, Raijmakers PG, Harms HJ et al. 2014. Effect of cardiac hybrid 15O-water PET/CT imagingon downstream referral for invasive coronary angiography and revascularization rate. *Eur Heart J Cardiovasc Imaging*. 15:170–179

Danad I, Raijmakers PG, Knaapen P. 2013. Diagnosing coronary artery disease with hybrid PET/CT: It takes two to tango. *J Nucl Cardiol*. 20:874–890.

De Bruyne B, Pijls NH, Kalesan B et al. 2012. Fractional flow reserve-guided PCI versus medical therapy in stable coronary disease. *N Engl J Med*. 367:991–1001.

Dilsizian V, Pohost GM, Nieman K. 2010. MSCT coronary imaging. In: *Cardiac CT, PET and MR*. Oxford: Wiley-Blackwell. 246–258.

Einstein AJ, Johnson LL, Bokhari S et al. 2010. Agreement of visual estimation of coronary artery calcium from low-dose CT attenuation correction scans in hybrid PET/CT and SPECT/CT with standard Agatston score. *J Am Coll Cardiol*. 56:1914–1921.

Fallavollita JA, Heavey BM, Luisi AJ, Jr et al. 2014. Regional myocardial sympathetic denervation predicts the risk of sudden cardiac arrest in ischemic cardiomyopathy. *J Am Coll Cardiol*. 63:141–149.

Fiechter M, Gebhard C, Ghadri JR et al. 2013. Myocardial perfusion imaging with 13N-ammonia PET is a strong predictor for outcome. *Int J Cardiol*. 167:1023–1026.

Figueroa AL, Abdelbaky A, Truong QA et al. 2013. Measurement of arterial activity on routine FDG PET/CT images improves prediction of risk of future CV events. *JACC Cardiovasc Imaging*. 6:1250–1259.

Fukushima K, Bravo PE, Higuchi T et al. 2012. Molecular hybrid positron emission tomography/computed tomography imaging of cardiac angiotensin II type 1 receptors. *J Am Coll Cardiol*. 60:2527–2534.

Goldman LW. 2007. Principles of CT and CT technology. *J Nucl Med Technol*. 35:115–128.

Goldman LW. 2007. Principles of CT: Radiation dose and image quality. *J Nucl Med Technol*. 35:213–225.

Groves AM, Speechly-Dick ME, Kayani I et al. 2009. First experience of combined cardiac PET/64-detector CT angiography with invasive angiographic validation. *Eur J Nucl Med Mol Imaging*.36:2027–2033.

Health Quality Ontario. 2010. 64-slice computed tomographic angiography for the diagnosis of intermediate risk coronary artery disease: An evidence-based analysis. *Ont Health Technol Assess Ser*. 10:1–44.

Herzog BA, Husmann L, Buechel RR et al. 2011. Rapid cardiac hybrid imaging with minimized radiation dose for accurate non-invasive assessment of ischemic coronary artery disease. *Int J Cardiol*. 153:10–13.

Herzog BA, Husmann L, Valenta I et al. 2009. Long-term prognostic value of 13N-ammonia myocardial perfusion positron emission tomography added value of coronary flow reserve. *J Am Coll Cardiol*. 54:150–156.

Janssen SP, Comans EH, Voskuyl AE, Wisselink W, Smulders YM. 2008. Giant cell arteritis: Heterogeneity in clinical presentation and imaging results. *J Vasc Surg*. 48:1025–1031.

Joshi NV, Vesey AT, Williams MC et al. 2014. 18F-fluoride positron emission tomography for identification of ruptured and high-risk coronary atherosclerotic plaques: A prospective clinical trial. *Lancet*. 383:705–713.

Kajander S, Joutsiniemi E, Saraste M et al. 2010. Cardiac positron emission tomography/computed tomography imaging accurately detects anatomically and functionally significant coronary artery disease. *Circulation*. 122:603–613.

Kalender W. 2005. *Computed Tomography: Fundamentals, System Technology, Image Quality, Applications*. 2nd ed. Publicis Corporate Publishing: Erlangen.

Kaufmann PA, Gnecchi-Ruscone T, di Terlizzi M, Schafers KP, Luscher TF, Camici PG. 2000. Coronary heart disease in smokers: Vitamin C restores coronary microcirculatory function. *Circulation*. 102:1233–1238.

Kjaer A, Meyer C, Nielsen FS, Parving HH, Hesse B. 2003. Dipyridamole, cold pressor test, and demonstration of endothelial dysfunction: A PET study of myocardial perfusion in diabetes. *J Nucl Med*. 44:19–23.

Ko BS, Cameron JD, Defrance T, Seneviratne SK. 2011. CT stress myocardial perfusion imaging using multi-detector CT—A review. *J Cardiovasc Comput Tomogr*. 5:345–356.

Labounty TM, Leipsic J, Min JK et al. 2010. Effect of padding duration on radiation dose and image interpretation in prospectively ECG-triggered coronary CT angiography. *AJR Am J Roentgenol*. 194:933–937.

Lee SG, Ryu JS, Kim HO et al. 2009. Evaluation of disease activity using F-18 FDG PET-CT in patients with Takayasu arteritis. *Clin Nucl Med*. 34:749–752.

Ling LF, Marwick TH, Flores DR et al. 2013. Identification of therapeutic benefit from revascularization in patients with left ventricular systolic dysfunction: Inducible ischemia versus hibernating myocardium. *Circulation Cardiovasc Imaging*. 6:363–372.

Logan J. 2000. Graphical analysis of PET data applied to reversible and irreversible tracers. *Nucl Med Biol*. 27:661–670.

Lopaschuk GD, Ussher JR, Folmes CD, Jaswal JS, Stanley WC. 2010. Myocardial fatty acid metabolism in health and disease. *Physiol Rev*. 90:207–258.

Lupi A, Buffon A, Finocchiaro ML, Conti E, Maseri A, Crea F. 1997. Mechanisms of adenosine-induced epicardial coronary artery dilatation. *Eur Heart J*. 18:614–617.

Lusic H, Grinstaff MW. 2013. X-ray-computed tomography contrast agents. *Chem Rev*. 113:1641–1666.

Mark DB, Berman DS, Budoff MJ et al. 2010. ACCF/ACR/AHA/NASCI/SAIP/SCAI/SCCT 2010 Expert consensus document on coronary computed tomographic angiography: A report of the American College of Cardiology Foundation Task Force on Expert Consensus Documents. *Circulation*. 121:2509–2543.

Matsufuji N, Tomura H, Futami Y et al. 1998. Relationship between CT number and electron density,scatter angle and nuclear reaction for hadron-therapy treatment planning. *Phys Med Biol*. 43:3261–3275.

Mc Ardle BA, Birnie DH, Klein R et al. 2013. Is there an association between clinical presentation and the location and extent of myocardial involvement of cardiac sarcoidosis as assessed by 18F-fluorodeoxyglucose positron emission tomography? *Circulation Cardiovasc Imaging*. 6:617–626.

Mc Ardle BA, Dowsley TF, deKemp RA, Wells GA, Beanlands RS. 2012. Does rubidium-82 PET have superior accuracy to SPECT perfusion imaging for the diagnosis of obstructive coronary disease?: A systematic review and meta-analysis. *J Am Coll Cardiol*. 60:1828–1837.

Mc Ardle BA, Leung E, Ohira H et al. 2013. The role of F18-fluorodeoxyglucose positron emission tomography in guiding diagnosis and management in patients with known or suspected cardiac sarcoidosis. *J Nucl Cardiol*. 20:297–306.

Min JK, Koo BK, Erglis A et al. 2012. Usefulness of noninvasive fractional flow reserve computed from coronary computed tomographic angiograms for intermediate stenoses confirmed by quantitative coronary angiography. *Am J Cardiol*. 110:971–976.

Miyashiro JK, Feigl EO. 1993. Feedforward control of coronary blood flow via coronary beta-receptor stimulation. *Circ Res*. 73:252–263.

Moorees J, Bezak E. 2012. Four dimensional CT imaging: A review of current technologies and modalities. *Australas Phys Eng Sci Med*. 35:9–23.

Mylonas I, Kazmi M, Fuller L et al. 2012. Measuring coronary artery calcification using positron emission tomography-computed tomography attenuation correction images. *Eur Heart J Cardiovasc Imaging*. 13:786–792.

Naghavi M, Libby P, Falk E et al. 2003. From vulnerable plaque to vulnerable patient: A call for new definitions and risk assessment strategies: Part I. *Circulation*. 108:1664–1672.

Nam CW, Hur SH, Cho YK et al. 2011. Relation of fractional flow reserve after drug-eluting stent implantation to one-year outcomes. *Am J Cardiol*. 107:1763–1767.

Naya M, Murthy VL, Foster CR et al. 2013. Prognostic interplay of coronary artery calcification and underlying vascular dysfunction in patients with suspected coronary artery disease. *J Am Coll Cardiol*. 61:2098–2106.

Oikawa M, Kagaya Y, Otani H et al. 2005. Increased [18F]fluorodeoxyglucose accumulation in right ventricular free wall in patients with pulmonary hypertension and the effect of epoprostenol. *J Am Coll Cardiol.* 45:1849–1855.

Pan X, Sidky EY, Vannier M. 2009. Why do commercial CT scanners still employ traditional, filtered backprojection for image reconstruction? *Inverse Probl.* 25:1230009.

Parker MW, Iskandar A, Limone B et al. 2012. Diagnostic accuracy of cardiac positron emission tomography versus single photon emission computed tomography for coronary artery disease: A bivariate meta-analysis. *Circulation Cardiovasc Imaging.* 5:700–707.

Ploux S, Riviere A, Amraoui S et al. 2011. Positron emission tomography in patients with suspected pacing system infections may play a critical role in difficult cases. *Heart Rhythm.* 8:1478–1481.

Rudd JH, Myers KS, Bansilal S et al. 2007. (18)fluorodeoxyglucose positron emission tomography imaging of atherosclerotic plaque inflammation is highly reproducible: Implications for atherosclerosis therapy trials. *J Am Coll Cardiol.* 50:892–896.

Saby L, Laas O, Habib G et al. 2013. Positron emission tomography/computed tomography for diagnosis of prosthetic valve endocarditis: Increased valvular 18F-fluorodeoxyglucose uptake as a novel major criterion. *J Am Coll Cardiol.* 61:2374–2382.

Schenker MP, Dorbala S, Hong EC et al. 2008. Interrelation of coronary calcification, myocardial ischemia, and outcomes in patients with intermediate likelihood of coronary artery disease: A combined positron emission tomography/computed tomography study. *Circulation.* 117:1693–1700.

Silva AC, Morse BG, Hara AK, Paden RG, Hongo N, Pavlicek W. 2011. Dual-energy (spectral) CT: Applications in abdominal imaging. *Radiographics.* 31:1031–1046.

Stanley WC, Recchia FA, Lopaschuk GD. 2005. Myocardial substrate metabolism in the normal and failing heart. *Physiol Rev.* 85:1093–1129.

Tahara N, Tahara A, Nitta Y et al. 2010. Heterogeneous myocardial FDG uptake and the disease activity in cardiac sarcoidosis. *JACC Cardiovasc Imaging.* 3:1219–1228.

Techasith T, Cury RC. 2011. Stress myocardial CT perfusion: An update and future perspective. *JACC Cardiovasc Imaging* 4:905–916.

Thackeray JT, Bengel FM. 2013. Assessment of cardiac autonomic neuronal function using PET imaging. *J Nucl Cardiol.* 20:150–165.

Tonino PA, De Bruyne B, Pijls NH et al. 2009. Fractional flow reserve versus angiography for guiding percutaneous coronary intervention. *N Engl J Med.* 360:213–224.

Velazquez EJ, Lee KL, Deja MA et al. 2011. Coronary-artery bypass surgery in patients with left ventricular dysfunction. *N Engl J Med.* 364:1607–1616.

Webb S. 1992. Historical experiments predating commercially available computed tomography. *Br J Radiol.* 65:835–837.

Wu C, Ma G, Li J et al. 2013. In vivo cell tracking via 18F-fluorodeoxyglucose labeling: A review of the preclinical and clinical applications in cell-based diagnosis and therapy. *Clin Imaging.* 37:28–36.

Wykrzykowska J, Lehman S, Williams G et al. 2009. Imaging of inflamed and vulnerable plaque in coronary arteries with 18F-FDG PET/CT in patients with suppression of myocardial uptake using a low-carbohydrate, high-fat preparation. *J Nucl Med.* 50:563–568.

Yoshinaga K, Chow BJ, Williams K et al. 2006. What is the prognostic value of myocardial perfusion imaging using rubidium-82 positron emission tomography? *J Am Coll Cardiol.* 48:1029–1039.

Yoshinaga K, Tamaki N. Imaging myocardial metabolism. 2007. *Curr Opin Biotechnol.* 18:52–59.

Youssef G, Leung E, Mylonas I et al. 2012. The use of 18F-FDG PET in the diagnosis of cardiac sarcoidosis: A systematic review and metaanalysis including the Ontario experience. *J Nucl Med.* 53: 241–248.

Zhang Y, Thorn S, DaSilva JN et al. 2008. Collagen-based matrices improve the delivery of transplanted circulating progenitor cells: Development and demonstration by ex vivo radionuclide cell labeling and in vivo tracking with positron-emission tomography. *Circulation Cardiovasc Imaging.* 1:197–204.

Zober TG, Mathews WB, Seckin E et al. 2006. PET imaging of the AT1 receptor with [11C]KR31173. *Nucl Med Biol.* 33:5–13.

Development of a second-generation whole-body small-animal SPECT/MR imaging system

BENJAMIN M.W. TSUI, JINGYAN XU, ANDREW RITTENBACH, JAMES W. HUGG, AND KEVIN B. PARNHAM

3.1 INTRODUCTION

Despite the initial uncertainty and skepticism, both positron emission tomography (PET)/computed tomography (CT) (Kinahan et al. 1998) and single photon emission (SPECT)/CT (Hasagawa et al. 1993) were quickly accepted as standard practice in both preclinical and clinical imaging. Since then, the importance of dual-modality imaging techniques has grown as new applications continue to emerge and their success continues to flourish. At the same time, their successes have spurred active research in other multimodality imaging techniques, including different combinations of PET, SPECT, CT, optical, and ultrasound.

Currently, the most exciting developments in multimodality imaging are dual-modality PET/magnetic resonance imaging (MRI) and SPECT/MRI. While x-ray CT provides anatomical information, it has several weaknesses. Foremost is the inability of CT to differentiate between soft tissues especially in the abdominal region. Second, especially in animal studies, extremely high doses of x-rays are needed to obtain high-resolution or diagnostic quality CT images. Third, although reasonable anatomical context can be obtained with CT, it is often limited in extracting subtle anatomical information. On the other hand, MRI provides a unique capability to differentiate between soft tissues and provide high-contrast-resolution anatomical images with no damaging effects from ionizing radiation. High-resolution magnetic resonance (MR) images and their ability to differentiate between soft tissues can be exploited in quantitative PET and SPECT image reconstruction when partial volume effect from differential radioactivity uptake in adjacent tissues is important. In addition, through the use of different contrast media and data acquisition pulse sequences, unique anatomical and functional information can be extracted from the MR images.

Multimodality images can be obtained from separate systems and fused using image registration. Despite the lower cost without the need of special instrumentation, there are several difficulties, including registration inaccuracy due to patient movement and incompatibility of scanning tables between different scanners and biological changes that may occur at different scanning times especially for functional studies. Modern PET/CT and SPECT/CT scanners, in a coplanar configuration, share the same gantry and the scanning table transports the patient between the different modalities. This provides a solution to the first two difficulties at some additional system cost, but the separate scanning time remains a limitation to these techniques. A multimodality scanner that allows simultaneous data acquisition from each component modality will provide the best coregistered images and truly multiplexed information, such as anatomical, functional, and biochemical, about the patient occurring at the same time.

The goal of simultaneous dual-modality PET/MR or SPECT/MRI is the development of a PET or SPECT subsystem or insert that can be operated inside an MRI system for simultaneous acquisition of PET or SPECT and MR image data from the same subject. The main challenges are the mutual interferences of the two subsystems that can significantly degrade the performance of the individual imaging systems and, if not taken care of, can generate significant image degradation, artifacts, and distortions. These challenges have been addressed in over 15 years of active research in PET/MRI, first in the preclinical and most recently in clinical areas.

The development of simultaneous SPECT/MRI lags behind the development of PET/MRI. In the early 2000s, Gamma Medica, Inc., first integrated a cadmium zinc telluride (CZT) solid-state detector with a low power application-specific integrated circuit (ASIC) and developed a compact detector module to replace the scintillator–photomultiplier-tube-based detector found in typical SPECT systems (Wagenaar et al. 2006a, 2006b). The performance characteristics of the CZT detector module were found to be minimally affected by the presence of a strong magnetic field. The feasibility of using the MR-compatible CZT detector modules in SPECT/MRI was evaluated (Meier et al. 2011; Nalciouglu et al. 2007). Recently, a small-bore SPECT/MRI system has been under development with spatial resolution down to ~500 microns (Meng, Tan, and Gu 2007; Tan, Cao, and Meng 2009). However, due to the high cost and system geometry, it is limited to image a mouse brain.

Our first-generation whole-body small-animal SPECT/MR insert was designed to be inserted inside a 12-cm-diameter small bore of a small-animal 9.4T MRI system (Tsui et al. 2011). Three rings of eight CZT detector modules formed a cylindrical SPECT insert. Each detector ring had an outer diameter of less than 12 cm and consisted of eight CZT detector modules connected seamlessly. A 24-pinhole cylindrical collimator provided pinhole projection images of a 2.5-mm-diameter common volume-of-view (CVOV) within the detector ring onto the 24 CZT detector modules. In addition, an accurate system calibration method and a sparse-view image reconstruction method that incorporated the pinhole response function were developed. The SPECT/MR insert demonstrated the feasibility of simultaneous SPECT/MRI with both phantom and small-animal studies. However, the first-generation SPECT/MR insert had several limitations, including the relatively small system geometry and relatively large pixel size of the detector modules. They resulted in limited pinhole magnification and a system resolution of ~3–5 mm that was insufficient for high-resolution SPECT imaging.

In this chapter, we present the development of a second-generation small-animal SPECT-MR insert that overcomes the shortcomings of the first-generation insert. We describe its design, fabrication, and evaluation using phantom and small-animal studies to demonstrate the feasibility of simultaneous SPECT/MRI and its potential applications in preclinical molecular imaging.

The second-generation stationary ring-type SPECT/MR insert consisted of five rings of 19 each seamlessly connected 2.54×2.54 cm^2, 16×16-pixel MR-compatible CZT detectors. It was designed to fit inside a standard gradient coil of a 4.7T MRI scanner with an inner core diameter of 20 cm. Two multipinhole (MPH) collimators were designed and constructed using a cylindrical shell filled with high-density tungsten powder and solid tungsten pinhole inserts to reduce eddy current pathways. Shielded birdcage (BC) quadrature transmit/receive radiofrequency (RF) coils were used for mouse imaging to fit inside the MPH collimator to maximize the MRI signal-to-noise ratio (SNR). Accurate system calibration and quantitative sparse-view three-dimensional (3-D) MPH reconstruction methods were developed for high-quality SPECT images.

The SPECT/MR insert was evaluated as a stand-alone high-performance small-animal SPECT system. The measured geometric efficiency and resolutions of both collimators agreed with their targeted designed values. With collimator-detector response modeling, the SPECT image resolution exceeded the target system resolution of the MPH collimators. Artifact-free high-quality MPH SPECT images were obtained from physical phantom studies and from small-animal studies using the SPECT/MR insert. The predicted Lorenz force effect on the CZT detectors in the presence of magnetic field was observed and corrected in the reconstruction. Placing the SPECT/MR insert inside a 3T clinical MRI system, we demonstrated its feasibility for simultaneous SPECT/MRI studies using physical phantoms. With minimized interactions between the two systems, e.g., SNR loss and distortions in MR images, and the use of less MR gradient demanding sequences, the simultaneously acquired SPECT, MR, and the fused SPECT/MR images were in good agreement. The SPECT/MR insert was further evaluated in simultaneous SPECT/MRI of mice in both static and dynamic acquisition studies. The results demonstrated the ability of the SPECT/MR insert to obtain small-animal whole-body dynamic studies and the feasibility of simultaneous small-animal SPECT/MRI.

3.2 METHODS

3.2.1 MR-COMPATIBLE SPECT INSERT

The basic detector module used in the second-generation SPECT-MRI system was the same as that was used in the first-generation SPECT-MRI system (Tsui et al. 2011), the same 5-mm-thick and 25.4 mm × 25.4 mm square CZT solid-state detector (Kromek/eV Products, Inc., Saxonburg, PA, USA) coupled with an ASIC shown in Figure 3.1a. It was digitized into a 16×16 pixel array with 1.6-mm pixel pitch. The widths of the row of pixels along the edges of each detector module were reduced by ~0.1 mm. This reduction ensured that the pitch of the detector pixels remained the same across detector modules when they were tiled to form large detector arrays or imaging camera.

A stationary cylindrical SPECT insert was designed to fit inside an MRI system with a minimum bore diameter of 20 cm, which was the diameter of the inner bore of a standard gradient coil of a 4.7T Bruker preclinical MRI system (Bruker BioSpin Corp., Billerica, MA, USA). It consisted of five rings of 19 CZT detector modules in each ring connected seamlessly as shown in Figure 3.1b. A cylindrical collimator with multiple pinholes, or MPH collimator, was placed inside the SPECT insert.

The small animal or object to be imaged was positioned inside the MPH collimator. The number and positions of the pinholes on the MPH collimator were designed to provide pinhole projections with a specific pattern onto the inner cylindrical surface of the SPECT insert. Figure 3.1c shows the unfolded cylindrical detector plane with the 5×19 CZT detector modules and a total of $5 \times 19 \times 16 \times 16 = 24{,}320$ detector pixels.

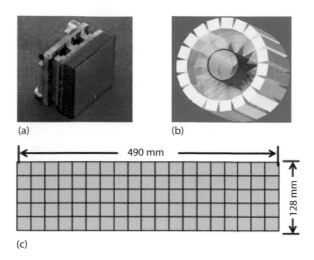

(a)   (b)

(c)

Figure 3.1 (a) The basic detector module used in the SPECT insert consisted of a 5-mm-thick and 25.4 mm × 25.4 mm square CZT solid-state detector module coupled with an ASIC. (b) The stationary cylindrical SPECT insert consisted of five rings of 19 CZT detector modules connected seamlessly. A cylindrical MPH collimator was placed inside the cylindrical detector ring. The small animal or object to be imaged was positioned within the MPH collimator. It was designed to provide multiple pinhole projections onto the cylindrical detector surface. (c) The unfolded detector surface showing the 5 × 19 CZT detector modules and a total of 5 × 19 × 16 × 16 = 24,320 detector pixels.

3.2.2 MPH COLLIMATOR DESIGN

The cylindrical-shaped SPECT insert and the continuous detector surface with the seamlessly tiled CZT detector modules allowed different MPH collimator designs for mouse and rat imaging. The first two MPH collimators were designed and fabricated for mouse imaging. Both provided a CVOV with 30 mm diameter and target system resolutions of 1 mm and 1.5 mm at the center of the CVOV. The MPH collimator designs were based on the theoretical formulation of the imaging characteristics of a MPH collimator, including the relationships between the pinhole diameter, the distances from the pinhole aperture to the detector plane and to the center of the CVOV, and the resolution and geometrical efficiency. Furthermore, the number of pinholes and their positions on the MPH collimator were optimized for maximum utilization of the detector surface shown in Figure 3.1c by the pinhole projections with less than 18% multiplexing or overlapping of the projection views. Among all possible MPH collimator designs with the targeted system resolution, the design that provided the highest geometric efficiency was chosen under the earlier

Table 3.1 Design parameters and theoretical predicted imaging characteristics of two MPH collimator designs

	Design 1	Design 2
Field-of-view	30 mm	30 mm
Target system resolution	1.0 mm	1.5 mm
Number of rows of pinholes	3	3
Total number of pinholes	18	36
Pinhole aperture size	0.3 mm	0.5 mm
Diameter of MPH collimator sleeve	49 mm	63 mm
Magnification of pinhole geometry	2.1	1.4
Geometric efficiency (%)[a]	0.026	0.069

[a] Based on Anger's formulation with tungsten inserts, 122 keV photons and 100% absorption by detector.

conditions. Table 3.1 shows the design parameters and theoretically predicted imaging characteristics of the two MPH collimators. The SPECT insert and the MPH collimator design allowed stationary SPECT data acquisition. The MPH collimator could rotate to two predetermined angular positions to obtain additional projections.

3.2.3 FABRICATION OF MR-COMPATIBLE MPH COLLIMATORS

To reduce the anticipated pulsed eddy currents resulting from the MR gradient field effects of the MPH collimator during simultaneous SPECT/MRI (Samoudi et al. 2015), special materials and methods were used in its fabrication. Figure 3.2 shows the components and the processes used in the fabrication of the MPH collimator. A stereolithographic model of an MPH sleeve with a hollow interior shown in Figure 3.2a was designed and fabricated using rapid prototyping technique. The openings allowed accurate placement of pinhole inserts designed to provide the pinhole projections as described in the previous section. The hollow interior was tightly packed with high-density, coated tungsten powder in an effect to reduce the eddy current that could be generated in solid metal material in a magnetic field. Individual pinhole inserts were made of solid tungsten by precision machining as shown in Figure 3.2b. They were inserted into the MPH collimator sleeve as shown in Figure 3.2c. Figure 3.2d shows the final MPH collimator with the collimator sleeves and all the pinhole inserts in place.

3.2.4 TRANSMIT/RECEIVE QUADRATURE BC RF COIL DESIGN AND FABRICATION

To acquire MR images, several iterations of transmit/receive shielded, quadrature (two feed cables) and linear (single feed cable) mode BC RF coils were used to fit inside the MPH collimator and to surround the small animal or the object to be imaged as shown in Figure 3.3a. The outer shield of the BC RF coil was later extended further and connected to the outer shield(s) of the coax cables to further reduce noise interference from the SPECT system in the MR acquisition. To preserve proper mode of operation for the BC, the linear mode BC coil with a single feed cable was later used for the dynamic SPECT mouse experiment (Rittenbach et al. 2012).

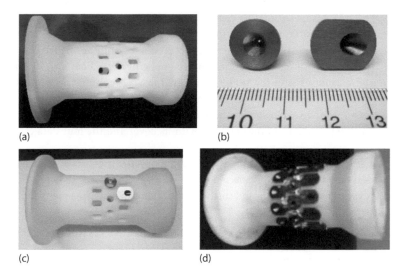

(a) (b) (c) (d)

Figure 3.2 (a) A stereo-lithographic rapid prototyping model of an MPH sleeve with a hollow cavity. Pinholes were inserted into the openings and provided projections of the small animal or the object to be imaged. The hollow cavity was tightly packed with coated high-density tungsten powder. (b) Samples of the individual pinhole inserts made of solid tungsten by precision machining. (c) Sample pinhole inserts placed into openings of the MPH collimator sleeve. (d) The final MPH collimator with the collimator sleeves and all the pinhole inserts.

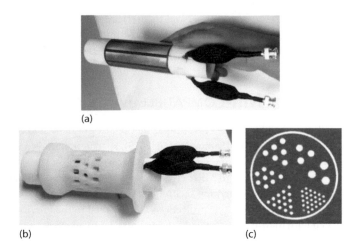

(a)

(b)   (c)

Figure 3.3 (a) A quadrature BC transmit/receive RF coil iteration with an RF shield and the drive/output cables and baluns. (b) The placement of the RF coil inside the MPH collimator sleeve. (c) Cropped MR image capture showing details of the hot-rod phantom obtained using the prototype quadrature BC transmit/receive RF coil shown in (a) that was placed inside the SPECT insert in a simultaneous SPECT/MR acquisition.

The configuration allowed operable MR image quality. The BC RF coil was fabricated by etching plastic sheets covered with a thin copper layer to minimize the attenuation of gamma-ray photon emissions from the object to be imaged. Figure 3.3a shows an iteration of the quadrature BC RF coil with RF shield and the drive/output cables with the two baluns. Figure 3.3b shows the placement of the RF coil inside the MPH collimator sleeve. Figure 3.3c shows a phantom image of the rod phantom acquired using a prototype quadrature transmit/receive BC RF coil in ~2.7 min with a prescribed slice thickness of 1.5 mm to initially demonstrate the image quality that can be obtained.

3.2.5 The completed SPECT/MR insert

The completed SPECT/MR insert without its cover is shown in Figure 3.4a. It shows the five rings of 19 CZT detector modules of the SPECT insert to the left and the data acquisition electronics to the right. The SPECT electronics were Faraday shielded to reduce radiation of noise to the RF coil and reciprocally to shield the SPECT data acquisition circuits from the magnetic field gradient and RF transmitter noise, which can produce false counts. Inside the SPECT insert are the MPH collimator and the transmit/receive quadrature BC RF coil described previously. The power consumption of each of the 95 low-power CZT detector modules was ~1 watt. To avoid overheating, cooling of the system was supplied by compressed air. Figure 3.4b shows the SPECT/MR from the back side with its cover and the four data cables and four air-cooling hoses.

An imaging bed attached to a 3-D scanning stage was constructed to position a small animal or object inside the SPECT/MR system. The assembly was mounted on an optical table as shown in Figure 3.4c for system evaluation and experimental phantom and small-animal studies outside of a magnetic field.

3.2.6 SPECT system calibration and correction

To obtain high-quality SPECT images, the radiation detection and imaging properties of the CZT detector modules, including the photopeak position and energy resolution of each of the 24,320 detector pixels, were experimentally determined using radionuclides with known photon energies. The photopeak position of each detector pixel was carefully adjusted to center within the preset energy window. The detection sensitivities of

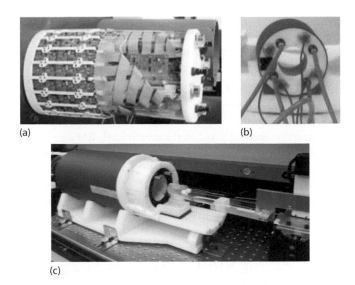

Figure 3.4 (a) The completed SPECT/MR insert without its cover. It shows the five rings of 19 CZT detector modules of the SPECT insert to the left and the data acquisition electronics to the right. (b) The back side of the SPECT/MR with its cover and the four data cable and four air-cooling hoses. (c) The SPECT/MR and an imaging bed attached to a 3-D scanning stage were mounted on an optical table for system evaluation and experimental phantom and small-animal studies outside of a magnetic field.

the detector pixels were then measured using a specially designed cylindrical shell phantom with an annular-ring cross-section that fit inside the cylindrical SPECT insert as shown in Figure 3.5. The precisely machined thin cylindrical shell with narrow width of the annular-ring was filled with liquid radioactivity (^{99m}Tc) to generate a flood image of the detector modules and pixels for the system uniformity measurement and detector non-uniformity correction.

Figure 3.5 The specially designed cylindrical shell phantom can be fit inside the SPECT insert and is used in uniformity correction of the detector pixels. The precisely machined thin cylindrical shell is filled with radioactive solution to provide a uniformity image of all the pixels of the cylindrical detector ring of the SPECT insert for nonuniformity correction.

The SPECT/MR insert was carefully calibrated to determine the pinhole aperture positions using a ^{57}Co point source. The estimated pinhole positions were used in a 3-D corrective MPH SPECT imaging reconstruction method described in the next section to provide high-resolution and artifact-free images (Rittenbach et al. 2012).

3.2.7 3-D CORRECTIVE AND SPARSE-VIEW MPH SPECT IMAGE RECONSTRUCTION METHOD

A 3-D MPH image reconstruction method based on the iterative ML-EM algorithm was developed for the SPECT insert. It was used to reconstruct the MPH projection data acquired from the SPECT insert after pixel energy and uniformity corrections. It incorporated the model of the measured pinhole response function and the SPECT system misalignment parameters for high-resolution and artifact-free MPH SPECT images. The corrective MPH SPECT image reconstruction method was evaluated for its use in sparse-view reconstruction for the same reconstructed image quality with a reduced number of projection views.

3.2.8 EVALUATION OF THE SPECT IMAGING CHARACTERISTICS OF THE SPECT/MR INSERT

After the SPECT system calibration and uniformity correction, the SPECT imaging characteristics of the SPECT/MR insert, including the spatial resolution and detection efficiency, were experimentally determined using point and line sources and compared to the corresponding theoretical predicted values (Rittenbach et al. 2012). The image quality was evaluated using an Ultra-Micro Hot Spot Phantom (Data Spectrum, Durham, NC, USA), which consisted of an insert with six groups of hollow cylindrical channels and a cylindrical phantom as shown in Figure 3.6. The phantom was used in experimental studies while a digitized version of the phantom was used in simulation studies to evaluate different aspects of the SPECT/MR inserts and the simultaneous SPECT/MRI techniques.

The 3-D corrective MPH SPECT image reconstructed method was used in the studies. The MPH SPECT images from the hot-rod phantom were compared to the known activity distribution in terms of accuracy and image artifacts. Comparison of results from the simulation study provided assessment of the standard 3-D MPH SPECT image reconstruction method with perfect system configuration. Comparison of results from the experimental phantom study will demonstrate the accuracy of the system calibration methods and effectiveness of the corrective 3-D MPH SPECT image reconstruction method.

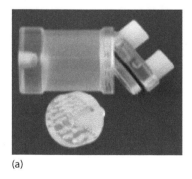
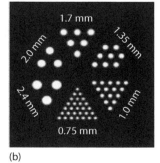

(a) (b)

Figure 3.6 (a) The Ultra-Micro Hot Spot Phantom (Data Spectrum, Durham, NC, USA) consists of an insert with six groups of hollow cylindrical channels with a channel height of 2.7 cm and inner diameters of 0.75, 1.0, 1.35, 1.7, 2.0, and 2.4 mm and a cylindrical phantom with an inside diameter of 2.8 cm and inside height of 2.8 cm. (b) A cross-section through the insert showing the configuration of the hollow channels. The phantom can be filled with water with specific concentration of radioactivity for use in SPECT imaging.

3.2.9 EVALUATION OF SPECT/MR INSERT FOR SIMULTANEOUS STATIC SPECT/MRI

The simultaneous SPECT/MRI capability of the SPECT/MR insert was evaluated using a clinical 3T MRI system (Rittenbach et al. 2013). Figure 3.7 shows the SPECT/MR insert with the phantom positioned on the scanning bed and imaged inside. The assembly was translated inside the bore of the MRI system for simultaneous SPECT/MRI.

The mouse RF coil of the SPECT/MR insert was connected to a special interface to the clinical MRI system. The interface has a transmit/receive switch and a signal amplification stage to maintain good SNR. The thick cable in the foreground is connected to the power supply placed in a low-magnetic-field region of the MRI room. At the back side of the SPECT/MR insert, the air hoses were connected to the compressed air outlet in the MRI room. To preserve the MR room shielding, the Ethernet data cables were interfaced to an optical communication module where fiber optic cable passed through a waveguide through a wall of the MRI room and connected to a personal computer (PC) in the control room for SPECT data acquisition.

The evaluation consisted of two stages. In the first stage, the mutual effects of the SPECT and MRI instrumentation and data acquisition on the SPECT and MR images during a static imaging mode were evaluated. In the second stage, the mutual effects were evaluated in a dynamic simultaneous SPECT/MRI mode.

In the experimental evaluation of the static simultaneous SPECT/MRI mode, the 36-pinhole collimator and the BC RF coil were placed inside the SPECT/MR insert. The ultra-micro-hot-spot phantom shown in Figure 3.5 was filled with a solution with ~3.5 mCi of ^{99m}Tc and Cu_2SO_4 and used in the simultaneous SPECT/MR acquisition. The ^{99m}Tc activity provided the 140 keV photons for SPECT imaging and the Cu_2SO_4 provided decreased T1 relaxation time contrast for MRI. The phantom was first imaged with the standalone SPECT/MR insert (outside of the magnetic field) and the MRI system separately. The separately obtained SPECT and MRI images of the phantom were used as the reference images. The SPECT/MR insert with the phantom inside was then placed inside the bore of a clinical 3T MRI system. A SPECT data acquisition

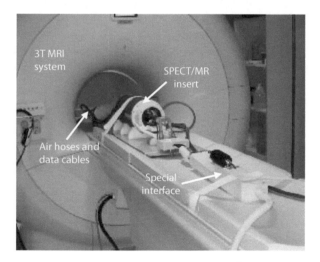

Figure 3.7 The SPECT/MR insert was positioned on the patient bed of a clinical 3T MRI system. The phantom or small animal to be imaged was placed inside the SPECT/MR insert and the assembly was translated inside the bore of the MRI system for simultaneous SPECT/MRI. The long white box in the foreground is a special interface that connects the RF coil to the data acquisition electronics of the clinical MRI system. The thick cable in the foreground is connected to the power supply placed in a low-magnetic-field region of the MRI room. At the back side of the SPECT/MR insert, the air hoses were connected to a compressed air outlet in the MRI room; the Ethernet data cables were connected to an optical communication interface where finer optic cables passed through a waveguide through a wall of the MRI room to the control room and were connected to a PC for SPECT data acquisition.

was obtained with the gradient field turned off. Then, a simultaneous SPECT/MR data acquisition was performed with the gradient field turned on. The SPECT data at a single collimator position were acquired for ~5 min while the MR images were acquired simultaneously using a 2-D gradient echo (GE) single-slice pulse sequence, which is not an aggressive gradient demanding sequence as compared to, for example, fast multislice turbo fast field with very short repetition time (TR) or the echo planar imaging sequences. The SPECT and MRI images obtained were compared with the corresponding reference images for possible image degradation and image artifact generation. They were also scaled and registered to each other for evaluation of image distortion.

3.2.10 Evaluation of SPECT/MR insert for simultaneous dynamic SPECT/MRI

In the experimental evaluation of the dynamic simultaneous SPECT/MRI mode, the same SPECT/MR insert configuration as in the static acquisition was used. A catheter was inserted into the tail vein of a ~30 g normal anesthetized mouse to extend the reach of the injection site. The animal was then placed inside the SPECT/MR insert, which in turn was positioned inside the MRI system with the kidneys of the mouse at the CVOV of the SPECT/MR insert.

The SPECT/MR insert was set to acquire list-mode SPECT data for 30 min. A multislice GE sequence was used to obtain dynamic MR images from five predetermined 1-mm-thick slices of the mouse's kidney with a 0.75-mm gap between slices. A bolus of ~3.5 mCi of ^{99m}Tc MAG3 (mercaptoacetyltriglycine) was injected into the mouse through the catheter line immediately after the SPECT data acquisition began. The dynamic MR data acquisition began soon after at 80-sec intervals for a total of 22 data sets during the 30 min of total acquisition time.

The dynamic MPH SPECT list-mode data were rebinned into 10-sec timeframes and each reconstructed using the corrective 3-D MPH SPECT image reconstruction method with 10 iterations. The dynamic SPECT and MRI images obtained were inspected separately for possible image degradation and image artifacts. The 10-sec frame dynamic 3-D MPH SPECT images were grouped into 80-sec time frames, from which five 1-mm-thick coronal image slices that closely matched the five acquired dynamic MR image slices were selected. The SPECT images were appropriately scaled and registered to the corresponding dynamic MR images for evaluation of possible image distortion. Finally, the dynamic SPECT and MR images were summed and registered for further evaluation.

3.3 RESULTS

3.3.1 SPECT system calibration and nonuniformity correction

The detection properties of the CZT detector modules within the cylindrical SPECT insert were measured using a cylindrical shell phantom filled with uniform ^{99m}Tc solution. Figure 3.8 shows a typical energy spectrum of a CZT detector pixel at a bias voltage of 500 volts and at 140 keV photon energy. The energy resolution was determined by fitting a Gaussian to the photopeak and calculating its full-width-at-half-maximum (FWHM). The average FWHM was found to be ~3.9% over 98% of the total 24,320 pixels.

The gain of each detector pixel was adjusted such that the photopeak was positioned at the center of a 10% energy window. The MPH projection data of the shell phantom was acquired to obtain a flood image for detector uniformity correction. Figure 3.9a shows a sample of the cross-sectional reconstructed image of a uniform cylinder without uniformity correction. Figure 3.9b shows the reconstructed image after uniformity correction of the projection data with the flood image. To further improve the uniformity, counts at detector pixels that were nonresponsive and were either hyperactive or hypoactive, i.e., the photopeak counts within the selected 10% energy window was either more than 175% or less than 25% of the average pixel count, were replaced by a 2-D interpolated value of its eight neighboring pixels. After this additional correction (of ~2.5% of total number of detector pixels), the reconstructed uniformity image is shown in Figure 3.9c.

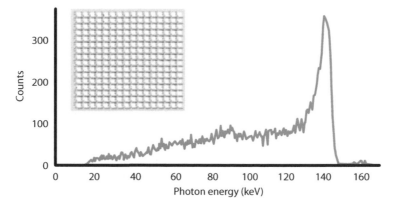

Figure 3.8 A typical energy spectrum of a CZT detector pixel. The energy resolution at 140 keV photon energy averaged over 98% of the total of 24,320 pixels is ~3.9%.

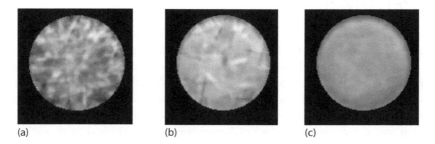

| (a) | (b) | (c) |

Figure 3.9 SPECT reconstructed images from the same sample slice through a cylindrical phantom filled with a uniform concentration of ^{99m}Tc solution obtained (a) without correction of detector pixel uniformity, (b) after uniformity correction using a flood image, and (c) with additional correction for nonresponsive, hyperactive, and hypoactive pixels.

3.3.2 IMAGING PERFORMANCE CHARACTERISTICS OF THE SPECT INSERT

After the energy calibration and uniformity correction of the CZT detector modules, the sensitivity and spatial resolution of the SPECT insert were determined using a reference ^{57}Co point source placed at the center of the CVOV and a ^{99m}Tc line source placed along the central axis, respectively (Rittenbach et al. 2013). Table 3.2 shows the measured and analytically calculated sensitivity and spatial resolution of the SPECT insert.

Table 3.2 Comparison of the measured and analytically calculated sensitivity and spatial resolution at the center of the CVOV of the SPECT insert

MPH collimator design	System detection sensitivity (cps/MBq)[a]		System spatial resolution in terms of FWHM (mm)[b]	
	Predicted	Measured	Predicted	Measured[b]
18-pinhole	197	172	1.0	1.1
36-pinhole	515	372	1.5	1.5

[a] Taking into account of the 75% absorption of the 122 keV photons by the 5-mm-thick CZT detector module.
[b] A ~0.5-mm inner diameter capillary tube was used as the line source; measured resolution was corrected for the finite width of the line source.

3.3.3 3-D CORRECTIVE AND SPARSE-VIEW MPH SPECT IMAGE RECONSTRUCTION METHOD

Figure 3.10 shows comparisons of MPH SPECT images obtained from the same sample slice through the hot-spot phantom (Figure 3.6) using the 3-D MPH SPECT image reconstruction methods with and without the pinhole response function modeling and with and without additional rotation of the 36-pinhole collimator used in the data acquisition. Comparison of images in rows (A) and (B) of Figure 3.10 shows the improvement of image quality in terms of both lowered noise and improved spatial resolution by incorporating the pinhole response functions in the image reconstruction method. Comparing images in columns (a) and (b) of Figure 3.10, additional collimator rotation provides improved image quality when no pinhole response function modeling was used, but minimal improvement was found when pinhole response function modeling was applied. The result indicates that the corrective image reconstruction method allows stationary imaging without collimator or system rotation.

3.3.4 EVALUATION OF SPECT/MR INSERT FOR SIMULTANEOUS STATIC SPECT/MRI

The hot-spot phantom shown in Figure 3.6 was used in an experimental study to evaluate the SPECT/MR insert for simultaneous static SPECT/MRI. Figure 3.11a and b shows a comparison between the separately and simultaneously acquired SPECT images. Depending on the RF coil used, MR SNR loss up to a factor of ~8 was observed due to the SPECT electronics interference. Visually it is also not evident that connecting the BC shield to the cable shield improved the general attained SNR. In addition, MR image geometrical distortions that can be attributed to gradient field distortions for fast survey multislice sequences were also recorded as well as slice position shifts for GE sequences. Despite the later mentioned shifts, targeted slices were retracked in the presence of the SPECT system and good-quality MR images were obtained and

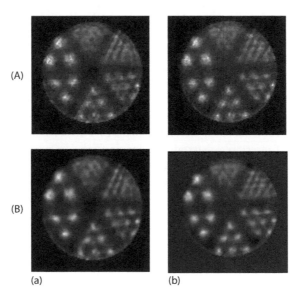

(A)

(B)

(a)   (b)

Figure 3.10 Comparisons of MPH SPECT images from the same sample slice through the hot-spot phantom shown in Figure 3.5, obtained using the 3D corrective MPH SPECT image reconstruction with 40 iterations: Row (A) without and row (B) with the pinhole response modeling, and column (a) without and column (b) with an additional rotational stop of the 36-pinhole collimator. All images have a reconstructed pixel size of 0.25 mm and image slice thickness of 2 mm.

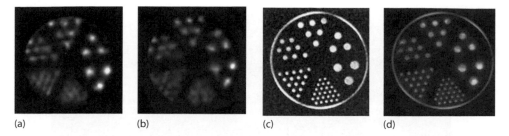

(a) (b) (c) (d)

Figure 3.11 (a) MPH SPECT image of a sample slice through the hot-spot phantom obtained from the SPECT/MR insert as a stand-alone system. The (b) MPH SPECT and (c) MR images of the same sample slice (with prescribed slice thickness of 1 mm) through the hot-spot phantom as in (a) but obtained simultaneously with the SPECT/MR insert placed inside the bore of a clinical 3T MRI system using a two feed BC RF coil with extended shield that was connected to the cables' shield. (d) The registered and fused SPECT and MR images from (b) and (c). [The Hot Iron color scale was used to display the MPH SPECT images from (b) in (d).]

registered to SPECT images. Figure 3.11c shows the simultaneously acquired MR image of the same sample slice through the hot-spot phantom. Minimal distortion and artifact were found in the MR image, indicating minimal effect of the SPECT/MR insert on the MR image. Figure 3.11d shows the registered and fused SPECT and MR images.

3.3.5 EVALUATION OF THE SPECT/MR INSERT FOR SIMULTANEOUS DYNAMIC SPECT/MRI

A dynamic ^{99m}Tc-MAG3 kidney function study of a mouse was conducted to evaluate the simultaneous dynamic SPECT/MRI capability of the SPECT/MR insert. Figure 3.12 shows the first four 80 sec/frame MR images (using 200 K-space line acquisition at TR = 200 ms with two averages per frame) of one of the four coronal slices through the mouse's kidney from the 30-min simultaneous dynamic SPECT/MR data acquisition. The dynamic MR images show the good image quality and SNR that could be initially obtained from the SPECT/MR insert using a linear mode BC RF coil with extended shielding. No noticeable MR geometric distortions were observed at the organ of interest (albeit slice offsets) in this case, indicating the feasibility of the SPECT/MR insert for simultaneous dynamic SPECT/MR data acquisition.

Figure 3.13 shows the first eight 10 sec/frame MPH SPECT images of the corresponding coronal slice shown in Figure 3.12 from the 30-min simultaneous dynamic SPECT/MR acquisition. The activity from the ~2.5 mCi of ^{99m}Tc-MAG3 injection went through the inferior vena cava to the heart and subsequently accumulated in the kidneys before excreting to the bladder. The dynamic MPH SPECT images show the

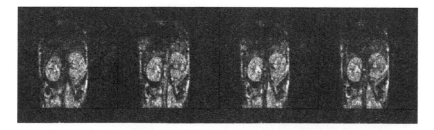

Figure 3.12 From left to right are the first four 80 sec/frame MR images of one of the four coronal slices through the mouse's kidneys from the 30-min simultaneous dynamic SPECT/MR data acquisition. All images have a reconstructed pixel size of 0.18 mm and prescribed image slice thickness of 1 mm. The signal loss in the upper part of the image may be attributed to distortions due to the presence of the SPECT system.

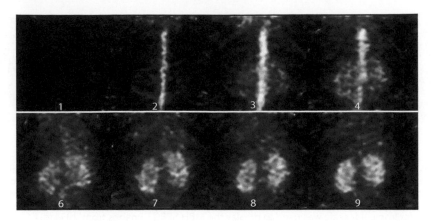

Figure 3.13 From left to right and top to bottom: the first eight 10 sec/frame MPH SPECT images of a sample coronal slice through the mouse's kidneys from the 30-min simultaneous dynamic SPECT/MR data acquisition. The 3-D MPH SPECT images were obtained using the 3-D ML-EM-based corrective MPH reconstruction method with 10 iterations and a reconstructed pixel size of 0.25 mm and image slice thickness of 1 mm.

high sensitivity and spatial resolution of the SPECT/MR insert. No noticeable distortion and artifacts were observed in the images, indicating minimal effect by the MRI system during simultaneous dynamic SPECT/MR data acquisition.

The 10 sec/frame MPH SPECT images shown in Figure 3.13 were combined to form 80 sec/frame images that matched the corresponding MR images in Figure 3.13. The resultant registered and fused SPECT/MR images of the first four 80 sec/frame are shown in Figure 3.14. These images demonstrate the feasibility of simultaneous SPECT/MRI with minimal mutual interactions between the SPECT and MRI instrumentation in the dynamic data acquisition mode.

The twenty-two 80 sec/frame dynamic MR images from each of the four image slices were summed to form four static images with 1-mm-thick and separated by 0.75 mm gaps. These images are shown in the top row of Figure 3.15. The 10 sec/frame dynamic 3-D MPH SPECT images over the entire 30-min acquisition were also summed to form a static 3-D MPH SPECT image. Four coronal image slices that correspond to the four MR image slices were selected and shown in the middle row of Figure 3.15. In the bottom row of Figure 3.15, the registered and fused corresponding SPECT and MR images are shown. These images demonstrate minimal distortion and mismatch between the SPECT and MR images during static data acquisition.

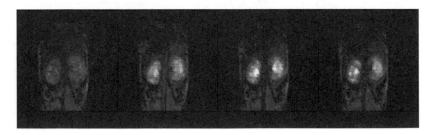

Figure 3.14 From left to right are the first four 80 sec/frame registered and fused SPECT and MR images of the corresponding coronal slice in Figures 3.11 and 3.12. (The color scale is used to display the MPH SPECT images.)

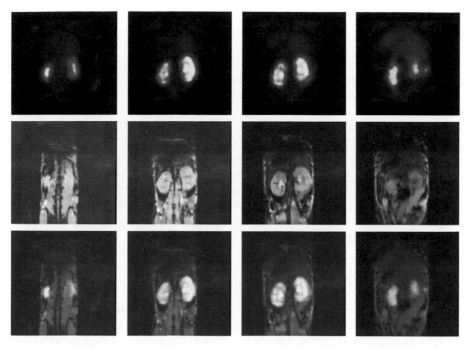

Figure 3.15 From left to right are the four corresponding 1-mm-thick coronal slices of static (top row) SPECT images, (middle row) MR images, and (bottom row) registered and fused SPECT/MR images that were summed over the entire 30-min acquisition time resulting in SNR improvements.

3.4 SUMMARY AND CONCLUSIONS

A second-generation small-animal SPECT/MR insert was designed and constructed for simultaneous SPECT/MRI of small animals, including mice and rats. It was designed to fit inside a MRI system with a minimum bore size of 20-cm diameter, e.g., the diameter of the inner bore of a standard gradient coil of a 4.7T Bruker preclinical MRI system. The same compact MR-compatible CZT detector modules used in the first-generation small-animal SPECT/MR system were connected seamlessly to form a stationary cylindrical SPECT insert.

Two cylindrical 18- and 36-pinhole collimators with optimized parameters and target system resolutions of 1.0 and 1.5 mm at the center of their CVOV were designed for mouse imaging. The MPH collimators can be rotated inside the SPECT to provide additional projection data. The MPH collimators were fabricated using a special MR-compatible high-density material with low conductivity to minimize their effects on the MRI system. A transmit/receive BC RF coil for MRI was designed to fit inside the cylindrical MPH collimator for good SNR imaging and was constructed using thin flexible sheets covered with thin copper layer to minimize its effect on the SPECT imaging. The SPECT insert with the RF coil forms the SPECT/MR insert.

A corrective sparse-view 3-D MPH image reconstruction method based on the iterative ML-EM algorithm and incorporating model of the pinhole response function was developed to provide quantitative SPECT imaging with improved image quality in terms of both improved spatial resolution and lower noise. Also it allowed a reduced number of collimator rotations without loss of image quality.

The SPECT/MR insert with the two MPH collimators and RF coils was evaluated as a stand-alone small-animal SPECT and inside an MR scanner using phantom studies. The results demonstrated its ability to provide MPH SPECT images with ~1 mm resolution and system sensitivity of ~170 cps/MBq or system sensitivity at ~370 cps/MBq and resolution of ~1.5 mm.

The SPECT/MR insert was also evaluated for its capability to provide simultaneous dynamic SPECT/MRI in a phantom and a mouse study. With minimal interference observed in terms of MR image geometric distortions and SNR loss due to the effect of the SPECT instrumentation on MRI, simultaneously acquired SPECT and MR images were possible. The results successfully demonstrated the feasibility of simultaneous small-animal SPECT/MRI using the SPECT/MR insert.

The preliminary results from our study are encouraging to further explore the full capability and applications of SPECT/MR instrumentation and simultaneous SPECT/MRI. Future instrumentation enhancements will include new MPH collimators for rat studies and for imaging radiolabel agents with medium- to high-energy photon emissions. Other important future studies include investigating methods to reduce MR SNR loss due to SPECT electronics as well as correcting for any MRI geometric distortions due to the possible deformation in the gradients' fields that may be eddy current dependent (Samoudi et al. 2015) and that may consequently enable using faster MRI sequences in the future. Most importantly, the search for potential applications of simultaneous SPECT/MR instrumentation and imaging techniques to biomedical and clinical research for better understanding of biochemical processes and disease will be the guide for future studies.

ACKNOWLEDGMENTS

The authors are indebted to AbdEl-Monem El-Sharkawy and the late William A. Edelstein for their contributions to the design and construction of the BC RF coils and the MR compatible multipinhole collimators. Also, the authors greatly appreciate their contributions to the design and implementation of the pulse sequences used in the MR experiments.

REFERENCES

Hasagawa BH, Lang TF, Brown JK, Gingold EL, Reily SM, Blankespoor SC, Liew SC, Tsui BMW, Ramanathan C. 1993. Object-specific attenuation correction of SPECT with correlation dual-energy x-ray CT. *IEEE Trans Nucl Sci* 40:1242–52.

Kinahan PE, Townsend DW, Beyer T, Sashin D. 1998. Attenuation correction for a combined 3D PET/CT scanner. *Med Phys* 25:2046–53.

Meier D, Wagenaar DJ, Chen S, Xu J, Yu J, Tsui BMW. 2011. A SPECT camera for combined MRI andSPECT for small animals. *Nucl Instrum Methods Phys Res A* 652:731–34.

Meng L-J, Tan J-W, Gu G. 2007. Design study of an MRI compatible ultra-high resolution SPECT for in vivomice brain imaging. *IEEE Nuclear Science Symposium Conference Record (NSS/MIC)*, Honolulu, HI: 2956–60.

Nalcioglu O, Muftuler L, Wagenaar DJ, Szawlowski M, Kapusta M, Pawlov N, Maehlum G, Patt B. 2007. Development of MR-compatible SPECT system: A feasibility study. *Proc Intl Soc Mag Reson Med* 15:920.

Rittenbach A, Xu J, El-Sharkawy A-M, Edelstein W, Liu A, Parnham K, Hugg J, Tsui BMW. 2012. System calibration method for a CZT detector based ring-type small animal SPECT system. Paper presented at the 59th Annual Meeting of the Society of Nuclear Medicine Miami, FL, 2012. *J Nucl Med* 53 (Suppl): 2392.

Rittenbach A, Xu J, El-Sharkawy A-M, Edelstein WA, Parnham K, Hugg JA, Tsui BMW. 2013. Continuing evaluation of an MR compatible SPECT insert for simultaneous SPECT-MR imaging of small animals. 2013. *Conference Record of the 2013 IEEE Nuclear Science Symposium and Medical Imaging Conference (NSS/MIC)*, Seoul, Korea: 1–5.

Samoudi AM, Van Audenhaege K, Vermeeren G, Poole M, Tanghe E, Martens L, Van Holen R, Joseph W. 2015. Analysis of eddy currents induced by transverse and longitudinal gradient coils in different tungsten collimators geometries for SPECT/MRI integration. *Magn Reson Med* 74:1780–89.

Tan J-W, Cao L, Meng L-J. 2009. A Prototype of the MRI-Compatible Ultra-High Resolution SPECT for in Vivo Mice Brain Imaging. *IEEE Nuclear Science Symposium Conference Record (NSS/MIC)*, Orlando, FL: 2800–5.

Tsui BMW, Hugg JW, Xu J, Chen S, Edelstein W, El-Sharkawy A-M, Wagenaar DJ, Patt BE. 2011. Design and development of MR-compatible SPECT systems for simultaneous SPECT-MR imaging of small animals. *Proc. SPIE, Medical Imaging 2011: Physics of Medical Imaging, Eds Pelc NJ and Nishikawa RM* 7961:79611Y-1-7.

Wagenaar DJ, Kapusta M, Li J, Patt BE. 2006a. Rationale for the combination of nuclear medicine with magnetic resonance for pre-clinical imaging. *Technol Cancer Res Treat* 5:343–50.

Wagenaar DJ, Nalcioglu O, Muftuler LT, Szawlowski M, Kapusta M, Pavlov N, Meier D, Maehlum G, Patt BE. 2006b. Development of MRI-compatible nuclear medicine imaging detectors. *IEEE Nuclear Science Symposium Conference Record (NSS/MIC)*, San Diego, CA: 1825–28.

Integrated PET and MRI of the heart

CIPRIAN CATANA AND DAVID E. SOSNOVIK

4.1 INTRODUCTION

Positron emission tomography (PET) and magnetic resonance imaging (MRI) are widely used in cardiovascular imaging and have very complementary attributes. MRI offers excellent anatomical and functional information, while PET provides highly sensitive readouts of molecular and metabolic processes. Some degree of overlap does exist between the modalities: molecular MRI of selected targets, such as fibrin and macrophage infiltration, has been performed in human patients *in vivo*; magnetic resonance (MR) spectroscopy provides important insights into myocardial metabolism; and physiological gating of PET studies is introducing a functional dimension into the technique. Nevertheless, a compelling case can be made to combine the attributes of MRI and PET into a single integrated instrument to further advance cardiovascular imaging.

Table 4.1 Commonly used PET tracers and analogous approaches in MR spectroscopy and imaging

	PET tracers	MRI techniques
Metabolism	^{18}FDG, ^{11}C-Acetate	Hyperpolarized ^{13}C, ^{31}P
Perfusion	^{13}N-Ammonia, ^{82}Rb	First pass Gd
Viability	^{18}FDG	Late Gd enhancement
Inflammation	^{18}FDG	T2W MRI
		Magnetic nanoparticles
Vascular/valve calcification	NaF	Ultra short TE (UTE) MRI

In the recent years, integrated scanners capable of simultaneously acquiring MRI and PET data have been developed, first for small-animal imaging and, more recently, for whole-body studies in humans. Simultaneous acquisition opens new opportunities for cardiovascular research and clinical applications. First, the quantification of PET data can be improved using the temporally and spatially correlated MR information. Second, various PET and MR methods can be cross-calibrated in an integrated scanner (Table 4.1). For example, fluorodeoxyglucose (FDG) PET and late gadolinium (Gd) enhancement MRI both provide extremely accurate assessments of myocardial viability via highly different mechanisms. While qualitative myocardial perfusion is being increasingly used clinically, quantitative flow measurement by MRI is still challenging and would benefit from the validation provided by a simultaneous PET measurement. Since perfusion studies involve a pharmacological challenge and a highly dynamic signal, simultaneous acquisition of the PET and MRI data is needed to allow a true comparison of the techniques to be performed under identical conditions (Rischpler et al. 2013). Third, by providing both molecular MRI and PET readouts, PET-MRI could become the modality of choice for molecular imaging applications. Examples include the characterization of myocardial regeneration, thrombosis, and atherosclerotic plaque using dual-labeled PET-MRI probes (Ciesienski et al. 2013; Uppal et al. 2011).

The applications that could potentially benefit from this new imaging modality are discussed in more detail elsewhere in the book. Here, we focus on the technical challenges that need to be overcome to integrate these two imaging modalities and the approaches that have been proposed for this purpose. Significant attention is given to the need to obtain accurate attenuation maps of the heart and thorax directly from the MR data. Finally, we discuss several methodological opportunities, such as MR-assisted motion and partial volume effects correction and image-based radiotracer arterial input function estimation. These topics highlight both the challenges and opportunities of simultaneous PET-MRI.

4.1.1 Motivation for simultaneous PET and MRI data acquisition

The most obvious motivation for integrating these two powerful imaging modalities is that they provide complementary data that are routinely analyzed concurrently in research and clinical applications. PET noninvasively images the distribution *in vivo* of biomolecules (small molecules, peptides, antibodies, nanoparticles, stem cells) labeled with radionuclides that undergo positron decay. Because of the high sensitivity of radioactive assays, PET can measure picomolar concentrations of labeled biomolecules. A wide variety of molecular targets and pathways have been imaged using PET radiotracers. However, the spatial resolution of PET is limited by physical factors due to positron physics and the difficulty of acquiring sufficient counting statistics. Furthermore, PET images often lack definitive anatomic information, thus making interpretation of the precise location of radiotracer accumulation difficult.

MRI can provide high spatial resolution anatomic images with exquisite soft tissue contrast by exploiting the differences in relaxation times of protons in different biochemical environments. The combination of high spatial resolution and contrast allows the anatomic consequences (e.g., cardiac wall motion abnormalities) of many disease processes to be visualized in patients and in animal models. More advanced MRI techniques can measure important physiologic parameters, including diffusion and permeability. The addition of passive contrast agents based on Gd or iron-oxide nanoparticles can further enhance contrast. MR spectroscopy, which measures the shift in frequency at which protons in different chemical environments resonate, allows the relative

Table 4.2 Maximum positron energies (E_{max}) for several radioisotopes relevant for PET imaging

Radionuclide	E_{max} (MeV)
^{18}F	0.63
^{11}C	0.96
^{13}N	1.20
^{15}O	1.73
^{68}Ga	1.89
^{62}Cu	2.93
^{82}Rb	2.60, 3.38

concentrations of abundant metabolites and some drugs administered at mass levels to be measured. However, the sensitivity of MRI for different metabolites and tracers is micromolar to nanomolar and thus many orders of magnitude lower than that of PET, imposing significant restrictions on the kinds of targets that can be visualized.

The initial technical motivation for simultaneous PET-MRI was the potential reduction of the positron range, one of the two physical factors that limit the spatial resolution in PET, in a magnetic field. The positron range, defined as the distance from the site of positron emission to the line between the detectors in which the two annihilation photons were detected, is dependent on the energy of the emitted positrons because higher energy positrons travel a longer distance before losing their energy and undergoing annihilation. Since the positron's trajectory (in the plane perpendicular to the field) becomes helical when moving in a magnetic field, it was shown that it annihilates closer to the place where it was emitted (Iida et al. 1986; Raylman 1991). Based on the early Monte Carlo simulations and subsequent experimental measurements of the positron range for different positron emitters (Raylman, Hammer, and Christensen 1996; Wirrwar et al. 1997) it appears that magnetic fields higher than 7 T, high-energy positron emitters (e.g., ^{15}O, ^{82}Rb) (Table 4.2), and scanners with small-diameter and submillimeter scintillation crystals are needed for this effect to be relevant.

One alternative to combined PET-MRI hardware is to acquire data separately and then use sophisticated software registration algorithms to merge them (Hayakawa et al. 2000; Minoshima et al. 1992; Woods et al. 1998). This approach has been widely and successfully applied in human imaging for tissues that can be approximated by rigid-body transformations (e.g., brain). However, in the case of nonrigid body structures such as the heart, nonlinear warping approaches are required, making accurate software registration a very difficult task. Even if this could be achieved, dynamic changes can occur between the two scans that would confound the spatial relationship of the two datasets. Another major drawback of this approach is that it removes the possibility of temporally correlating dynamic MR and PET signals or of comparing functional MRI with PET physiological measures.

4.1.2 TECHNICAL CHALLENGES IN INTEGRATING PET AND MRI

The major obstacle to PET inside an MRI scanner (aside from the limited space) is the presence of the strong magnetic field, which completely degrades the performance of photomultiplier tubes (PMTs). On the other hand, solid-state photon detectors, such as avalanche photodiodes (APDs) and Geiger mode APDs (also called solid-state photomultipliers, silicon photomultipliers [SiPMs], or multiphoton pixel counters), are magnetic field insensitive and can be used as alternatives to the more popular PMTs for this application. However, one challenge in the case of these semiconductor-based detectors is that their performance is a strong function of temperature, which affects their gain (changes of up to 5%/°C has been reported [Kolb et al. 2010]), breakdown voltage, and capacitance (Conradi 1974). In addition to the heat produced by the PET electronics, there is another aspect that has to be considered in an integrated PET-MRI scanner. The switching MR gradients induce eddy currents in the PET detector shielding material that can lead to temperature increases in the detector enclosure (as well as image artifacts and spatial distortions in the MR images).

Mutual electromagnetic interference has to be considered whenever two devices work next to each other. Signals in the radiofrequency (RF) part of the spectrum, where the MR scanner operates, are of special concern in this case as they could interfere with the operation of the PET and MR electronics, leading to spurious signals, artifacts, or decreased signal-to-noise ratio in the resulting images. For this reason, PET electronics are usually placed inside shielding enclosures (Faraday cages).

One requirement whenever materials have to be placed inside the MR scanner is that they minimally disturb the main magnetic field homogeneity. This could, for example, happen when diamagnetic or paramagnetic materials are placed inside the bore. Since at least the scintillator arrays have to be placed inside the magnet for simultaneous PET data acquisition, their magnetic susceptibility had to be evaluated. Fortunately, both bismuth germanium oxide and lutetium oxyorthosilicate (LSO) demonstrated similar susceptibility to that of human tissue and only minimally affected the MR images. On the other hand, lutetium Gd oxyorthosilicate caused significant artifacts and distortion in the MRI images (Yamamoto et al. 2003).

4.2 INSTRUMENTATION—INTEGRATED PET-MRI SCANNERS

Recognizing the potential to combine the strengths of the two modalities, while mitigating some of their limitations, approaches to combine PET and MRI in a single scanner have been proposed for many years, but the progress in the field has been slower than that of the PET/computed tomography (CT) technology in the early stages of development (Townsend et al. 2004). In this section, we discuss some of those efforts that lead to the development of integrated PET-MRI systems capable of simultaneous acquisition.

4.2.1 EARLY APPROACHES

The first attempts to combine PET and MRI were made in the 1990s. At the time, PMTs were the photon detectors of choice in PET and the early efforts focused on overcoming their magnetic field sensitivity. This was initially accomplished by using ~4-meter-long optical fibers to couple LSO scintillator elements placed inside the magnet to multichannel PMTs and electronics placed outside a 0.2-T open magnet MRI system. The first prototype, called McPET I, was used for proof-of-principle simultaneous data acquisition in phantoms (Shao et al. 1997a). The second-generation prototype (McPET II) had a larger transaxial diameter (54 mm instead of 38 mm) (Shao et al. 1997b) and was used for simultaneously acquiring PET images and 31P NMR spectra from isolated, perfused hearts (Garlick et al. 1997) and performing *in vivo* rat brain studies. Subsequently, several designs have been proposed either for improving the sensitivity or spatial resolution uniformity by using multiple layers of crystals in the transaxial direction (Mackewn et al. 2005, 2010), increasing the axial coverage by bending the optical fiber bundles (Raylman et al. 2006, 2007) or using slanted light guides (Tatsumi et al. 2012; Yamamoto et al. 2010). Alternatively, minimally modified PET detectors were positioned in the gap of a 1-T MR split-magnet (Lucas et al. 2006) or integrated into a field-cycled MRI system (Bindseil et al. 2011; Peng et al. 2010).

4.2.2 CURRENT STATE-OF-THE-ART

It was evident even from the early days that it would be impossible to use PMTs for building an integrated PET-MR scanner for routine human use. Fortunately, APDs have emerged as replacements for PMTs for this purpose and have been demonstrated to work inside MRI scanners at fields as high as 9.4 T (Pichler et al. 1997). An APD consists of a very thin, high-purity silicon wafer doped with impurities that creates regions with different properties across the material. Electrical contacts are placed on the two surfaces and a reverse bias voltage is applied across the device. When a photon with enough energy interacts in the absorption region, electron-hole pairs are produced. Under the influence of the applied electric field, the charges are separated and the electron drifts toward the anode and the hole toward the cathode. In the multiplication region, where the applied field is significantly higher, the electrons and holes gain enough energy to create other electron-hole pairs through impact ionization in a repeating process so that in the end, an avalanche of charge carriers results.

When compared with PMTs, the APDs have lower gains (10^2–10^3 vs. 10^6), higher noise levels, and consequently lower signal-to-noise ratio (SNR). Furthermore, the gain of an APD is strongly dependent on the temperature and bias voltage. However, they have similar timing and energy properties because of the higher quantum efficiency (60%–80% vs. 15%–25% for PMTs). As already mentioned, the main advantage of APDs in the context of PET-MRI is that they are insensitive to magnetic fields.

Several APD-based PET-MRI scanners have been developed and used for *in vivo* small-animal studies. The first one was developed at University of California, Davis (Catana et al. 2006, 2008). It used position-sensitive APDs (PSAPDs; RMD, Watertown, Massachusetts) as the photon detector and short bent optical fiber bundles to channel the light outside of the 7-T magnet's imaging field of view where the PSAPDs were located, reducing the potential for interference in this way (Figure 4.1). In parallel, researchers from the University of Tubingen built an APD-based insert for a 7-T small-animal MR scanner (ClinScan, Bruker, Germany) (Judenhofer et al. 2007) without using optical fibers between the photon detectors and the scintillator arrays (Judenhoper et al. 2008). Similarly, a PET insert for a Bruker 9.4-T MRI scanner was built at the Brookhaven National Laboratory (Woody et al. 2007) and proof-of-principle *in vivo* studies were performed in small animals, including rat gated mouse heart PET-MRI exams (Maramraju et al. 2011).

The first major equipment manufacturer to design and build an integrated PET-MRI scanner for human use was Siemens (Schlemmer et al. 2008). This prototype device, called BrainPET, is a head-only PET insert that fits into the standard 3-T MRI commercially available scanner (Magnetom Trio, Siemens, Erlangen, Germany) (Figure 4.2a). The first BrainPET scanner was installed in 2007 at the University of Tubingen, and three of these prototype devices are still in use. Numerous studies have been performed using these scanners for assessing the mutual interference (Chonde et al. 2013; Kolb et al. 2012), the performance of novel methods developed for addressing some of the specific challenges (Catana et al. 2010) or taking advantage of unique opportunities (Catana et al. 2011), and for exploring potential research and clinical applications (Catana et al. 2012; Catana, Guimaraes, and Rosen 2013; Frullano et al. 2010; Sander et al. 2013; Uppal et al. 2011).

Siemens also introduced the first fully integrated whole-body MR-PET scanner, called Biograph mMR, in 2010 (Figure 4.2c). Similar to the BrainPET prototype, this scanner also uses APD technology, but the PET and MRI hardware and software are now fully integrated. The PET detectors are placed between the body RF coil and the gradient set of a modified Vario 3-T MR scanner and the two scanners share the same gantry. As an improvement over the BrainPET prototype, which was air-cooled, chilled water is used to control the temperature of the detectors and associated electronics. In terms of performance, the PET component of the Biograph mMR is similar to that of the commercially available PET/CT scanner (Siemens mCT) (Delso et al. 2011). Numerous reports that presented the initial experiences with this scanner at several institutions

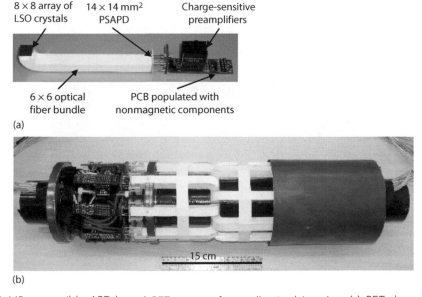

(a)

(b)

Figure 4.1 MR-compatible APD-based PET scanner for small-animal imaging: (a) PET detector module; (b) complete PET insert consisting of 16 detector modules, shielding material (removed on the left side) and carbon fiber tube used for support. (From Catana C et al., *J Nucl Med*, 47, 1968–1976, 2006.)

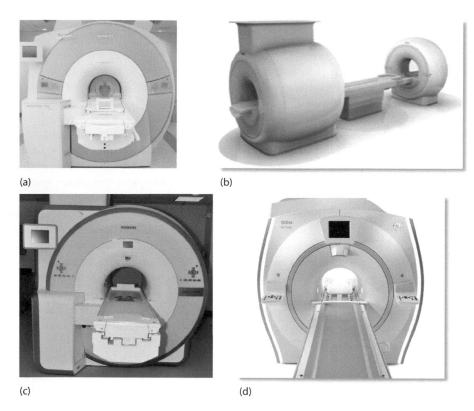

(a)

(b)

(c)

(d)

Figure 4.2 Integrated PET-MR scanners currently available for human use: (a) Siemens MR-BrainPET prototype, (b) Philips sequential TF PET/MR whole-body scanner, (c) Siemens Biograph mMR whole-body scanner, and (d) General Electric Signa PET/MR whole-body scanner. ([a–c] From Catana C et al., *J Nucl Med*, 54, 815–824, 2013.)

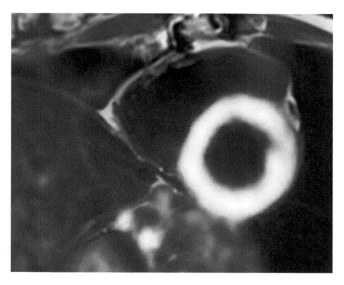

Figure 4.3 Integrated cardiac MRI and PET image in a human patient. The PET isotope injected was FDG. The MRI image was acquired with a turbo spin echo (RARE) sequence with a dual inversion (black blood) prepulse. Image courtesy of the authors.

around the world and the potential applications have already been published. An example of an FDG-PET-MRI cardiac study in a healthy volunteer using the scanner installed at our center is shown in Figure 4.3.

4.2.3 FUTURE DEVELOPMENTS

Recently, Geiger-mode APDs (GAPDs, a.k.a. SiPMs) have been proposed and tested for PET applications (Kolb et al. 2010; Roncali and Cherry 2011). These devices can be thought of as two-dimensional (2-D) arrays of very small APDs, called microcells, operated in Geiger-mode. This means that an avalanche occurs whenever an incident visible light photon produces an electron-hole pair in that microcell. The output signal of each microcell is not proportional to the number of these incident photons. However, the total output of the device is proportional to the number of microcells that fired and, under the assumption that the number of incident photons is significantly smaller than the number of microcells, the output signal is thus proportional to the energy deposited in the scintillator (Roncali and Cherry 2011). Some of their characteristics (e.g., gain comparable to that of PMTs, moderate bias voltage requirements, excellent timing properties, insensitivity to magnetic fields, etc.) make SiPMs very promising candidates for replacing APDs as the photon detector of choice for simultaneous PET-MRI.

Several groups have built prototype scanners using this technology and reported promising results in phantom and small-animal studies (Seong Jong et al. 2008; Yamamoto et al. 2011, 2012; Yoon et al. 2012). In a continued effort to minimize the interference with the MR, long flexible flat cables have been proposed for coupling the GAPDs to amplifiers located outside the 5-gauss line of the 7-T MRI scanner (Kang et al. 2011) or ~3-cm-long fibers were used to introduce gaps between the PET modules to allow the use of the standard body transmit coil of a 3-T MRI scanner (Hong et al. 2012).

In December 2013, General Electric introduced the first whole-body integrated PET-MRI scanner that uses digital SiPM technology, called SIGNA PET-MR (Figure 4.2d). The PET detector gantry has transaxial and axial fields of view of 60 cm and 25 cm, respectively. The PET detectors (Lutetium-based scintillators coupled to arrays of SiPMs) have excellent timing resolution (<400 ps was reported [Levin et al. 2013]), which enables time-of-flight PET image reconstruction. They were positioned between the body RF coil and the MR gradient set of the GE Discovery 750w 3-T MR scanner and each consists of five rings of 112 detector blocks. Interestingly, the SIGNA PET-MR scanner was designed to be fully upgradable from the stand-alone MR device on which it is based. The scanner includes active and passive mechanisms for controlling the temperature. Preliminary testing demonstrated interference-free operation both on the MR and PET side and very high PET sensitivity (21 cps/kBq).

4.3 TECHNIQUES FOR MR-ASSISTED PET DATA CORRECTIONS

4.3.1 MR-BASED PHOTON ATTENUATION CORRECTION

Since the early methods used in stand-alone PET or integrated PET/CT scanners for accounting for the reduction in the number of detected annihilation photons through interactions with the subject cannot be used in integrated PET-MRI scanners, for obvious reasons (e.g., limited space inside the MR bore, engineering challenges in integrating a rotating source mechanism given the MR environment, etc.), methods for obtaining similar information from the MR data had to be developed. Unlike x-ray CT, this is not a trivial task because the MR signal depends on proton density and tissue relaxation times, properties that are not directly related to tissue electron density, which is relevant for PET attenuation correction. Furthermore, imaging bone or lung tissue with conventional MR sequences is challenging, truncation artifacts are present in larger patients due to the limited field of view of MR, the annihilation photons can be absorbed in the RF MR coils located in the PET field of view, and subject motion leads to mismatches between the MR-derived attenuation map and the emission data. Several methods have been proposed for MR-based attenuation correction, ranging from those aimed at segmenting the body into a limited number of tissue classes to those that attempt to generate CT-like images from the MR data.

4.3.1.1 METHODS TO GENERATE DISCRETE-VALUED ATTENUATION MAPS

In this class of methods, a limited number of tissue classes (e.g., soft tissue, fat, bone, lung, air) are segmented and a single linear attenuation coefficient is assigned to each of these classes. These coefficients usually correspond to the average values measured from the CT- or transmission-based attenuation maps for all the segmented classes. Hence, the errors in the segmentation procedure and those introduced by the assumption that a predefined value represents the attenuation in the whole tissue class affect the accuracy of these methods. Furthermore, additional approximations need to be made because not all the tissue classes can easily be segmented from the MR data (e.g., bone segmentation is still very challenging) and the attenuation properties of tissue changes due to physiological factors (e.g., the lung attenuation is different in expiration versus inspiration), to give just a couple of examples.

In one of the first approaches proposed for whole-body MR-based attenuation correction, which was subsequently implemented on the Siemens Biograph mMR scanner, the body is segmented into four compartments (i.e., water, fat, lung, and air) (Martinez-Moller et al. 2009). From the MR data acquired in the coronal orientation using a two-point Dixon-VIBE sequence, the in-phase and out-phase images as well as water and fat images are obtained at each bed position. An air threshold is first determined from the in-phase images. Next, the lungs are segmented using connected component analysis. The voxels corresponding to soft and fat tissue are then obtained by applying thresholds to the water and fat images. Morphological operations are finally applied to remove noise and predefined linear attenuation coefficients (0, 0.022, 0.085, and 0.100 cm^{-1}) are assigned to the four tissue classes (background air, lung, fat, and soft tissue, respectively). An example of an attenuation map generated using this method is shown in Figure 4.4. Bone, lung, and neck regions showed the largest standardized uptake value (SUV) changes (up to 13%) using this approach. While these changes did not affect the clinical interpretation, they would have to be carefully considered when interpreting longitudinal studies. Although no cardiac lesions were included, only minimal changes in the SUV of node metastases in the mediastinum were reported, suggesting that the method performs reasonably well in the thorax as well (Martinez-Moller et al. 2009).

Only three components (i.e., water, lung, and air) were segmented in the procedure implemented for the Philips Ingenuity TF PET-MRI sequential scanner (Figure 4.2b) (Schulz et al. 2011). The MR data used for segmentation were acquired with a free-breathing 3-D T1-weighted spoiled gradient echo sequence in 6-cm-long slabs. To account for the drops in signal intensity at the edge of the 3-D slabs, the images were first smoothed in the axial direction. Next, a Laplace-weighted histogram (Wiemker and Pekar 2001) of the whole volume is used to automatically determine the thresholds to be used for segmenting the tissue classes. This step was required because using a fixed value (similar to the procedure used for segmenting CT images) would not work with MR data because of the larger intersubject variability in MR signal intensity compared to CT. Starting from these automatically determined thresholds and using region growing techniques, the body mask can be determined for each slice and it includes soft and bone tissue as well as air cavities. To separately segment the lungs, the coronal

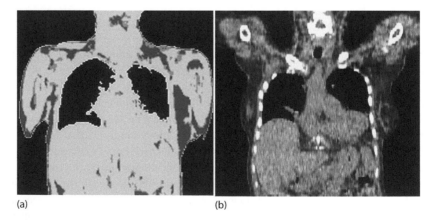

(a)   (b)

Figure 4.4 Segmented attenuation map of the torso obtained using the method currently implemented on the Biograph mMR scanner: (a) Dixon-VIBE- and (b) the corresponding CT-based attenuation maps. Note the different positioning of the arms of the subject for the two examinations and the fact that a separate bone tissue class is not included in the MR-based attenuation map. Images courtesy of the authors.

slice most likely to contain lung tissue (e.g., that with the maximum body cross section) is automatically selected. Starting from this slice and using a clustering algorithm combined with size criteria (based on expected values for the lungs) and 3-D region growing techniques, the lungs can be extracted. The linear attenuation coefficients assigned to soft and lung tissue in this method are 0.094 and 0.024 cm^{-1}, respectively. As expected, a tendency for underestimating the SUVs was again reported and it was attributed to the misclassification of bone as soft tissue (Sensakovic and Armato 2008).

Since segmenting additional tissue classes increases the complexity of the algorithms and the data acquisition times and reduces the reliability of the segmentation procedures, simulation studies have been performed to estimate the number of classes that need to be identified (Akbarzadeh et al. 2013; Keereman et al. 2011). For example, errors below 5% were reported when liver and adipose tissues were treated as soft tissue and only air, lung, soft tissue, spongeous, and cortical bone tissue classes were included separately in the attenuation map. Misclassifying lung tissue as air caused 45% bias in lung tissue and 17% in the thoracic spine. Not including cortical bone in the segmentation caused 20% errors. And a 30% bias in bone and adjacent tissue was reported when bone voxels were misclassified in a different study (Akbarzadeh et al. 2013).

Truncation artifacts in large patients are particularly relevant in PET-MRI for two reasons. First, the MR examinations are usually performed with the patients holding their arms down (as opposed to the arms-up procedure used for CT scanning). Second, the MR transaxial field of view of 50–60 cm is significantly smaller than that of stand-alone PET scanners. As a result, parts of the arms are missing from the MR-derived attenuation maps, which leads to 10%–20% biases in the resulting PET images (Delso et al. 2010). To address this issue, a pair of "soft tissue" cylinders approximating the missing arms were added to the attenuation maps using the nonattenuation corrected PET images for guidance (Delso et al. 2010) or a more advanced algorithm was used to estimate the contour of the body from the same images (Kalemis et al. 2013). A more elegant solution to this problem involves iterative reconstruction algorithms and will be discussed later in this chapter.

4.3.1.2 METHODS TO GENERATE CONTINUOUS-VALUED ATTENUATION MAPS

Atlas- or template-based methods have been successfully used for obtaining continuous-valued attenuation maps for brain imaging (Montandon and Zaidi 2005; Rota Kops and Herzog 2008), but they have proven more difficult to implement for whole-body applications given the anatomical variability and the larger susceptibility to errors in the image registration procedure. As a solution to address these issues, local pattern recognition combined with atlas registration was proposed first for brain (Hofmann et al. 2008) and subsequently for whole body imaging (Hofmann et al. 2011). The images in the atlas database (i.e., pairs of coregistered CT and MR images acquired in the same patients) are registered to the MR images of the patient of interest using a five-class segmentation. The transformations derived from these registrations are applied to the corresponding CT images and a set of pseudo-CT predictions are generated. Instead of averaging these values, a Gaussian process regression is used to predict the pseudo-CT for each voxel based on the subvolume that most likely matches the subvolume of interest, thus including the information about the local anatomy in the prediction process. Finally, two additional steps are applied automatically for incorporating the additional information obtained from the Dixon sequence regarding fat and water voxels and for increasing the chance of segmenting voxels corresponding to gas pockets. When using this method, the relative change in PET quantification with respect to the CT-based attenuation correction was, on average, 7.7% ± 8.4%. For comparison, the average errors reported when using the continuous-valued approach in the same patients was almost double. The largest difference between the two methods was observed in the pelvis, followed by the abdomen. Interestingly, their performance was similar in the thorax, likely because of the larger lung density variability that was not sufficiently captured in the atlas generated from MR data acquired with conventional sequences not optimized for lung imaging.

A registration-based approach was also suggested for including bone tissue in the segmented MR-based attenuation maps. This was accomplished by selecting the most similar CT from a database based on various patient specific similarity metrics such as gender, height, age, etc., and nonrigidly coregistering it to the subject's MRI. Only the voxels with Hounsfield units above 80 from the CT were selected and included in the pseudo-CT maps. Although the bias in volumes of interest that contained bone was dramatically reduced, the authors reported several limitations of their method that have to be addressed before it can be used routinely (Marshall et al. 2013).

4.3.1.3 METHODS USING THE EMISSION DATA TO ITERATIVELY IMPROVE THE ATTENUATION MAP

Relying only on the MR data or predefined atlases, it is extremely difficult to account for the large intersubject variability, particularly for whole-body applications. However, the PET emission data inherently contain information about the photon attenuation that could be used to overcome this limitation (Censor et al. 1979). The usefulness of these methods that involve the simultaneous estimation of both the emission and attenuation data was initially limited because of the cross-talk between the attenuation and emission estimation. However, this can be minimized by using the anatomical information provided by the MR in integrated PET-MR scanners or the time-of-flight information (Censor et al. 1979; Mollet et al. 2012; Panin, Aykac and Casey 2013; Rezaei et al. 2012; Salomon et al. 2011).

Such an approach was proposed for completing the truncated attenuation maps using the emission data (Nuyts et al. 1999). The method implemented on the Siemens Biograph mMR scanner was shown to reduce the error in the SUV estimation from 15%–50% (Delso et al. 2010) to less than 5% (Nuyts et al. 2010). Alternatively, the time-of-flight information can be used to improve the estimation of the linear attenuation coefficients assigned to the different regions segmented from the MR data (Salomon et al. 2011).

4.3.1.4 ADDITIONAL CONSIDERATIONS

There are other aspects that have to be considered for implementing an accurate whole-body MR-based attenuation correction method, particularly when focusing on the heart.

Estimating the lung attenuation map is very difficult because imaging the lungs with conventional MR sequences is very challenging. However, this is relevant because of the large variability in lung attenuation between subjects (Martinez-Moller et al. 2009; Schulz et al. 2011) as well as intrasubject, during the respiratory cycle, and in pathological conditions (Marshall et al. 2012).

Foreign objects present in the body often cause susceptibility artifacts, leading to signal voids in the MR images and the corresponding attenuation maps. As a simple solution, linear attenuation coefficients corresponding to the implant material could be manually assigned to the misclassified voxels (Ladefoged et al. 2013). It was also suggested that atlas-based techniques are more robust in the presence of this type of artifacts than segmentation-based approaches because of the implicit assumptions made about the values expected at a specific location (Hofmann et al. 2011). Using the time-of-flight information might also help minimize these artifacts. On the MR side, several MR sequences that are less sensitive to susceptibility artifacts (Ai et al. 2012; Sutter et al. 2012) could be used to minimize the signal voids in the MR images.

The attenuation properties of several MR contrast agents (e.g., ferumoxil, gadobutrol, ferumoxytol) have been investigated, and none of the ones used routinely for cardiovascular applications has been shown to bias the resulting attenuation maps (Borra et al. 2013; Lois et al. 2012).

Given the fact that the MR data used for generating the attenuation maps are acquired in tens of seconds while the PET data acquisition takes minutes or longer, a misregistration between the emission volume and the attenuation map is introduced due to motion. This mismatch can, in principle, be reduced in an integrated PET-MR scanner by using the MR signal for tracking the motion of the subject dynamically *during* the PET acquisition and for correcting the PET data retrospectively, as discussed later in this chapter.

Another aspect specific to integrated PET-MR scanners is the annihilation of photons in the RF coils that are located between the subject and the PET detectors. The attenuation properties of the standard and fixed RF coils (e.g., head and neck, spine) that are delivered with the commercially available scanners have been characterized by the scanner manufacturer and the corresponding attenuation maps are automatically selected during the PET data processing and image reconstruction step for each table position (Kalemis, Delattre, and Heinzer 2013). However, this is not the case for the body flexible arrays that are currently used for cardiac studies on these scanners. Fortunately, these coils have been designed to minimally attenuate the annihilation photons and can be ignored for routine applications. For advanced quantitative studies, fiducial markers can be used for determining the position of the coil in the PET field of view so that predefined coil attenuation maps can be included in the attenuation correction procedure (Paulus et al. 2012). Even regular MR coils such as the flexible anterior part of the cardiac coil used with the Philips scanner have been shown to introduce minimal bias when their position and attenuation properties are known and properly accounted for (Eldib et al. 2014).

4.3.2 NONRIGID BODY MR-ASSISTED PET MOTION CORRECTION

Organ deformations related to the physiologic cardiac and respiratory motion (in addition to voluntary and involuntary subject motion) affect PET data quality and quantification in cardiac studies. The high-temporal-resolution MR data acquired simultaneously with the PET data in integrated PET-MRI scanners can be used for deriving the information required for retrospectively performing PET motion correction. As opposed to the case of rigid-body motion estimation where characterizing the displacement of three noncollinear points is sufficient to describe the motion of the whole volume (i.e., head) (Catana et al. 2011), the displacements of all the voxels in the PET field of view have to be measured or estimated in the case of nonrigid body motion. The first demonstration of nonrigid body PET motion correction was done using a PMT-based preclinical MR-compatible PET scanner (Mackewn et al. 2010) and a two-section phantom (Tsoumpas et al. 2010). While the central section of the phantom was stationary, the outer one could be rotated, which generated nonrigid transformations between the two groups of rod sources located in each section. The MR images acquired at 10 positions were coregistered to the reference image using the Image Registration Toolkit Software (Rueckert et al. 1999) and the displacement vectors were applied to the corresponding PET images. Although artifacts were observed in the motion corrected images, these images were comparable to those obtained in the absence of motion.

Although both respiratory and cardiac motions are relevant and have to be addressed for quantitative cardiac PET studies, next we will discuss the solutions proposed for solving these issues separately.

4.3.2.1 RESPIRATORY MOTION CORRECTION

First attempts to characterize the respiratory motion in the context of PET-MRI have focused on the abdomen (Guerin et al. 2011). This is not surprising because this is a region of the body where the effects of respiratory motion can be largely disentangled from those of cardiac motion. Simple techniques (e.g., respiratory bellows, 1-D MR navigators for monitoring the position of the diaphragm, etc.) are currently used for respiratory gating in PET and MRI studies performed on stand-alone instruments. However, although these methods are useful for deriving a respiratory signal, they are clearly not sufficient for completely characterizing nonrigid organ deformations that accompany respiratory motion and more advanced MR techniques have been proposed for this purpose, such as phase contrast MRI, tagged-MRI, displacement encoding with stimulated echoes, and harmonic phase imaging (HARP) (Ozturk, Derbyshire, and McVeigh 2003).

In the context of PET-MRI, tagged-MRI has been suggested for tracking the motion in the abdomen (Chun et al. 2012; Guerin et al. 2011; Ozturk, Derbyshire, and McVeigh 2003). The MR tags can be thought of as fiducial markers that are temporarily superimposed on the tissue of interest. The change in the tag shape (e.g., a grid is most commonly used) at different time points is directly related to the motion of the tissue. Two sequences are required to estimate the motion using this procedure: one for tagging the tissue of interest and another one for deriving the motion fields. The first task can be accomplished using a delay alternating with nutation for tailored excitation sequence in the case of small-animal imaging (Mosher and Smith 1990) or complementary spatial modulation of magnetization (C-SPAMM) sequence (Axel and Dougherty 1989) in the case of whole-body human studies. For the second task, a regularized HARmonic phase (r-HARP) (Osman, McVeigh, and Prince 2000) or B-spline nonrigid image registration algorithm can be used.

Once derived from the MR data, these motion fields are used to update the system matrix and the attenuation map. A motion corrected list mode maximum likelihood expectation maximization algorithm is used to "register" each of the frames to the reference frame. Thus, a single reconstruction is performed for the whole data set, which means the motion of the organ of interest is "frozen" in the reference frame and all the recorded counts contribute to the final image, as opposed to standard gating techniques in which a large fraction (i.e., 7/8th if eight gates are defined) of the events are discarded. Simulations and early proof-of-principle PET-MRI studies demonstrated that the PET image variance is considerably reduced after motion correction, which leads to improvements in SNR and contrast recovery coefficients.

One obvious disadvantage of the tagged-MRI-based approach is that the MRI data collected for tracking the motion cannot be used for clinical purposes. Recognizing the need to run other sequences simultaneously with the PET data acquisition, a method to model the subject-specific motion has been proposed (King et al. 2012).

Once defined based on 3-D dynamic data set, a simple 1-D or 2-D navigator sequence can be used to drive the model. Forming the model is not a trivial task because of the intracycle (the motion paths during inspiration are different from those observed during expiration) and intercycle (the motion paths are different from one respiration to the next) variability. Sufficient 3-D dynamic MR data have to be acquired to characterize the motion of the thorax during each respiratory cycle and across different cycles. The motion fields with respect to a reference frame are then derived from these data. The continuously acquired 2-D navigator is used not only to apply the model but also to decide when a new set of dynamic 3-D MRI data is required to update the model.

Unfortunately, applying these methods in the thorax is still very challenging because of the reduced SNR of the lungs. Until MR methods for improving lung imaging are developed, the lung tissue could either be masked out to avoid introducing bias in adjacent regions (Guerin et al. 2011) or the lung motion fields could be interpolated from those estimated at the boundaries (King et al. 2012).

4.3.2.2 CARDIAC MOTION CORRECTION

The motion of the heart is considerably more complex than the motion of the lungs or internal organs in the abdomen. MR tagging can again be used to characterize it and proof-of-principle studies were performed on the BrainPET scanner using a cardiac beating phantom (Petibon et al. 2013). To build this phantom, the space between two inflatable balloons was filled with a radioactive gel and a methylcellulose background surrounded the outer one. Additionally, cold gel inserts were placed at different locations within the "myocardial wall." A

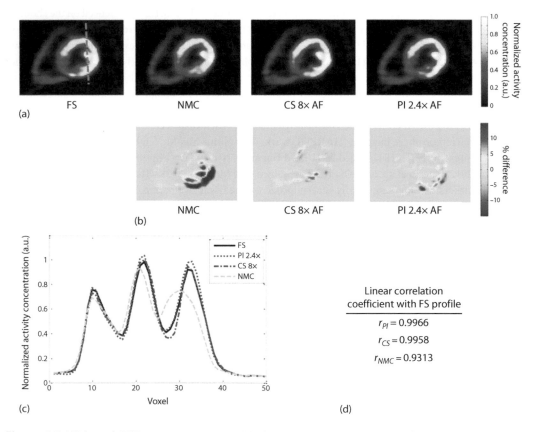

Figure 4.5 MR-based PET motion correction: (a) The PET images reconstructed using motion estimates derived from fully sampled (FS) and tagged-MR images accelerated using compressed sensing (CS) and parallel imaging (PI) techniques compared to those reconstructed without motion correction (NMC); (b) difference map using the FS case as reference; (c) line profiles for the four cases for the line shown in the first image; and (d) the linear correlation coefficients of the lines profiles using the FS case as reference. (From Huang C et al., *Med Phys*, 42, 1087–1097, 2015. Reprinted with permission.)

ventilator was used to inflate/deflate the innermost balloon, which mimicked the motion of the wall. The MR tagging was achieved with a SPAMM sequence and a multislice/multiphase gradient recalled echo sequence was run to acquire multiple representative volumes during the cycle. The motion fields were estimated from the tagged-MRI volumes using a nonrigid B-spline registration algorithm (Ledesma-Carbayo et al. 2008) and were incorporated into a list-mode iterative reconstruction framework. The defect/myocardium contrast recovery and lesion detectability were improved after MR-based motion correction when compared to the motion uncorrected case by 34%–206% and 62%–235%, respectively. Further significant improvements were noted in both metrics when the point-spread function of the scanner was included in the reconstruction and partial volume effects were minimized (Petibon et al. 2014).

Accelerated tagged-MRI using either parallel imaging or compressed sensing techniques has been recently suggested for minimizing the acquisition time of the sequences used for motion characterization. The generalized autocalibrating partially parallel acquisitions (GRAPPA) algorithm with different acceleration factors and the k-t focal underdetermined system solver (kt-FOCUSS) algorithm were used for parallel imaging and compressed sensing, respectively. The authors showed that the PET image quality corrected using motion fields obtained using this approach was similar to that obtained from fully sampled MR data for up to 4 times acceleration (Figure 4.5). Respiratory gating was performed in this study to minimize the influence of respiratory motion (Huang et al. 2015).

Standard tagged-MRI techniques cannot be used in more difficult situations, such as imaging the coronary arteries. Instead, a three-point Dixon MR sequence was proposed for imaging the surrounding fat (in addition to water). In simulation studies, the cardiac and respiratory motion fields were estimated from the gated fat-MR volumes and incorporated in the reconstruction. Substantial increases in plaque-to-background contrast after motion correction were predicted for all the plaque-to-background concentration ratios considered (Petibon et al. 2014).

4.3.2.3 COMBINED CARDIAC/RESPIRATORY MOTION CORRECTION

Respiratory and cardiac motion could, in principle, be simultaneously estimated from tagged-MR data. For example, the cardiac and respiratory cycles can each be divided into eight phases. Using one of these phases as the reference, the motion fields for the other 63 phases could be estimated using the HARP or B-spline nonrigid image registration algorithms discussed previously. For each ECG trigger, a 1-D navigator tracking the position of the diaphragm would first be used to determine the respiratory phase. Next, tagging followed by the rapid acquisition of the data corresponding to each of the eight cardiac phases would be performed. Because acquiring the data corresponding to all the respiratory and cardiac phases would take a prohibitively long time, many of the 64 phases would be undersampled. Compressed sensing approaches could be used to obtain the missing MR volumes or interpolation techniques could be used to derive the motion fields for the missing phases from the existing data (Ouyang, Li, and El Fakhri 2013).

4.3.3 PARTIAL VOLUME EFFECT CORRECTION

Partial volume effects lead to an underestimation of the radiotracer concentration in structures of interest (e.g., left ventricle myocardium wall) that are smaller than 2–3 times the full width at half maximum spatial resolution of the PET scanner as well as an overestimation of the activity in the adjacent structures (e.g., left ventricle chamber), the so-called spill-over and spill-in effects. In the case of cardiac imaging, this problem is further complicated by the fact that the location and thickness of the myocardium change as a result of cardiac and respiratory motion (as well as involuntary motion). Because of these issues, cardiac partial volume effects are both spatially and temporally variable. Particularly relevant to dynamic PET studies, required for accurate quantification of myocardial viability or blood flow, is that the relationship between the activity concentration in the myocardium and adjacent blood pool changes in time. For example, the activity concentration in the blood pool is significantly higher than that in the myocardium at the early time points, while the proportion is reversed at the later ones. A very comprehensive review of the myriad of partial volume effects correction methods, including examples of their use in cardiology, has been published elsewhere (Erlandsson et al. 2012).

In one of the first attempts to address the partial volume effects with the goal of improving the estimation of the absolute myocardial blood flow, the tissue fraction ("defined as the fraction of tissue mass in the volume of the region of interest") was included as an additional parameter in the kinetic model used to analyze the PET data in each of the regions of interest (Iida et al. 1988). A similar approach has been proposed by Hutchins et al. (1992). In this case, the assumption was that the measured PET signal in a specific region of interest is a combination of the activity concentration spilled over from the blood pool and true myocardial concentration multiplied by the fraction of the volume occupied by myocardial tissue. Although a strategy for placing the region of interest was proposed, this was obviously an oversimplification because it did not account for spill-over in other adjacent regions (e.g., right ventricle, lung, liver, etc.) and it ignored the spill-in effect from these regions.

Alternatively, methods that apply the correction before kinetic modeling have been developed. In an approach similar to the ones used for brain gray/white matter partial volume effects correction, an idealized image of the myocardial wall obtained directly from the PET images was convolved with the point spread function of the scanner to derive the spill-over and recovery coefficients (Nuyts et al. 1996). In a more recent implementation, high-resolution contrast-enhanced CT images were used to define the volumes of interest (Du et al. 2013). Using simulations, the authors also showed that the most accurate estimates of the activity are obtained when partial volume effect correction is performed independently for the images obtained from each gate (i.e., corresponding to the different phases of the cardiac cycle). This, however, requires a gated contrast-enhanced CT exam, which significantly increases the patient's radiation exposure. An interesting aspect revealed by this study was that the contrast between diseased and healthy tissue is actually affected by partial volume effects, with the highest accuracy being achieved when the defect is considered as a separate volume of interest. This highlights the need for a high-resolution high-contrast imaging technique, such as late Gd contrast-enhanced MRI, that is capable of providing anatomical and physiological information for accurately defining these volumes of interest for improving quantification in cardiac PET.

In the context of PET-MRI, those approaches that use the anatomical information derived from MRI for performing the correction, either post or during the image reconstruction, are particularly relevant. At this time, no one method has been accepted or is routinely used for research applications and even less so in clinical practice. This is probably because the performance of these methods is task specific and there is no optimal method that can be used in all situations. More importantly, the performance also depends on the accuracy of the anatomical segmentation procedure and the precision of the spatial coregistration of the PET and MRI volumes. The nonrigid-body motion correction methods discussed in the previous section would eliminate the image coregistration problem in integrated PET-MRI scanners (Petibon et al. 2013).

4.3.4 IMAGE-BASED RADIOTRACER ARTERIAL INPUT FUNCTION ESTIMATION

The radiotracer arterial input function is required for estimating biologically relevant parameters of interest from the dynamic PET data by kinetic modeling. Because the gold standard technique (i.e., manual blood sampling from the radial artery) is invasive and cannot be used in all the patients, image-based methods have been proposed as alternatives. The placement of the region of interest to obtain the most accurate input function from the PET images is still the subject of debate, the left atrium or ventricle, ascending or descending aorta, and even femoral arteries being suggested for this purpose (Hove et al. 2004; Vasquez, Johnson, and Gould et al. 2013; van der Weerdt et al. 2001). Intuitively, deriving the input function from a region of interest drawn on the left ventricle (Gambhir et al. 1989) should give the most accurate estimates given the fact that this is the largest pool of blood in the body and should be minimally affected by partial volume effects. However, the left ventricle myocardial activity increases in time and spills over into the left ventricle. Because of the size of the left ventricle, relatively large regions of interest can be defined at a safe distance from the wall to minimize the bias introduced by spill-out effects. However, if the region of interest is too small, statistical fluctuations can bias the estimates. In an integrated PET-MRI scanner, the anatomical information provided by MRI could help with the definition of the region of interest from which these data could be derived. Furthermore, in the absence of corrections for cardiac and respiratory motion, the values derived from such a region of interest is either underestimated due to spillover effects at the early time points or spill-in from the myocardium at the late time points. The MR-assisted motion correction method discussed previously would solve this problem.

4.4 CONCLUSION AND OUTLOOK

The integration of MRI and PET holds great promise but, as described earlier, still poses many technical challenges. Robust solutions to these challenges, however, are likely to be developed in the near future. At that time, it will be vital for rigorous studies to be performed to establish the value added by integrated MRI and PET in clinical practice. The full promise of the technology, however, will require new radiotracers and new indications to be developed. For instance, the imaging of coronary atherosclerotic plaque with current radiotracers and instrumentation is limited and remains a large unmet need in cardiovascular medicine. The full realization of PET-MRI will thus require an extensive and ongoing collaboration between molecular biologists, chemists, physicists, cardiologists, and industry.

REFERENCES

Ai T, Padua A, Goerner F et al. SEMAC-VAT and MSVAT-SPACE sequence strategies for metal artifact reduction in 1.5T magnetic resonance imaging. *Invest Radiol.* 2012;47:267–276.

Akbarzadeh A, Ay MR, Ahmadian A et al. MRI-guided attenuation correction in whole-body PET/MR: Assessment of the effect of bone attenuation. *Ann Nucl Med.* 2013;27:152–162.

Axel L, Dougherty L. MR imaging of motion with spatial modulation of magnetization. *Radiology.* 1989;171:841–845.

Bindseil GA, Gilbert KM, Scholl TJ et al. First image from a combined positron emission tomography and field-cycled MRI system. *Magn Reson Med.* 2011;66:301–305.

Borra R, Bowen SL, Attenberger U et al. Effects of ferumoxytol on quantitative accuracy of PET in simultaneous PET/MR imaging—A validation study. *RSNA.* 2013.

Catana C, Benner T, van der Kouwe A et al. MRI-assisted PET motion correction for neurologic studies in an integrated MR-PET scanner. *J Nucl Med.* 2011;52:154–161.

Catana C, Drzezga A, Heiss WD et al. PET/MRI for neurologic applications. *J Nucl Med.* 2012;53:1916–1925.

Catana C, Guimaraes AR, Rosen BR. PET and MR imaging: The odd couple or a match made in heaven? *J Nucl Med.* 2013;54:815–824.

Catana C, Procissi D, Wu Y et al. Simultaneous in vivo positron emission tomography and magnetic resonance imaging. *Proc Natl Acad Sci U S A.* 2008;105:3705–3710.

Catana C, Wu Y, Judenhofer MS et al. Simultaneous acquisition of multislice PET and MR images: Initial results with a MR-compatible PET scanner. *J Nucl Med.* 2006;47:1968–1976.

Catana C, van der Kouwe A, Benner T et al. Toward implementing an MRI-based PET attenuationcorrection method for neurologic studies on the MR-PET brain prototype. *J Nucl Med.* 2010;51:1431–1438.

Censor Y, Gustafson DE, Lent A et al. A new approach to the emission computerized tomography problem: Simultaneous calculation of attenuation and activity coefficients. *IEEE Trans Nucl Sci.* 1979;26:2775–2779.

Ciesienski KL, Yang Y, Ay I et al. Fibrin-targeted PET probes for the detection of thrombi. *Mol Pharm.* 2013;10:1100–1110.

Chonde DB, Abolmaali N, Arabasz G et al. Effect of MRI acoustic noise on cerebral fludeoxyglucose uptake in simultaneous MR-PET imaging. *Invest Radiol.* 2013;48:302–312.

Chun SY, Reese TG, Ouyang J et al. MRI-based nonrigid motion correction in simultaneous PET/MRI. *J Nucl Med.* 2012;53:1284–1291.

Conradi J. Temperature effects in silicon avalanche photodiodes. *Solid-State Electron.* 1974;17:99–106.

Delso G, Furst S, Jakoby B et al. Performance measurements of the Siemens mMR integrated wholebody PET/MR scanner. *J Nucl Med.* 2011;52:1914–1922.

Delso G, Martinez-Moller A, Bundschuh RA et al. The effect of limited MR field of view in MR/PET attenuation correction. *Med Phys.* 2010;37:2804–2812.

Du Y, Madar I, Stumpf MJ et al. Compensation for spill-in and spill-out partial volume effects in cardiac PET imaging. *J Nucl Cardiol.* 2013;20:84–98.

Eldib M, Bini J, Calcagno C et al. Attenuation correction for flexible magnetic resonance coils in combined magnetic resonance/positron emission tomography imaging. *Invest Radiol.* 2014;49:63–69.

Erlandsson K, Buvat I, Pretorius PH et al. A review of partial volume correction techniques for emission tomography and their applications in neurology, cardiology and oncology. *Phys Med Biol.* 2012;57:R119–R159.

Frullano L, Catana C, Benner T et al. Bimodal MR-PET Agent for quantitative pH imaging. *Angew Chem-Int Ed.* 2010;49:2382–2384.

Garlick PB, Marsden PK, Cave AC et al. PET and NMR dual acquisition (PANDA): Applications to isolated, perfused rat hearts. *NMR Biomed.* 1997;10:138–142.

Gambhir SS, Schwaiger M, Huang SC et al. Simple noninvasive quantification method for measuring myocardial glucose utilization in humans employing positron emission tomography and fluorine-18 deoxyglucose. *J Nucl Med.* 1989;30:359–366.

Guerin B, Cho S, Chun SY et al. Nonrigid PET motion compensation in the lower abdomen using simultaneous tagged-MRI and PET imaging. *Med Phys.* 2011;38:3025–3038.

Hayakawa N, Uemura K, Ishiwata K et al. A PET-MRI registration technique for PET studies of the rat brain. *Nucl Med Biol.* 2000;27:121–125.

Hofmann M, Bezrukov I, Mantlik F et al. MRI-based attenuation correction for whole-body PET/MRI: Quantitative evaluation of segmentation- and atlas-based methods. *J Nucl Med.* 2011;52:1392–1399.

Hofmann M, Steinke F, Scheel V et al. MRI-based attenuation correction for PET/MRI: A novel approach combining pattern recognition and atlas registration. *J Nucl Med.* 2008;49:1875–1883.

Hong SJ, Kang HG, Ko GB et al. SiPM-PET with a short optical fiber bundle for simultaneous PET-MR imaging. *Phys Med Biol.* 2012;57:3869–3883.

Hove JD, Iida H, Kofoed KF et al. Left atrial versus left ventricular input function for quantification of the myocardial blood flow with nitrogen-13 ammonia and positron emission tomography. *Eur J Nucl Med Mol Imaging.* 2004;31:71–76.

Huang C, Petibon Y, Ouyang J et al. Accelerated acquisition of tagged MRI for cardiac motion correction in simultaneous PET-MR: Phantom and patient studies. *Med Phys.* 2015;42:1087–1097.

Hutchins GD, Caraher JM, Raylman RR. A region of interest strategy for minimizing resolution distortions in quantitative myocardial PET studies. *J Nucl Med.* 1992;33:1243–1250.

Iida H, Kanno I, Miura S et al. A simulation study of a method to reduce positron-annihilation spread distributions using a strong magnetic-field in positron emission tomography. *IEEE Trans Nucl Sci.* 1986;33:597–600.

Iida H, Kanno I, Takahashi A et al. Measurement of absolute myocardial blood-flow with (H2O)-O-15 and dynamic positron-emission tomography—Strategy for quantification in relation to the partialvolume effect. *Circulation.* 1988;78:104–115.

Judenhofer MS, Catana C, Swann BK et al. PET/MR images acquired with a compact MR-compatible PET detector in a 7-T magnet. *Radiology.* 2007;244:807–814.

Judenhofer MS, Wehrl HF, Newport DF et al. Simultaneous PET-MRI: A new approach for functional and morphological imaging. *Nat Med.* 2008;14:459–465.

Kalemis A, Delattre BMA, Heinzer S. Sequential whole-body PET/MR scanner: Concept, clinical use, and optimisation after two years in the clinic. The manufacturer's perspective. *Magn Reson Mater Phys Biol Med.* 2013;26:5–23.

Kang J, Choi Y, Hong KJ et al. A small animal PET based on GAPDs and charge signal transmission approach for hybrid PET-MR imaging. *J Instrum.* 2011;6.

Keereman V, Van Holen R, Mollet P et al. The effect of errors in segmented attenuation maps on PET quantification. *Med Phys.* 2011;38:6010–6019.

King AP, Buerger C, Tsoumpas C et al. Thoracic respiratory motion estimation from MRI using a statistical model and a 2-D image navigator. *Med Image Anal.* 2012;16:252–264.

Kolb A, Lorenz E, Judenhofer MS et al. Evaluation of Geiger-mode APDs for PET block detector designs. *Phys Med Biol.* 2010;55:1815–1832.

Kolb A, Wehrl HF, Hofmann M et al. Technical performance evaluation of a human brain PET/MRI system. *Eur Radiol.* 2012;22:1776–1788.

Ladefoged CN, Andersen FL, Keller SH et al. PET/MR imaging of the pelvis in the presence of endoprostheses: Reducing image artifacts and increasing accuracy through inpainting. *Eur J Nucl Med Mol Imaging.* 2013;40:594–601.

Ledesma-Carbayo MJ, Derbyshire JA, Sampath S et al. Unsupervised estimation of myocardial displacement from tagged MR sequences using nonrigid registration. *Magn Reson Med.* 2008;59:181–189.

Levin C, Glover G, Deller T et al. Prototype time-of-flight PET ring integrated with a 3T MRI system for simultaneous whole-body PET/MR imaging. *J Nucl Med.* 2013;54:148.

Lois C, Bezrukov I, Schmidt H et al. Effect of MR contrast agents on quantitative accuracy of PET in combined whole-body PET/MR imaging. *Eur J Nucl Med Mol Imaging.* 2012;39:1756–1766.

Lucas AJ, Hawkes RC, Ansorge RE et al. Development of a combined microPET((R))-MR system. *Technol Cancer Res Treat.* 2006;5:337–341.

Mackewn JE, Halsted P, Charles-Edwards G et al. Performance evaluation of an MRI-compatible preclinical PET system using long optical fibers. *IEEE Trans Nucl Sci.* 2010;57:1052–1062.

Mackewn JE, Strul D, Hallett WA et al. Design and development of an MR-compatible PET scanner for imaging small animals. *IEEE Trans Nucl Sci.* 2005;52:1376–1380.

Maramraju SH, Smith SD, Junnarkar SS et al. Small animal simultaneous PET/MRI: Initial experiences in a 9.4 T microMRI. *Phys Med Biol.* 2011;56:2459–2480.

Marshall HR, Patrick J, Laidley D et al. Description and assessment of a registration-based approach to include bones for attenuation correction of whole-body PET/MRI. *Med Phys.* 2013;40:082509.

Marshall HR, Prato FS, Deans L et al. Variable lung density consideration in attenuation correction of whole-body PET/MRI. *J Nucl Med.* 2012;53:977–984.

Martinez-Moller A, Souvatzoglou M, Delso G et al. Tissue classification as a potential approach for attenuation correction in whole-body PET/MRI: Evaluation with PET/CT data. *J Nucl Med.* 2009;50:520–526.

Minoshima S, Berger KL, Lee KS et al. An automated method for rotational correction and centering of three-dimensional functional brain images. *J Nucl Med.* 1992;33:1579–1585.

Mollet P, Keereman V, Clementel E et al. Simultaneous MR-compatible emission and transmission imaging for pet using time-of-flight information. *IEEE Trans Med Imaging.* 2012;31:1734–1742.

Montandon ML, Zaidi H. Atlas-guided non-uniform attenuation correction in cerebral 3D PET imaging. *Neuroimage.* 2005;25:278–286.

Mosher TJ, Smith MB. A DANTE tagging sequence for the evaluation of translational sample motion. *Magn Reson Med.* 1990;15:334–339.

Nuyts J, Dupont P, Stroobants S et al. Simultaneous maximum a posteriori reconstruction of attenuation and activity distributions from emission sinograms. *IEEE Trans Med Imaging.* 1999;18:393–403.

Nuyts H, Maes A, Vrolix M et al. Three-dimensional correction for spillover and recovery of myocardial PET images. *J Nucl Med.* 1996;37:767–774.

Nuyts J, Michel C, Fenchel M et al. Completion of a Truncated Attenuation Image from the Attenuated PET Emission Data. *2010 IEEE Nuclear Science Symposium Conference Record.* New York: IEEE; 2010:2123–2127.

Osman NF, McVeigh ER, Prince JL. Imaging heart motion using harmonic phase MRI. *IEEE Trans Med Imaging.* 2000;19:186–202.

Ouyang JS, Li QZ, El Fakhri G. Magnetic resonance-based motion correction for positron emission tomography imaging. *Semin Nucl Med.* 2013;43:60–67.

Ozturk C, Derbyshire JA, McVeigh ER. Estimating motion from MRI data. *Proc IEEE Inst Electr Electron Eng.* 2003;91:1627–1648.

Panin VY, Aykac M, Casey ME. Simultaneous reconstruction of emission activity and attenuation coefficient distribution from TOF data, acquired with external transmission source. *Phys Med Biol.* 2013;58:3649–3669.

Paulus DH, Braun H, Aklan B et al. Simultaneous PET/MR imaging: MR-based attenuation correction of local radiofrequency surface coils. *Med Phys.* 2012;39:4306–4315.

Peng H, Handler WB, Scholl TJ et al. Proof-of-principle study of a small animal PET/field-cycled MRI combined system using conventional PMT technology. *Nucl Instrum Methods Phys Res Sect A-Accel Spectrom Dect Assoc Equip.* 2010;612:412–420.

Petibon Y, El Fakhri G, Nezafat R et al. Towards coronary plaque imaging using simultaneous PET-MR: A simulation study. *Phys Med Biol.* 2014;59:1203–1222.

Petibon Y, Huang C, Ouyang J et al. Relative role of motion and PSF compensation in whole-body oncologic PET-MR imaging. *Med Phys.* 2014;41:042503.

Petibon Y, Ouyang J, Zhu X et al. Cardiac motion compensation and resolution modeling in simultaneous PET-MR: A cardiac lesion detection study. *Phys Med Biol.* 2013;58:2085.

Pichler B, Lorenz E, Mirzoyan R et al. Performance test of a LSO-APD PET module in a 9.4 Tesla magnet. Paper presented at: Nuclear Science Symposium, 1997, IEEE, 1997.

Raylman RR. *Reduction of Positron Range Effects by the Use of a Magnetic Field: For Use in Positron Emission Tomography.* Ann Arbor, MI: University of Michigan; 1991.

Raylman RR, Hammer BE, Christensen NL. Combined MRI-PET scanner: A Monte Carlo evaluation of the improvements in PET resolution due to the effects of a static homogeneous magnetic field. *IEEE Trans Nucl Sci.* 1996;43:2406–2412.

Raylman RR, Majewski S, Lemieux SK et al. Simultaneous MRI and PET imaging of a rat brain. *Phys Med Biol.* 2006;51:6371–6379.

Raylman RR, Majewski S, Velan SS et al. Simultaneous acquisition of magnetic resonance spectroscopy (MRS) data and positron emission tomography (PET) images with a prototype MR-compatible, small animal PET imager. *J Magn Reson.* 2007;186:305–310.

Rezaei A, Defrise M, Bal G et al. Simultaneous reconstruction of activity and attenuation in time-offlight PET. *IEEE Trans Med Imaging.* 2012;31:2224–2233.

Rischpler C, Nekolla SG, Dregely I et al. Hybrid PET/MR imaging of the heart: Potential, initial experiences, and future prospects. *J Nucl Med.* 2013;54:402–415.

Rota Kops E, Herzog H. Template-based attenuation correction of PET in hybrid MR-PET scanners. *Society of Nuclear Medicine Annual Meeting Abstracts.* 2008;49:162P-c.

Roncali E, Cherry SR. Application of silicon photomultipliers to positron emission tomography. *Ann Biomed Eng.* 2011;39:1358–1377.

Rueckert D, Sonoda L, Hayes C et al. Nonrigid registration using free-form deformations: Application to breast MR images. *IEEE Trans Med Imaging.* 1999;18:712–721.

Salomon A, Goedicke A, Schweizer B et al. Simultaneous reconstruction of activity and attenuation for PET/MR. *IEEE Trans Med Imaging.* 2011;30:804–813. doi: 810.1109/TMI.2010.2095464. Epub 2092010 Nov 2095429.

Sander CY, Hooker JM, Catana C et al. Neurovascular coupling to D2/D3 dopamine receptor occupancy using simultaneous PET/functional MRI. *Proc Natl Acad Sci U S A.* 2013;30:30.

Schlemmer H-PW, Pichler BJ, Schmand M et al. Simultaneous MR/PET imaging of the human brain: Feasibility study. *Radiology.* 2008;248:1028–1035.

Schulz V, Torres-Espallardo I, Renisch S et al. Automatic, three-segment, MR-based attenuation correction for whole-body PET/MR data. *Eur J Nucl Med Mol Imaging.* 2011;38:138–152.

Sensakovic WF, Armato SG. *Magnetic Resonance Imaging of the Lung: Automated Segmentation Methods.* Vol 2. Dordrecht: Springer; 2008.

Seong Jong H, In Chan S, Ito M et al. An investigation into the use of Geiger-mode solid-state photomultipliers for simultaneous PET and MRI acquisition. *IEEE Trans Nucl Sci.* 2008;55:882–888.

Shao Y, Cherry SR, Farahani K et al. Simultaneous PET and MR imaging. *Phys Med Biol.* 1997a;42:1965–1970.

Shao Y, Cherry SR, Farahani K et al. Development of a PET detector system compatible with MRI/NMR systems. *IEEE Trans Nucl Sci.* 1997b;44:1167–1171.

Sutter R, Ulbrich EJ, Jellus V et al. Reduction of metal artifacts in patients with total hip arthroplasty with slice-encoding metal artifact correction and view-angle tilting MR imaging. *Radiology.* 2012;265:204–214.

Tatsumi M, Yamamoto S, Imaizumi M et al. Simultaneous PET/MR body imaging in rats: Initial experiences with an integrated PET/MRI scanner. *Ann Nucl Med.* 2012;26:444–449.

Tsoumpas C, Mackewn J, Halsted P et al. Simultaneous PET-MR acquisition and MR-derived motion fields for correction of non-rigid motion in PET. *Ann Nucl Med.* 2010;24:745–750.

Townsend DW, Carney JP, Yap JT et al. PET/CT today and tomorrow. *J Nucl Med.* 2004;45 Suppl 1:4S–14S.

Uppal R, Catana C, Ay I et al. Simultaneous MR-PET imaging of thrombus with a fibrin-targeted dual MR-PET probe: A feasibility study. *Radiology.* 2011;258:812–820.

Vasquez AF, Johnson NP, Gould KL. Variation in quantitative myocardial perfusion due to arterial input selection. *JACC Cardiovasc Imaging.* 2013;6:559–568.

van der Weerdt AP, Klein LJ, Boellaard R et al. Image-derived input functions for determination of MRGlu in cardiac (18)F-FDG PET scans. *J Nucl Med.* 2001;42:1622–1629.

Wiemker R, Pekar V. Fast computation of isosurface contour spectra for volume visualization. In: Lemke HU, Vannier MW, Inamura K, Farman AG, Doi K, eds. *Cars 2001: Computer Assisted Radiology and Surgery.* Vol 1230. Amsterdam: Elsevier Science Bv; 2001:372–377.

Wirrwar A, Vosberg H, Herzog H et al. 4.5 Tesla magnetic field reduces range of high-energy positrons-potential implications for positron emission tomography. *IEEE Trans Nucl Sci.* 1997;44:184–189.

Woods RP, Grafton ST, Holmes CJ et al. Automated image registration: I. General methods and intrasubject, intramodality validation. *J Comput Assist Tomogr.* 1998;22:139–152.

Woody C, Schlyer D, Vaska P et al. Preliminary studies of a simultaneous PET/MRI scanner based on the RatCAP small animal tomograph. *Nucl Instrum Methods Phys Res Sect A-Accel Spectrom Dect Assoc Equip.* 2007;571:102–105.

Yamamoto S, Imaizumi M, Kanai Y et al. Design and performance from an integrated PET/MRI system for small animals. *Ann Nucl Med.* 2010;24:89–98.

Yamamoto S, Kuroda K, Senda M. Scintillator selection for MR-compatible gamma detectors. *IEEE Trans Nucl Sci.* 2003;50:1683–1685.

Yamamoto S, Watabe H, Kanai Y et al. Interference between PET and MRI sub-systems in a siliconphotomultiplier-based PET/MRI system. *Phys Med Biol.* 2011;56:4147–4159.

Yamamoto S, Watabe T, Watabe H et al. Simultaneous imaging using Si-PM-based PET and MRI for development of an integrated PET/MRI system. *Phys Med Biol.* 2012;57:N1.

Yoon HS, Ko GB, Il Kwon S et al. Initial results of simultaneous PET/MRI experiments with an MRIcompatible silicon photomultiplier PET scanner. *J Nucl Med.* 2012;53:608–614.

CT-MRI

JAMES BENNETT AND GE WANG

5.1 INTRODUCTION

X-ray computerized tomography (CT) and cardiac magnetic resonance imaging (MRI) are emerging as the most promising complementary imaging modalities for coronary atherosclerotic disease detection (Fayad et al. 2002). The goal of hybrid cardiovascular imaging is combining complementary information from multiple modalities to improve diagnostic accuracy and treatment efficacy. Each current clinical imaging modality typically has at least one fundamental deficit when utilized individually: functional modalities, such as single photon emission computed tomography (SPECT) and positron emission tomography (PET), lack high spatial and/or temporal resolution, while structural modalities, such as CT and ultrasound, commonly lack tissue-specific functional imaging. The hybridization of CT and MRI can address the challenge of simultaneously providing both functional and structural images of the patient's cardiovascular system.

Hybrid CT-MRI cardiovascular imaging is an ideal application for the evaluation of chronic ischemic heart disease (CIHD). CIHD is a prevalent disease with a high mortality and can be technically challenging to diagnose. Conventional hybrid modalities used in CIHD diagnosis and treatment planning (e.g., SPECT-CT and PET-CT) involve significant ionizing radiation. High patient radiation exposure is an increasing concern in the medical community particularly due to the increasing use of diagnostic imaging. Fusion of cardiac CT and cardiac MRI (CMR) images delivers comparable diagnostic results to cardiac SPECT-CT and PET-CT with significantly reduced radiation dose (Scheffel et al. 2010; Stolzmann et al. 2011). Furthermore, there is the potential for sub-mSv dose exam with a dedicated CT-MRI hybrid scanner (Wang et al. 2013).

Hybrid CT-MRI has become a viable imaging option due to recent technical advances in multidetector computed tomography (MDCT) and CMR. Low-dose coronary CT angiography (CCTA) allows for three-dimensional (3-D) imaging of the coronary arteries with spatial and temporal resolution that approaches the current "gold standard" of coronary catheter angiography (CCA) (Donati et al. 2010). Common CCTA dose reduction techniques use prospective electrocardiogram (ECG) gating of the x-ray source and increased longitudinal detector coverage. Myocardial perfusion CMR and late gadolinium enhancement (LGE) methods now rival or surpass SPECT and PET for analysis of stress/rest myocardial viability without the radiation dose and special facilities associated with radionuclide imaging (Schwitter et al. 2008).

A "proof of concept" paper in 2002 by White et al. described coregistration of postreconstruction CCTA and CMR images to aid diagnosis and treatment planning in CIHD and coronary artery disease (CAD). A detailed description of this technique was first proposed in this paper. "Integrated imaging, using MDCT–based CCTA and MRI myocardial-viability maps, can help to noninvasively provide information about the morphologic and physiologic significance of obstructive and non-obstructive coronary lesions;" furthermore, "lesion characteristics and the presence or absence of collaterals beyond an occlusive coronary arterial lesion can be assessed in relation to the size and distribution of the resulting myocardial necrosis" (White and Setser 2002).

This trend was noted by another group in a review article describing "CT and CMR identify flow-limiting coronary stenoses and calcified plaques, directly image the atherosclerotic lesions, measure atherosclerotic burden, and characterize the plaque components;" "CT and CMR provide unique information that may predict cardiovascular risk, facilitate further study of the mechanisms of atherosclerosis progression and its response to therapy, and allow for assessment of subclinical disease" (Fayad et al. 2002). This study is followed by an excellent review and outlook article by the same group (Nikolaou et al. 2003). Images from Donati et al. in Figure 5.1 provide a striking example of the potential of hybrid cardiac CT-MRI.

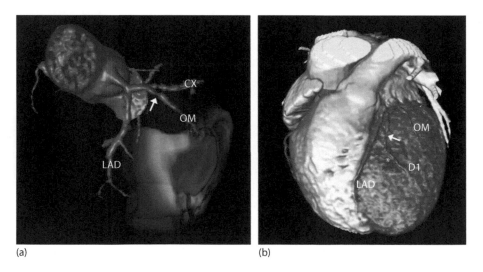

(a)   (b)

Figure 5.1 Two retrospective combinations of CT coronary angiography and CMR scans. (a) A fused surface rendering of the segmented coronary artery tree from low-dose CT with the mask representing the myocardium of the left ventricle and the segmented data from LGE. Lateral LGE can be assigned to a high-grade stenosis (arrow) of the obtuse marginal (OM) branch of the circumflex artery (CX). Extent of transmurality of LGE is displayed in an intensity-dependent manner for each region. No stenosis can be seen in the segmented left anterior descending (LAD) artery and proximal CX. (b) Segmented myocardial defects from CMR integrated with a volume rendering based on low-dose CT. Low-dose CT demonstrates the LAD and OM branch of circumflex artery and significant stenosis (arrow) of the first diagonal branch (D1). The fused images show an anterolateral perfusion deficit (asterisk) in relation to the D1 stenosis (arrow). (Adapted from Donati OF et al., *Invest Radiol*, 46, 331–340, 2011.)

A hybrid CT-MRI scanner does not yet exist; therefore, current applications are restricted to retrospective coregistration or fusion of separate CT and MRI scan datasets (reviewed in Section 5.2). Advances in CT and MRI technology have enabled the promise of low-dose, high-resolution CT-MRI hybrid scanner for combined morphological and functional imaging. Section 5.3 will include a discussion of the CT and MRI methodologies that can enable low-dose, high-resolution hybrid CT-MRI. Section 5.4.2 will describe the world's first hybrid CT-MRI scanner design. The breakthrough technology enabling this design comes from the unique attributes of a multimodality image reconstruction framework called "omni-tomography" (Wang et al. 2012). This chapter will conclude with a discussion of future hybrid CT-MRI scanner design, image reconstruction, and image fusion.

5.2 APPLICATIONS FOR CT-MRI FUSION AND COREGISTRATION

There are two research groups who have published the majority of studies on cardiovascular CT-MRI coregistration/fusion. White and Setser at the Cleveland Clinic published the first research papers on this topic in 2002 and 2005 using coregistered CCTA and CMR images. Subsequently, another research group based in Switzerland published a series of six papers between 2008 and 2011 describing their hybrid method which fused on CCTA and coronary calcium scoring (CCS) reconstructions and postreconstruction CMR images (Donati et al. 2010, 2011; Scheffel et al. 2008, 2010; Stolzmann et al. 2010, 2011). While most studies focused on diagnostic methods, White et al. also performed a retrospective study on the efficacy of coregistered CCTA/CMR for cardiac revascularization treatment. Another group from California performed an analysis of aortic CT and MRI registration accuracy using anatomic landmarks, which is highly relevant given the non-isotropic nature of 3-D MRI image reconstructions (Dey et al. 2006). Finally, a group from Finland published a study on the diagnostic accuracy of CCTA and correlated it with CMR on 14 patients (Holmström et al. 2006).

The first papers published in each series on cardiovascular CT-MRI begin with a feasibility/pilot study targeting CAD/CIHD detection/treatment planning. As previously mentioned, White and Setser (2002) were the first group to directly address hybrid cardiovascular CT-MRI with a research study on CAD and CIHD. Previous studies on hybrid cardiovascular imaging modalities demonstrated similar successful clinical approaches toward CAD/CIHD diagnosis using SPECT or PET for myocardial perfusion (functional) information combined with CT or MRI (structural) reconstructions (Faber et al. 1991). SPECT and PET became the gold standard for noninvasive 3-D functional cardiac imaging. Yet CMR continued to advance, yielding four current advantages over cardiac SPECT and/or PET:

- No ionizing radiation
- Higher spatiotemporal resolution
- Superior diagnostic accuracy versus SPECT (Schwitter et al. 2008)
- Detection of nontransmural myocardial necrosis via LGE (Kim et al. 1999)

Similarly, after decades of technical advances, CCTA has proven a viable alternative to the enduring gold standard for morphologic cardiovascular imaging, CCA. Advantages of CCTA versus CCA are

- Noninvasive procedure
- 3-D imaging of the myocardium and coronary artery tree
- Improved lesion characterization

Given these factors, White et al. proposed that suspected CAD/CIHD patients should follow a diagnostic CCTA/CMR protocol. A brief summary of the protocol is that patients should first undergo CCTA to identify and classify atherosclerotic lesions and to rule out other coronary lesions. Any atherosclerotic coronary lesions should be characterized according to "severity of stenosis, plaque composition, remodeling and the presence or absence of collaterals beyond [an occlusion]" (White and Setser 2002). Next, CMR with LGE should be performed to produce a myocardial viability map. Finally, the two datasets should be compared either via coregistration or side-by-side evaluation. The cross-referenced datasets can be used to evaluate the specific coronary morphopathology to determine appropriate treatment options.

5.2.1 REVASCULARIZATION PLANNING

The first clinical study on the efficacy of hybrid CT-MRI was performed by Setser et al. It targeted a pool of 26 patients meeting the following criteria:

- Diagnosed with CIHD via CCA
- Referred for surgical revascularization
- Cardiac surgeon requested CMR and CCTA for preoperative planning

The goals of the study were (a) to evaluate the assignment of left ventricular (LV) myocardial segments to coronary arterial territories by using co-registered MR imaging and MDCT displays, (b) to assess the accuracy of coregistered displays in determining the presence and distribution of clinically important CAD and the regional effect of CAD on LV myocardium, and (c) to determine the potential utility of coregistered displays in optimizing surgical revascularization planning (Setser et al. 2005). Goal (a) would be accomplished by comparing the accuracy of assigning CMR perfusion data to an American Heart Association standardized 17-segment LV model (Cerqueira et al. 2002) with and without coregistered CCTA data. The LV model relates a standard coronary arterial tree to 17 segments in the LV myocardium; however, individual anatomical variations can deviate from the model and may cause a mismatch between LV segment and the specific coronary artery supplying it. Indeed, the study found that 17% of segments were incorrectly assigned when CCA was used side by side with the CMR perfusion data, whereas coregistered CMR/CCTA datasets had no erroneous assignments.

Goal (b) was evaluated by comparing the CAD diagnostic accuracy using coregistered CCTA/CMR perfusion data versus CCA combined with standard functional imaging modalities (echocardiography and PET). The latter combination is the current gold standard for structural and functional cardiac imaging and thus was considered as the reference. Results of the study showed that the coregistered CCTA/CMR datasets identified CAD with the same accuracy as CCA plus echocardiography or PET. Finally, Goal (c) was assessed by having two experienced cardiac surgeons evaluate their treatment planning with CCA plus echocardiography or PET versus coregistered CCTA/CMR. The surgeons determined that coregistered CCTA/CMR improved surgical planning in 83% of the patients versus CCA plus echocardiography or PET. Thus, it was shown that the hybrid CCTA/CMR approach for evaluating CAD/CIHD patient treatment options was successful and merits further investigation in other areas of cardiovascular imaging.

5.2.2 CAD DIAGNOSIS

As a natural extension of the previous CIHD/CAD treatment planning study, Scheffel et al. (2010) evaluated "the diagnostic performance of low-dose CCTA and CMR and combinations thereof for the diagnosis of significant coronary stenoses." The performance of these two noninvasive modalities was tested against invasive CCA for detection of CAD abnormalities. Thus, the study was admittedly biased toward structural rather than functional abnormalities due to CCA as the reference standard. Furthermore, when CCTA and CMR data were used in combination to detect CAD, they were merely displayed side by side rather than coregistered or fused. These two deficits limit the broader applicability of the study toward hybrid CT-MRI cardiovascular imaging; however, there are two key observations: (1) low-dose CCTA delivers comparable diagnostic images of coronary stenoses as CCA with a radiation dose between 1 mSv and 4 mSv and (2) CMR provides valuable functional information to confirm abnormalities from CCTA. Another study by the same team found the diagnostic accuracy of CCTA and CCA for detecting CAD was comparable when combined with perfusion CMR (Donati et al. 2010), thereby confirming the observations in Scheffel's paper.

In yet another study published by the same group in 2010, Stolzmann describes a method for fusing subsequent CCTA and CMR images for diagnosis of CAD (Stolzmann et al. 2010). The fusion process represents a significant technical advance relative to the coregistration method previously utilized by White, Setser, and previous papers from their group. Scan datasets from a small cohort of five CAD patients who underwent both CCTA and CMR procedures were fused using a commercial software package developed by Fraunhofer MEVIS (Bremen, Germany). As this pilot study demonstrated, it is a reasonably straightforward process to coregister the CCTA and CMR image space using semiautomatic feature registration and nonrigid transformations. Once the image space is coregistered, the fusion process requires choosing appropriate coloration and 3-D rendering schemes. The CCTA/CMR acquisition protocols, image fusion process, and resulting reconstructions/renderings are explained further in this paper and should be referenced for details and images (Stolzmann et al. 2010).

The methodologies and tools in the previously described papers laid the foundation for a seminal study by Donati et al. (2011) who evaluated the accuracy and added diagnostic value of 3-D image fusion of CCTA and functional CMR for assessing hemodynamically relevant CAD. There were two goals to reach this aim: (a) evaluate spatial accuracy of CCTA and CMR image fusion and (b) determine added value to the diagnostic interpretation of CAD using CCTA/CMR image fusion versus side-by-side comparison of the datasets. In this subsection, the diagnostic implications of this study [Goal (b)] will be discussed, and the fusion accuracy [Goal (a)] will be covered in the next subsection of this chapter.

The study began with a population of 55 patients with suspected CAD and without prior cardiac revascularization. Diagnostic evaluation for CAD proceeded with CCTA and CMR examinations of each patient on the same day. The CCTA and CMR results were separately evaluated for coronary artery stenosis and myocardial perfusion/LGE defects, respectively. Patients with normal coronary artery CCTA or without CMR perfusion/LGE defect were excluded from further evaluation, as it was determined that fusion would not provide any benefit over side-by-side analysis. The remaining population of 27 patients had their CCTA and CMR datasets processed and evaluated using the same image fusion technique as described in a previous study (Stolzmann et al. 2010). Additionally, each patient subsequently had a CCA procedure. The results are summarized in the following excerpt:

> The fusion of morphologic data derived from CCTA and functional data from CMR allows for the simultaneous noninvasive evaluation of coronary artery stenosis along with myocardial perfusion and viability at a low radiation dose. Most importantly, CCTA/CMR fusion provides added diagnostic information for the integrative evaluation of patients with hemodynamically relevant CAD as compared with side-by-side analysis...the combined noninvasive approach of CCTA/CMR 3D image fusion may act as a gatekeeper and could aid in the planning of surgical or interventional revascularization procedures in patients having a high prevalence of CAD. (Donati et al. 2011)

In yet another permutation, combined CMR and CCS was evaluated for diagnostic performance in detecting CAD by Stolzmann et al. Sixty patients with suspected CAD underwent CCS and CMR; CCA was separately performed and served as the reference for CAD in coronary artery segments for each patient. Cardiac MR was evaluated individually and in combination with CCS by experienced readers with only clinical history to aid CAD diagnosis in each coronary segment. The resulting CAD detection accuracy was 87% for CMR with CCS (calcium score ≥495) and 82% for CMR alone (Stolzmann et al. 2011). It should be noted that a slight bias may be due to using CCA as the reference in this study; however, CCA is currently a gold standard for CAD diagnosis and prognosis. Furthermore, this study did not utilize any fusion or coregistration of the CMR and CCS images and thus may improve results further with these methods.

All of the previous studies described earlier provide a solid basis for the advantages of a combined CT and CMR approach toward CAD detection, diagnosis, prognosis, and treatment planning.

5.2.3 REAL-TIME SURGICAL INTERVENTION

As catheter-guided interventional surgery continues its advances, e.g., transcatheter aortic-valve implantation, hybrid CT-MRI may be uniquely positioned to complement these procedures. There have been several advances in MRI- and CT-guided cardiovascular interventional techniques, both in preclinical and clinical studies (Knecht et al. 2008; Lederman 2005; Raval et al. 2005; Serfaty et al. 2003; Sra et al. 2007). CT-guided real-time imaging (i.e., 4DCT, CT fluoroscopy) for interventional procedures has the downside of high radiation dose; however, advances in real-time reconstruction may reduce radiation dose for such procedures (Flach et al. 2013). These studies focused on technical improvements for the individual modalities (e.g., either CT or MRI), which proved to be quite powerful.

A study by Gutiérrez et al. (2007) went one step further to integrate CMR data with real-time conventional X-ray fluoroscopy in an article entitled "X-Ray Fused with MRI (XFM) Catheter Roadmaps." "In this early clinical experience, MRI-derived roadmaps added information to a range of invasive cardiovascular procedures including graft coronary arteriography, cardiac tumor biopsy, and peripheral artery intervention;" additionally, "X-ray fusion with MRI roadmaps is feasible and may prove a valuable adjunct to invasive cardiovascular procedures. With further development, XFM or related adjunctive imaging may improve the safety and success rate of a range of complex endovascular interventions."

5.2.4 IMAGE REGISTRATION ACCURACY

Current CT-MRI methods involve retrospective image coregistration and/or fusion as the scans must be acquired and reconstructed separately. Many diagnostic cardiovascular features require high spatial accuracy, such as coronary artery stenosis, which may be in the submillimeter range. Temporal accuracy is essential for the compatibility of CT and MRI datasets when the reconstructions must be aligned to the ECG cycle, such as the evaluation of ventricular motility. Thus, the accuracy of image coregistration/fusion is critical from both a spatial and temporal perspective. "Reproducible and interchangeable functional analysis based on CMR and CT examinations is crucial for the complementary and synergistic application of these imaging techniques" (Boll et al. 2006).

Two of the studies listed in Table 5.1 measure the spatial accuracy of CCTA and CMR fusion and a third addresses the temporal accuracy. The first study to directly address spatial registration error between CT and MRI for cardiovascular applications was performed by Dey et al. Calcified atherosclerotic lesions in the thoracic aorta were identified in 12 patients who underwent noncontrast electron beam CT (EBCT) and contrast-enhanced MRI examinations of the thoracic aorta. The mean 3-D geometric distance between landmarks manually identified by experienced readers in the CT and MRI images was 2.42 mm ± 1.65 mm. The study also performed a volumetric comparison of the calcified atherosclerotic plaque in the CT and MRI images. CT artifacts ("blooming") appeared to cause a significant distortion, such that the "average calcified plaque volume, circumferential extent, and maximal radial width by MRI were significantly smaller compared to CT (35%, 68%, and 53%, respectively; $p < 0.05$)" (Dey et al. 2006). Details of this method can be referenced in the cited paper.

Stolzmann et al. (2011) performed an analysis of the 3D CT-MRI fusion accuracy in their study of 27 patient datasets. The accuracy is calculated based on the distance between cardiac blood pool surfaces that were semiautomatically registered based on landmarks from an experienced reader. The Dice coefficient was also calculated for coregistered CT and MRI slices where a perfect resemblance between the images equals 1. "The mean surface distance of blood pools was 4.1 mm ± 1.3 mm (range, 2.4 mm–7.1 mm) and the mean Dice similarity coefficient was 0.78 ± 0.08 (range, 0.51–0.86). Dice similarity coefficients demonstrated high accuracy of CCTA/CMR image fusion in 25/27 datasets (93%)."

Finally, the study by Boll et al. characterized the temporal accuracy of cine cardiac MR and CT scan datasets. The results showed that "comparing MR and CT image data sets, no statistically significant differences were identified" for patients with normal heart rates; patients with tachycardia (>100 beats per minute) had temporal artifacts in their CT data sets (Boll et al. 2006).

Table 5.1 Cardiovascular CT-MRI research article survey

Year	Author	Journal	Title	Patients in study	Target pathology	CT-MRI visualization
2002	White et al.	American Journal of Cardiology	Integrated Approach to Evaluating Coronary Artery Disease and Ischemic Heart Disease	1	CAD	Co-registration
2005	Setser et al.	Radiology	Coregistered MR Imaging Myocardial Viability Maps and Multi–Detector Row CT Coronary Angiography Displays for Surgical Revascularization Planning: Initial Experience	26	CAD	Co-registration
2006	Dey et al.	Journal of Magnetic Resonance Imaging	Direct Quantitative in Vivo Comparison of Calcified Atherosclerotic Plaque on Vascular MRI and CT by Multimodality Image Registration	12	Aortic lesion	Co-registration
2006	Boll et al.	American Journal of Roentgenology	Synergy of MDCT and Cine MRI for the Evaluation of Cardiac Motility	20	Cardiac motility	Side-by-side
2006	Holmström et al.	Academic Radiology	Noninvasive Analysis of Coronary Artery Disease with Combination of MDCT and Functional MRI	14	CAD	Side-by-side
2010	Scheffel et al.	International Journal of Cardiovascular Imaging	Low-dose CT and Cardiac MR for the Diagnosis of Coronary Artery Disease: Accuracy of Single and Combined Approaches	43	CAD	Side-by-side
2010	Stolzmann et al.	European Radiology	Image Fusion of Coronary CT Angiography and Cardiac Perfusion MRI: A Pilot Study	5	CAD	Fusion
2010	Donati et al.	American Journal of Roentgenology	Combined Cardiac CT and MRI for the Comprehensive Workup of Hemodynamically Relevant Coronary Stenoses	47	CAD	Side-by-side
2011	Donati et al.	Investigative Radiology	3D Fusion of Functional Cardiac Magnetic Resonance Imaging and Computed Tomography Coronary Angiography: Accuracy and Added Clinical Value	27	CAD	Fusion
2011	Stolzmann et al.	International Journal of Cardiovascular Imaging	Combining Cardiac Magnetic Resonance and Computed Tomography Coronary Calcium Scoring: Added Value for the Assessment of Morphological Coronary Disease?	60	CAD	Side-by-side

5.3 CURRENT TECHNOLOGY

Current developments and trends in CT and CMR imaging will be briefly discussed in this section. The various scanner attributes and protocols for the studies described in Table 5.1 are displayed in Tables 5.2 and 5.3 for CT and in Table 5.4 for CMR.

5.3.1 CARDIAC CT

There have been tremendous advances in CT technology during the past decade. In this subsection, three current technologies that can enable low-dose, high-resolution cardiac CT imaging will be briefly covered.

Firstly, improvements in multislice CT, and subsequently MDCT, have greatly improved cardiovascular image quality and temporal resolution. Increased longitudinal coverage and/or MDCT allows for high pitch and even complete volumetric coverage for cardiac scans. Sub-100 msec temporal resolution is now possible with MDCT, which is quickly approaching the gold standard of 33 msec of CCA (30 frames/sec). Faster gantry rotation times have also helped improve temporal resolution, along with countless subcomponent advancements that are beyond the scope of this chapter. In many scenarios, CTA has challenged and in some cases surpassed CCA for cardiovascular imaging applications.

Secondly, ionizing radiation dose has become a greater clinical concern and is relatively high for cardiac imaging using CT and especially MDCT. Many CTA routines commonly deliver over 10 mSv per scan with retrospective gating. Fortunately, advances in x-ray source modulation deliver substantial reductions in radiation dose by providing prospective gating over a specified R–R interval. Iterative reconstruction techniques have also delivered similarly impressive reductions in radiation dose by decreasing the number of required projections for equivalent image quality compared with conventional filtered backprojection (FBP). A recent study has shown that combing these techniques to perform CCTA scans with patient radiation dose under 0.1 mSv is feasible (Ropers et al. 2012).

Finally, digital photon-counting (spectral) x-ray detectors have been developed for preclinical and clinical applications. The advent of spectral detectors promises the next revolution for CT by enabling quantitative material decomposition with a single scan. Numerous preclinical studies have demonstrated the potential efficacy of spectral CT for cardiovascular applications (Cormode et al. 2010; Feuerlein et al. 2008; Pan et al. 2009; Schlomka et al. 2008). Spectral CT fused with CMR may represent the next advancement in hybrid CT-MRI.

5.3.2 CMR

Advances in CMR commonly fall into three general categories: cine, LGE, and perfusion stress/rest studies. Each study utilized one or more of these techniques, which are cataloged in Table 5.4. The specific details of each study's CMR protocol are beyond the scope of this chapter and can be found within the relevant manuscripts. Here, we present a high-level summary of the current CMR technology.

Cine CMR performs several sequential ECG-gated acquisition sequences to produce a 4D (x, y, z, time) volume over the cardiac R–R cycle. The ECG-gating allows for higher-resolution images because it increases the sampling duration relative to real-time acquisitions, although artifact can occur with irregular patient heartbeat. Blood, which is typically bright, is well defined against the myocardium and provides an excellent window into the cardiac functionality. While most MRI scanner manufacturers have their own proprietary pulse sequences (e.g., TrueFISP, FIESTA, etc.), they are typically derivatives of balanced steady-state free precession (Chavhan et al. 2008). MRI scanners used for cine CMR are commonly 1.5 T with a cardiac coil.

LGE is another common technique that utilizes the preferential binding of gadolinium-based contrast agents with scar tissue to diagnose myocardial infarct (Moon et al. 2004). LGE is most commonly performed in conjunction with another contrast MRI study because both can be performed with a single injection. Approximately 5 to 20 min post contrast injection, depending on the protocol, the patient is imaged using a special inversion recovery pulse sequence. Areas of myocardial scarring, and therefore infarct, will display

Table 5.2 CT scanner attributes from studies in Table 5.1

Year	Author	CT scanner	X-ray source	Detector rows	Slice thickness	Gantry rotation time	Temporal resolution
2002	White et al.						
2005	Setser et al.	Siemens Sensation	Single	4 (n = 17) 16 (n = 9)	1 mm	500 msec (n = 17) 420 msec (n = 9)	125–250 msec (n = 17) 105–210 msec (n = 9)
2006	Dey et al.	GE Imatron eSpeed GE Lightspeed	Electron beam/ single	2 (n = 11) (EBCT) 16 (n = 1) (CT)	3 mm (n = 11) 1.25 mm (n = 1)	100 msec (n = 11) 500 msec (n = 1)	n/a
2006	Boll et al.	Philips Mx8000IDT	Single	16	0.75 mm	420 msec	n/a
2006	Holmström et al.	GE Lightspeed Ultra	Single	8	1.25	500 msec	n/a
2010	Scheffel et al.	Siemens Somatom Definition	Dual	64 (flying z-spot, 32 native)	0.6 mm	330 msec	83 msec
2010	Stolzmann et al.	Siemens Somatom Definition	Dual	64 (flying z-spot, 32 native)	0.6 mm	330 msec	83 msec
2010	Donati et al.	Siemens Somatom Definition	Dual	64 (flying z-spot, 32 native)	0.6 mm	330 msec	83 msec
2011	Donati et al.	Siemens Somatom Definition	Dual	64 (flying z-spot, 32 native)	0.6 mm	330 msec	83 msec
2011	Stolzmann et al.	Siemens Somatom Definition	Dual	64 (flying z-spot, 32 native)	0.6 mm	330 msec	83 msec

Table 5.3 CT scan protocol parameters from studies in Table 5.1

Year	Author	ECG gating (R-R interval)	Radiation dose	Tube voltage	Tube current	Reconstruction kernel	Contrast agent	Beta-blocker
2002	White et al.	n/a	n/a	n/a	n/a	n/a	n/a	n/a
2005	Setser et al.	55%; retrospective (n = 26)	n/a	120 kVp (n = 26)	300 mA (n = 17) 370 mA (n = 9)	Medium sharp body (B30f)	100–150 mL Iopromide (Ultravist 300)	Intravenous metoprolol (n = 4) None (n = 22)
2006	Dey et al.	45%–60%; prospective (n = 11) retrospective (n = 1)	n/a	n/a	n/a	n/a	Noncontrast (n = 11) 100 mL iohexol (Omnipaque 350) (n = 1)	n/a
2006	Boll et al.	0%–100%	7.2–9.1 mSv	140 kVp	550 mA with adaptive dose modulation	n/a	100 mL ioversol (Optiray 300)	None
2006	Holmström et al.	20%–80%	n/a	120 kVp	135 mA	n/a	120 mL Iopromide (Ultravist 370)	5 mL intravenous metoprolol (n = 10)
2010	Scheffel et al.	70%; prospective	n/a	120 kVp (n = 30) 100 kVp (n = 13)	Attenuation-based modulation; 190 mAs/rotation	Medium smooth tissue convolution (B30f) and sharp-tissue convolution (B45f)	80–100 mL Iopromide (Ultravist 370)	2.5 mg sublingual isosorbide dinitrate (Isoket)

(Continued)

Table 5.3 (Continued) CT scan protocol parameters from studies in Table 5.1

Year	Author	ECG gating (R-R interval)	Radiation dose	Tube voltage	Tube current	Reconstruction kernel	Contrast agent	Beta-blocker
2010	Stolzmann et al.	70%; prospective	2.2 ± 0.6 mSv	n/a	Attenuation-based modulation; 190 mAs/rotation	n/a	70–80 mL Iopromide (Ultravist 370)	2.5 mg sublingual isosorbide dinitrate (Isoket)
2010	Donati et al.	70%; prospective	2.5 mSv	120 kVp (n = n/a) 100 kVp (n = n/a)	Attenuation-based modulation; 190 mAs/rotation	Medium smooth tissue convolution (B30f) and sharp-tissue convolution (B45f)	80–100 mL Iopromide (Ultravist 370)	2.5 mg sublingual isosorbide dinitrate (Isoket)
2011	Donati et al.	70%; prospective	2.5 ± 0.8 mSv	120 kVp (n = ?) 100 kVp (n = ?)	Attenuation-based modulation; 190 mAs/rotation	Medium smooth-tissue convolution (B26f) and sharp-tissue convolution (B46f)	70–80 mL Iopromide (Ultravist 370)	5 to 10 mg intravenous metoprolol (Beloc) (n = 11) 2.5 mg sublingual isosorbide dinitrate (Isoket) (n = 27)
2011	Stolzmann et al.	70%; prospective	1.1 ± 0.3 mSv	120 kVp (n = 60)	100 mAs/rotation	Medium-soft-tissue convolution (B35f)	Noncontrast (n = 60)	None

Table 5.4 CMR scan protocol parameters for studies in Table 5.1

Year	Author	MRI scanner	MR coil(s)	Contrast agent	LGE delay time	Cine protocol	LGE protocol	Perfusion protocol
2002	White et al.	n/a	n/a	n/a	n/a	n/a	n/a	n/a
2005	Setser et al.	Siemens Sonata 1.5 T	n/a	0.2 mmol/kg gadopentetate dimeglumine (Magnevist)	20 min	Yes	Yes	n/a
2006	Dey et al.	Siemens Sonata 1.5 T	Anterior cardiac and posterior phased-array	Gadolinium-DTPA (Omniscan)	20 min	Yes	Yes	n/a
2006	Boll et al.	Siemens Magnetom Sonata 1.5 T	4-channel torso phased-array	None	n/a	Yes	n/a	n/a
2006	Holmström et al.	Siemens Magnetom Sonata 1.5 T	Body array	0.55 mmol/kg gadopentetate dimeglumine (Magnevist)	5–20 min	Yes	Yes	Yes
2010	Scheffel et al.	Philips Achieva 1.5 T	Cardiac phased-array receiver (five elements)	0.1 mmol/kg gadobutrolum (Gadovist 1.0) (once at stress and once at rest)	10 min	n/a	Yes	Yes
2010	Stolzmann et al.	Philips Achieva 1.5 T	Cardiac phased-array receiver (five elements)	0.1 mmol/kg gadobutrolum (Gadovist 1.0) (once at stress and once at rest)	10 min	n/a	Yes	Yes
2010	Donati et al.	Philips Achieva 1.5 T	Cardiac phased-array receiver (five elements)	0.1 mmol/kg gadobutrolum (Gadovist 1.0) (once at stress and once at rest)	10 min	n/a	Yes	Yes
2011	Donati et al.	Philips Achieva 1.5 T	Cardiac phased-array receiver (five elements)	0.1 mmol/kg gadobutrolum (Gadovist 1.0) (once at stress and once at rest)	10 min	n/a	Yes	Yes
2011	Stolzmann et al.	Philips Achieva 1.5 T	Cardiac phased-array receiver (five elements)	0.1 mmol/kg gadobutrolum (Gadovist 1.0) (once at stress and once at rest)	10 min	n/a	Yes	Yes

with contrast enhancement (bright) relative to normal myocardium in these images, hence the term "late gadolinium enhancement."

CMR perfusion imaging quantifies the perfusion of gadolinium-based contrast agent in the myocardium (Panting et al. 2002), and is very similar in mechanism to nuclear cardiac stress studies (e.g., Cardiolite/SPECT). Two contrast CMR scans are performed, once at rest and once under chemical stress (e.g., regadenoson, adenosine). Areas with transient myocardial infarction should display diminished contrast enhancement during the stress scan. Naturally, the timing of the contrast bolus into the left ventricle will be patient specific and is derived as part of the imaging protocol. More details can be found within the various manuscripts, as the specific protocol used with each CMR technique is highly site dependent.

5.4 FUTURE TECHNOLOGY

The future of hybrid CT-MR imaging will require significant advances beyond the pilot studies discussed in this chapter. Current technology can permit improved CAD detection results with lowered radiation dose via prospective gating and iterative image reconstruction techniques. However, two recent advances should enhance the synergy between CT and MRI that is missing with these previous studies.

First, conventional CT and MRI reconstruction algorithms operate independently of each other, and hence, mutual structural information between these two modalities is not utilized. A new framework, termed "omni-tomography," allows for high-resolution imaging with undersampled data by utilizing input from multiple modalities for simultaneous image reconstruction (Wang et al. 2012). Second, current studies utilize data from physically and temporally separate CT and MRI scans, which can introduce artifact. Retrospective image registration between CT and MRI is not a desirable alternative, because of registration errors due to nonrepeatable contrast dynamics, organ motion and deformation, MRI-induced geometric distortion, and signal nonlinearity, as well as inconsistent contrast mechanisms between CT and MRI (Wang et al. 2013). A novel hybrid CT-MRI architecture has been proposed based upon the aforementioned omni-tomography principle. These will be discussed in the following subsections of this chapter.

5.4.1 OMNI-TOMOGRAPHY

Omni-tomography was inspired and enabled by a recent theoretical breakthrough in interior tomography (Courdurier et al. 2008; Kudo et al. 2008; Wang et al. 2009; Wang and Yu 2010; Yang et al. 2010, 2011; Ye et al. 2007; Yu and Wang 2009), which is an approach initially developed for CT and is now considered a general tomographic imaging principle. Classic CT reconstruction theory targets theoretically exact reconstruction of a whole cross-section or volume from untruncated projections, yet real-world applications focus often on a specific region of interest (ROI). A long-standing barrier to interior tomography has been that conventional CT reconstruction techniques cannot exactly reconstruct an ROI solely from truncated x-ray projections through the ROI. In classic CT reconstruction, this is known as the "interior problem," which does not have a unique solution in an unconstrained setting.

The interior problem was extensively studied in the 1980s and 1990s without significant advance. That precise image reconstruction cannot be obtained from purely local data contributed to the longevity of CT and micro-CT scanner architectures whereby the detectors must be wide enough to cover a transaxial slice of a patient or animal. This problem also exists in other tomographic modalities, such as MRI, SPECT, and PET. In the late 2000s, the interior problem was revisited for theoretically exact image reconstruction over an ROI using practical conditions such as a known subregion or a piecewise constant or polynomial ROI model (Wang et al. 2009; Wang and Yu 2010; Yang et al. 2010, 2011; Ye et al. 2007; Yu and Wang 2009). This advancement means that the data acquisition system (e.g., detector array) can be made narrower and has been extended for other modalities such as SPECT and MRI (Yu et al. 2009; Zhang et al. 2009).

Thus, it has been recently proposed to transform each relevant imaging modality into a slim or compact "interior" imaging component for ROI-targeted data acquisition and image reconstruction (Wang et al. 2012). This is in contrast to the traditional global untruncated acquisition that requires full field-of-view coverage of the patient from all angles and results in the aforementioned physical conflict for grand fusion of multiple modalities in a single scanner/gantry. For example, SPECT-CT scanners have each imaging chain adjacent to each other, each with its own slip ring/gantry, because the full-size SPECT detectors cannot physically fit with the CT components (e.g., x-ray source, detectors, etc.) on a single slip ring/gantry. However, using interior acquisition, for example, compact "interior" detector array, system components from multiple modalities can be integrated into a single gantry for simultaneous data acquisition and composite interior reconstruction in a unified framework.

Although various modalities have different contrast mechanisms, the physiological process to be reconstructed is typically similar in nature. Consequently, there is strong correlation between reconstructed datasets from various modalities. The unified omni-tomographic approach is based on intermodality coherence to optimize the composite image quality. This is particularly important when image reconstruction is separately considered for each modality, and prior information is unavailable in the traditional reconstruction approach. In light of lowered sampling rate using temporal/spectral image coherence (Gao et al. 2011a, 2011b), it is hypothesized that intermodality coherence can be utilized for omni-tomography to reduce data requirements further, while giving the same image quality. In general, the multimodal imaging model is

$$P_i(A_i x_i) = y_i, \quad i \le N \tag{5.1}$$

where P_i, A_i, x_i, and y_i are an undersampling operator, an imaging system matrix, an image, and data, respectively, for the ith modality. The matrix model is

$$AX = Y \tag{5.2}$$

where the ith column of X (Y) corresponds to x_i (y_i), and A should be understood in terms of Equation 5.1. Next, the rank-sparsity decomposition (Chandrasekaran et al. 2009)

$$X = X_L + X_S \tag{5.3}$$

where X_L and X_S are low-rank and sparse components, respectively. Finally, the optimization problem

$$(X_L, X_S) = \underset{(X_L, X_S)}{\arg \min} \left\| A(X_L + X_S) - Y \right\|^2 + \lambda_* \left\| T_L(X_L) \right\|_* + \lambda_1 \left\| T_S(X_S) \right\|_1 \tag{5.4}$$

where $\left\| \cdot \right\|_*$ is the nuclear norm to enforce the inter-modality coherence of X_L after a transform T_L with a parameter λ_*, and $\left\| \cdot \right\|_1$ is the L_1 norm to promote the sparsity of X_S after a transform T_S with a parameter λ_1 (Cai et al. 2010). Here, T_L is the identity transform and T_S a linear framelet transform. Equation 5.4 was solved using the split Bregman method (Goldstein and Osher 2009).

In a pilot study by Wang et al., a set of MRI and CT head scans were selected from the NIH Visible Human Project (http://www.nlm.nih.gov/research/visible), containing MRI T1, T2, proton density, and CT images of a human cadaver. The data were undersampled with a factor of 8. The MR Cartesian k-space data were pseudo-randomly undersampled along the phase encoding direction (Bieri et al. 2005). The fan-beam CT data were undersampled using the dynamical strategy (Gao et al. 2011a). Figure 5.2 shows that the unified reconstruction improved the image quality significantly. Clearly, the unified reconstruction framework can be extended to cover more imaging modalities in support of omni-tomography.

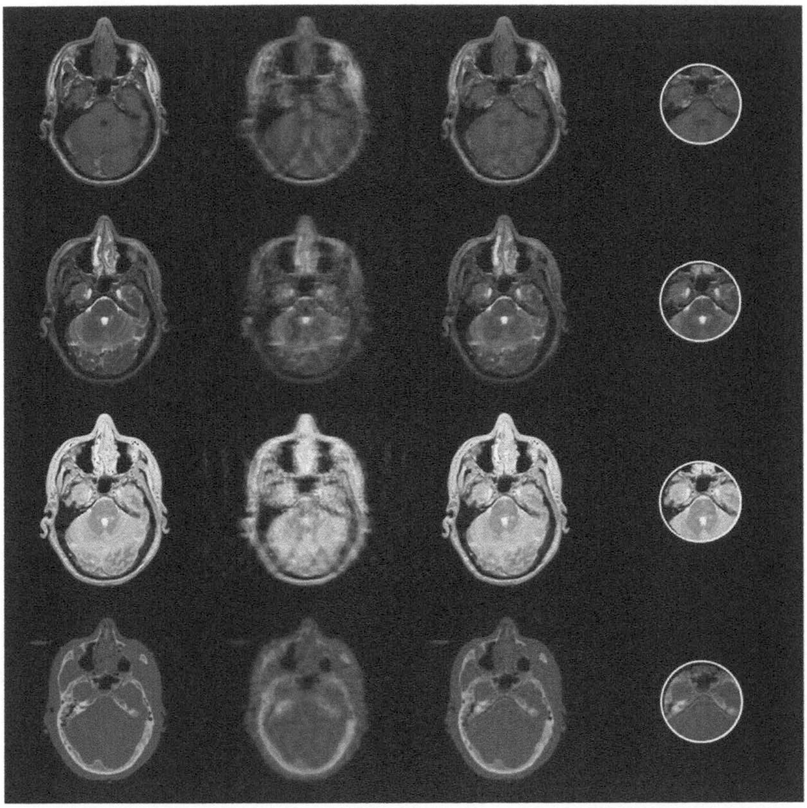

Figure 5.2 An MRI-CT head scan consisted of MR T1 (first column), T2 (second column), proton density images (third column), and a CT image (fourth column). The top row shows the phantom images; the middle row, the images separately reconstructed using the conventional FFT or FBP method; and the bottom row, the images simultaneously reconstructed in the unified rank-sparsity decomposition framework. (Adapted from Wang G et al., *PLoS One*, 7, e39700, 2012.)

5.4.2 CT-MRI SCANNER DESIGN

A CT-MRI system proposed by Wang et al. (2013) is illustrated in Figure 5.3. This design is based on the previously described omni-tomography principles.

5.4.2.1 CT SUBSYSTEM

The CT subsystem consists of nine pairs of x-ray tube and detector array, as shown in Figure 5.3. This design allows for near-instantaneous data acquisition of nine views (projections). Nine views may be sufficient for reconstruction of special cases, but such a low number of views is generally considered insufficient for ROI reconstruction with diagnostic image quality. Thus, it was proposed that the multisource data acquisition assembly can be rotated for three or five sets of nine views. Each data acquisition session can be ECG-gated with breath-holding. In an alternative design, 45 or more carbon-nanotube x-ray focal spots can be distributed along a half or full arc for rapidly multiplexed acquisitions in nine-view groups.

The x-ray tube electron beam(s) used for the CT in the proposed CT-MRI system will be under the influence of the MRI magnetic field (roughly 0.5 T background field in the FOV). The magnetic field will change the trajectories of both the primary electrons from the cathode and the backscattered electrons from the anode. The primary electrons can be deflected by the magnetic field without significant issues (Wen et al. 2007a), yet the deflection issue is more severe when the magnetic field is oblique to the cathode-anode electric

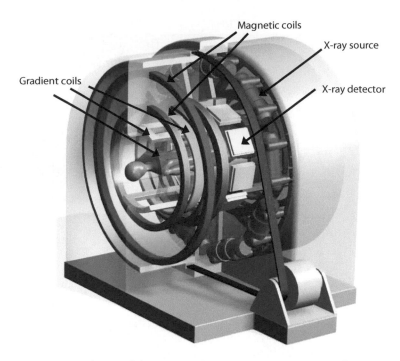

Figure 5.3 Cut-away view rendering of the proposed CT-MRI scanner. (Adapted from Wang G et al., Top-level design of the first CT-MRI scanner, In The 12th International Meeting on Fully Three-Dimensional Image Reconstruction in Radiology and Nuclear Medicine, Lake Tahoe, USA, 2013.)

field in the x-ray tube. Electron beam defocusing remains problematic even if the magnetic field is parallel to the cathode-anode electric field because electrons emitted from a thermionic cathode are omnidirectional and have a distribution of initial velocities. A significant portion of those electrons will be backscattered when the accelerated primary electrons strike the anode. More than 50% of the electrons incident on the target are backscattered for an X-ray tube working at 65 kV (Wen et al. 2007b). Only a small fraction of these backscattered electrons will return to the anode without an external magnetic field. However, the backscattered electrons will have an elevated probability of striking the anode again when a strong magnetic field exists. Thus, backscattered electrons result in an increased total x-ray output, slightly softening the x-ray spectrum and enlarging the focal spot size (Wen et al. 2007b).

The proposed CT-MRI system will have x-ray tubes that work in the static main magnetic field (main magnets) without complication of the magnetic fields from the gradient and radiofrequency (RF) coils. The main magnetic field is designed to be parallel to the longitudinal direction of the CT-MRI system. The cathode-anode axes in all the x-ray tubes can be made parallel to the direction of the magnetic field using a reflection-type anode at a suitable take-off angle. The simulation study in Wang et al. (2013) indicates that a 0.5 T static magnetic field is expected at the x-ray tube position. Active shielding can be used to minimize the magnetic field near the focal spot to prevent the backscattered electrons from returning to the focal spot area (Lillaney et al. 2012).

5.4.2.2 MRI SUBSYSTEM

The proposed MRI subsystem uses a pair of superconducting electromagnets to produce a 1.5 T static field with 1 part-per-million (peak–peak) variation within an interior cardiac ROI (25-cm sphere). The large central gap (more than 40 cm) for the CT subsystem requires consideration of the component's electromagnetic, mechanical, and thermal properties. As shown in Figure 5.4, the superconducting magnet has a two-layer coil configuration in that the inner layer (50 cm radius) provides a primary field in the field of view and the outer layer (80 cm radius) is mainly for shielding the stray magnetic field within the domain of interest (8 m in length and 5 m in radius). The wire layouts were determined using a nonlinear optimization scheme

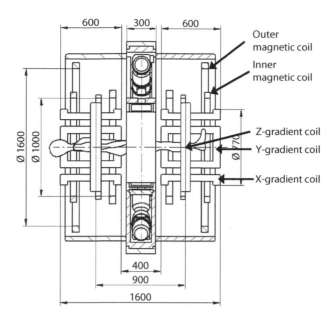

Figure 5.4 Geometric parameters of the proposed MRI subsystem with a split magnetic core and a two-layer coil configuration. (Adapted from Wang G et al., Top-level design of the first CT-MRI scanner, In The 12th International Meeting on Fully Three-Dimensional Image Reconstruction in Radiology and Nuclear Medicine, Lake Tahoe, USA, 2013.)

(Crozier and Liu 2008), followed by a quench simulation study to balance imaging performance and system cost. Field-based passive-shimming (Liu et al. 2011) and current-based active shimming techniques were used to improve field uniformity/harmonics and ROI homogeneity, respectively.

Three orthogonal gradient coils provide linear gradient fields, with paired saddle coils for the x- and y-gradient coils, and a Maxwell pair for the z-gradient coil. Coil patterns may be possibly optimized for advanced MRI applications (Zhu et al. 2012). Eddy current should be minimized with coil refinement, shield optimization, and gradient pre-emphasis. The RF coils are composed of aluminum to minimize interaction with the x-ray beam. Array coils may be arranged around the spherical FOV to transmit and acquire RF signals (Hong et al. 2011). The proposed design should have a temporal resolution at 20 msec–60 msec and an in-plane spatial resolution at 1 mm–2 mm. These capabilities should be sufficient for cardiac function studies, although higher spatial resolution may be possible with motion correction techniques.

5.4.3 CONCLUSION

Although the CT-MRI system design is based on physical and engineering principles, construction of a real system is not an easy task since it is significantly more complicated than any commercially available multimodality system. Thus, it would be prudent to start with proof-of-concept prototypes to show unique utilities and then gradually move to product-ready platforms. Vulnerable plaque characterization is the penultimate goal of hybrid CT-MRI cardiac imaging and would likely need 50 μm spatial resolution, 20 msec temporal resolution, and other cutting-edge quality indices. These specifications will require significant advances in technology but also represent unprecedented research opportunities.

While the multimodality reconstruction is currently performed modality-wise, it is hypothesized that omni-tomographic reconstruction can be implemented in a dictionary/atlas-based framework. The most detailed domain knowledge could be put into a composite dictionary/parameterized atlas for the specific imaging mechanisms involved. A hypothetical unified iterative reconstruction protocol follows:

1. Initial composite image is set based on the atlas.
2. Image is updated with an omni-tomographic scan.

3. Diffeomorphism is driven via flow (such as Ricci flow used to solve the Poincare conjecture) by the dictionary/atlas, yielding a Beltrami coefficient (BC) distribution (Wang 2012).

4. Image is refined to minimize the data discrepancy, BC-based sparsity, and other regularizing terms including cross-modality similarity. A scheme is to combine all the fidelity and penalty terms into one objective function, and minimize it using a split-Bregman-type scheme.

Omni-tomography offers biological, technical, physical, mathematical, and economic opportunities. Biologically, the "all-in-one" and "all-at-once" imaging power allows for observation of well-registered spatiotemporal features *in vivo*. Physically, multiphysics modeling suggests new imaging modes for synergistic information (such as photoacoustic imaging which combines ultrasound resolution and optical contrast). Technically, a paradigm shift of system engineering is required to integrate several different types of imaging components. Economically, a "one-stop-shop" for diagnosis and intervention may be realized that could be often more cost-effective than a full-fledged imaging center with independent modalities. Nevertheless, omni-tomography does have limitations due to an ROI-oriented restriction, increased complexity, and possible tradeoffs as more imaging contrast mechanisms are involved, whereas these are tractable with innovative technology and methods.

REFERENCES

Bieri O, Markl M, and Scheffler K. 2005. Analysis and compensation of eddy currents in balanced SSFP. *Magn Reson Med* 54:129–137.

Boll DT, Bossert AS, Aschoff AJ, Hoffmann MH, and Gilkeson RC. 2006. Synergy of MDCT and cine MRI for the evaluation of cardiac motility. *Am J Roentgenol* 186:S379–S386.

Cai J-F, Candès EJ, and Shen Z. 2010. A singular value thresholding algorithm for matrix completion. *SIAM J Optim* 20:1956–1982.

Cerqueira MD, Weissman NJ, Dilsizian V, Jacobs AK, Kaul S, Laskey WK, Pennell DJ, Rumberger JA, Ryan T, and Verani MS. 2002. Standardized myocardial segmentation and nomenclature for tomographic imaging of the heart: A statement for healthcare professionals from the Cardiac Imaging Committee of the Council on Clinical Cardiology of the American Heart Association. *Circulation* 105:539–542.

Chandrasekaran V, Sanghavi S, Parrilo PA, and Willsky AS. 2009. Sparse and low-rank matrix decompositions. In *Communication, Control, and Computing*, 2009 Allerton 2009 47th Annual Allerton Conference on. IEEE. 962–967.

Chavhan GB, Babyn PS, Jankharia BG, Cheng H-LM, and Shroff MM. 2008. Steady-state MR imaging sequences: Physics, classification, and clinical applications. *Radiographics* 28:1147–1160.

Cormode DP, Roessl E, Thran A, Skajaa T, Gordon RE, Schlomka J-P, Fuster V, Fisher EA, Mulder WJM, Proksa R, and Fayad ZA. 2010. Atherosclerotic plaque composition: Analysis with multicolor CT and targeted gold nanoparticles. *Radiology* 256:774–782.

Courdurier M, Noo F, Defrise M, and Kudo H. 2008. Solving the interior problem of computed tomography using a priori knowledge. *Inverse Probl* 24:065001.

Crozier S, and Liu F. 2008. Shielded, asymmetric magnets for use in magnetic resonance imaging. Google Patents.

Dey D, Slomka P, Chien D, Fieno D, Abidov A, Saouaf R, Thomson L, Friedman JD, and Berman DS. 2006. Direct quantitative in vivo comparison of calcified atherosclerotic plaque on vascular MRI and CT by multimodality image registration. *J Magn Reson Imaging* 23:345–354.

Donati OF, Alkadhi H, Scheffel H, Kuehnel C, Hennemuth A, Wyss C, Azemaj N, Plass A, Kozerke S, and Falk V. 2011. 3D fusion of functional cardiac magnetic resonance imaging and computed tomography coronary angiography: Accuracy and added clinical value. *Invest Radiol* 46:331–340.

Donati OF, Scheffel H, Stolzmann P, Baumüller S, Plass A, Leschka S, and Alkadhi H. 2010. Combined cardiac CT and MRI for the comprehensive workup of hemodynamically relevant coronary stenoses. *Am J Roentgenol* 194:920–926.

Faber TL, McColl RW, Opperman RM, Corbett JR, and Peshock RM. 1991. Spatial and temporal registration of cardiac SPECT and MR images: Methods and evaluation. *Radiology* 179:857–861.

Fayad ZA, Fuster V, Nikolaou K, and Becker C. 2002. Computed tomography and magnetic resonance imaging for noninvasive coronary angiography and plaque imaging: Current and potential future concepts. *Circulation* 106:2026–2034.

Feuerlein S, Roessl E, Proksa R, Martens G, Klass O, Jeltsch M, Rasche V, Brambs H-J, Hoffmann MHK, and Schlomka J-P. 2008. Multienergy photon-counting k-edge imaging: Potential for improved luminal depiction in vascular imaging. *Radiology* 249:1010–1016.

Flach B, Kuntz J, Brehm M, Kueres R, Bartling S, and Kachelrieß M. 2013. Low dose tomographic fluoroscopy: 4D intervention guidance with running prior. *Med Phys* 40:101909.

Gao H, Cai JF, Shen ZW, and Zhao HK. 2011a. Robust principal component analysis-based four-dimensional computed tomography. *Phys Med Biol* 56:3181–3198.

Gao H, Yu HY, Osher S, and Wang G. 2011b. Multi-energy CT based on a prior rank, intensity and sparsity model (PRISM). *Inverse Probl* 27:115012.

Goldstein T, and Osher S. 2009. The split Bregman algorithm for l1 regularized problems. *SIAM J Imaging Sci* 2:323–343.

Gutiérrez LF, Silva Rd, Ozturk C, Sonmez M, Stine AM, Raval AN, Raman VK, Sachdev V, Aviles RJ, and Waclawiw MA. 2007. Technology preview: X-ray fused with magnetic resonance during invasive cardiovascular procedures. *Catheter Cardiovasc Interv* 70:773–782.

Holmström M, Vesterinen P, Hänninen H, Sillanpää MA, Kivistö S, and Lauerma K. 2006. Noninvasive analysis of coronary artery disease with combination of MDCT and functional MRI. *Acad Radiol* 13:177–185.

Hong M, Yu Y, Wang H, Liu F, and Crozier S. 2011. Compressed sensing MRI with singular value decomposition-based sparsity basis. *Phys Med Biol* 56:6311.

Kim RJ, Fieno DS, Parrish TB, Harris K, Chen E-L, Simonetti O, Bundy J, Finn JP, Klocke FJ, and Judd RM. 1999. Relationship of MRI delayed contrast enhancement to irreversible injury, infarct age, and contractile function. *Circulation* 100:1992–2002.

Knecht S, Skali H, O'neill MD, Wright M, Matsuo S, Chaudhry GM, Haffajee CI, Nault I, Gijsbers GH, and Sacher F. 2008. Computed tomography–fluoroscopy overlay evaluation during catheter ablation of left atrial arrhythmia. *Europace* 10:931–938.

Kudo H, Courdurier M, Noo F, and Defrise M. 2008. Tiny a priori knowledge solves the interior problem in computed tomography. *Phys Med Biol* 53:2207–2231.

Lederman RJ. 2005. Cardiovascular interventional magnetic resonance imaging. *Circulation* 112:3009–3017.

Lillaney P, Shin M, Conolly SM, and Fahrig R. 2012. Magnetostatic focal spot correction for x-ray tubes operating in strong magnetic fields using iterative optimization. *Med Phys* 39:5567.

Liu F, Zhu J, Xia L, and Crozier S. 2011. A hybrid field-harmonics approach for passive shimming design in MRI. *IEEE Trans Appl Supercond* 21:60–67.

Moon JC, Reed E, Sheppard MN, Elkington AG, Ho S, Burke M, Petrou M, and Pennell DJ. 2004. The histologic basis of late gadolinium enhancement cardiovascular magnetic resonance in hypertrophic cardiomyopathy. *J Am Coll Cardiol* 43:2260–2264.

Nikolaou K, Poon M, Sirol M, Becker CR, and Fayad ZA. 2003. Complementary results of computed tomography and magnetic resonance imaging of the heart and coronary arteries: A review and future outlook. *Cardiol Clin* 21:639–655.

Pan D, Williams TA, Senpan A, Allen JS, Scott MJ, Gaffney PJ, Wickline SA, and Lanza GM. 2009. Detecting vascular biosignatures with a colloidal, radio-opaque polymeric nanoparticle. *J Am Chem Soc* 131:15522–15527.

Panting JR, Gatehouse PD, Yang G-Z, Grothues F, Firmin DN, Collins P, and Pennell DJ. 2002. Abnormal subendocardial perfusion in cardiac syndrome X detected by cardiovascular magnetic resonance imaging. *N Engl J Med* 346:1948–1953.

Raval AN, Telep JD, Guttman MA, Ozturk C, Jones M, Thompson RB, Wright VJ, Schenke WH, DeSilva R, and Aviles RJ. 2005. Real-time magnetic resonance imaging—Guided stenting of aortic coarctation with commercially available catheter devices in swine. *Circulation* 112:699–706.

Ropers D, Layritz C, Eisentopf J, Daniel W, Lell M, and Pflederer T. 2012. Ultra-low dose coronary CT angiography with less than 0.1 mSv radiation exposure using prospectively ECG-triggered high-pitch spiral acquisition and iterative reconstruction. *J Am Coll Cardiol* 59:E1354.

Scheffel H, Alkadhi H, Leschka S, Plass A, Desbiolles L, Guber I, Krauss T, Gruenenfelder J, Genoni M, Luescher TF, Marincek B, and Stolzmann P. 2008. Low-dose CT coronary angiography in the step-and-shoot mode: Diagnostic performance. *Heart* 94:1132–1137.

Scheffel H, Stolzmann P, Alkadhi H, Azemaj N, Plass A, Baumueller S, Desbiolles L, Leschka S, Kozerke S, Falk V, Boesiger P, Wyss C, Marincek B, and Donati O. 2010. Low-dose CT and cardiac MR for the diagnosis of coronary artery disease: Accuracy of single and combined approaches. *Int J Cardiovasc Imaging* 26:579–590.

Schlomka JP, Roessl E, Dorscheid R, Dill S, Martens G, Istel T, Bäumer C, Herrmann C, Steadman R, Zeitler G, Livne A, and Proksa R. 2008. Experimental feasibility of multi-energy photon-counting K-edge imaging in pre-clinical computed tomography. *Phys Med Biol* 53:4031.

Schwitter J, Wacker CM, van Rossum AC, Lombardi M, Al-Saadi N, Ahlstrom H, Dill T, Larsson HBW, Flamm SD, Marquardt M, and Johansson L. 2008. MR-IMPACT: Comparison of perfusion-cardiac magnetic resonance with single-photon emission computed tomography for the detection of coronary artery disease in a multicentre, multivendor, randomized trial. *Eur Heart J* 29:480–489.

Serfaty JM, Yang X, Foo TK, Kumar A, Derbyshire A, and Atalar E. 2003. MRI-guided coronary catheterization and PTCA: A feasibility study on a dog model. *Magn Reson Med* 49:258–263.

Setser RM, O'Donnell TP, Smedira NG, Sabik JF, Halliburton SS, Stillman AE, and White RD. 2005. Coregistered MR imaging myocardial viability maps and multi–detector row CT coronary angiography displays for surgical revascularization planning: Initial experience. *Radiology* 237:465–473.

Sra J, Narayan G, Krum D, Malloy A, Cooley R, Bhatia A, Dhala A, Blanck Z, Nangia V, and Akhtar M. 2007. Computed tomography-fluoroscopy image integration-guided catheter ablation of atrial fibrillation. *J Cardiovasc Electrophysiol* 18:409–414.

Stolzmann P, Alkadhi H, Scheffel H, Hennemuth A, Kuehnel C, Baumueller S, Kozerke S, Falk V, Marincek B, and Donati O. 2010. Image fusion of coronary CT angiography and cardiac perfusion MRI: A pilot study. *Eur Radiol* 20:1174–1179.

Stolzmann P, Alkadhi H, Scheffel H, Plass A, Leschka S, Falk V, Kozerke S, Wyss C, and Donati O. 2011. Combining cardiac magnetic resonance and computed tomography coronary calcium scoring: Added value for the assessment of morphological coronary disease? *Int J Cardiovasc Imaging* 27:969–977.

Wang G. 2012. Omni-Tomographic Imaging for interior reconstruction using simultaneous data acquisition from multiple imaging modalities. Google Patents.

Wang G, Liu F, Liu F, Cao G, Gao H, and Vannier MW. 2013. Top-level design of the first CT-MRI scanner. In The 12th International Meeting on Fully Three-Dimensional Image Reconstruction in Radiology and Nuclear Medicine, Lake Tahoe, USA.

Wang G, Yu H, and Ye Y. 2009. A scheme for multi-source interior tomography. *Med Phys* 36:3575–3581.

Wang G, and Yu HY. 2010. Can interior tomography outperform lambda tomography? *Proc Natl Acad Sci U S A* 107:E92–E93.

Wang G, Zhang J, Gao H, Weir V, Yu H, Cong W, Xu X, Shen H, Bennett J, Furth M, Wang Y, and Vannier M. 2012. Towards omni-tomography—Grand fusion of multiple modalities for simultaneous interior tomography. *PLoS One* 7:e39700.

Wen Z, Fahrig R, Conolly S, and Pelc NJ. 2007a. Investigation of electron trajectories of an x-ray tube in magnetic fields of MR scanners. *Med Phys* 34:2048.

Wen Z, Pelc NJ, Nelson WR, and Fahrig R. 2007b. Study of increased radiation when an x-ray tube is placed in a strong magnetic field. *Med Phys* 34:408.

White RD, and Setser RM. 2002. Integrated approach to evaluating coronary artery disease and ischemic heart disease. *Am J Cardiol* 90:L49–L55.

Yang J, Yu H, Jiang M, and Wang G. 2011. High order total variation minimization for interior SPECT. *Inverse Probl* 28:015001.

Yang JS, Yu HY, Jiang M, and Wang G. 2010. High-order total variation minimization for interior tomography. *Inverse Probl* 26:035013.

Ye Y, Yu H, Wei Y, and Wang G. 2007. A general local reconstruction approach based on a truncated Hilbert transform. *Int J Biomed Imaging* 2007:63634.

Yu H, and Wang G. 2009. Compressed sensing based Interior tomography. *Phys Med Biol* 54:2791–2805.

Yu H, Yang J, Jiang M, and Wang G. 2009. Interior SPECT—Exact and stable ROI reconstruction from uniformly attenuated local projections. *Commun Numer Methods Eng* 25:693–710.

Zhang J, Yu H, Corum C, Garwood M, and Wang G. 2009. Exact and stable interior ROI reconstruction for radial MRI. 72585G-72585G-72588.

Zhu M, Xia L, Liu F, Zhu J, Kang L, and Crozier S. 2012. A finite difference method for the design of gradient coils in MRI—An initial framework. *IEEE Trans Biomed Eng* 59:2412–2421.

Hybrid x-ray luminescence and optical imaging

RAIYAN T. ZAMAN, MICHAEL V. MCCONNELL, AND LEI XING

6.1 PRINCIPLE BEHIND X-RAY IMAGING TECHNIQUE

X-ray diagnostic imaging is a major worldwide activity. In the United States, approximately 250 million x-ray examinations are performed annually, and in Europe, a similar level of radiological activity is undertaken. This results in the fact that the largest contribution to radiation exposure to the population as a whole is known to be from manmade radiation sources arising in the form of diagnostic x-ray (UNSCEAR 2012). For diagnostic radiology, the image is generated by the interaction of x-ray photons, which have transmitted through the patient, with a photon detector. These photons can either be primary photons, which have passed through the tissue without interacting, or secondary photons, which result from an interaction along their path through the patient. The secondary photons will in general be deflected from their original direction and result in scattered radiation. The basic principles of projection radiography are that

1. X-rays are produced in an x-ray tube.
2. The energy distribution of the photons is modified by inherent and additional filtration.
3. The x-rays are attenuated differently by the various body tissues.
4. Scattered radiation, which impairs image contrast, is reduced.
5. The transmitted photons are detected.

The x-ray image is a two-dimensional (2-D) projection of the attenuating properties of all the tissues along the paths of the x-rays. The photons emitted by the x-ray tube are collimated by a beam-limiting device. Then they enter the patient, where they may be scattered, absorbed, or transmitted without interaction. The primary photons recorded by the image receptor form the image. The secondary photons create a certain amount of background radiation which degrades contrast. If necessary, the majority of the scattered photons can be removed by placing an antiscatter device between the patient and the image receptor. This device can simply be an air gap or a so-called antiscatter grid formed from a series of parallel metal strips.

6.2 PRINCIPLE BEHIND OPTICAL IMAGING TECHNIQUE

Optical imaging has gradually found its way into clinical applications over the last couple of decades. The advantages of optical imaging are that it does not use ionizing radiation (unlike x-rays) and that it can have higher spatial and temporal resolution than magnetic resonance. The main disadvantage of optical imaging is that tissue in general is highly scattering and absorbing. Optical imaging methods can be broadly divided into two categories: linear and nonlinear imaging—based on the dependence of the signals on the incident light intensity. Most imaging methods are implemented in the linear regime due to the low cost and ease of use associated with operation. Some of the most common methods include bioluminescence, fluorescence, absorption, and reflectance. The problem with linear imaging is that depth resolved information is difficult to achieve due to light scattering. Nonlinear imaging can partly circumvent this problem since the detected signals depend nonlinearly on the incident intensity and as a result are only generated at the focus of the light beam. This strong localization allows 3-D optical sectioning in images based on a nonlinear contrast. Transmitted light or back-scattered light can be collected, knowing it must have originated in the voxel being interrogated. This works even in heavily scattering samples, such as tissue. Another advantage of nonlinear imaging methods is that they encompass two-photon processes whereby lower-energy photons (in the near infrared [NIR] spectral range) may be used to probe the sample, thereby taking advantage of the therapeutic window shown in Figure 6.1. Some nonlinear optical imaging methods include two-photon fluorescence, two-photon absorption, and self-phase modulation.

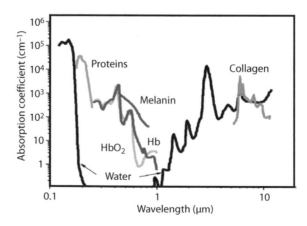

Figure 6.1 Optical absorption spectra of various tissue components in the ultraviolet to infrared frequency range.

6.3 WHY HYBRID IMAGING MODALITIES ARE NEEDED

Optical imaging promises increased sensitivity and specificity to disease compared to traditional anatomical imaging modalities. The information gained from optical imaging has the potential to provide patient-specific selection of therapy, improved prediction of outcomes, and increased treatment efficacy. X-ray radiography and computed tomography (CT) are commonly used anatomical imaging modalities; however, although they provide invaluable information in the clinic, they have been largely unsuccessful for biological imaging. This deficiency is due to their lack of sensitivity to low concentrations of contrast agents; x-ray imaging is many orders of magnitude less sensitive than optical or radionuclide imaging. This poor sensitivity arises from the low x-ray stopping power of diluted contrast agents, which necessitates high concentrations compared to other imaging modalities.

Current imaging modalities include magnetic resonance imaging (MRI), x-ray CT, positron emission tomography (PET), single-photon emission CT (SPECT), and optical (bioluminescence and fluorescence) and ultrasound imaging. Each modality cannot provide fully comprehensive information due to their inherent limitations. Therefore, it is desirable to combine different imaging modalities into a single imaging modality, leveraging complementary features. Radionuclide imaging has a limited spatial resolution, CT has low sensitivity to molecular targets, MRI is often semiquantitative, and optical imaging cannot penetrate more than ~5–8 cm of tissue (Nahrendorf et al. 2009a). Combining modalities can compensate for some of these limitations while building on their respective strengths. This is particularly important for the heart, a challenging imaging target with fast-moving, small structures. A modality with high resolution and good soft-tissue contrast can provide anatomical structures while a second, more sensitive modality may sample molecular information. Hybrid imaging provides intuitive integration of information based on functional and structural imaging techniques without requiring a detailed understanding of the technologies needed to produce them.

6.4 RECENT DEVELOPMENT IN HYBRID IMAGING

Although the main focus of this chapter is x-ray luminescence and optical hybrid imaging, some recent hybrid imaging modalities will be discussed briefly in this section. Subsequent developments have seen the advent of PET-CT designed as an integrated hardware and software platform. Due to such success of hybrid PET-CT scanners, none of the major manufacturers currently offers stand-alone PET scanners for commercial sale. Recent advances include incorporating 64-slice multidetector CT (MDCT), new detector technologies, and development of time-of-flight PET scanners (Muehllehner and Karp 2006; Surti et al. 2007). The rapid popularity of PET-CT in clinical practice has generated significant fall in manufacturing cost and availability of more technologically advanced systems at comparable prices. Following the success of PET-CT scanners with integrated diagnostic MDCT, Philips and Siemens both subsequently introduced SPECT-CT scanners that also included a diagnostic MDCT. The superior diagnostic quality of the CT on these scanners, particularly for soft tissues in the abdomen, and the much shorter CT acquisition times offer significant advantages. However, these systems are more expensive and involve higher radiation doses (O' Connor and Kemp 2006). Despite the success and popularity of PET-CT and, more recently, of SPECT-CT, there are some shortcomings in the use of CT as a complementary anatomical imaging modality. First, CT adds radiation dose to the overall examination, particularly if used in a full diagnostic role (Wu et al. 2004). Second, CT provides relatively poor soft tissue contrast in the absence of oral and intravenous iodinated contrast, particularly if low-dose acquisition protocols are utilized to minimize incremental radiation exposure. These two theoretical limitations do not apply to MRI, which does not involve ionizing radiation and provides soft tissue imaging with high spatial resolution and superior contrast compared to CT. MRI can also provide more advanced functional techniques such as diffusion and perfusion imaging as well as spectroscopy, which may be complementary to functional information obtained by PET. Furthermore, the high sensitivity of PET may also complement the poor signal strength inherent in current functional MRI. The combination of PET and MRI into a single scanner may therefore prove to be the ultimate hybrid imaging modality, combining the metabolic and molecular information of PET with the excellent anatomical detail of MRI, while offering new potential applications with respect to functional MRI techniques (Seemann 2005).

6.5 DEVELOPMENT OF X-RAY/OPTICAL HYBRID IMAGING SYSTEMS

6.5.1 PRECLINICAL HYBRID IMAGING SYSTEMS

6.5.1.1 X-RAY/OPTICAL LUMINESCENCE IMAGING WITH NANOPHOSPHOR

A study by Carpenter et al. first introduced a combined x-ray/optical instrument to investigate the use of nanosized inorganic phosphors as potential biological contrast agents for medical imaging (Carpenter et al. 2010). The emission from this contrast agent is evaluated to determine the practicality of this new modality. The implications for this x-ray-activated contrast agent are discussed with regard to its potential to enable molecular imaging during fluoroscopy, x-ray CT, or projection x-ray imaging. The hybrid system consists of an x-ray source, x-ray detector, and a charge-coupled device (CCD) camera. This system was tested with a tissue emulating phantom made from agar with embedded gadolinium oxysulfides (GOS) doped with either trivalent europium (Eu) or terbium (Tb) phosphors (GOS:Tb and GOS:Eu) or lanthanum oxysulfides (Figure 6.2). A major concern of these x-ray-excitable phosphor contrast agents is that they require ionizing dose to activate. Thus, lower concentrations of phosphors will necessitate higher doses. Although this work demonstrated the potential of x-ray luminescence imaging for imaging a superficial object, imaging of lesions centimeters deep should be possible, with contrast-resolution limitations dependent on tissue properties, concentration, and nonspecific uptake.

The development of deep-tissue x-ray luminescence imaging will require the incorporation of optical tomographic models. With x-ray luminescent imaging, the x-ray source must be modeled in tissue to give

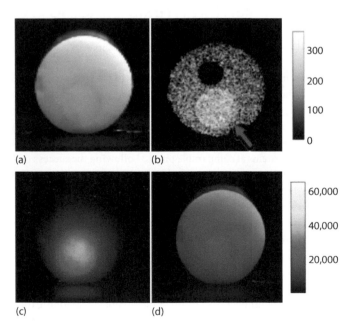

Figure 6.2 Agar imaging phantom with embedded phosphors and tissue emulating optical properties. (a) White-light optical image. (b) Projection fluoroscopy image (note the distinction between the phosphor inclusion—indicated by the arrow—around 300 units, compared to the black circle caused by a screw hole in the optical table supporting the phantom). (c) Optical emission from the phantom. (d) Overlay of the white light image (a) and the light emission (c). (From Carpenter CM et al., *Med Phys*, 37, 4011–4018, 2010. With permission.)

an accurate description of dose. There are many sophisticated tools to model dose, such as Monte Carlo or analytical models, which have been shown to be accurate (e.g., within 4% in biological tissues). Concurrent x-ray structural imaging will further improve these calculations. After dose distribution is calculated, tomographic imaging may be performed with a reconstruction model that uses a model of the light propagation in tissue to minimize the difference between calculated and optical measurements. This is very similar to the fluorescence molecular imaging problem. Once again, the knowledge of anatomical information will aid the optical reconstruction problem by providing structural detail, which may be used to improve optical modeling and reconstruction.

6.5.1.2 X-RAY/OPTICAL LUMINESCENCE IMAGING WITH X-RAY LUMINESCENCE CT

Recently, Pratx and coworkers developed x-ray luminescence CT (XLCT), which detects the presence and location of x-ray phosphors in optically diffusive medium (Pratx et al. 2010a). Using 1–100 cGy of radiation dose, they achieved a spatial resolution of 1 mm through tissue-mimicking material based on the selective excitation of subpicomolar of 50 nm phosphor and optical detection of their luminescence. XLCT is proposed as a new hybrid molecular imaging modality based on the selective excitation and optical detection of x-ray-excitable phosphor nanoparticles. These nanosized particles can be fabricated to emit NIR light when excited with x-rays, and because both x-rays and NIR photons propagate long distances in tissue, they are particularly well suited for in vivo biomedical imaging. In XLCT, tomographic images are generated by irradiating the subject using a sequence of programmed x-ray beams, while sensitive photodetectors measure the light diffusing out of the subject. By restricting the x-ray excitation to a single, narrow beam of radiation, the origin of the optical photons can be inferred regardless of where these photons were detected and how many times they scattered in tissue. This study presents computer simulations exploring the feasibility of imaging small objects with XLCT, such as research animals. The accumulation of 50 nm phosphor nanoparticles in a 2-mm-diameter target can be detected and quantified with subpicomolar sensitivity using less than 1 cGy of radiation dose. Provided sufficient signal-to-noise ratio, the spatial resolution of the system can be made as high as needed by narrowing the beam aperture. In particular, 1-mm spatial resolution was achieved for a 1-mm-wide x-ray beam. By including an x-ray detector in the system, anatomical imaging is performed simultaneously with molecular imaging via standard x-ray CT. The molecular and anatomical images are spatially and temporally coregistered, and if a single-pixel x-ray detector is used, they have matching spatial resolution. Pratx et al. recently showed that XLCT could image the cross-sectional distribution of microsize phosphor particles in 1 cm of an agar tissue phantom (Pratx et al. 2010b).

6.5.1.3 X-RAY/OPTICAL LUMINESCENCE IMAGING WITH SCINTILLATORS

Sensors for small chemical analytes in thick tissue were demonstrated by Chen et al. using scanning x-ray excited optical luminescence (Chen et al. 2011). This study reports a high spatial resolution imaging technique to measure optical absorption and detect chemical and physical changes on surfaces embedded in thick tissue. Developing sensors to measure chemical concentrations on implanted surfaces through tissue is an important challenge for analytical chemistry and biomedical imaging. Tissue scattering dramatically reduces the resolution of optical imaging. In contrast, x-rays provide high spatial resolution imaging through tissue but do not measure chemical concentrations. It describes a hybrid technique that uses a scanning x-ray beam to irradiate Gd_2O_2S scintillators and detect the resulting visible luminescence through the tissue. The amount of light collected is modulated by optical absorption in close proximity to the luminescence source. By scanning the x-ray beam and measuring total amount of light collected, one can measure the local absorption near scintillators at a resolution limited by the width of luminescence source (i.e., the width of the x-ray excitation beam). For proof of principle, a rectangular 1.7-mm scanning x-ray beam was used to excite a single layer of 8-mm Gd_2O_2S particles and detect the absorption of 5-nm-thick silver island film through 10 mm of pork. Lifetime and spectroscopic measurements, as well as changing the refractive index of the surroundings, indicate that the silver reduces the optical signal through attenuated total internal reflection. The technique was used to image the dissolution of regions of the silver island film, which were exposed to 1 mM of H_2O_2 through 1 cm of pork tissue.

6.5.1.4 X-RAY/FLUORESCENCE MICROSCOPY

Metals and other trace elements play an important role in many physiological processes in all biological systems. Characterization of precise metal concentrations, their spatial distribution, and chemical speciation in individual cells and cell compartments will provide much needed information to explore the metallome in health and disease. Ralle et al. described in their review article synchrotron-based x-ray fluorescent microscopy (SXRF) to be the ideal tool to quantitatively measure trace elements with high sensitivity at high resolution (Ralle and Lutsenko 2009). SXRF is based on the intrinsic fluorescent properties of each element and is therefore element specific. Recent advances in synchrotron technology and optimization of sample preparation have made it possible to image metals in mammalian tissue with submicron resolution.

SXRF is a rapidly emerging high-resolution method to image elements in biological samples. The principle of SXRF is based on the intrinsic fluorescent properties of elements. High-energy x-rays (primary radiation) are used to emit inner shell electrons (such as K- or L-shell) from atoms of an object into the continuum. Outer shell electrons fill the inner shell vacancies, thereby emitting fluorescence (secondary radiation). The energy of these secondary x-rays depends on the properties of the nucleus and the electron shell and is different for each element and/or oxidation state. X-ray fluorescent measurements of biological samples yield quantitative information about the spatial distribution of multiple elements simultaneously with high sensitivity and low background (Yun et al. 1999). No dyes are needed, and when carried out in energy scanning mode, SXRF can be used to determine the oxidation state of an element. While many x-ray synchrotron sources have focusing optics that achieve resolution in the low micrometer range (Miller et al. 2006; McCrea et al. 2008), only third-generation synchrotron sources can provide high brilliance at high photon energy that is needed to acquire 2-D elemental maps at submicron (~200 nm) resolution. To date, SXRF is the only available method for quantitative imaging of whole cells.

SXRF is exceptionally sensitive; attomolar amounts (concentration of 10^{-18} moles per liter) of trace elements can be detected in single cells or 10-μm tissue sections (Twining et al. 2003), so even slight losses or redistribution of elements during tissue preparation and sectioning may lead to incorrect conclusions. In addition, to increase the information content of the SXRF generated data, one needs to increase the size of the measured sample (areas to be scanned) beyond a few cells. Similarly, the number of imaged sections should be sufficiently high to obtain statistically significant data. This number is largely determined by the heterogeneity (different cell types) of the scanned tissue.

SXRF imaging is the ideal and, so far, the only tool to reliably determine elemental distribution in tissue samples at the subcellular level with high sensitivity and low background. With optimized protocols for sample preparation and correlative techniques in place, all necessary tools are now available to investigate fundamental problems related to metal content of tissues. It is certain that SXRF will significantly contribute to the study of metallomes in different organisms, from plants to humans. However, SXRF is not a routine spectroscopic method and is not suitable for analysis of large numbers of samples because of the scarcity of available synchrotron sources. The amount of data that can be collected during a typical user cycle is limited by the scan speed and desired resolution. The scan speed (the time the detector spends at each point on a sample) is determined by the time it takes to collect a sufficient amount of photons. The resolution is dependent on the available focusing optics. However, often, the deciding factor when choosing the resolution for a scan is the total scan time (higher resolution = longer scan time). Users have to choose appropriate scan size in order to have adequate resolution and scan speed that would result in sufficient data collection given the limitation of beam time. The time it takes to scan a sample at a given resolution is ultimately determined by the brilliance of the focused beam at high photon energy and will significantly improve only if the brilliance of the beam is improved. Thus, SXRF is likely to help when important biological questions about metal ion distribution, concentration, and oxidation state need to be answered and only small amounts of materials are available (such as biopsy samples) while the resolution of individual cells (or their compartments) is critical.

6.6 NONINVASIVE X-RAY/OPTICAL HYBRID SYSTEM FOR CARDIOVASCULAR MOLECULAR IMAGING

The development and refinement of noninvasive cardiovascular imaging over the past two decades have provided new tools for the identification of preclinical disease that reach beyond identification of flow-limiting stenoses. Molecular imaging, unlike anatomic imaging, focuses on the immunobiology hidden behind the endothelium and may therefore be able to identify prospective culprit lesions in coronary arteries. Once an inflamed plaque at risk of rupture can be located, the goal is to employ systemic or local measures to prevent myocardial infarction (Fuster et al. 2005). Imaging biology could also better triage treatment: which lesion should be treated, and which should be left alone? A decision that a stent may not be needed could avoid unnecessary complications and reinterventions. Thus, there is significant need for next-generation imaging strategies that build on the increased knowledge in vascular biology. Molecular imaging of coronary arteries should be able to assess the regional risk that is specific to a lesion, which can then be used in concert with global risk factors to personalize the therapeutic strategy (Matter, Stuber, and Nahrendorf 2009).

6.6.1 MULTISPECTRAL X-RAY/CT IMAGING

Another hybrid imaging is multispectral CT imaging, with excellent spatial resolution. Spectral CT imaging, used with carefully chosen contrast agents such as Au-HDL, can be used to detect macrophages in atherosclerotic plaque while also imaging plaque composition (e.g., calcified vs. noncalcified plaques) at the same time (Cormode et al. 2010). This hybrid system has the potential to be used with other types of contrast agents to probe other biologic processes and diseases.

A preclinical spectral CT system in which incident x-rays are divided into six different energy bins was used for multicolor imaging (Cormode et al. 2010). Au-HDL, an iodine-based contrast agent, and calcium phosphate were imaged in a variety of phantoms (Figure 6.3). Apolipoprotein E knockout (apo E-KO) mice were used as the model for atherosclerosis. Gold nanoparticles targeted to atherosclerosis (Au-HDL) were

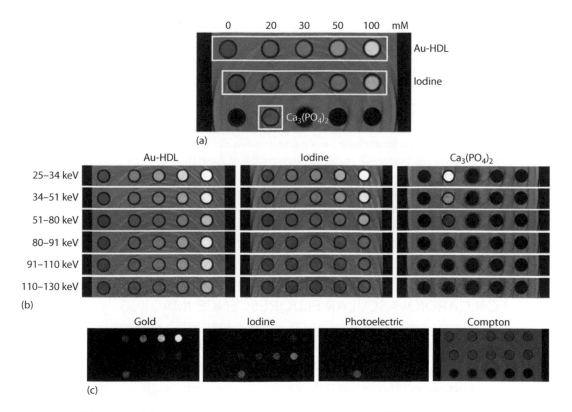

Figure 6.3 CT images of phantom containing various concentrations of Au-HDL, an iodinated contrast agent, and calcium phosphate powder, $Ca_3(PO_4)_2$, to simulate calcium-rich tissue. (a) Labeled conventional CT image; (b) spectral CT energy bin images; and (c) gold, iodine, photoelectric, and Compton images derived from energy bins are shown. (From Nahrendorf M et al., *Circ Cardiovasc Imaging*, 2, 56–70, 2009.)

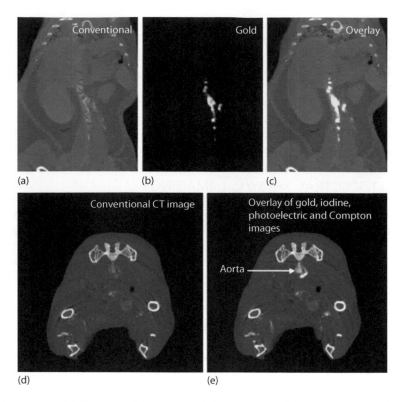

Figure 6.4 (a)–(c) Spectral CT images of thorax and abdomen in apo E–KO mouse injected 24 hours earlier with Au-HDL. (d), (e) Spectral CT images near bifurcation of aorta in apo E–KO mouse injected with Au-HDL and an iodinated emulsion contrast agent (Fenestra VC) for vascular imaging. (From Cormode DP et al., *Radiology* 256: 774–782, 2010. With permission.)

intravenously injected at a dose of 500 mg per kilogram of body weight. Iodine-based contrast material was injected 24 hours later, after which the mice were imaged. Wild-type mice were used as controls. Macrophage targeted by Au-HDL was further evaluated using transmission electron microscopy and confocal microscopy of aorta sections. Multicolor CT enabled differentiation of Au-HDL, iodine-based contrast material, and calcium phosphate in the phantoms. Accumulations of Au-HDL were detected in the aortas of the apo E-KO mice, while the iodine-based contrast agent and the calcium-rich tissue could also be detected and thus facilitated visualization of the vasculature and bones (skeleton), respectively, during a single scanning examination. Microscopy revealed Au-HDL to be primarily localized in the macrophages on the aorta sections (Figure 6.4); hence, the multicolor CT images provided information about the macrophage burden.

6.7 NONINVASIVE X-RAY/OPTICAL HYBRID SYSTEM FOR CARDIOVASCULAR FLUORESCENCE IMAGING

6.7.1 X-RAY CT/FLUORESCENCE LIFETIME IMAGING MICROSCOPY

Before introducing the hybrid x-ray CT/fluorescence lifetime imaging microscopy (FLIM) imaging modality, the next four paragraphs will describe how FLIM is in use in cardiovascular imaging as a standalone modality. Autofluorescence spectroscopy has the potential to quantify the elastin, collagen, and lipid contents in plaques (Arakawa et al. 2002). Although steady-state fluorescence spectroscopy is much simpler to implement, it is less robust and less sensitive than time-resolved spectroscopy (Jo et al. 2006). Time-resolved autofluorescence spectroscopy has the potential to distinguish between intimal thickening, fibrotic, and fibro-lipid plaques (Marcu et al. 2009); however, point spectroscopy has limited spatial resolution and slow acquisition speed. Both spatial resolution and acquisition speed can be significantly improved by implementing time-resolved fluorescence spectroscopy in imaging mode, most commonly known as FLIM (Lakowicz 2006). Two recent studies have shown the potential of FLIM for biochemical imaging of atherosclerotic plaques; however, its ability to distinguish intimal-thickening, fibrotic and fibro-lipid plaques (as in point-spectroscopy) has not yet been established (Thomas et al. 2010; Phipps et al. 2011). Recently, luminal multispectral FLIM imaging of coronary arteries demonstrated practical resolution and acquisition speed to perform biochemical imaging of coronary arteries with sufficient sensitivity and specificity to distinguish plaques with high content of either collagen or lipids (Park et al. 2012). Endogenous multispectral FLIM imaging was performed on the lumen of 58 segments of postmortem human coronary arteries. Briefly, the scanning FLIM system was implemented following a direct pulse-recording scheme, in which the pixel rate could be equal to the laser repetition rate.

A frequency tripled Q-switched Nd:YAG laser is used as the excitation source (355 nm, 30 kHz max. repetition rate, 1 ns pulse FWHM) (Park et al. 2012). The fluorescence emission is separated into three bands using a set of dichroic mirrors and filters (390 ± 20 nm for collagen, 452 ± 22.5 nm for elastin, and 550 ± 20 nm for lipids). Each band has launched into a fiber with different lengths (1 m, 10 m, and 19 m, respectively) chosen to provide ~45 ns intervals between each emission-band-decay. The three consecutive decays can be detected with a micro channel plate-photomultiplier tube (rise time: 150 ps) and sampled with a high-bandwidth digitizer (1.5 GHz, 4 GS/s). The system lateral resolution is measured to be 100 m. Each multispectral FLIM image (field of view: 2 mm × 2 mm at 60 × 60 pixels) is acquired in ~7 s.

The fluorescence normalized intensity and average lifetime from each emission band could be used to classify each pixel of an image as either "high collagen," "high lipids," or "low collagen/lipids" via multiclass Fisher's linear discriminant analysis (LDA) (Ripley 2007). Classification of plaques as either "high collagen," "high lipids," or "low collagen/lipids" based on the endogenous multispectral FLIM could be achieved with a sensitivity/specificity of 96/98%, 89/99%, and 99/99%, respectively, where histopathology served as the gold standard.

Clinically, x-ray CT is presently used in cardiovascular management to measure the amount of calcifications in the coronary vessels and, when combined with iodinated contrast agents, for noninvasive coronary angiography. CT images are based on measuring x-ray attenuation in small volumes at various different angles within the coronary artery by a mathematical reconstruction technique called filtered back projection, or the Radon transform, the x-ray attenuation in a single volume element (voxel) can be calculated in order to generate 3-D images of the x-ray absorption (Berman et al. 2006). Based on this x-ray CT imaging, a coronary calcium score can be calculated to give an indication of the global atherosclerotic burden and have strong predictive power for long term outcome of coronary artery disease. de Weert and Phillips et al. have shown that x-ray CT/FLIM can be used to differentiate between lipids and fibrous tissue in the carotid artery (de Weert et al. 2006; Phipps et al. 2011, 2012). In the latter study, plaques were analyzed from 18 patients (9 female, 9 male, average age 65 ± 2.8 years) undergoing carotid endarterectomy (Phipps et al. 2011). To correlate CT measurements with autofluorescence of carotid endarterectomy specimens three optical filters, F377: 377/50 nm, F460: 460/66 nm, and F510: 510/84 nm (center wavelength/bandwidth), were used. A Laguerre deconvolution technique was used for the evaluation of fluorescence decay dynamics (Pande and Jo 2011; Phipps et al. 2012). The resulting decay parameters (average fluorescence lifetime and four Laguerre coefficients at each of the recorded bandwidths) were used for sample characterization. LDA was used to classify each image into collagen or lipid-rich regions based on these parameters (Balakrishnama and Ganapathriraju 1998). In addition, de Weert et al. performed histology study on endarterectomy samples to correlate with the CT measurement, and they found that the x-ray absorption in Houndsfield units (HU) of calcifications, lipids, and fibrous tissue was 657 ± 416 HU, 88 ± 18 HU, and 25 ± 19 HU, respectively.

6.7.2 X-RAY/CT WITH FLUORESCENCE MOLECULAR TOMOGRAPHY

Fluorescence molecular tomography (FMT) is emerging as a quantitative modality for noninvasive whole mouse imaging (Besterman et al. 1981) and can be used to detect and quantitate inflammatory protease activity (Weissleder and Ntziachristos 2003; Vengrenyuk et al. 2006). High-throughput capabilities with scan times of <5 minutes, cost-effectiveness, high sensitivity, absence of radiation, and simultaneous imaging of multiple biomarkers in spectrally resolved channels are advantages that promote the use of FMT (Pratten et al. 1982).

As atherosclerotic plaques are small and vary in distribution, their accurate localization in such data sets is challenging. To overcome this limitation, hybrid FMT-CT protocols were developed, similarly to a combined PET-CT imaging (Marsh and Helenius 1980; Pratten and Lloyd 1983; Selby et al. 1995). An example of how advantages can be gained from fusing modalities is the combination of x-ray CT with FMT (Nahrendorf et al. 2009b, 2010; Panizzi et al. 2010). Like PET imaging, FMT is quantitative and sensitive (Nahrendorf et al. 2010) but provides limited anatomical information. Stand-alone FMT imaging is sufficient in large targets such as subcutaneously implanted tumors. Vascular and myocardial FMT, however, benefits from adding a modality that provides precise anatomic information. Two strategies have been proposed. An integrated FMT-CT device that combines laser, CCD cameras, and x-ray source on one gantry (Schulz et al. 2010) facilitates the use of CT priors for improved image reconstruction of FMT data sets (Ntziachristos 2010). An alternative approach uses separate devices. Here, the animal is transferred from the FMT imager to the CT or MRI scanner and image fusion is based on fiducials incorporated in an imaging cassette holding the anesthetized mouse. The advantage of the second approach lies in its versatility. It allows integration of any modality (FMT with PET, CT, MRI, or multiple combinations). Spectrally resolved FMT provides several channels for simultaneous quantification of molecular markers. In the commercially available system, the excitation/emission wavelengths are 635/655 nm, 680/700 nm, 750/780 nm, and 785/815 nm. Theoretically, PET and MRI could be added to yield a total of six channels. The most complex study to date combined three FMT channels with PET and CT and indicates that hybrid imaging of multiple targets is feasible (Nahrendorf et al. 2010). For instance, an infarct imaging study could use both the commercially available FMT system and probes, to quantify myocyte apoptosis, monocyte/macrophage numbers, protease activity, and integrin expression, and MRI, to measure collagen content (Helm et al. 2008), in one single study. These parameters could then be followed over time and correlated to the left ventricular function and myocardial infarct size. Such a multimodal protocol would provide insight into the generation of heart failure on a systems level by following a network of interdependent biomarkers. The use of quantum dots (QDs) as fluorescence reporters may add even more channels because these materials' emission profiles are narrow (Michalet et al. 2005). Another interesting development is the quantitation of NIR reporter proteins by FMT (Shu et al. 2009), which could be used to investigate gene expression or stem cell survival in the cardiovascular system in concert with the biomarkers noted previously.

6.7.2.1 DUAL-MODALITY FMT-CT IMAGING WITH A CONTRAST AGENT

Macrophage infiltration in atherosclerotic plaques plays a vital role in the progression of atherosclerosis (Zhang et al. 2006). As a result, macrophages have become widely recognized as a key target for atherosclerosis imaging, since they contribute significantly to the progression of atherosclerosis (Nahrendorf et al. 2008). FMT-CT hybrid imaging system can be used to characterize atherosclerotic plaque by targeting macrophages in inflamed coronary plaque using noninvasive molecular marker.

Proteases are emerging biomarkers of inflammatory diseases. In atherosclerosis, these enzymes are often secreted by inflammatory macrophages, digest the extracellular matrix of the fibrous cap, and destabilize atheromata. Protease function can be monitored with protease activatable imaging probes and quantitated in vivo by FMT. There are two major constrains currently associated with imaging of murine atherosclerosis: lack of highly sensitive probes and absence of anatomic information. The FMT-CT hybrid imaging system could be a robust and observer-independent tool for noninvasive assessment of inflammatory murine atherosclerosis (Nahrendorf et al. 2009b). A study by Nahrendorf et al. achieved coregistration of FMT and CT using a multimodal imaging cartridge containing fiducial markers detectable by both modalities (Figure 6.5). A high-resolution CT angiography protocol accurately localized fluorescence to the aortic root of atherosclerotic apoE$^{-/-}$ mice. FMT-CT imaging is performed at 680/700 nm excitation/emission wavelength using an FMT 2500 system (VisEn Medical Inc., Bedford, Massachusetts, USA) with an isotropic resolution of 1 mm. Each sensor is injected into apoE$^{-/-}$ mice with normal diet or high-cholesterol diet. Total imaging time for FMT acquisition is typically 5 to 8 minutes. Data are post processed using a normalized Born forward equation to calculate 3-D fluorochrome concentration distribution (Liner 1991). CT angiography immediately followed FMT to robustly identify the aortic root as the region of interest. The imaging cartridge lightly compressed the anesthetized mouse between optically translucent windows and thereby prevented motion during transfer to the CT (Inveon PET-CT, Siemens). The CT x-ray source operated at 80 kVp and 500 µA

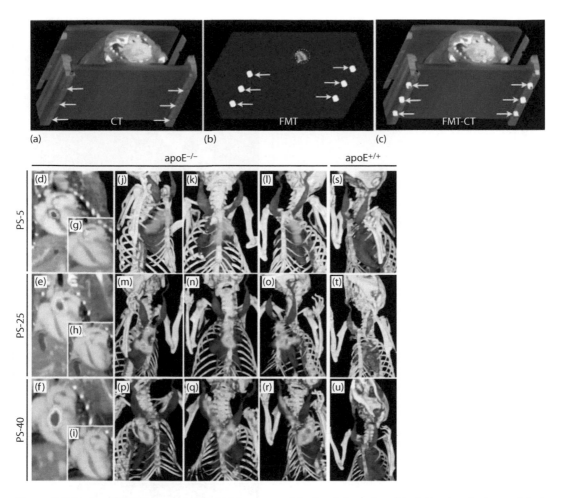

Figure 6.5 In vivo FMT-CT imaging. (a)–(c) Image coregistration is based on fiducial landmarks (arrows) that are incorporated into the animal holder and are identifiable on CT (a) and FMT (b). The software coaligns these fiducials to create a hybrid data set (c). Fluorescence signal in the aortic root of an apoE$^{-/-}$ mouse is encircled. (d)–(f) 2-D FMT-CT long-axis views of apoE$^{-/-}$ mice injected with respective protease sensors. Fluorescence signal is observed in the aortic root and arch, regions with high plaque load and high ex vivo fluorescence signal (as shown in Figure 6.2). (g)–(i) CT-only views of D through F. Arrow heads depict vascular calcification, likely colocalizing with plaques. (j)–(r) 3-D maximum-intensity projection of hybrid data sets show skeletal and vascular anatomy and the distribution of fluorescence signal. Most signal is observed in the root and arch; however, Q and R show additional activation of the protease sensor in the carotid artery, also a region predisposed to atheroma build-up in this model. (s)–(u) FMT-CT after injection of respective sensor into wild-type mice. (From Nahrendorf M et al., *Arterioscler Thromb Vasc Biol*, 29, 1444–1451, 2009. With permission.)

with an exposure time of 370 to 400 ms to acquire 360 projections. The effective 3-D CT resolution is 80 μm isotropic. During CT acquisition, Isovue-370, a radiopaque contrast agent, is infused at 55 μL/min through a tail vein catheter. The CT reconstruction protocol performed bilinear interpolation, used a Shepp-Logan filter, and scaled pixels to Hounsfield units (Tu, Shaw, and Chen 2006). Then, data need to be imported into OsiriX (The OsiriX foundation) to coregister FMT and CT images. Fiducials on the imaging cartridge can be visualized and tagged in FMT and CT images with point markers to define their XYZ coordinates. Using these coordinates, data can be resampled, rotated, and translated to match the image matrices and finally displayed in one hybrid image.

Hydrophobic QDs embedded in iodinated oil subsequently dispersed in water to form the oil-in-water nanoemulsion of 80 nm can be used as a FMT-CT dual-modal contrast agent (Ding et al. 2013). QDs have great potential in disease diagnosis because of complementary combination of the high spatial resolution of CT with

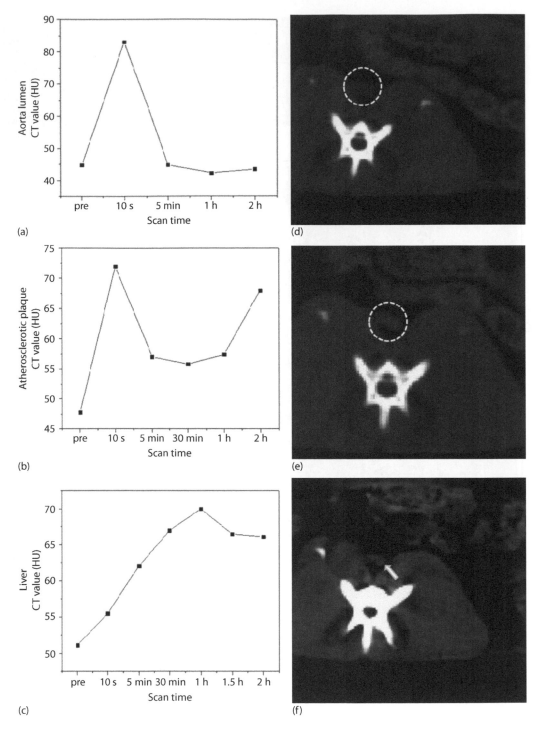

Figure 6.6 Kinetics and distribution of QD-iodinated oil nanoemulsion in atherosclerotic rabbits. (a)–(c) CT values of different organs for preinjection and at different time points for postinjection of the nanoemulsion. (a) Aortas lumen; (b) atherosclerotic plaques, showing that the atherosclerotic plaques were significantly higher 2 h postinjection; (c) liver. (d)–(f) CT images of the abdominal aorta (white circle) of NZW rabbit. The atherosclerotic plaques were indicated by arrows. (d) Before, (e) 10 s, and (f) 2 h after the injection of QD-iodinated oil nanoemulsion. Before injection, the atherosclerotic plaques could not be differentiated from the surrounding tissues, whereas a strong enhancement was detected in the plaques 2 h after the injection of the nanoemulsion. (From Ding J et al., *Biomaterials*, 34, 209–216, 2013. With permission.)

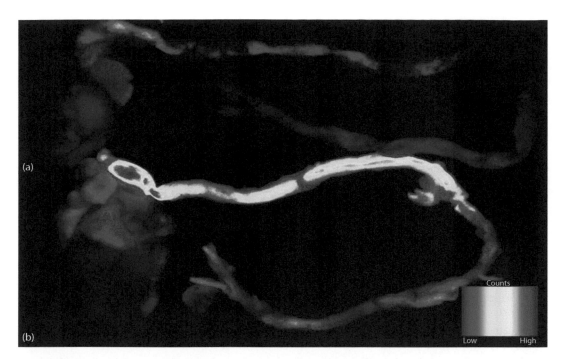

Figure 6.7 Fluorescence imaging of excised aortas with QD-iodinated oil nanoemulsion. The whole aortas were excised (a) without injection of the nanoemulsion; (b) 2 h after the injection. The signal intensity was significantly higher in the atherosclerotic plaques, meaning the accumulation of the nanoemulsion. (From Ding J et al., *Biomaterials*, 34, 209–216, 2013. With permission.)

the high sensitivity of optical imaging. In this study, atherosclerotic rabbits were used to detect macrophages with CT after administration of the nanoemulsion. This animal model has been widely used to study the effects of contrast agents on atherosclerotic plaques due to its high levels of macrophage infiltration and similar to human coronary atherosclerotic plaques (Kosuge et al. 2011). After ear vein injection of the nanoemulsion, serial CT imaging of the rabbits was performed to observe the enhancement in atherosclerotic plaques to detect the macrophage infiltration (Figure 6.6). Next, the aortas of rabbits were excised and scanned using an NIR fluorescence (NIRF) imaging system to estimate uptake of the nanoemulsion in aortas. The aortic arch and the abdominal aorta had higher signal intensity, meaning these segments were plaque-loaded areas (Figure 6.7). Therefore, it is further demonstrated that the QDs-iodinated oil nanoemulsion has a great potential targeting to macrophage-rich region to evaluate atherosclerosis with CT/fluorescence imaging.

6.7.3 INVASIVE CORONARY HYBRID MOLECULAR IMAGING

During the past decades, there has been considerable progress in understanding the pathophysiology of coronary artery disease, while devices, imaging techniques, and therapeutic strategies have been developed to optimize the treatment of patients with ischemic heart disease. These advances have made feasible the treatment of complex lesions and high-risk patients, improving their prognosis and quality of life. However, these developments have also created the need for more detailed imaging of coronary anatomy and pathology. It is apparent that contrast coronary angiography, which is the traditional method for the visualization of coronary artery disease, has significant limitations in assessing the extent and severity of atherosclerosis, as it permits only a projection evaluation of luminal dimensions and is unable to provide information regarding the vessel wall.

Atheroma burden and its composition affect prognosis, as there is evidence that cardiovascular outcomes and the occurrence of acute coronary events depend not only on the severity of luminal narrowing but also on plaque characteristics and inflammation (Yusuf et al. 2001). To address these limitations and study in more detail the natural evolution of atherosclerosis, considerable effort has been made in developing new imaging modalities that would permit the precise evaluation of coronary pathobiology. Advances in signal processing and the miniaturization of medical devices have allowed numerous invasive intravascular molecular imaging techniques to emerge (e.g., intravascular ultrasound [IVUS], optical coherence tomography [OCT], NIR spectroscopic imaging, intravascular magnetic spectroscopy, intravascular MRI, Raman spectroscopy, intravascular photoacoustic imaging, NIRF imaging, and time resolved fluorescence spectroscopic imaging), which have enriched our knowledge of coronary atherosclerosis by providing detailed visualization of luminal and plaque morphology and reliable quantification of the atheroma burden and its composition. Although these modalities offer a plethora of new data, each also has inherent limitations that do not allow complete evaluation of the coronary arteries. To address this drawback, hybrid imaging has been proposed.

The aims of this section is to describe the currently available invasive intravascular-based molecular hybrid imaging modalities, cite the advantages of the provided images and models, stress the methodological limitations of each approach, and discuss their potential value in the study of coronary atherosclerosis.

6.7.3.1 X-RAY CORONARY ANGIOGRAPHY-OCT HYBRID IMAGING

The fusion of coronary x-ray angiography (XA) and OCT aims to merge information regarding vessel geometry and plaque type in a single hybrid model. Two methodologies are currently available. The first approach uses the method applied to reconstruct the coronary arteries from IVUS and angiographic data, while the other, which was introduced by Tu et al. (2011), can integrate quantitative coronary angiography (QCA) and OCT/IVUS (Bourantas et al. 2012). The latter method uses biplane QCA to reconstruct the vessel and then uses anatomical landmarks seen on both QCA and OCT/IVUS to register the frames onto the reconstructed vessel by QCA.

Despite the fact that the 3-D angiographic reconstruction has value, the foremost limitation of XA-based systems remains the inability to image beyond the vessel lumen. In other words, the 3-D reconstructed vessel remains a "lumenogram," though with better 3-D capabilities. Thus, early stages of plaque formation may not be evident with XA due to the occurrence of coronary artery remodeling, and vulnerable plaques cannot be recognized. These limitations have been well addressed by intravascular tomography-based imaging techniques, among which grey scale IVUS is a well-established and validated modality. IVUS provides a wealth of information including vessel wall composition, which is crucial to the assessment of coronary atherosclerosis. Later on, the role of intravascular tomography-based imaging techniques was greatly enhanced by the radiofrequency data analysis for plaque characterization and OCT for the assessment of the thin fibrous cap atheromas and malapposition of stent struts. These new imaging techniques have extended the capabilities in the assessment of coronary artery disease. However, the fact that intravascular tomography-based imaging does not preserve the global topology information could lead to erroneous interpretations. Although a longitudinal view (L-View) is available in most IVUS/OCT consoles to provide an overview of the pullback series, the presentation of the L-View by stacking cross-sectional images along a straightened version of the transducer pullback trajectory is a very unnatural way of conceptualization. As a result, the interpretation can be quite challenging.

Given the different but complementary perspectives provided by XA and IVUS/OCT, the fusion/integration of the two imaging modalities by using XA as a roadmap while exploiting detailed vessel wall information from IVUS/OCT will benefit the interpretation of coronary artery disease and the guidance of coronary interventions. Currently, if IVUS/OCT is performed in the preintervention stage, the treatment planning is determined to a great extent by the IVUS/OCT interpretation. However, since XA fluoroscopy is still the only imaging tool available during stent deployment and positioning, the interventionalist must mentally establish the correspondence between XA and IVUS/OCT images (Tu et al. 2011). This spatial corresponding process is not always easy, especially for diseases of early stages or long diffused lesions where lumen narrowing is not clearly evident and no side branch is present in the neighborhood of the lesion borders. Thus, XA-IVUS/OCT integrated systems are currently requested in the market to better support coronary interventions. The clinical applicability of such fused/integrated systems depends to a great extent on the reliability and robustness of the coregistration approach. Once a reliable correspondence between angiographic and IVUS/OCT images is established, the issue of fusing/integrating information from the two image modalities becomes relevant.

The approach starts with standard QCA of the vessel of interest in the two angiographic views (either biplane or two monoplane views). Next, the vessel of interest is reconstructed in 3-D and registered with the corresponding IVUS/OCT pull-back series by a distance mapping algorithm (Danielsson 1980). An example of combing QCA and OCT imaging is given by Figure 6.8. The accuracy of the registration was retrospectively evaluated on 12 silicone phantoms with coronary stents implanted and on 24 patients who underwent both coronary angiography and IVUS examinations of the left anterior descending artery. Stent borders or side branches were used as markers for the validation. While the most proximal marker was set as the baseline position for the distance mapping algorithm, the subsequent markers were used to evaluate the registration error. The correlation between the registration error and the distance from the evaluated marker to the baseline position was analyzed. The XA-IVUS registration error for the 12 phantoms was 0.03 ± 0.32 mm ($p = 0.75$). One OCT pullback series was excluded from the phantom study, since it did not cover the distal stent border. The XA-OCT registration error for the remaining 11 phantoms was 0.05 ± 0.25 mm ($p = 0.49$). For the in vivo validation, two patients were excluded due to insufficient image quality for the analysis. In total, 78 side branches were identified from the remaining 22 patients and the registration error was evaluated on 56 markers. The registration error was 0.03 ± 0.45 mm ($p = 0.67$). The error was not correlated to the distance between the evaluated marker and the baseline position ($p = 0.73$). In conclusion, the new XA-IVUS/OCT coregistration approach is a straightforward and reliable solution to combine XA and IVUS/OCT imaging for the assessment of the extent of coronary artery disease. It provides the interventional cardiologist with detailed information about vessel size and plaque size at every position along the vessel of interest, making this a suitable tool during the actual intervention.

In contrast to the approach proposed by Bourantas et al. (2012), this method can process frequency domain OCT data, but it is unable to estimate the orientation of the OCT frames onto the 3-D vessel. IVUS has helped us study the natural evolution of atherosclerosis, but its low resolution did not allow detailed

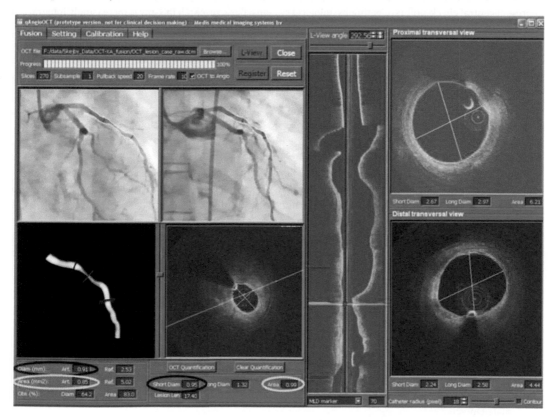

Figure 6.8 The XA-OCT coregistration and quantification. A sidebranch was manually identified from both XA and OCT images to be used as baseline position to register the two imaging modalities. After the registration, the markers superimposed in the OCT longitudinal view were synchronized with the same markers in the XA views. The OCT measurements could be compared with 3D QCA at the same position. (From Tu S et al., *Int J Cardiovasc Imaging*, 27, 197–207, 2011. With permission.)

examination of the distribution of culprit lesions and investigation of the association between plaque rupture and local hemodynamic status. Some investigators have advocated that high shear and tensile stress may contribute to plaque destabilization and rupture (Slager et al. 2005; Fukumoto et al. 2008). It is expected that the integration of OCT with coronary angiography will provide the substrate to study in vivo the impact of blood flow on acute coronary syndromes.

6.7.3.2 OCT-NIRF HYBRID IMAGING

The fusion of molecular imaging with conventional intravascular imaging techniques such as IVUS or OCT is expected to provide a more thorough evaluation of plaque vulnerability, as it would permit the simultaneous evaluation of vessel anatomy and plaque inflammation. A dual-modality OCT-NIRF catheter has recently been presented that allows the acquisition and coregistration of OCT and NIRF data (Yoo et al. 2011). In vitro and in vivo studies have confirmed the feasibility and efficacy of this device and demonstrated the great potential of the combined microstructural and molecular functional imaging. However, further evaluation of the safety of NIRF imaging is required before combined OCT and NIRF imaging can be used in humans.

6.8 CONCLUSION

Motivated by the potential to transform preclinical research and clinical care, cardiovascular hybrid imaging has made advances toward targeting coronary atherosclerosis and heart failure. In the next decade, our rapidly growing knowledge in cardiovascular biology promises identification of advanced targets, which will likely represent central check-points of pathophysiological systems. Basic research in x-ray is increasingly integrating imaging data, while optical techniques hold particular promise for dissemination into nonimaging laboratories. The synergy created by multimodal imaging will unite molecular, physiologic, and anatomic information.

Imaging is already an integral part of clinical trials, in which imaging biomarkers may serve as surrogate endpoints. However, formidable hurdles to clinical translation of advanced imaging systems still exist. To overcome these hurdles, the cardiovascular community can learn from the field of cancer imaging, which is quickly adopting hybrid imaging for both bench work and clinical translation. Dedicated grants have funded large imaging centers that combine expertise across medical physics, cancer biology, and probe chemistry. Basic scientists without extensive imaging expertise profit from this infrastructure, using high-end imaging tools on a collaborative base. Similar ongoing efforts in cardiovascular science could replicate this success.

Cardiovascular research has focused in recent years on the study of coronary atherosclerosis, aiming to predict its evolution and identify future culprit lesions. Consequently, considerable effort has been made to thoroughly visualize the coronary anatomy and pathobiology and understand in detail the effect of local and systemic factors on plaque progression. Hybrid imaging—both noninvasive and intravascular—has considerable future potential. Further effort is expected to be undertaken in upcoming years in advancing hybrid imaging techniques, to reveal novel aspects of plaque pathophysiology. Hybrid imaging has a bright future and is anticipated to constitute a valuable ally to tackle this challenging disease.

REFERENCES

Arakawa K, Isoda K, Ito T, Nakajima K, Shibuya T et al. (2002) Fluorescence analysis of biochemical constituents identifies atherosclerotic plaque with a thin fibrous cap. *Arterioscler Thromb Vasc Biol* 22: 1002–1007.

Balakrishnama SG, Ganapathriraju A (1998) Linear Discriminant Analysis—A Brief Tutorial. Institute for Signal and Information Processing, Department of Electrical and Computer Engineering Mississippi State University.

Berman DS, Hachamovitch R, Shaw LJ, Friedman JD, Hayes SW et al. (2006) Roles of nuclear cardiology, cardiac computed tomography, and cardiac magnetic resonance: Noninvasive risk stratification and a conceptual framework for the selection of noninvasive imaging tests in patients with known or suspected coronary artery disease. *J Nucl Med* 47: 1107–1118.

Besterman JM, Airhart JA, Woodworth RC, Low RB (1981) Exocytosis of pinocytosed fluid in cultured cells: Kinetic evidence for rapid turnover and compartmentation. *J Cell Biol* 91: 716–727.

Bourantas CV, Papafaklis MI, Naka KK, Tsakanikas VD, Lysitsas DN et al. (2012) Fusion of optical coherence tomography and coronary angiography—In vivo assessment of shear stress in plaque rupture. *Int J Cardiol* 155: e24–e226.

Carpenter CM, Sun C, Pratx G, Rao R, Xing L (2010) Hybrid x-ray/optical luminescence imaging: Characterization of experimental conditions. *Med Phys* 37: 4011–4018.

Chen H, Longfield DE, Varahagiri VS, Nguyen KT, Patrick AL et al. (2011) Optical imaging in tissue with x-ray excited luminescent sensors. *Analyst* 136: 3438–3445.

Cormode DP, Roessl E, Thran A, Skajaa T, Gordon RE et al. (2010) Atherosclerotic plaque composition: Analysis with multicolor CT and targeted gold nanoparticles. *Radiology* 256: 774–782.

Danielsson P-E (1980) Euclidean distance mapping. *Comput Graphic Image Process* 14: 21.

de Weert TT, Ouhlous M, Meijering E, Zondervan PE, Hendriks JM et al. (2006) In vivo characterization and quantification of atherosclerotic carotid plaque components with multidetector computed tomography and histopathological correlation. *Arterioscler Thromb Vasc Biol* 26: 2366–2372.

Ding J, Wang Y, Ma M, Zhang Y, Lu S et al. (2013) CT/fluorescence dual-modal nanoemulsion platform for investigating atherosclerotic plaques. *Biomaterials* 34: 209–216.

Fukumoto Y, Hiro T, Fujii T, Hashimoto G, Fujimura T et al. (2008) Localized elevation of shear stress is related to coronary plaque rupture: A 3-dimensional intravascular ultrasound study with in-vivo color mapping of shear stress distribution. *J Am Coll Cardiol* 51: 645–650.

Fuster V, Fayad ZA, Moreno PR, Poon M, Corti R et al. (2005) Atherothrombosis and high-risk plaque: Part II: Approaches by noninvasive computed tomographic/magnetic resonance imaging. *J Am Coll Cardiol* 46: 1209–1218.

Helm PA, Caravan P, French BA, Jacques V, Shen L et al. (2008) Postinfarction myocardial scarring in mice: Molecular MR imaging with use of a collagen-targeting contrast agent. *Radiology* 247: 788–796.

Jo JA, Fang Q, Papaioannou T, Baker JD, Dorafshar AH et al. (2006) Laguerre-based method for analysis of time-resolved fluorescence data: Application to in-vivo characterization and diagnosis of atherosclerotic lesions. *J Biomed Opt* 11: 021004.

Kosuge H, Sherlock SP, Kitagawa T, Terashima M, Barral JK et al. (2011) FeCo/graphite nanocrystals for multi-modality imaging of experimental vascular inflammation. *PLoS One* 6: e14523.

Lakowicz JR (2006) *Principles of Fluorescence Spectroscopy*. New York: Springer.

Liner CL (1991) Born theory of wave-equation dip moveout. *Geophysics* 56: 7.

Marcu L, Jo JA, Fang Q, Papaioannou T, Reil T et al. (2009) Detection of rupture-prone atherosclerotic plaques by time-resolved laser-induced fluorescence spectroscopy. *Atherosclerosis* 204: 156–164.

Marsh M, Helenius A (1980) Adsorptive endocytosis of Semliki Forest virus. *J Mol Biol* 142: 439–454.

Matter CM, Stuber M, Nahrendorf M (2009) Imaging of the unstable plaque: How far have we got? *Eur Heart J* 30: 2566–2574.

McCrea RP, Harder SL, Martin M, Buist R, Nichol H (2008) A comparison of rapid-scanning x-ray fluorescence mapping and magnetic resonance imaging to localize brain iron distribution. *Eur J Radiol* 68: S109–S113.

Michalet X, Pinaud FF, Bentolila LA, Tsay JM, Doose S et al. (2005) Quantum dots for live cells, in vivo imaging, and diagnostics. *Science* 307: 538–544.

Miller LM, Wang Q, Telivala TP, Smith RJ, Lanzirotti A et al. (2006) Synchrotron-based infrared and x-ray imaging shows focalized accumulation of Cu and Zn co-localized with beta-amyloid deposits in Alzheimer's disease. *J Struct Biol* 155: 30–37.

Muehllehner G, Karp JS (2006) Positron emission tomography. *Phys Med Biol* 51: R117–R137.

Nahrendorf M, Zhang H, Hembrador S, Panizzi P, Sosnovik DE et al. (2008) Nanoparticle PET-CT imaging of macrophages in inflammatory atherosclerosis. *Circulation* 117: 379–387.

Nahrendorf M, Sosnovik DE, French BA, Swirski FK, Bengel F et al. (2009a) Multimodality cardiovascular molecular imaging, Part II. *Circ Cardiovasc Imaging* 2: 56–70.

Nahrendorf M, Waterman P, Thurber G, Groves K, Rajopadhye M et al. (2009b) Hybrid in vivo FMT-CT imaging of protease activity in atherosclerosis with customized nanosensors. *Arterioscler Thromb Vasc Biol* 29: 1444–1451.

Nahrendorf M, Keliher E, Marinelli B, Waterman P, Feruglio PF et al. (2010) Hybrid PET-optical imaging using targeted probes. *Proc Natl Acad Sci U S A* 107: 7910–7915.

Ntziachristos V (2010) Going deeper than microscopy: The optical imaging frontier in biology. *Nat Methods* 7: 603–614.

O'Connor MK, Kemp BJ (2006) Single-photon emission computed tomography/computed tomography: Basic instrumentation and innovations. *Semin Nucl Med* 36: 258–266.

Pande P, Jo JA (2011) Automated analysis of fluorescence lifetime imaging microscopy (FLIM) data based on the Laguerre deconvolution method. *IEEE Trans Biomed Eng* 58: 172–181.

Panizzi P, Swirski FK, Figueiredo JL, Waterman P, Sosnovik DE et al. (2010) Impaired infarct healing in atherosclerotic mice with Ly-6C(hi) monocytosis. *J Am Coll Cardiol* 55: 1629–1638.

Park J, Pande P, Shrestha S, Clubb F, Applegate BE et al. (2012) Biochemical characterization of atherosclerotic plaques by endogenous multispectral fluorescence lifetime imaging microscopy. *Atherosclerosis* 220: 394–401.

Phipps J, Sun Y, Saroufeem R, Hatami N, Fishbein MC et al. (2011) Fluorescence lifetime imaging for the characterization of the biochemical composition of atherosclerotic plaques. *J Biomed Opt* 16: 096018.

Phipps JE, Sun Y, Fishbein MC, Marcu L (2012) A fluorescence lifetime imaging classification method to investigate the collagen to lipid ratio in fibrous caps of atherosclerotic plaque. *Lasers Surg Med* 44: 564–571.

Pratten MK, Cable HC, Ringsdorf H, Lloyd JB (1982) Adsorptive pinocytosis of polycationic copolymers of vinylpyrrolidone with vinylamine by rat yolk sac and rat peritoneal macrophage. *Biochim Biophys Acta* 719: 424–430.

Pratten MK, Lloyd JB (1983) Effect of suramin on pinocytosis by resident rat peritoneal macrophages: An analysis using four different substrates. *Chem Biol Interact* 47: 79–86.

Pratx G, Carpenter CM, Sun C, Xing L (2010a) X-ray luminescence computed tomography via selective excitation: A feasibility study. *IEEE Trans Med Imaging* 29: 1992–1999.

Pratx G, Carpenter CM, Sun C, Rao RP, Xing L (2010b) Tomographic molecular imaging of x-ray excitable nanoparticles. *Opt Lett* 35: 3345–3347.

Ralle M, Lutsenko S (2009) Quantitative imaging of metals in tissues. *Biometals* 22: 197–205.

Ripley BD (2007) *Pattern Recognition Neural Networks*. Cambridge, New York Cambridge University Press.

Schulz RB, Ale A, Sarantopoulos A, Freyer M, Soehngen E et al. (2010) Hybrid system for simultaneous fluorescence and x-ray computed tomography. *IEEE Trans Med Imaging* 29: 465–473.

Seemann MD (2005) Whole-body PET/MRI: The future in oncological imaging. *Technol Cancer Res Treat* 4: 577–582.

Selby DM, Singer DF, Anderson RW, Coligan JE, Linderman JJ et al. (1995) Antigen-presenting cell lines internalize peptide antigens via fluid-phase endocytosis. *Cell Immunol* 163: 47–54.

Shu X, Royant A, Lin MZ, Aguilera TA, Lev-Ram V et al. (2009) Mammalian expression of infrared fluorescent proteins engineered from a bacterial phytochrome. *Science* 324: 804–807.

Slager CJ, Wentzel JJ, Gijsen FJ, Thury A, van der Wal AC et al. (2005) The role of shear stress in the destabilization of vulnerable plaques and related therapeutic implications. *Nat Clin Pract Cardiovasc Med* 2: 456–464.

Surti S, Kuhn A, Werner ME, Perkins AE, Kolthammer J et al. (2007) Performance of Philips Gemini TF PET/CT scanner with special consideration for its time-of-flight imaging capabilities. *J Nucl Med* 48: 471–480.

Thomas P, Pande P, Clubb F, Adame J, Jo JA (2010) Biochemical imaging of human atherosclerotic plaques with fluorescence lifetime angioscopy. *Photochem Photobiol* 86: 727–731.

Tu SJ, Shaw CC, Chen L (2006) Noise simulation in cone beam CT imaging with parallel computing. *Phys Med Biol* 51: 1283–1297.

Tu S, Holm NR, Koning G, Huang Z, Reiber JH (2011) Fusion of 3D QCA and IVUS/OCT. *Int J Cardiovasc Imaging* 27: 197–207.

Twining BS, Baines SB, Fisher NS, Maser J, Vogt S et al. (2003) Quantifying trace elements in individual aquatic protist cells with a synchrotron x-ray fluorescence microprobe. *Anal Chem* 75: 3806–3816.

United Nations Scientific Committee on the Effects of Atomic Radiation. SOURCES, Effects and Risks of Ionizing Radiation, UNSCEAR 2012 Report.

Vengrenyuk Y, Carlier S, Xanthos S, Cardoso L, Ganatos P et al. (2006) A hypothesis for vulnerable plaque rupture due to stress-induced debonding around cellular microcalcifications in thin fibrous caps. *Proc Natl Acad Sci U S A* 103: 14678–14683.

Weissleder R, Ntziachristos V (2003) Shedding light onto live molecular targets. *Nat Med* 9: 123–128.

Wu TH, Huang YH, Lee JJ, Wang SY, Wang SC et al. (2004) Radiation exposure during transmission measurements: Comparison between CT- and germanium-based techniques with a current PET scanner. *Eur J Nucl Med Mol Imaging* 31: 38–43.

Yoo H, Kim JW, Shishkov M, Namati E, Morse T et al. (2011) Intra-arterial catheter for simultaneous microstructural and molecular imaging in vivo. *Nat Med* 17: 1680–1684.

Yun WLB, Cai Z, Maser J, Legnini D, Gluskin E, Chen Z, Krasnoperova A, Valdimirsky Y, Cerrina F, Di Fabrizio E, Gentili M (1999) Nanometer focusing of hard x-rays by phase zone plates. *Rev Sci Instrum* 70: 2238–2241.

Yusuf S, Reddy S, Ounpuu S, Anand S (2001) Global burden of cardiovascular diseases: Part I: General considerations, the epidemiologic transition, risk factors, and impact of urbanization. *Circulation* 104: 2746–2753.

Zhang Z, Machac J, Helft G, Worthley SG, Tang C et al. (2006) Non-invasive imaging of atherosclerotic plaque macrophage in a rabbit model with F-18 FDG PET: A histopathological correlation. *BMC Nucl Med* 6: 3.

X-ray fluoroscopy–echocardiography

R. JAMES HOUSDEN AND KAWAL S. RHODE

7.1 INTRODUCTION

Minimally invasive interventions are becoming an increasingly common alternative to open-heart surgery. The minimally invasive approach has many benefits, including reduced mortality rates, reduced risk of infection, and faster recovery for the patient. These complex procedures require the surgeon to remotely guide a catheter or surgical device to specific locations inside the heart. This has largely been made possible by the use of imaging modalities, particularly x-ray fluoroscopy, to provide visual feedback on the catheter location.

X-ray fluoroscopy is a live imaging modality with frame rates of up to 30 frames per second. The catheters and interventional devices are designed to be highly visible in x-ray, and since x-ray is a projection modality, it is able to visualize these catheters along their whole length within the field of view. X-ray fluoroscopy therefore provides excellent visualization of the interventional devices and is used in almost all minimally invasive interventions. However, x-ray displays soft tissue with considerably less contrast than the catheters, making the cardiac anatomy difficult to identify, as illustrated in Figure 7.1. This, along with the two-dimensional (2-D) nature of x-ray images, makes it difficult to navigate an interventional device to the correct anatomy using x-ray guidance alone. Consequently, procedure times are lengthened

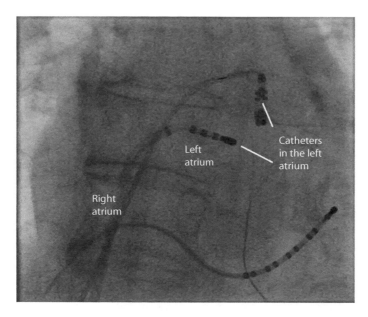

Figure 7.1 A typical x-ray image from a minimally invasive intervention. The catheters are easily seen, but the soft-tissue anatomy has lower contrast, making it difficult to determine the exact location of a catheter on the anatomy.

and x-ray use is increased, ultimately leading to more exposure of patients and staff to the ionizing x-ray radiation. This is particularly a concern for pediatric patients where exposure carries a significant risk (Tzifa et al. 2010).

In order to improve the image guidance and simplify the intervention, other imaging modalities are often introduced. These include computed tomography (CT), magnetic resonance imaging (MRI), and echocardiography. Each of these is valuable for its ability to visualize cardiac anatomy, thereby providing the information missing from the x-ray images. However, these modalities do not represent a replacement for x-ray since they have limitations of their own. CT and MRI are usually acquired prior to the intervention, sometimes by several weeks. Although intraprocedure CT or MRI is possible, the images are always static views of the heart. Apart from the problems of anatomical motion due to breathing, heartbeat, and the interventional devices themselves, these static images are unable to convey the necessary live location information of the interventional devices. Echocardiography is a live modality capable of visualizing both anatomy and devices, but due to its small field of view and changing orientation relative to the patient, it can be difficult to interpret without a skilled operator to communicate the context of the images to the surgeon.

The solution is therefore to combine the x-ray and one of the other modalities in a hybrid visualization, so that the complementary aspects of each modality are available to the surgeon in a single view. The usual approach is to display the x-ray projection image in its standard view because surgeons are familiar with x-ray images and how they relate to the patient. One of the other modalities is then registered to x-ray space and is usually displayed as a projection overlaid on the x-ray. This directly relates anatomy and devices in a single view and displays the additional modality in a context easily related to the patient. An example using echocardiography is shown in Figure 7.2.

The requirement common to all such techniques is the need to register the additional modality to x-ray space so that it can be projected into the x-ray image. A further requirement for clinical utility is that this be done as quickly as possible, with minimal modification of the standard clinical workflow. This chapter will review the various methods used to achieve this registration for the echocardiography imaging modality.

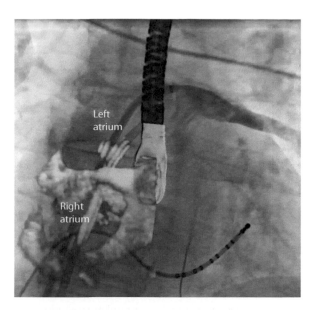

Figure 7.2 An x-ray image overlaid with 3-D echo. The echo adds anatomical information to the x-ray image, allowing the catheters to be more easily related to the anatomy. The hybrid view is presented in the familiar coordinates of the x-ray system, allowing easy interpretation of the combined images.

7.2 IMAGING EQUIPMENT

7.2.1 X-RAY FLUOROSCOPY SYSTEMS

X-ray fluoroscopy systems for use during cardiac surgery comprise a rotatable gantry containing an x-ray source and detector on opposite sides of the patient (see Figure 7.3). These are known as C-arm systems because of the shape of the gantry around the patient. The live x-ray images are streamed to a screen in the operating room, which provides feedback on the catheter location as the surgeon manipulates its position.

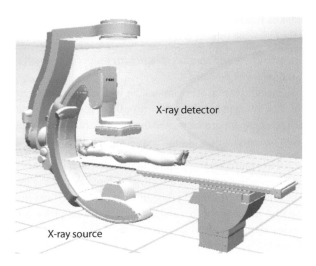

Figure 7.3 A cardiac interventional C-arm x-ray system. The x-ray source and detector are mounted on a gantry that can rotate around the patient and image different projections.

Typically, the field of view is large enough to cover two or three heart chambers. The frame rate can be up to 30 frames per second but is often less than this to reduce the radiation exposure. The gantry is usually capable of rotating about two axes: one parallel to the patient's head-to-foot direction and one in the patient's left-to-right direction. Most often, only the rotation about the head-to-foot axis is varied, to maintain the optimum projection view of the catheter motion.

The projection nature of x-rays can be restrictive for use in surgical guidance, and some systems provide a second source and detector that can be rotated independently. Two x-ray streams are then acquired simultaneously from different angles, which together provide additional information to the surgeon on the 3-D location of the catheter. These biplane systems are more expensive and so are less common in clinical practice.

7.2.2 Echocardiographic systems

Cardiac echocardiographic (echo) systems are currently available in three varieties depending on the positioning of the ultrasound probe: transthoracic echo (TTE), transesophageal echo (TEE), and intracardiac echo (ICE). Transthoracic systems (Rabben 2011) are rarely used for interventional guidance because it is impractical to manually hold the ultrasound probe in place against the patient for the duration of the procedure. More importantly, it is impossible for the sonographer to hold the ultrasound probe in place while simultaneous x-ray is applied in the region of the ultrasound probe.

TEE (Salgo 2011) is a recent development in which a miniaturized ultrasound probe is placed in the patient's esophagus. This requires the patient to be under general anesthesia. Precise positioning of the ultrasound transducer is possible via controls on the shaft of the probe outside the patient. These allow the sonographer to remotely steer the head of the ultrasound probe inside the esophagus. The placement of the ultrasound probe relative to the heart produces images of better resolution and clarity compared to TTE but requires a skilled sonographer to be used effectively. Both TTE and TEE systems are capable of producing 2-D and live 3-D images.

ICE is a more invasive catheter-based approach that images the heart from the inside, producing very clear and useful images. It is expensive because the ultrasound catheter cannot be reused and it is usually able to produce only 2-D images.

7.3 REGISTRATION OF ECHOCARDIOGRAPHIC IMAGES AND X-RAY FLUOROSCOPY

A variety of precalibration and image-based methods have been developed for x-ray-echo registration. The registration is limited by the following echo-specific constraints:

- X-ray and echo images have almost no common features or properties, which means that a direct image-to-image registration is not possible.
- Echo imaging differs significantly from other modalities by the requirement for very flexible placement and manual control of the ultrasound probe to ensure optimal images. The construction of ultrasound scanners means that the component producing the images, the ultrasound probe, is not stationary and is not tracked by any built-in mechanism of the imaging equipment.

These constraints have led to some common features of the various approaches to x-ray-echo registration. First, since the echo images themselves cannot be directly registered to an x-ray image, all hybrid x-ray-echo systems rely instead on tracking the ultrasound probe in x-ray space. This requires precalibration to locate the echo imaging space relative to the probe. Second, as the ultrasound probe is not tracked automatically, some physical or image-based method is required that locates the probe in 3-D x-ray space and then to the 2-D x-ray image space via the known C-arm geometry of the x-ray scanner. Figure 7.4 shows the coordinate systems involved.

Among the variety of methods are those that locate the probe via physical sensing equipment that tracks both the 3-D position of the probe and the position of the x-ray scanner. A second class of methods, designed

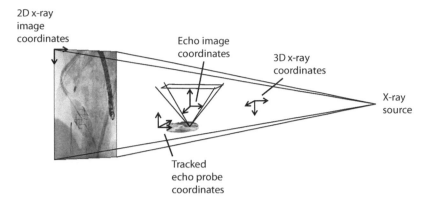

Figure 7.4 The echo image is registered to x-ray by tracking the probe in 3-D x-ray space. The echo image position is located relative to the probe by calibration. The echo image is then projected onto the 2-D x-ray image via the known geometry of the C-arm system.

specifically for transesophageal imaging, utilize the fact that the TEE probe is visible in the accompanying x-ray image. The probe position can be tracked by an image-based method and registered to 3-D x-ray space via the 2-D x-ray view. These methods can be either feature based, tracking fiducial markers attached to the probe, or intensity based, tracking the probe image itself.

7.3.1 PRECALIBRATION OF ECHO PROBE TO ECHO IMAGE SPACE

The objective of the precalibration stage is to determine the offset between the ultrasound image coordinate space and the coordinate space of the tracked device. The tracked coordinate space has an origin at a predefined point, either on the probe itself or equivalently on an object rigidly attached to the probe (see Figure 7.5). The position of the probe or attachment is tracked by one of the techniques described in Sections 7.3.2 through 7.3.4 and the position and orientation of the ultrasound images are known via the precalibrated offset.

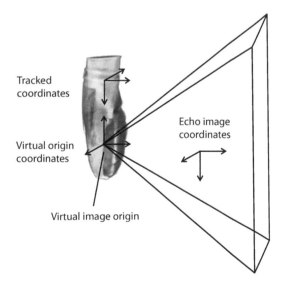

Figure 7.5 Coordinate systems of echo calibration. The tracked coordinates and virtual origin coordinates are both at fixed points on the echo probe and their offset can be found by calibration. The image coordinate system varies depending on the geometry of the scanned region. Their offset from the tracked coordinates must be determined at the time of acquisition from scanner settings.

Modern ultrasound probes use array technology (Rabben 2011), which allows the imaged region to be varied electronically without changing the probe position. Examples include changes in the sector width, the lateral steering, and in the case of 3-D ultrasound probes, rotation of the imaging volume about the central vertical axis of the images. While this is useful for scanning flexibility, it also means that the offset between the tracked and imaging coordinate systems is not necessarily constant.

In order to resolve this complication, the offset is divided into two stages that utilize a virtual origin point of the ultrasound images (Figure 7.5). This origin is fixed regardless of the varying imaging parameters and so is at a fixed offset from the tracked coordinates. This stage of the offset can be precalibrated. The remaining offset must be determined during live imaging as a function of the various scanner settings (sector width, steering angle, etc.). Knowledge of how these parameters affect the image location is built into the ultrasound scanner but is usually inaccessible from outside. Clinical systems designed for hybrid imaging therefore require a modified interface to the ultrasound scanner—see Section 7.3.5 for further discussion of these requirements.

The fixed part of the offset is determined by precalibration. Many different methods have been proposed for this kind of calibration (Mercier et al. 2005) and only a few examples will be given here. The most popular methods generally consist of imaging a simple phantom containing identifiable point targets with the ultrasound system, in a range of poses. The ultrasound probe is tracked during the imaging using the tracking method to be calibrated, and the point targets are also located by the same tracking system. At this point, the physical target points are at a known location relative to the probe for each ultrasound view and they are also located in ultrasound image space. Calibration then consists of registering corresponding targets in the two spaces, ultrasound image space and tracked probe space, to determine the unknown fixed offset.

For example, a method of calibration for tracking using physical sensing equipment involves also tracking the tip of a needle using the same sensing system (Jain et al. 2009). The needle tip is imaged by the ultrasound in a range of poses while simultaneously tracking both the ultrasound probe and the needle tip. Alternatively, for an image-based approach in which the probe is tracked directly in x-ray space, the target points are then also located using the same x-ray system (Gao et al. 2012). This is possible by imaging a fixed set of target points with x-ray from two different angles and backprojecting (Hawkes et al. 1987) to find their 3-D positions in x-ray space.

7.3.2 PHYSICAL PROBE TRACKING METHODS

Physical probe tracking refers to methods that attach a physical position-sensing device to the ultrasound probe. A similar device is also attached to the x-ray equipment, so that both systems are tracked in a common coordinate system. Through the tracking of the two devices, the probe is located relative to the x-ray scanner coordinate space. For this to be possible, both attachments require calibration: the ultrasound by the methods discussed in the previous section and the x-ray by a similar approach of imaging a phantom (Rhode et al. 2005).

The majority of hybrid x-ray-echo cardiac systems use the TEE variety of ultrasound because of the difficulty, discussed earlier, of using TTE in combination with x-ray. However, the need to safely insert and manipulate a TEE probe inside the esophagus significantly restricts the devices that can be attached to it. In fact, the only position sensing device that has plausibly been attached to a TEE probe is the Aurora Electromagnetic (EM) Tracking System (NDI, Waterloo, Ontario, Canada). This system uses very small sensors, which can be attached to the TEE probe with minimal change to its profile. An example attachment is shown in Figure 7.6. The probe and sensor are covered in a liquid plastic layer to achieve a rigid attachment (Jain et al. 2009). In this example, the registration accuracy between the two imaging modalities was measured as 2 mm in a phantom environment.

A second example of physical probe tracking involves a TTE probe. Although TTE is usually not suitable for combination with x-ray, there is one approach that has achieved a clinically feasible hybrid imaging system (Ma et al. 2009, 2010). The approach is to control the TTE probe via a robotic interface (Figure 7.7a), which avoids the sonographer being required to stand near the path of the x-ray while imaging. Instead, the robot is controlled remotely via a haptic feedback controller that relays force sensor information to the control device (Figure 7.7b). This ensures that the patient cannot be injured by excessive probe pressure and also enables the probe position to adjust automatically to maintain constant contact pressure and hence image quality.

Figure 7.6 An electromagnetic tracker attached to a TEE probe. The light colored material is a liquid plastic that rigidly holds the sensor to the outside of the probe. (With kind permission from Springer Science+Business Media: *Functional Imaging and Modeling of the Heart–FIMH*, 3D TEE registration with x-ray fluoroscopy for interventional cardiac applications, 5528, 2009, 321–329, Jain, A, Gutierrez, L, Stanton, D, Figure 3, Image copyright © Springer.)

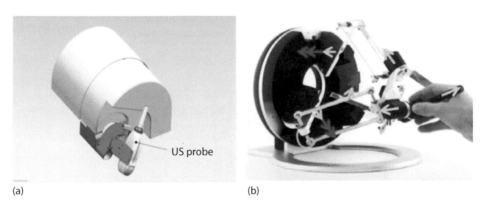

(a)   (b)

Figure 7.7 (a) The robotic probe holder that is controlled remotely by (b) a haptic feedback device. (Reprinted with permission from Ma, Y et al., Evaluation of a robotic arm for echocardiography to x-ray image registration during cardiac catheterization procedures, in *Proceedings of the Annual International Conference of the IEEE Engineering in Medicine and Biology Society–EMBC*, 5829–5832. © 2009 IEEE. All rights reserved.)

The robot is rigidly attached to the x-ray table and the probe rigidly attached to the moving end of the robot. The position of its moving end relative to this attachment is determined automatically through the known geometry of the robot. Both the x-ray system and the rigid part of the robot are tracked by an Optotrak Optical Tracking System (NDI, Waterloo, Ontario, Canada). In many respects, the system uses the same registration concept as the EM tracking of a TEE probe, the only difference being that the probe tracking is via a two-stage tracking system comprising a robot that is itself optically tracked. As with all such systems, precalibration is required to determine the offset between the x-ray scanner and its tracker and also between the ultrasound image and the probe attachment (in this case the moving end of the robot).

While these types of sensor-based tracking are potentially a reliable and highly accurate means to register the two modalities, there are some obstacles to clinical use. These are mainly the additional equipment needed to operate the systems. Optical sensors require cameras with a line of sight to the tracked targets and the targets themselves can be bulky. Magnetic sensors are more compact but have a small region of operation outside of which they rapidly become inaccurate. They are also susceptible to distortion of their magnetic field by nearby metallic objects, which introduces further errors. In either case, the use of a sensor system introduces further constraints into an already complex operating procedure. In addition, when using TEE, there are concerns over the safety of attaching additional devices to a probe that is designed to enter the esophagus.

The current limitations are understandable given that the two systems described here are both research rather than commercial systems. In order to overcome these limitations, it will be necessary to develop fully integrated systems in which the user does not need to be aware of the sensors. In this regard, EM sensors can more easily be integrated into existing devices, as they do not require a line of sight. For example, a recent development is an EM sensor fully integrated into a TEE probe (Moore et al. 2010). Ultimately, integrated sensor systems, which to the user appear the same as the original hardware, may prove to be the simplest means of registration available.

7.3.3 FIDUCIAL-BASED TEE TRACKING

Fiducial-based tracking is an image-based registration technique that bypasses the need for physical sensors and x-ray calibration by directly tracking the ultrasound probe in the x-ray images. This approach is designed specifically for TEE probe tracking, as a TEE probe is often within the x-ray field of view during interventional x-ray guidance. An example is shown in Figure 7.8. Although the registration will eventually be used to overlay the echo image on the x-ray projection, the initial requirement is to locate the probe head in the 3-D coordinate space of the x-ray system. Given that the registration is being done via the projection image, this may seem like a backward step. However, knowing the 3-D location means that the registration will still be valid if the angle of the C-arm system is changed. A new overlay can then be generated simply by reprojecting through the echo image at the new angle.

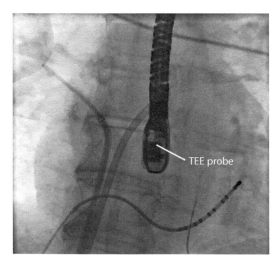

Figure 7.8 A typical x-ray image from an intervention involving TEE. The TEE probe is clearly visible in the x-ray.

The key feature of fiducial-based approaches is to add an external attachment to the TEE probe that contains point markers that are easily identified in x-ray. These markers, rather than the probe itself, are tracked by estimating their pose from the x-ray image and thus determining their position in the 3-D coordinates of the x-ray system. With a biplane x-ray system, it is possible to completely determine the 3-D position of a point marker by backprojecting from the two views (Hawkes et al. 1987). However, due to the limited availability of biplane x-ray systems compared to single-view monoplane systems, the registration must be able to work on monoplane images. A single point backprojected from a monoplane image could be anywhere along the x-ray projection line, but when an object comprises at least three noncollinear points, it is possible to uniquely determine their 3-D location from a single image. This includes their positions along the x-ray projection lines, because the magnification of the object varies with distance from the x-ray source: objects closer to the source appear larger. The primary example of a fiducial-based tracking system using monoplane x-ray is reported by Lang et al. (2012).

As with sensor-based tracking, there is a limit to what can be attached to the TEE probe without risking injury to the patient. One possibility is a closely fitting sleeve that covers the probe head (Lang et al. 2012) as shown in Figure 7.9a. In this example, the fiducial points are seven 1-mm-diameter tantalum beads embedded in a 2-mm-thick layer of silicone, arranged around the head of the probe. The fiducials appear as dark spots in the x-ray image and are distinct enough to be detected automatically (Figure 7.9c).

The registration process (Lang et al. 2012) requires a 3-D model of the fiducial locations. This is obtained from a high-resolution micro-CT scan of the probe head and attachment. It is this 3-D model of the fiducial points that is registered to the x-ray image. Registration involves four steps as follows:

1. Obtain an initial estimate of the model position in 3-D x-ray space.
2. Identify projected fiducial markers in the x-ray image.
3. Determine correspondence between model points and image points.
4. Optimize model pose to match its projected image.

The initial estimate is obtained by approximately matching a reference image, like that in Figure 7.9b, to the x-ray image by intensity-based registration. This requires that the probe be in a similar orientation to the reference image. In a sequence of registrations, this initialization is necessary for the first registration, but subsequent registrations are initialized by the result of earlier registrations in the sequence. Given this initial

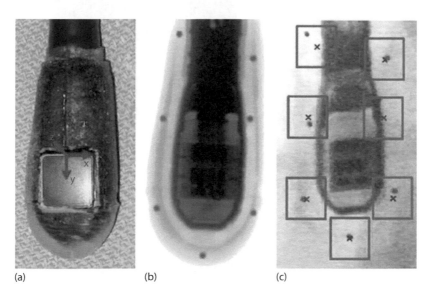

(a)   (b)   (c)

Figure 7.9 (a) Silicone sleeve containing x-ray visible beads. (b) Model x-ray image used for initialization. (c) Example of fiducial detection. (Reprinted with permission from Lang P et al. *IEEE Trans Biomed Eng*, 59, 1444–1453, 2012. Images copyright © 2012 IEEE. All rights reserved.)

estimate of the pose of the 3-D model, Step 2 is achieved by projecting the fiducial locations on the model into the 2-D image. This gives a starting point for finding the fiducial locations in the x-ray image, which are then detected automatically by examining the image in the region of the projected points (Lang et al. 2012). Step 3 requires a feature-pairing algorithm (Pilu 1997). In Step 4, the pose of the 3-D model is then iteratively adjusted so that the projected fiducials from the 3-D model closely match the detected fiducials in the image (Habets et al. 2009). This algorithm, including initialization, is fully automated.

The registration is complicated by the possibility of fiducials being obscured in the x-ray, either by other overlapping devices or by the TEE probe itself. While a minimum of three fiducials is needed to register the 3-D model, more than three will improve the registration accuracy, particularly in the rotation and out-of-plane position (Lang et al. 2012). Additionally, there is the possibility of other objects being incorrectly identified as a fiducial in the x-ray image, which will cause large registration errors. The reported mean alignment error of this system is approximately 1.2 mm in vivo, with tracking failures occurring in less than 1% of images.

Compared to sensor-based systems, the changes in clinical setup required by this approach are minimal, the main addition being the fiducial attachment on the ultrasound probe. The close-fitting sheath with embedded beads was designed as a prototype for a system with beads incorporated into the probe housing. Such a system would operate similarly to the one described herein and would avoid any safety concerns over attachments to the probe inside the esophagus. Furthermore, the registration algorithm requires only minimal variations from the standard interventional protocol, making it a good candidate for clinical integration.

7.3.4 INTENSITY-BASED TEE TRACKING

For each of the systems described so far, the conclusion has been that clinical integration requires as few changes as possible to the imaging hardware, particularly to the ultrasound probe inside the esophagus. This reasoning has led to the final tracking approach, which uses an unmodified TEE probe. This relies on the internal structure of the probe having enough distinct features that its pose can be determined from its projection in the x-ray image.

Similar to the fiducial-based approach, intensity-based tracking requires a 3-D model of the tracked object. This is acquired as a high-resolution nano-CT scan of the TEE probe head (resolution 0.2 mm voxels) shown in Figure 7.10. CT is particularly appropriate in this case because the intensities in the 3-D model are

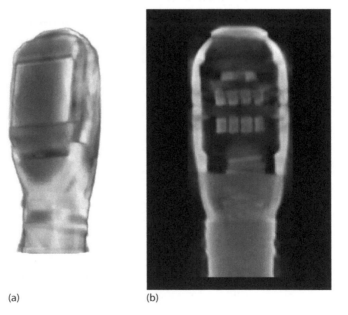

(a)   (b)

Figure 7.10 3-D model of the TEE probe acquired by nano-CT. The figure shows (a) a volume rendering of the model and (b) a slice through the model. The internal structure and electronics are clearly visible in the model.

closely related to the projected intensities of the x-ray image. As with the fiducial-based approach, the model can be registered to either biplane or monoplane images, but monoplane is preferred due to its greater clinical availability.

The registration involves the following steps (Gao et al. 2012):

1. Initialize the 3-D model to a location in 3-D x-ray space.
2. Iteratively adjust the model's pose to match its projected image by
   a. Generating a simulated projection through the 3-D model.
   b. Calculating the similarity of the simulated projection and actual projection.

Step 2 is therefore an iterative optimization process aiming to maximize the similarity of a simulated projection with the real projection. The simulated projection is called a digitally reconstructed radiograph (DRR) and is generated by casting rays through the 3-D CT probe model (Penney et al. 1998) according to the known projection parameters of the x-ray system. The CT image values represent the attenuation coefficient at each point in the tissue. Integrating the values along the rays is a close approximation to the x-ray image formation process and the resulting DRR is similar in appearance to an x-ray image. A significant difference between the x-ray and DRR image generation is the presence of other structures in the x-ray that do not appear in the 3-D CT probe image. These include some variations due to soft tissue, but also other interventional devices that may overlap with the probe projection. The most appropriate similarity measures are those that are robust to both slowly varying soft tissue differences and small-scale differences caused by narrow interventional devices (Penney et al. 1998).

The generation of DRRs is a computationally expensive process. In an iterative optimization, many DRRs will need to be generated as different poses of the probe model are examined. The registration process would take several minutes on an ordinary central processing unit, which is inadequate for tracking real-time probe movements. A solution is to implement the algorithm in a graphical processing unit, which allows certain computations to be speeded up significantly. In this way, the generation of a DRR is reduced to less than 10 ms (Gao et al. 2012), and in current systems, the complete registration takes approximately 0.5–1 s (Housden et al. 2012).

The initialization (Step 1 of the registration algorithm above) is the main obstacle to clinical convenience. The algorithm (Gao et al. 2012) currently requires an approximate manual initialization of the probe pose, within 10 mm and 10° of its correct pose. In a real-time sequence of images, subsequent registrations are initialized automatically by the most recent registration result, on the assumption that changes in probe position will be small from one image to the next. More recent developments in this area have extended the initialization range to 40 mm and 40° by registering at multiple resolution levels, so that the probe position is initially estimated only approximately and then refined (Kaiser et al. 2013).

While manual initialization in this range is certainly possible in clinical practice, this is not ideal and has led to the development of an alternative, fully automated algorithm (Mountney et al. 2012; Heimann et al. 2013). The key feature of this algorithm is to determine the registration in several steps rather than as a global optimization of the six degrees of freedom of the pose. This process is illustrated in Figure 7.11. First, the two in-plane position parameters are detected, followed by the in-plane rotation, then the out-of-plane position, and finally the two out-of-plane rotation parameters. All but the last of these steps is determined using a discriminative learning approach, with the classifiers becoming gradually more discriminative at each step as the pose parameters become known. The out-of-plane rotations have a more complex effect on the projection and are instead determined by matching the projection image to a collection of template images for different out-of-plane rotations. The registration takes 0.5 s, a similar time to the DRR approach.

The average accuracy of echo image to x-ray alignment using the DRR approach is 2.9 mm for in vivo images (Housden et al. 2012). The fully automated approach is a more recent development and it remains to be seen how it will perform clinically. In either case, by needing no modification of the TEE probe, these methods are the easiest to integrate into a clinical procedure.

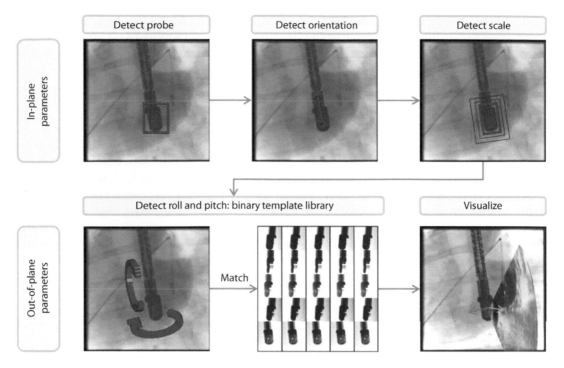

Figure 7.11 Fully automatic detection of TEE probe position and orientation in x-ray. (With kind permission from Springer Science+Business Media: *Medical Image Computing and Computer-Assisted Intervention–MICCAI*, Ultrasound and fluoroscopic images fusion by autonomous ultrasound probe detection, 7511, 2012, 544–551, Mountney P, Ionasec R, Kaizer M et al., Figure 1, Image copyright © Springer.)

7.3.5 CLINICAL INTEGRATION

In addition to the probe modification and hardware requirements mentioned previously, each of the methods has certain requirements in order to integrate them into a clinical intervention. Since the objective is to provide an additional visualization with minimal change to the existing clinical protocol, the requirements are mainly in data flow rather than hardware. Common to all methods is the need for a personal computer (PC) separate to the x-ray and echo systems on which to receive data from both modalities, perform the registration, and generate the hybrid visualization. Each modality is therefore required to stream live images to the PC. The echo modality may be imaging in 3-D, which will require a high data rate. One solution is to stream the line data of the echo image, which is sampled less densely than the final voxel array, and then reconstruct the echo volume on the PC. However, the reconstruction requires knowledge of the echo system that will likely only be available to the manufacturer.

Apart from the image data itself, the data streams need to contain geometrical information on the imaging modalities. Specific requirements are the current angle of the x-ray C-arm, the x-ray projection geometry, the position of the x-ray table if it moves during the procedure, and the current size and steering settings of the echo volume for calibration (Section 7.3.1). Accessing the images and geometric information may require modification of the standard imaging hardware.

In terms of suitability of the various registration techniques for clinical integration, they have been presented in Sections 7.3.2 through 7.3.4 in order of decreasing modification of the hardware, which is likely to simplify clinical integration and acceptance. This is particularly important when considering modifications to the TEE probe, because safety, as well as convenience, is a concern. For these reasons, the only commercial system currently available for hybrid x-ray-echo imaging (EchoNavigator, Philips Healthcare) uses the intensity-based DRR approach with an unmodified TEE probe. However, while the intensity-based approaches have the benefit of the simplest clinical integration, it should be noted that there is a different trend in achievable accuracy: while the intensity-based accuracy is acceptable, the other methods do better, with the best accuracy obtained using fiducial registration (Lang et al. 2010). For this reason, future developments may produce clinical systems from the other registration approaches. The fiducial method in particular is a good candidate, providing superior accuracy and requiring almost as little hardware modification as the intensity-based methods.

7.4 ACCURACY AND SOURCES OF ERROR

Ultimately, a hybrid x-ray-echo system aims to provide a visualization in which cardiac structures are perfectly aligned in the two modalities and the visualization is up to date with the moving heart. Potential sources of error in this alignment include registration inaccuracy, low imaging frame rates, and temporal synchronization errors between the two modalities.

Temporal errors arise from low frame rate imaging used to image the continuous movement of the heart. This can mean that the information displayed at any moment is slightly out of date. Fortunately, x-ray and echo are both real-time modalities able to run at a sufficiently high frame rate for cardiac imaging. As such, guidance errors of this type are minimal and are no worse than those encountered in established x-ray-only or echo-only guidance. Synchronization of the two modalities is potentially a greater problem unique to hybrid imaging in which the two image streams experience different lags. One example of measuring the relative lag has found it to be not more than 100–200 ms, or 10–20% of the cardiac period (Housden et al. 2012).

Regarding spatial alignment of the two modalities, the required accuracy depends to some extent on the application (Linte et al. 2012), although alignment within 5 mm is usually taken as an acceptable target for a useful interventional guidance system. Each of the systems described in this chapter meets this target, in some cases even when measuring the overall system accuracy in a clinical setting. This is promising for the eventual use of these systems in routine clinical practice.

7.5 APPLICATION EXAMPLES

Hybrid x-ray-echo imaging is potentially useful in any intervention where a device or catheter must be directed to specific locations on the anatomy. The anatomy is not easily seen in x-ray, making the navigation difficult using x-ray alone. Echo is capable of imaging the anatomy and therefore enhances the navigation of devices through complex vasculature. It is also a live modality and so is suited to applications requiring navigation to moving targets, particularly in the heart.

A commonly cited application of hybrid imaging is to guide aortic valve replacement, in which a prosthetic aortic valve is implanted in place of the original valve. This procedure is necessary in patients suffering from severe aortic stenosis, due to calcification of the valve. In patients where standard surgery has a high risk, a minimally invasive alternative is transcatheter aortic valve implantation (TAVI) (Thomas 2009; Walther et al. 2009; Wendler et al. 2010). The procedure is to deliver the prosthetic valve via a catheter, either along the aorta or through an incision in the apex of the heart. The main risk in this procedure is in positioning the new valve; if placed incorrectly, it could leak or block the coronary arteries. The aortic root anatomy is difficult to see in x-ray (Figure 7.12a), making it very difficult to correctly align the valve to the anatomy using x-ray alone. Usually, TEE imaging is also available and is directed at the aortic root and existing valve for additional guidance. It is therefore a small step to introduce hybrid x-ray-echo imaging using one of the previous techniques (Gao et al. 2012; Housden et al. 2012; Lang et al. 2011, 2012). The benefit of the hybrid view is the visualization of the device and anatomy in echo, but guided by the x-ray image, which provides both a standard coordinate system in which to interpret the echo and a clearer image of the prosthetic valve (Figure 7.12).

Similar benefits have also been suggested in electrophysiology procedures involving the left atrium (Gao et al. 2012; Housden et al. 2012). Electrophysiology procedures aim to measure and correct abnormalities in the electrical activity of the heart. When the procedure involves catheters in the left atrium, the catheters must first enter the right atrium via the vena cava and then pass through the interatrial septum, via a trans-septal

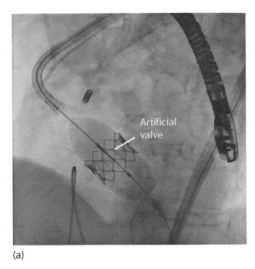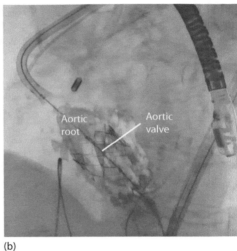

(a)   (b)

Figure 7.12 Example overlays from a TAVI intervention. (a) The x-ray clearly shows the interventional device, in this case an artificial valve. (b) The overlay adds the anatomy of the aortic root to the view.

puncture, to access the left atrium. Figure 7.2 is an example of the hybrid visualization imaging devices across the septum. The trans-septal puncture is usually guided by TEE to reduce the risk of puncturing the wrong part of the heart (Earley 2009). Accurate registration of x-ray and echo images is therefore expected to assist in this step by enhancing the echo with x-ray containing a clear image of the puncture needle.

In addition to imaging structural information, echo is useful for measuring cardiac function. A proposed use for hybrid x-ray-echo is therefore to observe the effect of a procedure to improve cardiac function. A specific example is pacing wire placement in pacemaker implantation (Ma et al. 2010). The location of the pacing wires affects the cardiac output. Cardiac output is measured using real-time 3-D echo, which provides live feedback on the effect of pacing wire location, and this allows the pacing wire placement to be optimized during the procedure.

These three examples highlight some potential application areas of hybrid x-ray-echo, but the primary area of interest is in repair of structural heart disease, of which aortic valve replacement is an example. Although current research has concentrated on aortic valve replacement, the hybrid guidance system is potentially useful in many structural heart disease procedures where a device must be guided to a specific part of the anatomy. It is likely that future trials will begin to look at using hybrid x-ray-echo in a range of these applications.

REFERENCES

Earley MJ. 2009. How to perform a transseptal puncture. *Heart* 95 (1): 85–92.

Gao G, G Penney, Y Ma et al. 2012. Registration of 3D trans-esophageal echocardiography to x-ray fluoroscopy using image-based probe tracking. *Med Image Anal* 16 (1): 38–49.

Habets DF, SI Pollmann, X Yuan, TM Peters, and DW Holdsworth. 2009. Error analysis of marker-based object localization using a single-plane XRII. *Med Phys* 36 (1): 190–200.

Hawkes DJ, ACF Colchester, and CR Mol. 1987. The accurate 3-D reconstruction of the geometric configuration of vascular trees from x-ray recordings. In *Physics and Engineering of Medical Imaging*, ed. by R. Guzzardi, 119:250–256. NATO ASI Series.

Heimann T, P Mountney, M John, and R Ionasec. 2013. Learning without labeling: Domain adaptation for ultrasound transducer localization. In *Medical Image Computing and Computer-Assisted Intervention–MICCAI*, ed. by K. Mori, I. Sakuma, Y. Sato, C. Barillot, and N. Navab, 8151:49–56. Lecture Notes in Computer Science.

Housden RJ, A Arujuna, Y Ma et al. 2012. Evaluation of a real-time hybrid three-dimensional echo and x-ray imaging system for guidance of cardiac catheterisation procedures. In *Medical Image Computing and Computer-Assisted Intervention–MICCAI*, ed. by N. Ayache, H. Delingette, P. Golland, and K. Mori, 7511:25–32. Lecture Notes in Computer Science.

Jain A, L Gutierrez, and D Stanton. 2009. 3D TEE registration with x-ray fluoroscopy for interventional cardiac applications. In *Functional Imaging and Modeling of the Heart–FIMH*, ed. by N. Ayache, H. Delingette, and M. Sermesant, 5528:321–329. Lecture Notes in Computer Science.

Kaiser M, M John, A Borsdorf, A Nöttling, and T Neumuth. 2013. A hybrid optimization approach for 2D–3D registration based fusion of ultrasound and x-ray. *Biomed Eng/Biomed Tech* 58 (Sl-1 Track-L).

Lang P, P Seslija, D Bainbridge et al. 2011. Accuracy assessment of fluoroscopy-transesophageal echocardiography registration. In *Proceedings of SPIE Medical Imaging: Visualization, Image-Guided Procedures, and Modeling*, ed. By K. H. Wong and D. R. Holmes III, vol. 7964, 79641Y.

Lang P, P Seslija, MWA Chu et al. 2012. US-fluoroscopy registration for transcatheter aortic valve implantation. *IEEE Trans Biomed Eng* 59 (5): 1444–1453.

Lang P, P Seslija, DF Habets, MWA Chu, DW Holdsworth, and TM Peters. 2010. Three-dimensional ultrasound probe pose estimation from single-perspective x-rays for image-guided interventions. In *Medical Imaging and Augmented Reality–MIAR*, ed. by H. Liao, P. J. Edwards, X. Pan, Y. Fan, and G.-Z. Yang, 6326:344–352. Lecture Notes in Computer Science.

Linte CA, P Lang, ME Rettmann et al. 2012. Accuracy considerations in image-guided cardiac interventions: Experience and lessons learned. *Int Comput Assist Radiol Surg* 7 (1): 13–25.

Ma Y, GP Penney, D Bos et al. 2009. Evaluation of a robotic arm for echocardiography to x-ray image registration during cardiac catheterization procedures. In *Proceedings of the Annual International Conference of the IEEE Engineering in Medicine and Biology Society–EMBC*, 5829–5832.

Ma Y, GP Penney, D Bos et al. 2010. Hybrid echo and x-ray image guidance for cardiac catheterization procedures by using a robotic arm: A feasibility study. *Phys Med Biol* 55 (13): N371–N382.

Mercier L, T Langø, F Lindseth, and DL Collins. 2005. A review of calibration techniques for freehand 3-D ultrasound systems. *Ultrasound Med Biol* 31 (4): 449–471.

Moore JT, AD Wiles, C Wedlake et al. 2010. Integration of transesophageal echocardiography with magnetic tracking technology for cardiac interventions. In *Proceedings of SPIE Medical Imaging: Visualization, Image-Guided Procedures, and Modeling*, ed. by K. H. Wong and M. I. Miga, vol. 7625, 76252Y.

Mountney P, R Ionasec, M Kaizer et al. 2012. Ultrasound and fluoroscopic images fusion by autonomous ultrasound probe detection. In *Medical Image Computing and Computer-Assisted Intervention–MICCAI*, ed. by N. Ayache, H. Delingette, P. Golland, and K. Mori, 7511:544–551. Lecture Notes in Computer Science.

Penney GP, J Weese, JA Little, P Desmedt, DLG Hill, and DJ Hawkes. 1998. A comparison of similarity measures for use in 2-D-3-D medical image registration. *IEEE Trans Med Imaging* 17 (4): 586–595.

Pilu M. 1997. A direct method for stereo correspondence based on singular value decomposition. In *Proceedings of IEEE Computer Society Conference on Computer Vision and Pattern Recognition–CVPR*, 261–266.

Rabben SI. 2011. Technical principles of transthoracic three-dimensional echocardiography. Chap. 2 in *Textbook of Real-Time Three Dimensional Echocardiography*, ed. by L. Badano, R. M. Lang, and J. L. Zamorano, 9–24. Springer London.

Rhode KS, M Sermesant, D Brogan et al. 2005. A system for real-time XMR guided cardiovascular intervention. *IEEE Trans Med Imaging* 24 (11): 1428–1440.

Salgo IS. 2011. 3D transesophageal echocardiographic technologies. Chap. 3 in *Textbook of Real-Time Three Dimensional Echocardiography*, ed. by L. Badano, R. M. Lang, and J. L. Zamorano, 25–32. Springer London.

Thomas M. 2009. Trans-catheter aortic valve implantation in the United Kingdom: NICE guidance. *Heart* 95 (8): 674–675.

Tzifa A, GA Krombach, N Krämer et al. 2010. Magnetic resonance-guided cardiac interventions using magnetic resonance-compatible devices: A preclinical study and first-in-man congenital interventions. *Circ Cardiovasc Interv* 3 (6): 585–592.

Walther T, T Dewey, MA Borger et al. 2009. Transapical aortic valve implantation: Step by step. *Ann of Thorac Surg* 87 (1): 276–283.

Wendler O, T Walther, P Nataf et al. 2010. Trans-apical aortic valve implantation: Univariate and multivariate analyses of the early results from the SOURCE registry. *Eur Cardio thorac Surg* 38 (2): 119–127.

Combined ultrasound and photoacoustic imaging

DOUG YEAGER, ANDREI KARPIOUK, NICHOLAS DANA,
AND STANISLAV EMELIANOV

8.1 INTRODUCTION

The use of ultrasound (US) for medical diagnostics has become a diverse tool for clinicians since its initial introduction in the early twentieth century. While Karl Dussik first applied ultrasonography for medical diagnoses in the 1940s (Dussik 1942), it was a decade later before Edler and Hertz performed the first successful echocardiogram (Edler and Hertz 2004), followed by the clinical introduction of Doppler US as a means of assessing blood flow by the mid-1960s (Franklin, Schlegel, and Rushmer 1961). Today, US imaging is a widely utilized clinical imaging modality, used to assess cardiovascular anatomical features and hemodynamic function, using both noninvasive and minimally invasive US transducer probes. Common noninvasive applications include imaging of the intima-media thickness of carotid arteries and their Doppler flow characteristics, as well as transthoracic echocardiography for assessment of the heart. Invasive imaging procedures include transesophageal echocardiography, for improved imaging of the heart and Doppler flow through its different chambers, and intravascular US (IVUS), during which a US catheter is navigated into coronary arteries to image atherosclerotic plaques and guide percutaneous interventions. While US imaging has proven to be a powerful diagnostic modality in the field of cardiovascular medicine, with the exception of Doppler assessment of hemodynamic function, conventional US imaging techniques are ill-suited for molecular or functional imaging, due in large part to its inherently low contrast between soft-tissue constituents. To that end, recent biomedical research has seen a dramatic expansion in the field of combined US and photoacoustic (PA) imaging, two modalities that utilize shared hardware to provide complimentary information about target tissues.

PA imaging, also referred to as optoacoustic imaging or, more generally, thermoacoustic imaging, relies on the detection of pressure waves generated as a result of rapid thermal expansion following absorption of optical pulses, an effect analogous to thunder and lightning. Although the PA effect was initially reported by Alexander Graham Bell in 1880 (Bell 1880), widespread research into its applications did not commence until after the development of pulsed optical sources, particularly the extensive development of laser-based technologies, in the 1960s and 1970s. Even then, early PA applications predominantly focused on spectral characterization of gases or thin film solids and liquids (Kinney and Staley 1982; Rosencwaig 1978). However, it was more than a decade later before the PA effect was first applied for soft-tissue imaging (Bowen 1981) and began to be introduced as a biomedical imaging modality (Kruger et al. 1995; Oraevsky et al. 1994). Since that time, basic and preclinical research within the field of PA imaging has rapidly expanded, now encompassing numerous techniques and applications of relevance to cardiovascular medicine.

PA imaging offers several unique characteristics that make it an attractive modality and have prompted the recent efforts for its development as both a clinical and preclinical modality. First, PA imaging is capable of enabling signal detection with optical-based contrast at comparable resolution and depth to that which is achieved using US imaging. Most optical imaging techniques are extremely sensitive to optical scattering of photons, and as a result, the achievable imaging depths within highly scattering biological tissues are typically less than 1 mm. However, PA imaging requires only one-way travel of light and is dependent only on optical absorption for signal generation. Therefore, while optical scattering impacts the penetration and distribution of light in tissues, PA imaging is less sensitive to scattering than other optical techniques, thus enabling imaging depths up to several centimeters. Additionally, due to the dependence of PA signal generation on optical absorption, the modality is capable of providing endogenous tissue characterization through the careful selection of one or more optical wavelengths to differentiate chromophores of interest. Spectroscopic PA (sPA) imaging is often utilized as a method to differentiate oxy- and deoxy-hemoglobin or lipid-rich tissue based on their unique optical absorbance characteristics. Functional imaging is also feasible by determining oxygen saturation based on relative ratios of oxygenated and deoxygenated hemoglobin, again utilizing differences in the optical absorbance of the two states of hemoglobin. Finally, PA imaging is an attractive tool for expanded cellular and molecular-specific imaging, which can be achieved using targeted exogenous contrast agents, including a wide range of plasmonic metal nanoparticles, dyes, and carbon nanotubes, each with unique optical properties. Each of these techniques and applications will be discussed in more detail throughout this chapter.

Due to the fact that the PA effect relies on the generation of pressure waves, the resulting signal detection can be achieved using conventional US transducers. Therefore, while the information gained from US and PA imaging modalities is fundamentally different—US signals arising from backscatter generated by adjacent tissues with acoustic impedance mismatches and PA signals arising from optical absorption—PA and US imaging can be performed using shared hardware. To that end, combined US and PA imaging has found particular merit as a hybrid imaging modality for cardiovascular applications, with PA imaging enabling optical absorption-based imaging of tissue compositional and functional properties to compliment US-based anatomical assessment of tissue.

This chapter provides an overview of fundamental principles and available hardware for PA imaging. Different combined US and PA imaging techniques and applications that have been implemented for cardiovascular imaging will then be introduced. Finally, remaining challenges that must be addressed before US and PA imaging will be translated into the clinic as a hybrid technique are discussed.

8.2 PRINCIPLES OF PA IMAGING

At a high level, PA imaging is achieved through the delivery of temporally modulated light to induce measureable pressure waves within the targeted tissue. PA imaging therefore utilizes an optical source and a transducing receiver. The interplay between optical excitation and US detection permits a wide range of PA imaging characteristics in terms of the obtained informational content (e.g., tissue composition, temperature maps) and achievable spatial and temporal resolution. This section provides an overview of PA signal generation and the attainable resolution of PA imaging.

8.2.1 PA PRESSURE GENERATION

PA imaging relies on absorption of optical energy, subsequent thermal expansion of the absorber, and the generation of a resultant pressure wave, which can be detected and localized using acoustic imaging techniques. The application of modulated electromagnetic radiation, such as pulsed laser illumination, can lead to the generation of acoustic waves via several distinct processes, including dielectric breakdown, vaporization, thermal expansion, electrostriction, and radiation pressure (Sigrist 1986; Tam 1986). The vast majority of biomedical applications of PA imaging utilize the mechanism of thermal expansion, a process by which optical absorption from a pulsed laser source induces a mechanical expansion and relaxation of the absorbing chromophore (Beard 2011; Bowen 1981).

In order to achieve highly efficient PA signal generation, both the thermal and stress confinement conditions should be satisfied.

$$\tau_0 \ll \tau_{th} = \frac{d_c^2}{4\alpha_{th}} \tag{8.1}$$

$$\tau_0 \ll \tau_s = \frac{d_c}{v_s} \tag{8.2}$$

Thermal and stress confinement conditions are defined by Equations 8.1 and 8.2, respectively, where τ_0 is the optical pulse duration, τ_{th} is the thermal diffusion time, τ_s is the stress relaxation time, d_c is a characteristic dimension of the excited region (determined by the size of an absorber or the penetration depth of light within an absorber), α_{th} is the thermal diffusivity, and v_s is the speed of sound (Gusev and Karabutov 1993). In effect, under thermal confinement, the laser pulse duration should be significantly shorter than the thermal diffusion time, thus meaning that thermal dissipation from the absorbing region is negligible during the excitation pulse and PA transient generation process. Similarly, under stress confinement, propagation of the acoustic wave, which is generated as a result of the mechanical expansion and relaxation, can be considered to be negligible over the duration of the optical pulse. While α_{th} and v_s are tissue-dependent and temperature-dependent parameters, typical values for soft tissues are on the order of approximately 1.3×10^{-3} cm²/s and 1.5 mm/s, respectively (Walsh 2011). As an example, to resolve an absorber with a characteristic dimension of 100 μm, τ_{th} and τ_s would be approximately 19 ms and 67 ns, respectively, meaning that efficient PA signal generation is achieved if the pulse duration is less than 67 ns. In practice, the confinement conditions are satisfied for absorbers on this size scale through the use of nanosecond pulsed laser sources (<10 ns/pulse) for PA imaging of endogenous biological tissues. Examples and descriptions of such sources are provided in Section 8.3.

Under thermal confinement, τ_{th}, the PA pressure, p, generated within a nonviscous and acoustically homogenous medium is described by Equation 8.3 (Diebold and Sun 1994, Li and Wang 2009):

$$\nabla^2 p - \frac{1}{v_s^2}\frac{\partial^2}{\partial t^2}p = -\frac{\beta}{C_p}\frac{\partial}{\partial t}H(r,t) \tag{8.3}$$

where the heating function, $H(r,t)$, represents a spatial position, r, and time, t, dependent conversion of the applied electromagnetic radiation to thermal energy; β represents the isobaric volume expansion coefficient; and C_p represents the isobaric specific heat capacity. Considering a pulsed optical illumination heating source, the heating function, $H(r,t)$, can be expressed in terms of the local optical absorption coefficient of the medium, μ_a, and the local laser fluence, Φ, by

$$H(r,t) = \mu_a(r)\Phi(r,t) \tag{8.4}$$

Further assuming that the optical source satisfies the stress confinement condition, τ_s, the temporal component of the heating function can be approximated as a delta function, δ, effectively isolating the spatial dependency of the heating profile by

$$H(r,t) = H(r)\delta(t) \tag{8.5}$$

Therefore, given sufficiently short laser pulses and a large enough spot size to achieve both thermal and stress confinement conditions, the peak PA pressure, p_o, is generated as defined by Equation 8.6,

$$p_0 = \frac{\beta v_s^2}{C_p}H(r) = \frac{\beta v_s^2}{C_p}\mu_a(r)\Phi(r) = \Gamma\mu_a(r)\Phi(r) \tag{8.6}$$

It can be seen that the generated pressure is defined by only the local optical fluence, Φ, the local optical absorption coefficient, μ_a, and the Grüneisen parameter, Γ, which is itself comprised of the isobaric volume expansion coefficient, the speed of sound, and the isobaric specific heat capacity (Grüneisen 1912).

In the case of imaging biological tissues with nanosecond pulsed lasers, Equation 8.6 is widely utilized as the fundamental, albeit simplified, PA equation, whereby the generated PA signal magnitude is dependent on only three local parameters: light delivery, optical absorption, and efficiency of converting the absorbed energy into a pressure wave. More comprehensive derivations of PA pressure generation are beyond the scope of this chapter but can be found elsewhere (Diebold and Sun 1994; Diebold 2009). The light delivery can be controlled, in part, by adjusting the delivered light energy. Typically, PA differentiation of specific absorbers of interest is achieved through the careful selection of imaging wavelength(s), which correspond to optical absorption peaks of the desired imaging target while reducing background signal and light attenuation. Such optical wavelength selection is highly tissue and target specific, as discussed in Section 8.4.

While PA imaging of endogenous tissues has several key applications within cardiovascular medicine, preclinical research also relies heavily on the utilization of exogenous absorbers for improved molecular imaging capabilities and enhanced contrast. PA contrast agents under investigation include dyes, carbon nanotubes, and metallic nanoparticles, each of which is discussed in more detail in Section 8.4.4 of this chapter. In the case of utilizing these nanometer-sized agents for PA imaging, however, it is important to note that the previous equations describing generated PA pressure must be reconsidered because the thermal and stress confinement conditions are often not maintained using optical sources most commonly used for PA imaging. This ultimately means that, despite the high absorption, both the thermal confinement and stress confinement conditions may not be valid when using a nanosecond pulsed laser for nanoparticle-enhanced PA imaging applications. Under these conditions, it has been shown that the applied laser pulse induces heat within the absorbing particle, the heat rapidly diffuses throughout the nanoparticle–tissue interface into the surrounding environment where the acoustic pressure wave is induced (Chen et al. 2012). This indicates that both the optical properties of the contrast agent and the efficiency of the heat transfer into the surrounding environment will significantly affect the generated acoustic pressure. To account for the lack of stress and thermal confinement in the case of imaging nanoparticle contrast agents, the simplified PA pressure equation can be modified as indicated in Equation 8.7:

$$p_0 \propto \eta \Gamma_{eff} \sigma N \Phi \tag{8.7}$$

where η is the heat transfer efficiency from the particle to its surroundings, Γ_{eff} is an effective Grüneisen parameter of the nanoparticle-environment mixture, σ is the absorption cross-section of the nanoparticle, and N represents the number of nanoparticles per unit volume (Chen et al. 2013).

Additionally, whereas variations in optical absorption spectra of either endogenous absorbers or exogenous contrast agents provide the most common means of PA imaging tissue characterization, it is important to note that the Grüneisen parameter offers a separate means for PA imaging and sensing. Because the terms comprising the Grüneisen parameter are tissue and temperature specific, PA imaging performed at a single optical wavelength, and at a constant optical fluence, can also be utilized both to differentiate tissue types and to monitor temperature change, thus providing a potential means of therapy guidance.

8.2.2 IMAGING DEPTH AND SPATIAL AND TEMPORAL RESOLUTION OF PA IMAGING

Presently, the temporal resolution or frame rates of PA imaging systems are generally limited by the pulse repetition frequencies of their optical sources, which are highly dependent on the characteristics of the laser and can range from tens of Hertz (Hz) for tunable laser systems to several kilo-Hertz (kHz) for fixed wavelength lasers. In US imaging, on the other hand, US pulsers are capable of acoustic pulse repetition frequencies of several kHz enabling imaging in real-time.

The majority of PA imaging applications within cardiovascular medicine utilize the delivery of optical excitation pulses with subsequent detection using focused US beam. This approach results in an achievable spatial resolution that is comparable to that of conventional US imaging. Due to strong light scattering in soft tissues, the optical spot size of the laser beam in deep tissues is larger than the acoustic beam profile, thus making the spatial resolution primarily determined by the beam profile of the US transducer used to receive the PA pressure waves. When using a nanosecond pulsed excitation source, broadband acoustic transients with frequency content up to several tens of mega-Hertz (MHz) can be induced, depending on the physical dimensions of the absorbing source. Therefore, the spatial resolution is primarily dependent on frequency-dependent acoustic attenuation (Beard 2011). In addition, specific values for spatial resolution vary depending on transducer characteristics, including whether it is a single element or array transducer, the frequency bandwidth, focusing and the numerical aperture. However, transducers utilized for US imaging of biological tissues typically provide resolution, both axial and lateral, on the order of tens to hundreds of micrometers.

One major exception in the determining factor of PA spatial resolution is optical resolution PA microscopy (OR-PAM). With this technique, the optical beam profile is tightly focused into a small PA excitation region, resulting in an improved achievable resolution. OR-PAM is restricted to shallow penetration depths, less than approximately 1 mm, in which the focused photons remain within the ballistic regime prior to the high optical scattering of biological tissues. Within these conditions, the lateral resolution is limited by

Table 8.1 Common US and PA imaging parameters

Parameter	Transducer frequency (MHz)	Axial resolution (μm)	Lateral resolution (μm)	Imaging depth (mm)
Ultrasound	5	150	300	70
	20	75	165	30
Macroscopic PA	5	150	300	40
Microscopic PA	50	15	50	3
Optical Resolution PA	75	15	5	0.7

Source: Reprinted from *Trends Biotechnol*, 29, Mallidi, S., Luke, G.P., Emelianov, S., Photoacoustic imaging in cancer detection, diagnosis, and treatment guidance, 213–221, Copyright 2011, with permission from Elsevier.

acoustic diffraction and can be as little as a few micrometers, while axial resolution remains limited by acoustic properties of the system (Beard 2011).

Spatial resolution and imaging depth represent a tradeoff for US and PA imaging applications, whereby higher-frequency transducers typically provide high-resolution US images at the cost of reduced imaging depth, and vice versa. However, US imaging depths are generally on the order of 1 to 10 cm. As a result, the imaging depth achievable for PA imaging is largely dependent on the sensitivity of the signal detection and the delivery of sufficient optical energy to deeper tissues to generate a measureable signal. To this end, light penetration is a critical parameter for PA imaging depth and can be improved by increasing the optical fluence and imaging at optical wavelengths in which the background provides minimal scattering and absorption, such as the near-infrared (NIR) tissue optical window. In fact, through careful selection of the excitation wavelength, PA imaging of vasculature has been demonstrated at a depth of up to 4 cm *in vivo* (Kruger et al. 2010), while *ex vivo* studies have achieved depths of greater than 5 cm with the help of exogenous contrast for signal enhancement (Ku and Wang 2005). Table 8.1 summarizes the tradeoffs between different US and PA imaging techniques, including transducer frequency, axial and lateral resolution, and imaging depth.

8.3 PA IMAGING INSTRUMENTATION AND SYSTEMS

Given the similarities in the generated signals between PA and US imaging, much of the instrumentation required for combined US/PA imaging can be shared by the two modalities. In terms of hardware, the fundamental difference between the two modalities is the need for a modulated optical source to generate PA pressure. Beyond that, all components associated with receiving and processing pressure waves can potentially be shared between both US and PA imaging, including transducers, preamplifiers, analog-to-digital converters (ADC), and signal processing software. This ability to share instrumentation results in relatively low overall costs for the hybrid imaging systems as well as straightforward means for obtaining coregistered US/PA images.

Applications for US/PA imaging within cardiovascular medicine have utilized several different types of systems, ranging from preclinical imaging systems capable of noninvasive full-body imaging to a more clinically translatable, catheter-based system for imaging arterial tissue. Regardless of the embodiment, each of the systems requires an optical source capable of generating PA signals within the tissue to be imaged and a transducer to receive the incoming signals. During PA imaging, the transducer can simply passively receive incoming signals; then, for US imaging, the same transducer can be used for pulse-echo signal detection. Once received, the US and PA signals are digitized, appropriately filtered, and processed for image reconstruction. The remainder of this section highlights requirements of PA instrumentation for systems of particular relevance to cardiovascular imaging.

8.3.1 OPTICAL EXCITATION

Currently, a variety of optical sources suitable for PA imaging is available. Potential sources include different types of pulsed lasers and other laser-based systems as well as nonlaser, broadband light sources such as light-emitting diodes and flash lamps. Depending on system- and application-specific parameters, requirements for optical delivery are often more stringent and limit the potential sources that can be used for PA imaging. Critical parameters that affect the efficacy and imaging frame rate for a specific PA imaging application and system include the pulse duration, the maximum pulse energy, the achievable wavelength or spectral range of optical output, and the pulse repetition frequency (PRF).

The appropriate choice of pulse duration for a PA light source is affected by limitations to the maximum fluence, which is itself wavelength dependent, that can safely be delivered as well as thermal and stress confinement requirements discussed in the previous Section 8.2 (American National Standards Institute [ANSI] 2007). Ideally, the pulse duration should be chosen as short as possible to ensure that both thermal and stress confinements are maintained during PA signal generation. However, excessive shortening of optical pulse duration would require a corresponding increase in light intensity in order to deliver the same energy, which could potentially cause damage to either the light delivery system or the tissue (Noack and Vogel 1999). Typically, 3–7-ns pulse durations are used for PA imaging. In the case of imaging endogenous tissue absorbers, both thermal and stress confinements are maintained. Note, however, that both stress and thermal confinements are not satisfied if nanoparticle-based contrast agents are used, but PA imaging with the same optical sources is still feasible as long as the modified equation describing the generated PA pressure is considered (Equation 8.4).

For applications other than OR-PAM, the light source utilized for PA signal generation should provide pulsed light with necessary energy per pulse to generate detectable PA transients throughout the field of view of the receiving transducer. Exact values necessary to satisfy this condition vary greatly depending on the imaging system and application but generally fall on the order of tens of millijoules (mJ) or less. It is also important to consider the effect that the anticipated imaging environment may have on the propagation of light. For example, when imaging from within or through major blood vessels or the heart, the presence of blood as well as other soft tissues will attenuate light due to significant light scattering and absorption, the extent of which is highly wavelength dependent (Jacques 2013; Roggan et al. 1999).

The output wavelength, or range of wavelengths, required of the optical source is also dependent on the PA signal source(s) of interest. Briefly, a few potential PA imaging targets that have been investigated are metallic stents, lipid-rich tissue, and customized contrast agents, each of which is discussed in later sections of this chapter. In the case of imaging the stent, the choice of a wavelength to use for PA imaging is relatively flexible due to the wide absorption spectrum of metals and their very high light absorption coefficient relative to endogenous tissue (Karlsson and Ribbing 1982). If attempting to image lipid, however, the wavelengths that are required of the optical source are limited to a few select optical bands, which correspond to regions of high absorption of lipid relative to other water-based tissue types. The spectral bands tend to be broad, on the order of tens of nanometers or more, meaning that the optical source could be either monochromatic or more broadband.

As with the pulse energy and output wavelength, the PRF of the optical source is also highly dependent on the other system components and the intended application. From a high level, if real-time PA imaging is desired, a PRF of at least 30 would be needed. This is indeed the case for PA imaging systems in which mechanical scanning is not required for two-dimensional imaging, provided that the illumination is powerful enough to excite PA signals throughout the imaging region. On the other hand, if mechanical scanning is required, then this requirement can quickly increase, to the point where lasers with a PRF of several kHz may be required to achieve real-time PA imaging. In reality, the majority of PA imaging systems developed to date do not yet provide real-time imaging, and the PRF of the optical sources is usually the limiting factor.

One of the most widely used laser sources is a diode-pumped solid state laser utilizing different crystal matrices doped by Nd^{3+} ions and operated in a Q-switched mode, the most common example being the yttrium aluminum garnet-doped matrix. Q-switched neodymium-doped yttrium aluminum garnet (Nd:YAG) lasers are common PA optical sources (Eleccion 1972). They emit with pulse durations of less than 10 ns at a fundamental wavelength of 1064 nm and can easily be frequency doubled, tripled, and quadrupled to obtain 532, 355, and 266 nm, respectively. To generate these short-duration and high-energy pulses at other wavelengths within the visible and NIR spectral ranges, Nd:YAG lasers can be used to pump, for example, optical parametrical oscillators (OPOs) (Nikogosyan 1991). Typically, OPOs are capable of generating output wavelengths across a wide spectral range (e.g., 680 nm to > 2000 nm), thus providing versatility for preclinical PA imaging systems in which wavelength tunability remains critical for differentiation of target chromophores.

8.3.2 ACOUSTIC DETECTION

While Equations 8.6 and 8.7 describe the PA pressure generated at the location of the source, attenuation of the US wave as it propagates through biological tissue toward the receiving transducer should also be considered. Acoustic attenuation, α, is described by

$$\alpha = \alpha f^n \, [\text{dB/cm}] \tag{8.8}$$

where a is a tissue-dependent constant and f is US frequency, expressed in MHz. For biological media, n is generally considered to be 1. At a frequency of 1 MHz, the acoustic attenuation ranges within 0.18–3.3 dB/cm for various soft tissues but is as high as 20 dB/cm and 40 dB/cm for bone and lung tissue, respectively. While fairly low for soft tissues at low frequency, the attenuation also increases with frequency. For each MHz increase in frequency, there is approximately 0.5 dB of additional attenuation per centimeter. Therefore, if a high-frequency transducer is used for US/PA imaging, the effective imaging depth could potentially be limited by acoustic attenuation rather than optical delivery. Notably, for a given frequency, attenuation is also approximately twice as significant for US imaging as it is for PA imaging due to the required two-way travel of the acoustic pressure, though US insonification pulses can often achieve sufficiently high-pressure amplitudes to allow for strong backscattered US signal at several centimeters of tissue depth. Frequency-dependent attenuation also means that the tissue itself serves as a low-pass filter, reducing the achievable resolution and thus providing a tradeoff between imaging depth and spatial resolution.

Given the scope of this chapter, discussion of transducers utilized for PA signal receiving, as well as US pulse-echo imaging, is limited to piezoelectric crystals. While other transducer types could be used for PA imaging, such as optical pressure sensors, piezoelectric transducers are currently the most commonly used transducers in biomedical PA imaging applications (Oraevsky and Karabutov 2000). In contrast to US wave generation, a mechanical strain at the surface of a piezoelectric crystal results in the generation of a proportional voltage, thus enabling the transduction of an acoustic pressure to an electrical signal. The transducers are designed to exhibit a specific resonant frequency, a function of the speed of sound within the crystal and its geometry, and to enable wideband detection so that they are sensitive to a wide range of frequencies around the resonant frequency of the transducer. Specific examples of piezoelectric transducer materials are lead zirconate titanate (PZT) or lead magnesium niobate (PMN) crystals or polyvinylidene fluoride films. The active elements of the transducer are also attached to a backing layer intended to transfer as much energy as is possible from the active element. Finally, a matching layer is used on the surface of the active element to match the acoustic impedances of the active element and water.

In practice, PA transducers with center frequencies ranging from 1 MHz to as much as 60 MHz have been utilized for cardiovascular applications. The PA imaging systems further range from the use of mechanically scanned single-element US transducers to array transducers composed of hundreds of piezoelectric elements in planar, cylindrical, or spherical orientations, which can be electronically scanned. As such, the selection of transducer(s) for PA imaging applications, as is the case with US imaging, is highly application specific. More details on the detector geometries used for preclinical and intravascular PA (IVPA) imaging systems are provided in Sections 8.3.4 and 8.3.5, respectively.

While piezoelectric transducers originally designed for US imaging applications have been widely utilized for PA detection, optimization of PA imaging quality may require more careful design that is specifically tailored for particular PA imaging applications. For example, most clinical US array transducers are capable of the detection of acoustic signals of only a few megahertz (MHz), whereas broadband PA transients can have a frequency content of several tens of MHz. As a result, such US transducers are ill-suited for efficient PA signal detection. To address this limitation, several custom transducer designs have been demonstrated for PA-imaging specific applications. The most notable example of such a design involves the utilization of optical-based US detection in lieu of a piezoelectric transducer. The design enables the detection of acoustically induced changes in the optical thickness of a Fabry-Perot polymer film interferometer, with an inherently broadband sensitivity compared to clinically available piezoelectric transducers (Zhang, Laufer, and Beard 2008). The design also offers a further potential advantage in that the mirrors required for the interferometer can be selected to be transparent at the desired PA excitation wavelength range, thus mitigating challenges associated with optical delivery around an opaque US array transducer. Within the preclinical setting, the utilization of such a sensor, custom designed for PA imaging, represents a tradeoff between the achievable PA image quality and the cost of system development. As PA imaging moves toward the clinic, the selection of acoustic receiver(s) for combined US and PA imaging systems will ultimately be driven by the desired application and associated requirements for both US and PA imaging resolution and sensitivity.

8.3.3 SIGNAL ACQUISITION AND IMAGE RECONSTRUCTION

The acquisition and the subsequent image reconstruction of generated PA signals are conceptually analogous to that of US imaging. In the most general sense, generated PA pressure waves propagate through tissue to the surface of US transducer, which is then used to convert the pressure into an electrical signal, with the arrival time and magnitude of the generated electrical signal corresponding to the location and intensity of the PA signal source, respectively. For PA imaging applications within the focus of this chapter, those in which the spatial resolution is determined by the characteristics of the receiving US transducer rather than by the size of narrowly focused region of optical delivery, the highly scattering nature of biological tissues enables the assumption of diffuse optical delivery throughout the imaged region. If necessary, gross variations in the optical delivery throughout the field of view can also be further accounted for by calibrating the PA imaging system and using simulations based on estimated optical properties of the imaged medium, such as the Monte Carlo method (Wang, Jacques, and Zheng 1995), and subsequently scaling the intensity of received PA signals to account for heterogeneity in the delivered optical fluence.

Following conversion of the acoustic pressure, the electric signal can be amplified before being digitized, using an ADC, for signal processing. The sampling rate of the data acquisition card used for digitization must be high enough to avoid aliasing of the incoming signals, effectively limiting them to tens or hundreds of mega-Hertz depending on the specifications of the US transducer. Once the signal is digitized and processed, image reconstruction can be started by using a back-projection technique to localize the origin of the generated PA signals. Given that light travel time is negligible relative to the time of flight for PA pressure propagation in tissue and assuming that all PA signals propagate spherically from their sources, the amplitude of the PA signal measured at each point in time can be considered the sum of all PA signal sources, which are generated equidistant from the transducer. Using the speed of sound, approximately 1540 m/s in soft tissue, multiplied by the time of flight for PA pressure arrival at the transducer, the origin of PA signals can be localized. This method represents a simple delay and sum beamforming technique, which can be used for image reconstruction (Beard 2011). While more sophisticated beamforming techniques have been developed and are described elsewhere in the literature (Kuchment and Kunyansky 2011), the delay and sum method provides an overview of the fundamental approach for localizing PA signal sources when using array transducers for signal detection.

Alternatively, if a single element transducer is utilized for PA signal acquisition, as has been the case to date for IVPA imaging, then image reconstruction is more straightforward. The time of flight can be used to detect the axial distance of a signal source from the transducer. The lateral or azimuthal location of signals is determined by hardware and based on the beamwidth of the US transducer—the region within the tissue from which it is sensitive to receive PA signals. The single-element US transducer can then be mechanically scanned to enable two- or three-dimensional image reconstruction.

For both single-element and array transducers, US and PA images reconstruction is performed using fundamental equivalent methods, with one notable variation. PA imaging involves only the one-way travel of pressure waves, whereas US imaging must account for the two-way travel of the acoustic waves. Therefore, the time of flight must be reduced by a factor of two to accurately measure axial position, after which PA and US image coregistration can be achieved.

The digitized signals are then band-pass filtered according to the central frequency of the transducer and then subjected to an envelope detection algorithm, such as the Hilbert transform. Lastly, the signals may also need to undergo a scan conversion process to convert from the image into the appropriate orientation. For example, if a single-element transducer was rotated, the reconstructed image must also be scan converted from polar to Cartesian system of coordinates. Typically, US and PA images are displayed on a logarithmic scale; although in some applications for PA imaging in particular, a linear scale may also be used to more accurately display image contrast. Image renderings comprised of an underlying US image with an overlaying PA mask are also commonly utilized to display the location of strong PA signals within the context of the overall tissue morphology provided by US imaging.

8.3.4 SMALL-ANIMAL PA IMAGING SYSTEMS

Multiple preclinical PA imaging systems are commercially available for small-animal imaging applications, shown in Figure 8.1 (Bayer, Luke, and Emelianov 2012). Among these preclinical systems, two fundamentally different approaches for the PA imaging of small animals has been adopted, namely, linear array-based US/PA imaging and curved array-based PA tomography. With the first approach (left panel of Figure 8.1), a linear US array transducer is coupled with optical fiber bundles, which deliver light approximately focused within the US plane and at the focus depth of the US transducer (Visualsonics Vevo LAZR system, Toronto, Canada). This approach enables complimentary US/PA imaging where high-frequency linear array transducers can be utilized to effectively increase the image resolution for small-animal applications, with reported resolutions of commercially available systems of less than 100 μm. Because the optical and US delivery are coupled, three-dimensional imaging can be achieved by translating either the image target or the coupled transducer. However, the ability to perform whole-body imaging is limited due to the challenge of maintaining consistent orientation of the coupled fiber relative to the surface of the body. As an alternative approach, lower center frequency (e.g., 5 MHz) bowl-shaped US detector arrays have been coupled with circumferential optical illumination to enable tomographic PA imaging, including two commercially available systems (iThera Medical, Munich, Germany, and Endra Life Sciences, Ann Arbor, Michigan, USA) shown in the center and right panels of Figure 8.1, respectively. In an analogous approach to the method utilized for computed tomography, curved US arrays, covering approximately 180° surrounding the imaging target, enables tomographic acquisition of a full 360° cross-section. Conventional US imaging has not been coupled within the system for this approach; instead, PA imaging is performed at multiple wavelengths to obtain both the target morphology and PA differentiation of particular absorbers of interest. Using a curved array therefore offers the advantage of a simplified approach for whole-body imaging at the expense of image resolution. With each approach, the utilization of tunable OPO lasers enables imaging over a broad optical wavelength range within the visible and NIR regions. Tunable wavelength output is particularly

Figure 8.1 Commercially available preclinical photoacoustic imaging systems. (From Bayer CL et al., *Acoustics Today*, 8, 15–23, 2012. With permission.)

useful for performing spectroscopic imaging to assess hemoglobin oxygenation as well as to identify exogenous contrast agents that are widely utilized for preclinical imaging applications.

8.3.5 IVUS AND IVPA IMAGING: CATHETER AND SYSTEM DESIGN

Endoscopic and intravascular US devices are commonly utilized within the clinic to assess the anatomical properties of the heart and surrounding arteries. Recently, IVPA imaging has been combined with IVUS imaging to enable combined IVUS and IVPA (IVUS/IVPA) imaging (Sethuraman et al. 2007). Briefly, combined IVUS/IVPA imaging offers the possibility for high-resolution characterization of arterial morphology and composition and is being investigated as a tool for enhanced diagnosis of atherosclerotic lesion stability. This section will overview requirements for the development of an integrated IVUS/IVPA imaging catheter and prototype designs that have been proposed to date, as well as provide an example of a complete IVUS/IVPA imaging system.

Fundamentally, an integrated IVUS/IVPA imaging catheter is no different from all combined US/PA imaging systems in that it must couple optical light delivery with US transducer and also digitize and process the received data for image reconstruction. Practically, however, there are significant challenges that must be addressed, particularly the need to maintain a small profile for the catheter while ensuring that the optical and US beams overlap and are projected sideways onto the arterial wall from within the lumen. This is achieved using a thin optical fiber as a waveguide to transfer optical energy, coupled from an external laser source at the proximal end of the integrated IVUS/IVPA catheter, to the distal end of the catheter, where it is reflected toward the artery wall. Also located at the distal wall is an IVUS transducer. Commercial IVUS catheter designs utilize either an annular array based transducer operating at a center frequency of approximately 20 MHz or a single element US transducer, which is mechanically rotated and typically have a center frequencies ranging within 30–60 MHz. While the array-based transducer design could potentially be modified to allow for optical delivery, only single element receivers have been reported for IVUS/IVPA imaging to date.

Integrated IVUS/IVPA catheter prototypes utilize multimode optical fibers with core diameters ranging from 200 to 600 μm (Jansen et al. 2011a; Karpiouk, Wang, and Emelianov 2010a; Wei et al. 2011). The choice of the thickness is based on a tradeoff between the energy of light pulses to be delivered and limitations on the sizes and flexibility of the catheters. The redirection and alignment of light with the US beam have been realized using both micro-optics such as mirrors (Hsieh et al. 2012a; Karpiouk, Wang, and Emelianov 2010b; Wei et al. 2011; VanderLaan et al. 2014) or prisms (Yin et al. 2011) and side-fire fibers utilizing total internal reflection (Jansen et al. 2011a; Karpiouk, Wang, and Emelianov 2010a), examples of which are shown in Figure 8.2a–c. The mirror-based optical reflection approach is suitable to redirect light with lower losses and offers greater variability in the reflected angle compared to the side-fire fiber approach. However, the small core diameter of the optical fibers results in high light intensity, which can damage the mirror. This problem can be partially overcome by increasing the distance between the fiber and the mirror and using the same mirror to redirect both the US and light beams (Wei et al. 2011), but the rigid part of the catheter becomes elongated using this approach, sacrificing the achievable catheter flexibility and bend radius. Additionally, an all optical IVUS/IVPA catheter approach has also been demonstrated using optical fibers to deliver the light for IVPA and to excite a microring resonator to generate US pulse and a separate optical fiber to transmit returned IVUS and IVPA signals (Hsieh et al. 2012b). Currently, most experimental achievements using integrated IVUS/IVPA catheters have employed side-fire fiber-based light delivery systems and single-element transducers. The smallest integrated imaging catheter demonstrated has an outer diameter of 1 mm, housed within a drive double-wound drive cable secured to a 5-mm-long rigid fixture at the distal tip (Figure 8.2b) (VanderLaan et al. 2014). While further miniaturization of integrated IVUS/IVPA catheters is likely to be achieved with future design modifications, to less than 1-mm outer diameter, their flexibility and rotational stability should also be demonstrated before they can be considered for clinical studies.

A schematic diagram of an IVUS/IVPA imaging set-up, used for interrogation of vessel mimicking phantoms and freshly excised artery sections, is provided in Figure 8.2d (Wang et al. 2010b). Briefly, the distal end of an integrated IVUS/IVPA imaging catheter is inserted into the lumen of the imaging target. A nanosecond pulsed laser was focused into a multimode optical fiber at the proximal end of an integrated IVUS/IVPA imaging catheter, which also housed a single-element IVUS transducer. During imaging, each pulse of the laser provided a trigger for collecting radiofrequency (RF) signals from a pulser/receiver connected to the IVUS

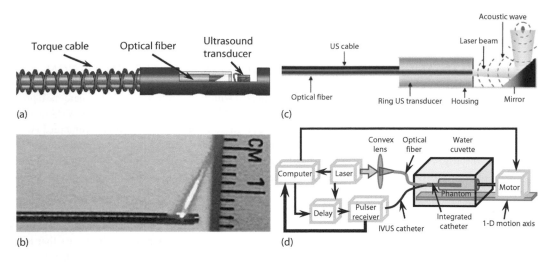

Figure 8.2 Integrated IVUS/IVPA imaging catheter and system designs. (a) Schematic and (b) photograph of the distal end of an integrated catheter utilizing a side-fire fiber illumination design (c) Schematic of an integrated IVUS/IVPA catheter utilizing a ring transducer and mirror to direct ultrasound and optical propagation. (d) Schematic diagram of an IVUS/IVPA imaging system. ([b] From VanderLaan D et al., System and integrated catheter for real-time intravascular ultrasound and photoacoustic imaging. Ultrasonics Symposium (IUS), 2014 IEEE International, September 3–6, 2014; [c] from Wei W et al., *J Biomed Opt*, 16, 106001, 2011. With permission; [d] from Wang B et al., *IEEE J Select Top Quantum Electron*, 16, 588–599, 2010.)

transducer. The RF signals were digitized, resulting in a single IVPA A-line signal. Following a user-defined delay, the pulser/receiver was then used to acquire a conventional IVUS A-line. The resulting acquired RF data was therefore composed of spatially coregistered and temporally consecutive IVPA and IVUS signals. For generation of two-dimensional images of a vessel cross section, the imaging target was rotated using a stepper motor. Additionally, for three-dimensional imaging, a second motor was utilized to translate the image target.

A 10-μm carbon fiber was used to approximate an IVUS/IVPA point-source in order to characterize the imaging system's resolution. The axial and lateral resolution was estimated based on the full width half maximum of the frequency response and beam profile from the carbon fiber, respectively. The IVUS and IVPA imaging axial resolutions were 54 μm and 38 μm, respectively, whereas the measured lateral resolutions were 3.2° and 5.5°, respectively. In a separate experiment, demonstrating the complimentary nature of combined IVUS/IVPA imaging, a vessel mimicking polyvinyl alcohol phantom with a highly optically absorbing graphite rod was imaged (Sethuraman et al. 2007). Figure 8.3 shows IVUS, IVPA, and combined IVUS/IVPA

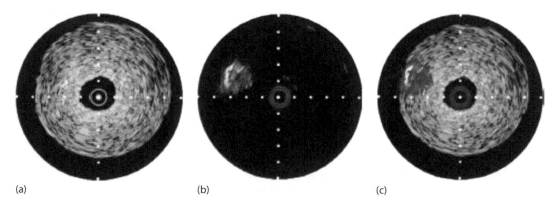

Figure 8.3 IVUS/IVPA imaging of vessel phantom with optically absorbing inclusion. (a) IVUS image showing the structure of the vessel phantom. (b) IVPA imaging revealing the location of the inclusion. (c) Combined IVUS/IVPA image. (From Sethuraman S et al., *IEEE Trans Ultrason Ferroelectr Freq Control*, 54, 978–986, 2007. With permission.)

images to demonstrate how IVPA imaging provides compositional information based on optical absorption properties and how it complements the overall morphology demarcated by the corresponding IVUS image. This highlights the feasibility and characteristics of combined IVUS/IVPA imaging. Specific applications of this technique are described in detail in Section 8.5 of this chapter.

8.4 PA IMAGING TECHNIQUES

Over the past two decades, PA imaging has been increasingly investigated, using a variety of techniques, as a modality capable of augmenting clinical cardiovascular imaging modalities. Taken together, the multitude of different techniques are indicative of the flexibility of PA imaging, whereby adjustments in the instrumentation set-up, imaging procedure, and targeted signal sources can all be tailored to specific imaging needs. Generalized imaging techniques that have been utilized for PA imaging can be categorized based on the physical parameter that is modified in order to induce PA signal changes, for example, adjustment of optical wavelength(s) or induced local temperature changes during PA imaging. Alternatively, the technique can be categorized by the nature of the PA signal sources being evaluated, specifically, relying on either endogenous chromophores or exogenous contrast agents. Each of these techniques is discussed in the following sections.

8.4.1 sPA IMAGING

The amplitude of a generated PA signal is proportional to the optical absorption, the local laser fluence, and the efficiency of thermo-acoustic transfer within the local environment (the Grüneisen parameter), as evident by Equations 8.6 and 8.7. Therefore, provided that the Grüneisen parameter does not vary during PA imaging, one major advantage of PA imaging is that it enables the detection of specific chromophores of interest based on analysis of PA signal intensity when the optical excitation wavelength is adjusted and the local laser fluence is held constant. Using this approach, wavelength-dependent optical absorption spectral characteristics are exploited as an independent variable to enable PA detection of specific absorbers of interest, including both endogenous tissues and exogenous contrast agents. Errors associated with the assumption of constant local fluence can be reduced by choosing a limited optical wavelength range for PA imaging because differences in optical scattering are minimized. The technique, termed sPA imaging or multispectral optoacoustic tomography (MSOT), is widely utilized for both identifying the location of one or more absorber(s) as well as estimating their relative concentrations.

If the goal is to detect the presence of a particular absorber rather than the relative concentration within the imaged tissue, it is possible to use a simple sPA approach in which the slope of the PA signal over two or more wavelengths is used to differentiate endogenous tissues based on unique optical absorption spectral trends relative to other signal sources. Using this approach, the PA signal obtained from each position within the image is normalized with respect to wavelength and compared to known optical absorption spectra or spectral trends of a potential signal source, such as deoxy- and oxy-hemoglobin, lipid, and exogenous contrast agents. In many cases, the region for PA imaging can be selected such that an absorber of interest can be expected to have a significantly different spectral curve from other signal sources. Therefore, differentiating the presence of the absorber can be achieved by assessing the slope(s) of the normalized PA signal versus wavelength(s). Indeed, this is the case for lipid, which decreases in optical absorption over the range of 1200–1230 nm, whereas water-based tissues exhibit a nearly constant optical absorption profile, thus allowing for sPA detection of lipid (Graf et al. 2012; Jansen et al. 2013; Wang et al. 2010a). Similarly, a ratiometric assessment of PA signal intensity at two different wavelengths has been utilized to assess the presence and status of an activatable contrast agent (Razansky et al. 2012).

While these simplified sPA approaches are adequate and even beneficial in terms of the required image acquisition and processing times, they are inherently limited in their ability to differentiate numerous absorbers. To this end, more complex spectral unmixing strategies have been deployed for sPA imaging applications, including least squares and cross-correlation algorithms. For example, an intraclass correlation algorithm has been used to simultaneously differentiate between multiple different absorbers (Mallidi et al. 2009).

More computational intensive approaches have been developed that enable each pixel to be treated as a linear combination of absorbers, thus allowing detection of the relative contribution of numerous absorbers to the generated sPA signal based on minimizing the mean squared error (Kim et al. 2011). A linear least squares algorithm has also been used to differentiate nanoparticle contrast agents from endogenous absorbers for spectroscopic IVPA (sIVPA) imaging applications (Yeager et al. 2012).

The diversity in the number of sPA algorithms that have been developed is indicative of the complexity that multiwavelength imaging adds to PA imaging. Taking into account wavelength dependencies and the presence of N distinct absorbers within an imaged tissue, Equation 8.6 must be modified to

$$p_0(\lambda) = \Gamma\Phi(\lambda)\left(\mu_{a_1}(\lambda) + \mu_{a_2}(\lambda) + \ldots + \mu_{a_N}(\lambda)\right) \tag{8.9}$$

If an estimate of the relative concentration of each absorber for each pixel is desired, it is also necessary to further consider the wavelength-dependent optical absorption of each absorber, μ_{a_i}, to be the product of the concentration, $C[i]$, and the molar absorption cross-section, $\sigma_{a_i}(\lambda)$. Equation 8.9 must therefore be rewritten as (Hsieh et al. 2012a):

$$p_0(\lambda) = \Gamma\Phi(\lambda)\left(C[1]\sigma_{\alpha_1}(\lambda) + C[2]\sigma_{\alpha_2}(\lambda) + \cdots + C[N]\sigma_{\alpha_N}(\lambda)\right) \tag{8.10}$$

If the Grüneisen parameter is assumed constant during image acquisition and the wavelength-dependent, local optical fluence can be reliably estimated, Equation 8.9 or 8.10 can be rewritten as a set of linear equations. Provided that the number of wavelengths utilized for sPA imaging is greater than the number of expected absorbers, N, the absorption coefficient or relative concentration vector from the system of equations can then be solved.

It is also important to note that the wavelengths selected for sPA imaging can have a significant impact on the ability to reliably differentiate between absorbers present within the imaged medium. In general, it is desirable to image over a short wavelength range to minimize the effects of wavelength-dependent changes in the optical fluence, which are difficult to estimate in heterogeneous media. Additionally, the selection of wavelength regions where absorbers of interest exhibit unique spectral properties will help improve the overall efficacy of an algorithm. An algorithm was recently developed to aid in the optimization of selected wavelengths for sPA imaging based on the spectra of absorbers known to be present within the imaged tissue (Luke, Nam, and Emelianov 2013).

In summary, sPA imaging offers the ability to use variations in known optical absorption spectra to assess the composition of the imaged tissue. Numerous algorithms have been developed for sPA imaging, ranging from a crude assessment of the presence of a single absorber to more sophisticated techniques that enable the calculation of the relative concentration of an array of absorbers. Regardless of the approach, however, sPA provides a means of capitalizing on the compositional specificity that makes PA imaging an attractive modality to combine with the more limited molecular contrast of US imaging.

8.4.2 THERMAL PA IMAGING

The intensity of a generated PA signal is proportional to the local laser fluence, optical absorption, and the Grüneisen parameter. Additionally, all components of the Grüneisen parameter, the thermal coefficient of volume expansion, the speed of sound, and the isobaric specific heat capacity, are temperature sensitive and tissue specific. Therefore, whereas sPA imaging techniques assume that the Grüneisen parameter remains constant during imaging, the deliberate modification of temperature in the imaged tissue provides an additional PA imaging technique, termed thermal PA (tPA) imaging. Using the tPA technique, optical fluence and wavelength are held constant and the obtained signal intensity is monitored over time as the temperature of the target tissue is modified. As a result, the magnitude of the tPA signal at each location within the tissue is linearly proportional to the local temperature within water-based tissue, thus allowing for temperature distribution mapping throughout the imaging field. Therefore, tPA has been investigated for applications in the noninvasive monitoring of the delivered thermal dosage during targeted hyperthermia or cryogenic treatments, including the use of additional laser sources for photothermal heating. For example, tPA imaging has been utilized for temperature mapping during continuous wave (CW) laser heating of highly optically absorbing contrast agents, as seen in Figure 8.4a and b (Chen et al. 2013). Using such approaches, the colinearity of PA signal intensity and temperature has been characterized (Figure 8.4c) and found to enable resolutions of temperature changes as fine as 0.15°C in *ex vivo* studies (Pramanik and Wang 2009).

Interestingly, it is also possible to differentiate between lipid-based and water-based tissues based on the PA signal change with temperature (Wang and Emelianov 2011). It has been demonstrated that the thermal coefficient of volume expansion and the speed of sound (two components of the Grüneisen parameter)

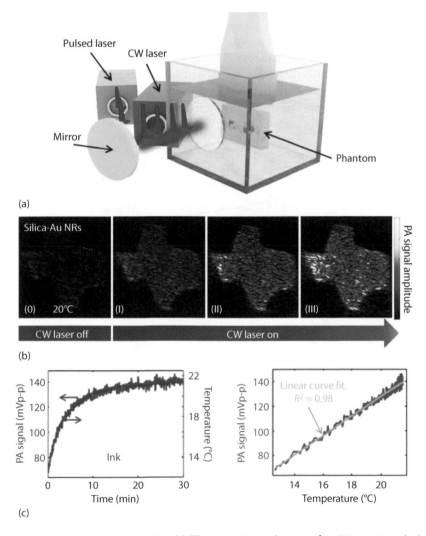

Figure 8.4 Thermal photoacoustic imaging. (a) The experimental set up for tPA-monitored photothermal experiments with both pulsed laser and CW laser. The pulsed laser is used to generate the PA signals, and the CW laser to heat a phantom containing strong optically absorbing contrast agents. (b) Initial photoacoustic signals without CW laser heating (left) and subsequent PA images with CW laser illumination at 1-minute intervals. (c) PA signal and the actual temperature of ink solution as cold solution was allowed to come to room temperature and PA signal versus temperature, which shows almost a linear relationship. ([b] From Chen Y-S et al., *J Biophotonics*, 6, 534–542, 2013. With permission; [c] from Pramanik M et al., *J Biomed Opt*, 14, 054024, 2009. With permission.)

exhibit opposite temperature-dependent trends between water and animal fat (Duck 1990; Fidanza, Keys, and Anderson 1953). Therefore, as an alternative to temperature mapping, tPA may also be used as a means of lipid detection by evaluating the change of PA signal intensity with temperature. As a result, both sPA and tPA offer the potential to differentiate lipid from surrounding water-based tissues, an application of significance to numerous cardiovascular pathologies where the buildup of lipid plays a critical role in the disease severity or progression.

8.4.3 PA IMAGING OF ENDOGENOUS TISSUE

Within biological media, unique optical absorption spectra of many different tissue components enable the characterization of the molecular composition using PA imaging. More specifically, sPA imaging is widely utilized to differentiate wavelength-dependent optical absorption characteristics of several different endogenous tissue chromophores, including oxy- and deoxy-hemoglobin, lipid, and collagen. Figure 8.5 provides optical absorption spectra for endogenous chromophores frequently utilized for molecular-specific PA imaging. For example, sPA imaging can be implemented to detect the presence of lipid-rich or collagen-rich regions within atherosclerotic plaques (Jansen et al. 2011a, 2013; Sethuraman et al. 2007; Wang et al. 2010a, 2012a, 2012b, 2012c) or to quantify the amount of oxy- and deoxy-hemoglobin and estimate oxygen saturation (sO_2) (Kim et al. 2011; Li et al. 2008). Details of specific sPA imaging applications that involve the detection of endogenous absorbers of particular interest within the field of cardiovascular medicine are introduced in Section 8.5 of this chapter.

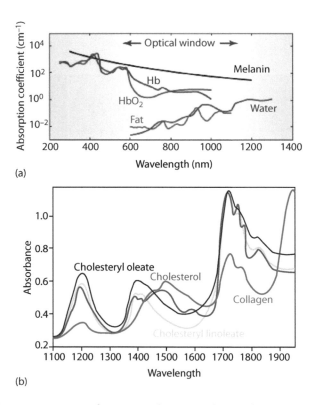

Figure 8.5 Optical absorption spectra of various endogenous chromophores. (a) Visible and near-infrared absorption spectrum of oxy- and deoxy-hemoglobin, melanin, water, and fat. (b) Near-infrared absorption spectra of various cholesterol composition and collagen. ([a] Reprinted from *Trends Biotechnol*, 29, Mallidi S, Luke GP, Emelianov S, Photoacoustic imaging in cancer detection, diagnosis, and treatment guidance, 213–221, Copyright 2011, with permission from Elsevier; [b] reprinted from *J Am Coll Cardiol*, 47, Caplan JD, Waxman S, Nesto RW, Muller JE, Near-infrared spectroscopy for the detection of vulnerable coronary artery plaques, C92–C96, Copyright 2006, with permission from Elsevier.)

In addition to sPA differentiation of one or more endogenous absorbers, *PA Doppler imaging* has also been demonstrated as a technique capable of quantifying fluid flow rates (Fang, Maslov, and Wang 2007a). PA Doppler imaging is analogous to Doppler US techniques with the notable distinction that the detected US signal during PA Doppler arises from the strong optical absorption of hemoglobin within red blood cells (RBCs) rather than US reflections by RBCs, which are relatively weak. The net result of this difference in the signal source is an improved signal-to-noise ratio (SNR), particularly at low flow rates (Beard 2011). PA Doppler imaging has been achieved using a variety of techniques, primarily differentiated based on the applied optical excitation. Initial experimental results demonstrated the ability to determine PA Doppler frequency shifts using CW laser excitation, with a major limitation being that the resulting measurements were not depth resolved (Fang, Maslov, and Wang 2007a, 2007b). This limitation was subsequently overcome using pulsed excitation and the spectral analysis of tone-burst Doppler PA signals, resulting in velocity resolutions ranging from 1.5 to 200 mm/s in a manner that is inversely proportional to the achievable spatial resolution (Sheinfeld, Gilead, and Eyal 2010a, 2010b). Lastly, an alternative approach relies on the measurement of changes in PA signal arrival time following successive PA excitation pulses using a cross-correlation algorithm, which enables the use of nanosecond pulsed laser excitation and yields high spatial and velocity resolution (Brunker and Beard 2010, 2011, 2013).

8.4.4 PA IMAGING OF EXOGENOUS CONTRAST AGENTS

PA imaging applications are limited by the number of endogenous absorbers within the body. Physiological parameters related to blood and a few biomarkers, including lipid and collagen, can be assessed using PA imaging based on the contrast that they inherently provide within certain optical wavelength ranges. The additional use of exogenous contrast agents with strong and unique optical absorption spectral characteristics has greatly expanded the role of PA imaging as a cellular and molecular-specific imaging modality. The introduction of exogenous contrast agents can also greatly improve PA image contrast if the utilized agent is appropriately selected. It is important that the contrast agents provide PA signal within an optical region where background signal from other tissues is minimized, such as the NIR tissue optical window from approximately 700–1100 nm. Special effort is also taken to ensure that the contrast agents produce limited toxicity and that their physical properties such as size, shape, and surface charge are appropriately tuned to allow the intended delivery. While US contrast agents have also been developed for similar applications, their use is limited by their large size or their inability to enable longitudinal imaging because of the destruction of the contrast. Despite these important variables, a wide array of contrast agents have been developed and evaluated in preclinical IVPA imaging applications, as outlined in Table 8.2 (Luke, Yeager, and Emelianov 2012). PA contrast agents that have been widely investigated to date fall into three main categories; dyes, single-walled carbon nanotubes (SWNTs), and noble metal nanoparticles. Each of these categories of PA contrast agents is discussed in this section.

Several different molecular dyes have been utilized as PA imaging contrast agents. Originally developed as more conventional optical imaging contrast agents, dyes are typically on the order of 1 nm in size and are rapidly cleared from circulation through the renal system. Interestingly, the majority of dyes were initially intended to be used based on their fluorescent properties, in which applied light is rapidly reemitted at a longer optical wavelength, a phenomenon that counteracts the efficiency of PA signal generation by reducing the ability of the molecule to convert absorbed light into heat during thermal expansion. However, many of the available dyes that absorb light within the NIR window exhibit a low quantum yield, meaning that they can be effectively repurposed as PA contrast agents. Examples of dyes that have been utilized for PA imaging include indocyanine green (ICG) (Kim et al. 2007) and methylene blue (Song et al. 2008), which are both already clinically approved for limited applications, along with IRDye800CW (Stantz et al. 2010), AlexaFluor750 (Bhattacharyya et al. 2008), and MMPSense680 (Razansky et al. 2012). Dyes can also be modified through the direct conjugation of molecular targeting moieties (Bhattacharyya et al. 2008; Stantz et al. 2010) or incorporation into aggregates to form larger particles to better control their ability to target specific molecules of interest during PA imaging (Altinoglu et al. 2008; Kim et al. 2007).

Table 8.2 Overview of different exogenous contrast agents utilized for PA imaging

Contrast agent	Description	Size (nm)	Peak absorption wavelength (nm)
Indocyanine green PEBBLEs	Fluorescent green dye encapsulated in ormosil spheres	100	790
IRDye800CW	Fluorescent dye	<2	774
AlexaFluor	Fluorescent dye	<2	350–790
Methylene blue	Clinically approved dye	<2	677
Gold nanospheres	Solid gold spheres	2–60	520–540
Gold nanorods	Solid rice-shaped gold particles	10 by 40–60	650–1100
Gold nanoshells	Spherical particles with a silica core and gold shell	50–500	700–1100
Silver nanoplates	Thin triangular silver discs	6–20 by 100–200	600–1200
Single-walled carbon nanotubes	One-atom-thick cylinders of graphene	1 by 50–300	690
Gold-coated single-walled carbon nanotubes	Single-walled carbon nanotubes coated with a thin layer of gold	11 by 100	900
Gold nanocages	Silver cubes coated with a porous layer of gold	40	800
Nanobeacons	Liposomes containing plasmonic nanoparticles	100–200	520–1100
Nanostars	Spherical iron oxide cores with a spiked gold coating	50	800–900
Nanoroses	Clusters of gold-coated iron oxide spheres	20–40	700–800
Pearl necklaces	Gold nanorods with iron oxide spheres attached to their surfaces	90	785
Nanowantons	Crescent-shaped cobalt cores coated with a layer of gold	30–90	400–800
Iron oxide – gold nanoshells	Spherical iron oxide cores with gold shells	30–40	660–900
Magnetic and plasmonic liposomes	Liposomes loaded with iron oxide spheres and gold nanorods	100–200	800
Photoacoustic nanodroplets	Superheated liquid perfluorocarbon droplets loaded with plasmonic nanoparticles	200	520–1200
Radiolabeled gold nanorods	Gold nanorods radiolabeled with [125I]	10 by 40–60	650–1100
Gold nanoclusters	4-nm nanospheres connected with a biodegradable polymer	50–100	700–900

Source: With kind permission from Springer Science+Business Media: *Ann Biomed Eng*, Biomedical applications of photoacoustic imaging with exogenous contrast agents, 40, 2012, 422–437, Luke, G, Yeager, D, Emelianov, S.

Beyond molecular dyes, SWNTs and noble metal nanoparticles have also been utilized as PA contrast agents. SWNTs exhibit a high optical absorption over a broad wavelength range, supporting their ability to efficiently generate PA signals (de la Zerda et al. 2008). Attachment of targeting agents to the SWNT surface has also been demonstrated (de la Zerda et al. 2010). Additionally, the development of hybrid SWNTs which incorporate metallic coating (Kim et al. 2009) or the attachment of dyes (de la Zerda et al. 2010) have been shown to increase the efficiency of PA signal generation.

Noble metal nanoparticles represent a widely investigated category of PA contrast agents. Metallic nanoparticles are of particular interest to researchers because they exhibit surface plasmon resonance (SPR) coupling, a process in which free electrons at the nanoparticle surface oscillate with the applied electromagnetic field, which enables very high, tunable optical absorption properties based on modifications to the nanoparticles size and shape (Grzelczak et al. 2008; Xia et al. 2009). As a result of the SPR effect, an individual nanoparticle exhibits very high optical absorption, up to five orders of magnitude greater than that exhibited by a dye molecule (Jain et al. 2006), resulting in efficient PA signal generation. Additionally, the size and shape of these plasmonic nanoparticles can be easily tailored, resulting in changes to their optical absorption spectrum and the ability to use sPA imaging for their detection over background endogenous tissues. For example, the location of the peak absorbance generated by rod-shaped gold nanoparticles can be adjusted within the NIR optical window by tuning the particles' aspect ratios (Nikoobakht and El-Sayed 2003). In fact, a significant number of gold nanoparticle PA contrast agents have been developed, including spherical gold nanoparticles (Mallidi et al. 2009; Wang et al. 2009), gold nanoshells (Liangzhong et al. 2006; Yang et al. 2007), gold nanorods (Bayer et al. 2011; Chen et al. 2011; Meng-Lin et al. 2008; Song et al. 2009; Yeager et al. 2012), gold nanoplates (Millstone et al. 2009), gold nanocages (Skrabalak et al. 2008), and aggregated gold nanoclusters (Yoon et al. 2010). Each of these gold particles has its own unique physical and optical absorption properties. Silver nanoparticles have also been developed and investigated for PA imaging applications (Homan et al. 2010). Similar to the other classes of PA contrast agents, the surface of plasmonic nanoparticles can be readily conjugated with targeting moieties for molecular imaging applications (Kumar, Aaron, and Sokolov 2008) or otherwise modified to improve their stability (Chen et al. 2010) and PA signal generation efficiency (Chen et al. 2011). As a result, plasmonic nanoparticle delivery followed by sPA imaging can be employed to greatly expand cellular and molecular imaging capabilities from that achieved based solely on endogenous absorbers.

Finally, several hybrid imaging contrast agents have also been developed based on the aforementioned PA contrast agents. A limited number of these hybrid particles have sought to provide both US and PA contrast within a single particle construct. This includes the loading of conventional US microbubble contrast agents with dye molecules to add PA contrast (Kim et al. 2010), as well as the development of PA nanodroplets (PAnDs) (Wilson, Homan, and Emelianov 2012). PAnDs are composed of a bovine albumin shell surrounding a mixture of liquid perfluorocarbon and plasmonic nanoparticles, which can be optically triggered to generate both US and PA contrast. The basis of this multimodal contrast agent has also been expanded, whereby the plasmonic nanoparticles are replaced with ICG dye, resulting in a particle that is more readily translatable to the clinical environment (Hannah et al. 2014). Together, the wide array of exogenous contrast agents available for PA imaging applications offers numerous options for expanding the ability of the modality to provide compositional assessment of targeted tissues that compliment the morphological information obtained from US imaging.

In summary, numerous PA imaging techniques can be utilized to provide complimentary information to traditional US imaging based on variations in either the nature of the PA systems, the means of PA signal characterization, or the targeted signal sources. The versatility of these PA imaging techniques opens the door for many new applications of particular relevance within the field of cardiovascular imaging.

8.5 APPLICATIONS OF COMBINED US AND PA IMAGING IN CARDIOVASCULAR MEDICINE

Over the past decade, combined US and PA imaging has been investigated as a tool for enhancing more conventional, clinically available imaging techniques across a wide array of applications within cardiovascular medicine (Beard 2011; Mallidi, Luke, and Emelianov 2011; Mehrmohammadi et al. 2013). The unique ability to obtain spatially coregistered morphological and physiological or compositional information at physiologically relevant imaging depths has motivated continued growth in this hybrid imaging modality. As PA imaging techniques continue to be refined, their utilization for diagnostic and therapy monitoring procedures will likely expand to include even more diverse preclinical and, ultimately, clinical applications. The subsequent sections overview the predominant categories of combined US and PA imaging applications that are of specific interest within cardiovascular medicine.

8.5.1 CHARACTERIZATION OF ATHEROSCLEROTIC PLAQUES

Despite significant advances in preventative measures and interventional procedures over the past few decades, cardiovascular disease remains the leading cause of death within industrialized nations, with coronary heart disease accounting for more than 7 million annual deaths. The high mortality rate can be attributed, in part, to an inability to reliably differentiate and appropriately treat rupture-prone atherosclerotic plaques, which can lead to acute coronary events, strokes, and peripheral vascular disease progression. These so-called vulnerable plaques are characterized by the presence of a thin fibrous cap (<65 μm) overlaying a large, lipid-rich pool and accompanied by an increased number of inflammatory macrophages (Davies and Thomas 1985; Falk 1983; Muller et al. 1994).

PA imaging has been increasingly utilized to better characterize atherosclerotic plaques, using a variety of the techniques introduced in Section 8.4 to assess the cellular and molecular composition within identified lesions (Jansen, van Soest, and van der Steen 2014). Combined IVUS/IVPA imaging was recently introduced as a potentially clinically translatable hybrid technique to help improve diagnostic accuracy of vulnerable atherosclerotic lesions by providing a means of supplementing the morphological information provided by IVUS with additional capability for assessing the composition of atherosclerotic lesions based on unique optical absorption properties of specific plaque components or delivered contrast agents (Sethuraman et al. 2007). Custom fabricated integrated IVUS/IVPA catheters and imaging systems, described in Section 8.3.5, enable the acquisition of spatially coregistered and temporally consecutive IVUS and IVPA images, therefore allowing for complimentary IVUS/IVPA imaging of arterial cross-sections with shared signal detection hardware.

sIVPA imaging has been utilized to identify the distribution of endogenous absorbers within arterial plaques such as collagen (Sethuraman et al. 2007; Wang et al. 2012c) and lipid (Jansen et al. 2011b, 2014; Wang et al. 2010a, 2012c). Due in large part to its correlation with atherosclerotic lesion instability, lipid has been a major focus of IVPA imaging. Optical absorption peaks present at approximately 1210 nm and 1720 nm have both been utilized for the localization of lipid within atherosclerotic plaques of excised arteries using combined IVUS/IVPA systems (Figure 8.6), including in the presence of luminal blood (Jansen et al. 2013; Wang et al. 2012a, 2012b). It has also been shown that it is possible to image lipid within atherosclerotic lesions using IVUS/IVPA imaging at a single optical wavelength of 1720 nm rather than using a spectroscopic analysis due to the low background signal from water-based tissues (Wang et al. 2012b). This single wavelength IVPA technique was also demonstrated *in vivo* in an animal model of atherosclerosis; however, a larger sample size should be imaged in the future to further evaluate the technique. US/sPA imaging for morphological assessment and lipid detection was also expanded to the carotid arteries (Graf et al. 2012). Additionally, US/PA imaging has been investigated as a tool for detection and staging of deep vein thrombosis based on age-dependent changes in the hemoglobin optical properties within the thrombus (Karpiouk et al. 2008). While further investigation is required, it is conceivable that this application may be expanded to the assessment of coronary arteries, using IVUS/IVPA imaging to help identify ruptured plaques.

While endogenous chromophores provide several relevant tissues for PA imaging localization to supplement the US assessment of arterial plaques, the introduction of exogenous contrast significantly increases the potential imaging targets. For example, macrophages, an indicator of the local inflammatory status of an atherosclerotic plaque, are scavenger cells which phagocytose foreign bodies, thus allowing the potential for their detection using PA following labeling with exogenous nanoparticles, as demonstrated in Figure 8.7a and b (Wang et al. 2009). The role of macrophages in disease progression makes them an important and relatively simple potential labeling target for contrast agents, provided that they are successfully delivered to the site of the atheroma. In an atherosclerotic rabbit model, arteries excised following a systemic injection of polymer-stabilized spherical (Wang et al. 2009) and rod-shaped (Yeager et al. 2012) plasmonic gold nanoparticles (Figure 8.7c and d), it has been shown that IVUS/sIVPA is capable of localizing regions of high nanoparticle concentration. Histological analysis subsequently revealed that these PA contrast agents tend to preferentially label

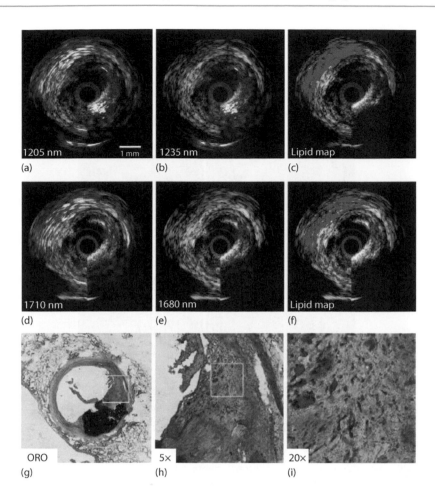

Figure 8.6 Lipid detection in an atherosclerotic human coronary artery using sIVPA at 1210 μm and 1720 μm. (a) 1205 nm and (b) 1235 nm combined IVPA/IVUS images (IVPA 25 dB, IVUS 40 dB). (c) Lipid map based on two-wavelength relative difference between the PA signal at 1205 nm and 1235 nm. (d) 1710 nm and (e) 1680 nm combined IVPA/IVUS images (IVPA 25 dB, IVUS 40 dB). (f) Lipid map resulting from the two-wavelength relative difference between the PA signal at 1710 nm and 1680 nm. Both lipid maps are shown overlaid on the corresponding IVUS image. (g) Lipid histology stain (Oil Red O); lipids are stained red; calcification is stained black. (h) 5× magnification of the part of the atherosclerotic plaque indicated as lipid rich by the lipid stains [area outlined in black in (g)] shows larger extracellular lipid droplets, while the lipids in all other parts of the lesion are intracellular or contained in small extracellular droplets. (i) 4× magnification of area outlined in black in (h). (Reprinted from *Photoacoustics*, 2, Jansen K, Wu M, van der Steen AF, van Soest G, Photoacoustic imaging of human coronary atherosclerosis in two spectral bands, 12–20, Copyright 2013, with permission from Elsevier.)

atherosclerotic lesions and colocalize with macrophages. In addition to macrophages, several other biomarkers that are overexpressed at various stages of atherosclerotic plaque progression have been targeted using PA contrast agents and preclinical imaging set-ups. One such example is the use of antibody conjugated nanoparticles targeted to vascular cell adhesion molecule-1 and subsequent PA detection in a mouse mode with an induced inflammatory response (O'Donnell et al. 2010). Matrix metalloproteinases (MMPs), another marker of plaque vulnerability, have also been targeted for PA imaging. Using an organic dye that exhibits a shift in absorption properties when cleaved by MMPs, PA imaging was used to localize MMP activity in excised carotid endarterectomy specimens (Figure 8.7e and f) (Razansky et al. 2012). As ongoing research continues to identify novel contrast agent designs, realization of the potential PA imaging targets is likely to continue to expand, opening the door for even greater ability for US/PA imaging systems to diagnose and characterize atherosclerotic lesion cellular and molecular composition. However, a cost–benefit analysis should be considered for each contrast agent to establish if it is suitable for preclinical and/or clinical applications.

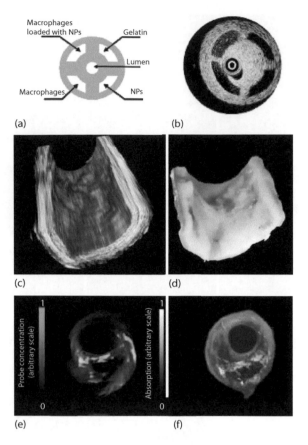

Figure 8.7 PA imaging using exogenous contrast agents. (a) Schematic diagram of the tissue mimicking phantom. (b) The combined IVUS/IVPA image of the phantom obtained using 680 nm wavelength illumination, indicating the origin of the photoacoustic responses in IVPA images. (c) Reconstructed three-dimensional integrated IVUS/IVPA renderings of gold nanorod-labeled atherosclerotic plaque. (d) photograph of the corresponding artery section revealing gold nanorods (AuNR) distribution at the luminal surface (red). (e–f) Localization of MMP activity in three carotid specimens. (e) Imaging results from intact plaques made using MSOT, revealing location of a molecular probe (MMPsense 680) activatable by local MMP activity. (f) The corresponding epi-fluorescent image from dissected plaque (in green) superimposed onto a color image of the corresponding cryosection. ([b] Reprinted with permission from Wang B, Yantsen E, Larson T, Karpiouk AB, Sethuraman S, Su JL, Sokolov K, Emelianov S, *Nano Lett*, 9, 2212–2217, Copyright 2009 American Chemical Society; [d] from, Yeager D, Karpiouk A, Wang B, Amirian J, Sokolov K, Smalling R, Emelianov S, *J Biomed Opt*, 17, 106016, 2012. With permission; [f] with kind permission from Springer Science+Business Media: *Mol Imaging Biol*, Multi-spectral optoacoustic tomography of matrix metalloproteinase activity in vulnerable human carotid plaques, 14, 2012, 277–285, Razansky D, Harlaar NJ, Hillebrands JL, Taruttis A, Herzog E, Zeebregts CJ, van Dam GM, Ntziachristos V.)

8.5.2 EVALUATION AND MONITORING OF THE MYOCARDIUM AND CARDIOVASCULAR FUNCTION

Beyond the direct detection and characterization of atherosclerosis, US/PA imaging is also being investigated for the assessment of hemodynamic function. For example, a rat infarct model was used to demonstrate the ability of PA imaging to detect myocardial ischemia and characterize intramyocardial fractional blood volume (Li et al. 2011). Similarly, noninvasive sPA imaging of small-animal models was also used to demonstrate an approach for characterizing myocardial infarction by localizing a PA contrast agent targeted to selectins (Taruttis et al. 2013).

Given the optical absorption spectral differences between oxy- and deoxy-hemoglobin, sPA is commonly used to detect the presence of each and to quantify oxygen saturation (sO_2). This application for PA imaging has been widely utilized for the characterization of blood in the microvasculature environment. In particular, PA imaging has been applied for monitoring physiological parameters of tumors, although PA has not yet been utilized for more direct applications in cardiovascular medicine. Also, as discussed in Section 8.4.3 of this chapter, PA Doppler techniques allow for the quantification of low flow rates (Sheinfeld, Gilead, and Eyal 2010a, 2010b). While PA Doppler imaging is currently at an early stage in experimental development, as the overall approach continues to be refined, the achievable sensitivity at low flow rates may provide new interesting clinical applications. One primary challenge facing the translation of the technique to *in vivo* flow measurements is that each of the described implementations treats RBCs as discrete particles, while it remains to be demonstrated whether whole blood can be treated as an optically heterogeneous medium or if the RBCs are packed so close that the blood appears optically homogenous. However, if this hurdle can be overcome, it is possible that the ability to measure RBC flow rates can be combined with the ability to detect sO_2 using sPA imaging techniques to enable the noninvasive estimation of oxygen delivery and consumption within a target tissue region, thereby providing otherwise difficult to measure physiological information (Beard 2011).

8.5.3 THERAPY GUIDANCE

US/PA imaging has also been demonstrated as a platform that is well suited for delivery of therapeutics, both during cardiac surgery and percutaneous coronary interventions. In a natural progression from the application for characterizing coronary atherosclerotic plaques, IVUS/IVPA has also been investigated as a means of guiding stent placement (Karpiouk et al. 2012; Su, Wang, and Emelianov 2009). The metallic stent struts provide very high IVPA signal relative to endogenous tissue, allowing the characterization of stent strut apposition, an important parameter in assessing the likelihood of postprocedural events, with high contrast relative to IVUS imaging alone. In an additional expansion of IVUS/IVPA imaging, the imaging system was modified to capitalize on the temperature sensitivity of PA signal intensity to enable monitoring of selective hyperthermia in atherosclerotic lesions in a feasibility study (Yeager et al. 2013). A CW laser source was incorporated as a second optical source through the optical fiber of the integrated catheter, serving as a means of heating highly absorbing plasmonic gold nanorods located within macrophages. Using this approach, the IVUS/IVPA imaging platform can be expanded to a theranostic platform, first used for plaque characterization and then for therapy delivery and monitoring. Beyond atherosclerosis, US/PA imaging has also been investigated for the three-dimensional assessment of cardiac ablation lesions, using spectroscopic analysis to differentiate hemoglobin-rich and ablated tissue, as demonstrated in Figure 8.8 (Dana et al. 2014). While the use of US/PA imaging as a tool for therapy guidance in each of these applications is currently in the early stages, it offers the potential to expand the modality beyond strictly diagnostic imaging.

8.6 CLINICAL TRANSLATION OF PA IMAGING: THE CHALLENGES AHEAD

While the complementary nature of the morphological and compositional information that can be obtained from combined US and PA imaging has motivated a rapid and diverse expansion in related preclinical research, PA imaging has not yet been translated into the clinic. Two major advantages of US over other clinical modalities have been that it is nonionizing and that it enables real-time imaging. Ideally, the coupling of PA imaging would retain those advantages while providing clinicians with complimentary functional or compositional information about targeted tissues. However, if these advantages are to remain true for hybrid US/PA imaging applications, significant hurdles remain with regard to both the need for further technical improvements and thorough demonstration of the overall safety of PA imaging. Furthermore, in the case of intravascular imaging, or potential future hybrid endoscopic techniques, the invasive nature of the required imaging procedure introduces further challenges that must be met prior to its widespread clinical acceptance.

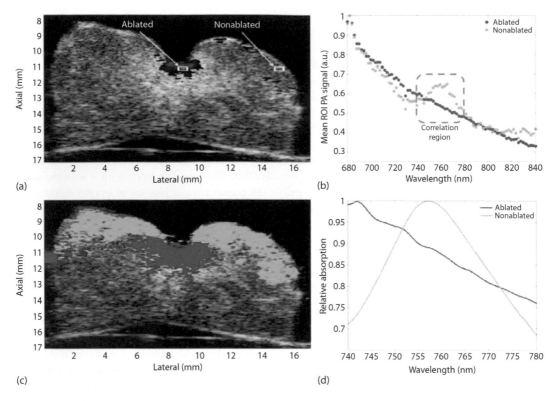

Figure 8.8 PA characterization of cardiac ablation lesions. (a) Single-wavelength photoacoustic image (710 nm) overlaying ultrasound image with ablated and nonablated regions of interest. (b) Mean region of interest photoacoustic signal plotted vs. wavelength. (c) Tissue characterization map. (d) Reference spectra for ablated and non-ablated tissue. (Reprinted from *Heart Rhythm*, 11, Dana N et al., In vitro photoacoustic visualization of myocardial ablation lesions, 150–157, Copyright 2014, with permission from Elsevier.)

8.6.1 CHALLENGES FOR REAL-TIME CLINICAL PA IMAGING

To date, PA imaging techniques have most commonly utilized commercially available US transducers coupled with illumination from a Q-switched Nd:YAG laser to pump an OPO or dye lasing medium, in order to select the desired wavelength (Nikogosyan 1991; Osterberg and Margulis 1986). Such optical sources are attractive in preclinical settings because they offer high pulse energies, on the order of tens of millijoules, over a broad range of wavelengths, allowing optical tuning for PA detection of a wide range of absorbers of interest. The versatility of these laser sources, however, is available at the expense of PRF, which is on the order of tens of hertz. Furthermore, dye lasers require manual switching of the dye material when imaging beyond a limited optical spectral band and are subject to declining energy during imaging as a result of photobleaching, thus limiting their clinical utility for diverse US/PA systems. If PA signal acquisition is required at multiple wavelengths for spectroscopic imaging or signal averaging is utilized to improve image SNR, real-time imaging (≥30 frames per second) may not be achieved using a single laser source. Whereas OPO lasers may be adequate for some US/PA applications that utilize array transducers to capture full two-dimensional images with a single pulse, mechanically scanned intravascular imaging provides an extreme case for optical source requirements if real-time imaging is to be maintained. Using an integrated IVUS/IVPA catheter that is mechanically rotated and pulled back to enable three-dimensional imaging, an optical source operating at several kilohertz and providing approximately 1 mJ/pulse would be required to maintain real-time IVPA imaging capabilities. Such sources, however, are becoming commercially available with output wavelengths desirable for specific PA imaging applications. Moving forward, as specific applications of US/PA imaging look to be translated into real-time clinical systems, the switch from the slow but highly adjustable OPO sources to the faster laser sources with limited spectral ranges may be required.

Additional technical considerations warrant further investigation as US/PA systems look to realize their full clinical potential, namely, the utilized PA signal receivers and signal processing algorithms. In the pre-clinical setting, most US/PA imaging applications have made use of commercially available US transducers for the US transmit-receive as well as the PA signal receiving. In general, however, generated PA pressure waves possess comparatively broadband spectral and lower amplitude characteristics than are present in conventional US imaging. Therefore, the development of more sensitive, broadband PA signal detectors could help improve the overall system sensitivity. Also, the improvement of algorithms for spectral unmixing and the calculation of local optical fluence will help further improve the overall accuracy and sensitivity of PA imaging (Beard 2011). An ultimate goal of PA imaging is to enable true quantification of absorbers, whether endogenous tissue components or cellular/molecular-targeted exogenous contrast agents. In the case of PA imaging and quantification of exogenous contrast agents, the intelligent design and selection based on the unique optical absorbance characteristics can help to improve PA contrast over background absorbers (Luke, Nam, and Emelianov 2013). For this to be realized, however, it is first necessary to develop more precise yet computationally efficient algorithms to account for local fluence and the related wavelength-dependent changes. These technical challenges are all being actively studied within the preclinical arena, and in general, none are likely to prevent the future translation of US/PA imaging systems into the clinic. The extent to which each is addressed, however, will likely change the overall impact which such systems have in clinical practice.

Finally, in the case of IVPA imaging applications, the overall catheter design should exhibit comparable physical dimensions and flexibility to existing clinical catheters. To date, the smallest diameter IVPA cath-eter that has been utilized is comprised of a rigid distal tip and an outer diameter of approximately 1.25 mm (Jansen et al. 2011a), although further minimization to less than 1 mm diameter is possible. Clinically used, mechanically rotated IVUS catheters, on the other hand, rotate within a sheath that has an outer diameter of approximately 700 µm and with a rigid segment surrounding the transducer of less than 1 mm. Combined IVUS/IVPA catheters must meet similar criteria prior to their clinical adoption. While such a miniaturiza-tion is conceptually feasible, it is yet to be demonstrated, and the effectiveness of light delivery for PA signal generation from the small optical fibers that would be required remains a potential obstacle.

8.6.2 SAFETY OF PA IMAGING

Beyond remaining technical challenges, the clinical acceptance of PA imaging will also require a more thor-ough investigation into the safety of each application. In particular, the desire for enhanced SNR should be balanced with unintentional laser–tissue interactions, which may lead to irreparable damage during PA imaging. The ANSI has established wavelength dependent laser fluence exposure limits which can be used as a benchmark for predicting laser safety during PA signal generation (ANSI 2011). Laser–tissue interactions may prove to be of particular interest for IVPA imaging, where the limited size of the optical fiber results in a high local fluence at the distal end of the catheter which can result in local heating during imaging (Sethuraman et al. 2006, 2008).

Additionally, while nanoparticle contrast agents provide great promise as tools for expanding the cellular/molecular sensitivity of PA imaging, the long-term safety of such contrast agents is not yet fully understood. It has been shown in systemically injected animal models, for example, that metallic nanoparticles larger than 6 nm do not ever fully clear the body and that high percentages of the initial injection are retained within the liver and spleen. However, particles smaller than 6 nm are quickly cleared from circulation and therefore may not enable enough time for sufficient accumulation within the desired targets (Longmire, Choyke, and Kobayashi 2008). In an effort to address these conflicting restrictions, biodegradable nanoparticle cluster PA contrast agents have been developed that are comprised of cross-linked 4-nm primary particles that form larger initial sizes until exposed to reduced pH values within intracellular lysosomes (Tam et al. 2010a, 2010b, Yoon et al. 2010). While such an approach that increases the likelihood of contrast agent clearance is ideal in the clinical setting, the overall safety of metallic nanoparticle contrast agents that are greater than 6 nm is also being studied in ongoing U.S. Food and Drug Administration clinical trials for applications beyond cardiovascular medicine (Bao, Mitragotri, and Tong 2013). While such studies will help guide future research utilizing metallic nanoparticle contrast agents by demonstrating their biological interactions and potential toxicity, such relationships will likely need to be further investigated to assess potential additional effects, such as local heating, which can result from the interaction with the nanoparticles and nanosecond-pulsed optical illumination during PA imaging.

8.7 CONCLUSIONS

Combined US/PA imaging offers the potential for acquisition of complimentary information regarding tissue anatomy, composition, and functional/physiological properties, capitalizing on shared US instrumentation to generate spatially coregistered images. The recent commercialization of numerous small-animal PA imaging systems opens the door for investigation of new applications of this hybrid modality within cardiovascular medicine as well as the characterization of novel contrast agents designed to help study disease progression. Furthermore, as US/PA imaging techniques are further refined and tailored to specific applications, the modality can be expected to enter clinical trials and begin to progress toward its true clinical potential.

ABBREVIATIONS

ADC	Analog-to-digital converter
ANSI	American National Standards Institute
AuNR	Gold nanorods
CW	Continuous wave
FDA	United States Food and Drug Administration
FWHM	Full width half maximum
IB-IVUS	Integrated backscatter intravascular ultrasound
ICC	Intraclass correlation
IMT	Intima-media thickness
IVPA	Intravascular photoacoustic
IVUS	Intravascular ultrasound
LDL	Low density lipoprotein
LED	Light-emitting diode
MMP	Matrix metalloproteinase
MSOT	Multispectral optoacoustic tomography
NIR	Near-infrared
NIRS	Near-infrared spectroscopy
Nd:YAG	Neodymium-doped yttrium aluminum garnet
OCT	Optical coherence tomography
OPO	Optical parametric oscillator
OR-PAM	Optical resolution photoacoustic microscopy
PA	Photoacoustic
PAnD	Photoacoustic nanodroplet
PRF	Pulse repetition frequency
PVA	Polyvinyl alcohol
RF	Radiofrequency
sIVPA	Spectroscopic intravascular photoacoustic
SNR	Signal-to-noise ratio
sO_2	Oxygen saturation
sPA	Spectroscopic photoacoustic
SPR	Surface plasmon resonance

SWNTs   Single-walled carbon nanotubes
tPA   Thermal photoacoustic imaging
US   Ultrasound
VCAM-1   Vascular cell adhesion molecule-1
VH-IVUS   Virtual histology intravascular ultrasound

REFERENCES

Altinoglu EI, TJ Russin, JM Kaiser, BM Barth, PC Eklund, M Kester, and JH Adair. 2008. Near-infrared emitting fluorophore-doped calcium phosphate nanoparticles for in vivo imaging of human breast cancer. *ACS Nano* 2 (10):2075–84.

American National Standards Institute (ANSI). 2007. American National Standard for Safe Use of Lasers. ANSI Z136.1-2007, Laser Institute of America, Orlando, FL.

American National Standards Institute (ANSI). 2011. American National Standard for Safe Use of Lasers in Healthcare. ANSI Z136.3-2011, Laser Institute of America, Orlando, FL.

Bao G, S Mitragotri, and S Tong. 2013. Multifunctional nanoparticles for drug delivery and molecular imaging. *Annu Rev Biomed Eng* 15:253–82.

Bayer CL, YS Chen, S Kim, S Mallidi, K Sokolov, and S Emelianov. 2011. Multiplex photoacoustic molecular imaging using targeted silica-coated gold nanorods. *Biomed Opt Express* 2 (7):1828–35.

Bayer CL, GP Luke, and SY Emelianov. 2012. Photoacoustic imaging for medical diagnostics. *Acoustics Today* 8 (4):15–23.

Beard P. 2011. Biomedical photoacoustic imaging. *Interface Focus* 1 (4):602–31.

Bell AG. 1880. Upon the production and reproduction of sound by light. *J Soc Telegraph Eng* 9 (34):404–26.

Bhattacharyya S, S Wang, D Reinecke, W Kiser, Jr, RA Kruger, and TR DeGrado. 2008. Synthesis and evaluation of near-infrared (NIR) dye-herceptin conjugates as photoacoustic computed tomography (PCT) probes for HER2 expression in breast cancer. *Bioconjug Chem* 19 (6):1186–93.

Bowen T. 1981. Radiation-induced thermoacoustic soft tissue imaging. 1981 Ultrasonics Symposium, October 14–16, 1981.

Brunker J, and P Beard. 2010. Pulsed photoacoustic Doppler flowmetry using a cross correlation method. *Proc. SPIE 7564, Photons Plus Ultrasound: Imaging and Sensing 2010*, 756426 (February 23, 2010); doi:10.1117/12.841760.

Brunker J, and P Beard. 2011. Pulsed photoacoustic Doppler flow measurements in blood-mimicking phantoms. *Proc. SPIE 7899, Photons Plus Ultrasound: Imaging and Sensing 2011*, 78991K (February 28, 2011); doi:10.1117/12.874469.

Brunker J, and P Beard. 2013. Acoustic resolution photoacoustic Doppler velocity measurements in fluids using time-domain cross-correlation. *Proc. SPIE 8581, Photons Plus Ultrasound: Imaging and Sensing 2013*, 85811U (March 4, 2013); doi:10.1117/12.2004742.

Chen Y-S, W Frey, S Aglyamov, and S Emelianov. 2012. Environment-dependent generation of photoacoustic waves from plasmonic nanoparticles. *Small* 8 (1):47–52.

Chen Y-S, W Frey, S Kim, P Kruizinga, K Sokolov, and S Emelianov. 2010. Enhanced thermal stability of silica-coated gold nanorods for photoacoustic imaging and image-guided therapy. *Opt Express* 18 (9):8867–78.

Chen Y-S, W Frey, S Kim, P Kruizinga, K Homan, and S Emelianov. 2011. Silica-coated gold nanorods as photoacoustic signal nanoamplifiers. *Nano Lett* 11 (2):348–54.

Chen Y-S, W Frey, C Walker, S Aglyamov, and S Emelianov. 2013. Sensitivity enhanced nanothermal sensors for photoacoustic temperature mapping. *J Biophotonics* 6 (6–7):534–42.

Dana N, L Di Biase, A Natale, S Emelianov, and R Bouchard. 2014. In vitro photoacoustic visualization of myocardial ablation lesions. *Heart Rhythm* 11 (1):150–7.

Davies MJ, and AC Thomas. 1985. Plaque fissuring—The cause of acute myocardial infarction, sudden ischaemic death, and crescendo angina. *Br Heart J* 53 (4):363–73.

de la Zerda A, Z Liu, S Bodapati, R Teed, S Vaithilingam, BT Khuri-Yakub, X Chen, H Dai, and SS Gambhir. 2010. Ultrahigh sensitivity carbon nanotube agents for photoacoustic molecular imaging in living mice. *Nano Lett* 10 (6):2168–72.

de la Zerda A, C Zavaleta, S Keren, S Vaithilingam, S Bodapati, Z Liu, J Levi, BR Smith, TJ Ma, O Oralkan, Z Cheng, X Chen, H Dai, BT Khuri-Yakub, and SS Gambhir. 2008. Carbon nanotubes as photoacoustic molecular imaging agents in living mice. *Nat Nanotechnol* 3 (9):557–62.

Diebold GJ. 2009. Photoacoustic monopole radiation. In *Photoacoustic Imaging and Spectroscopy*, edited by L. V. Wong, 3–17. CRC Press, Boca Raton, FL.

Diebold GJ, and T Sun. 1994. Properties of photoacoustic waves in one, two, and three dimensions. *Acta Acust United Acust* 80 (4):339–51.

Duck FA. 1990. *Physical Properties of Tissue: A Comprehensive Reference Book*. Michigan: Academic Press.

Dussik KT. 1942. Über die Möglichkeit, hochfrequente mechanische Schwingungen als diagnostisches Hilfsmittel zu verwerten. *Z Gesamte Neurologie Psychiatrie* 174 (1):153–68.

Edler I, and CH Hertz. 2004. The use of ultrasonic reflectoscope for the continuous recording of the movements of heart walls. 1954. *Clin Physiol Funct Imaging* 24 (3):118–36.

Eleccion M. 1972. The family of lasers: A survey. *IEEE Spectr* 9 (3):26–40.

Falk E. 1983. Plaque rupture with severe pre-existing stenosis precipitating coronary thrombosis. Characteristics of coronary atherosclerotic plaques underlying fatal occlusive thrombi. *Br Heart J* 50 (2):127–34.

Fang H, K Maslov, and LV Wang. 2007a. Photoacoustic Doppler effect from flowing small light-absorbing particles. *Phys Rev Lett* 99 (18):184501.

Fang H, K Maslov, and LV Wang. 2007b. Photoacoustic Doppler flow measurement in optically scattering media. *Appl Phys Lett* 91 (26).

Fidanza F, A Keys, and JT Anderson. 1953. Density of body fat in man and other mammals. *J Appl Physiol* 6 (4):252–6.

Franklin DL, W Schlegel, and RF Rushmer. 1961. Blood flow measured by Doppler frequency shift of back-scattered ultrasound. *Science* 134 (3478):564–5.

Graf IM, S Kim, B Wang, R Smalling, and S Emelianov. 2012. Noninvasive detection of intimal xanthoma using combined ultrasound, strain rate and photoacoustic imaging. *Ultrasonics* 52 (3):435–41.

Grüneisen E. 1912. Theorie des festen Zustandes einatomiger Elemente. *Annalen der Physik* 344:257–306.

Grzelczak M, J Perez-Juste, P Mulvaney, and LM Liz-Marzan. 2008. Shape control in gold nanoparticle synthesis. *Chem Soc Rev* 37 (9):1783–91.

Gusev VE, and AA Karabutov. 1993. *Laser optoacoustics, Lazernaia optoakustika. English*. New York: American Institute of Physics.

Hannah A, G Luke, K Wilson, K Homan, and S Emelianov. 2014. Indocyanine green-loaded photoacoustic nanodroplets: Dual contrast nanoconstructs for enhanced photoacoustic and ultrasound imaging. *ACS Nano* 8 (1):250–9.

Homan K, S Kim, YS Chen, B Wang, S Mallidi, and S Emelianov. 2010. Prospects of molecular photoacoustic imaging at 1064 nm wavelength. *Opt Lett* 35 (15):2663–5.

Hsieh BY, SL Chen, T Ling, LJ Guo, and PC Li. 2012a. All-optical scanhead for ultrasound and photoacoustic dual-modality imaging. *Opt Express* 20 (2):1588–96.

Hsieh B-Y, S-L Chen, T Ling, LJ Guo, and P-C Li. 2012b. All-optical scanhead for ultrasound and photoacoustic dual-modality imaging. *Opt Express* 20 (2):1588–96.

Jacques SL. 2013. Optical properties of biological tissues: A review. *Phys Med Biol* 58 (11):R37–61.

Jain PK, KS Lee, IH El-Sayed, and MA El-Sayed. 2006. Calculated absorption and scattering properties of gold nanoparticles of different size, shape, and composition: Applications in biological imaging and biomedicine. *J Phys Chem B* 110 (14):7238–48.

Jansen K, AF van der Steen, HM van Beusekom, JW Oosterhuis, and G van Soest. 2011a. Intravascular photoacoustic imaging of human coronary atherosclerosis. *Opt Lett* 36 (5):597–9.

Jansen K, M Wu, AF van der Steen, and G van Soest. 2013. Lipid detection in atherosclerotic human coronaries by spectroscopic intravascular photoacoustic imaging. *Opt Express* 21 (18):21472–84.

Jansen K, M Wu, AF van der Steen, and G van Soest. 2014. Photoacoustic imaging of human coronary atherosclerosis in two spectral bands. *Photoacoustics* 2 (1):12–20.

Jansen K, AFW van der Steen, HMM van Beusekom, JW Oosterhuis, and G van Soest. 2011b. Intravascular photoacoustic imaging of human coronary atherosclerosis. *Opt Lett* 36 (5):597–99.

Jansen K, G van Soest, and AFW van der Steen. 2014. Intravascular photoacoustic imaging: A new tool for vulnerable plaque identification. *Ultrasound Med Biol* 40(6):1037–48.

Karlsson B, and CG Ribbing. 1982. Optical constants and spectral selectivity of stainless steel and its oxides. *J Appl Phys* 53 (9):6340–6.

Karpiouk AB, SR Aglyamov, S Mallidi, J Shah, WG Scott, JM Rubin, and SY Emelianov. 2008. Combined ultrasound and photoacoustic imaging to detect and stage deep vein thrombosis: Phantom and ex vivo studies. *J Biomed Opt* 13 (5):054061.

Karpiouk AB, B Wang, J Amirian, RW Smalling, and SY Emelianov. 2012. Feasibility of in vivo intravascular photoacoustic imaging using integrated ultrasound and photoacoustic imaging catheter. *J Biomed Opt* 17 (9):96008–1.

Karpiouk AB, B Wang, and SY Emelianov. 2010a. Development of a catheter for combined intravascular ultrasound and photoacoustic imaging. *Rev Sci Instrum* 81 (1):014901.

Karpiouk A, B Wang, and S Emelianov. 2010b. Development of a catheter for combined intravascular ultrasound and photoacoustic imaging. *Rev Sci Instrum* 81 (1):014901.

Kim C, R Qin, JS Xu, LV Wang, and R Xu. 2010. Multifunctional microbubbles and nanobubbles for photoacoustic and ultrasound imaging. *J Biomed Opt* 15 (1):010510.

Kim G, SW Huang, KC Day, M O'Donnell, RR Agayan, MA Day, R Kopelman, and S Ashkenazi. 2007. Indocyanine-green-embedded PEBBLEs as a contrast agent for photoacoustic imaging. *J Biomed Opt* 12 (4):044020.

Kim JW, EI Galanzha, EV Shashkov, HM Moon, and VP Zharov. 2009. Golden carbon nanotubes as multimodal photoacoustic and photothermal high-contrast molecular agents. *Nat Nanotechnol* 4 (10):688–94.

Kim S, YS Chen, GP Luke, and SY Emelianov. 2011. In vivo three-dimensional spectroscopic photoacoustic imaging for monitoring nanoparticle delivery. *Biomed Opt Express* 2 (9):2540–50.

Kinney JB, and RH Staley. 1982. Applications of photoacoustic spectroscopy. *Annu Rev Mater Sci* 12 (1):295–321.

Kruger RA, RB Lam, DR Reinecke, SP Del Rio, and RP Doyle. 2010. Photoacoustic angiography of the breast. *Med Phys* 37 (11):6096–100.

Kruger RA, P Liu, YR Fang, and CR Appledorn. 1995. Photoacoustic ultrasound (PAUS)—Reconstruction tomography. *Med Phys* 22 (10):1605–9.

Ku G, and LV Wang. 2005. Deeply penetrating photoacoustic tomography in biological tissues enhanced with an optical contrast agent. *Opt Lett* 30 (5):507–9.

Kuchment P, and L Kunyansky. 2011. Mathematics of photoacoustic and thermoacoustic tomography. In *Handbook of Mathematical Methods in Imaging*, edited by Otmar Scherzer, 817–65. Springer, New York.

Kumar S, J Aaron, and K Sokolov. 2008. Directional conjugation of antibodies to nanoparticles for synthesis of multiplexed optical contrast agents with both delivery and targeting moieties. *Nat Protocols* 3 (2):314–20.

Li C, and LV Wang. 2009. Photoacoustic tomography and sensing in biomedicine. *Phys Med Biol* 54 (19):R59–97.

Li Z, H Li, H Chen, and W Xie. 2011. In vivo determination of acute myocardial ischemia based on photoacoustic imaging with a focused transducer. *J Biomed Opt* 16 (7):076011.

Li L, HF Zhang, RJ Zemp, K Maslov, and L Wang. 2008. Simultaneous imaging of a lacZ-marked tumor and microvasculature morphology in vivo by dual-wavelength photoacoustic microscopy. *J Innov Opt Health Sci* 1 (2):207–15.

Liangzhong X, X Da, G Huaimin, Y Diwu, Z Lvming, and Y Sihua. 2006. Gold nanoshell-based photoacoustic imaging application in biomedicine. International Symposium on Biophotonics, Nanophotonics and Metamaterials, Metamaterials, 2006, October 16–18, 2006.

Longmire M, PL Choyke, and H Kobayashi. 2008. Clearance properties of nano-sized particles and molecules as imaging agents: Considerations and caveats. *Nanomedicine (Lond)* 3 (5):703–17.

Luke GP, SY Nam, and SY Emelianov. 2013. Optical wavelength selection for improved spectroscopic photoacoustic imaging. *Photoacoustics* 1 (2):36–42.

Luke GP, D Yeager, and SY Emelianov. 2012. Biomedical applications of photoacoustic imaging with exogenous contrast agents. *Ann Biomed Eng* 40 (2):422–37.

Mallidi S, T Larson, J Tam, PP Joshi, A Karpiouk, K Sokolov, and S Emelianov. 2009. Multiwavelength photoacoustic imaging and plasmon resonance coupling of gold nanoparticles for selective detection of cancer. *Nano Lett* 9 (8):2825–31.

Mallidi S, GP Luke, and S Emelianov. 2011. Photoacoustic imaging in cancer detection, diagnosis, and treatment guidance. *Trends Biotechnol* 29 (5):213–21.

Mehrmohammadi M, SJ Yoon, D Yeager, and SY Emelianov. 2013. Photoacoustic imaging for cancer detection and staging. *Curr Mol Imaging* 2 (1):89–105.

Meng-Lin L, O Jung-Taek, X Xie, G Ku, W Wei, L Chun, G Lungu, G Stoica, and LV Wang. 2008. Simultaneous molecular and hypoxia imaging of brain tumors in vivo using spectroscopic photoacoustic tomography. *Proc IEEE* 96 (3):481–9.

Millstone JE, SJ Hurst, GS Metraux, JI Cutler, and CA Mirkin. 2009. Colloidal gold and silver triangular nanoprisms. *Small* 5 (6):646–64.

Muller JE, GS Abela, RW Nesto, and GH Tofler. 1994. Triggers, acute risk factors and vulnerable plaques: The lexicon of a new frontier. *J Am Coll Cardiol* 23 (3):809–13.

Nikogosyan DN. 1991. Beta barium borate (BBO): A review of its properties and applications. *Appl Phys A* 52:359–68.

Nikoobakht B, and MA El-Sayed. 2003. Preparation and growth mechanism of gold nanorods (nrs) using seed-mediated growth method. *Chem Mater* 15 (10):1957–62.

Noack J, and A Vogel. 1999. Laser-induced plasma formation in water at nanosecond to femtosecond time scales: Calculation of thresholds, absorption coefficients, and energy density. *IEEE J Quantum Electron* 35 (8):1156–67.

O'Donnell M, ER McVeigh, HW Strauss, A. Tanaka, BE Bouma, GJ Tearney, MA Guttman, and EV Garcia. 2010. Multimodality cardiovascular molecular imaging technology. *J Nucl Med* 51 Suppl 1:38s–50s.

Oraevsky AA, SL Jacques, RO Esenaliev, and FK Tittel. 1994. Laser based optoacoustic imaging in biological tissues. *Proc. SPIE* 0277-786X (2134A):122–8.

Oraevsky AA, and AA Karabutov. 2000. Ultimate sensitivity of time-resolved optoacoustic detection.

Osterberg U, and W Margulis. 1986. Dye laser pumped by Nd:YAG laser pulses frequency doubled in a glass optical fiber. *Opt Lett* 11 (8):516–8.

Pramanik M, and LV Wang. 2009. Thermoacoustic and photoacoustic sensing of temperature. *J Biomed Opt* 14 (5):054024.

Razansky D, NJ Harlaar, JL Hillebrands, A Taruttis, E Herzog, CJ Zeebregts, GM van Dam, and V Ntziachristos. 2012. Multispectral optoacoustic tomography of matrix metalloproteinase activity in vulnerable human carotid plaques. *Mol Imaging Biol* 14 (3):277–85.

Roggan A, M Friebel, K Do Rschel, A Hahn, and G Mu Ller. 1999. Optical properties of circulating human blood in the wavelength range 400–2500 nm. *J Biomed Opt* 4 (1):36–46.

Rosencwaig A. 1978. Photoacoustic spectroscopy. In *Advances in Electronics and Electron Physics*, edited by L. Marton, 207–311. Academic Press, New York.

Sethuraman S, SR Aglyamov, JH Amirian, RW Smalling, and SY Emelianov. 2007. Intravascular photoacoustic imaging using an IVUS imaging catheter. *IEEE Trans Ultrason Ferroelectr Freq Control* 54:978–86.

Sethuraman S, SR Aglyamov, RW Smalling, and SY Emelianov. 2008. Remote temperature estimation in intravascular photoacoustic imaging. *Ultrasound Med Biol* 34 (2):299–308.

Sethuraman S, A Rakalin, S Aglyamov, J Amirian, R Smalling, and S Emelianov. 2006. 6G-3 Temperature monitoring in intravascular photoacoustic imaging. Ultrasonics Symposium, 2006. IEEE, October 6, 2006.

Sheinfeld A, S Gilead, and A Eyal. 2010a. Photoacoustic Doppler measurement of flow using tone burst excitation. *Opt Express* 18 (5):4212–21.

Sheinfeld A, S Gilead, and A Eyal. 2010b. Simultaneous spatial and spectral mapping of flow using photoacoustic Doppler measurement. *J Biomed Opt* 15 (6):066010.

Sigrist MW. 1986. Laser generation of acoustic waves in liquids and gases. *J Appl Phys* 60 (7):R83–122.

Skrabalak SE, J Chen, Y Sun, X Lu, L Au, CM Cobley, and Y Xia. 2008. Gold nanocages: Synthesis, properties, and applications. *Acc Chem Res* 41 (12):1587–95.

Song KH, C Kim, K Maslov, and LV Wang. 2009. Noninvasive in vivo spectroscopic nanorod-contrast photoacoustic mapping of sentinel lymph nodes. *Eur J Radiol* 70 (2):227–31.

Song KH, EW Stein, JA Margenthaler, and LV Wang. 2008. Noninvasive photoacoustic identification of sentinel lymph nodes containing methylene blue in vivo in a rat model. *J Biomed Opt* 13 (5):054033.

Stantz KM, M Cao, B Liu, KD Miller, and L Guo. 2010. Molecular imaging of neutropilin-1 receptor using photoacoustic spectroscopy in breast tumors.

Su JL, B Wang, and SY Emelianov. 2009. Photoacoustic imaging of coronary artery stents. *Opt Express* 17 (22):19894–901.

Tam AC. 1986. Applications of photoacoustic sensing techniques. *Rev Mod Phys* 58 (2):381–431.

Tam JM, AK Murthy, DR Ingram, R Nguyen, KV Sokolov, and KP Johnston. 2010a. Kinetic assembly of near-IR-active gold nanoclusters using weakly adsorbing polymers to control the size. *Langmuir* 26 (11):8988–99.

Tam JM, JO Tam, A Murthy, DR Ingram, LL Ma, K Travis, KP Johnston, and KV Sokolov. 2010b. Controlled assembly of biodegradable plasmonic nanoclusters for near-infrared imaging and therapeutic applications. *ACS Nano* 4 (4):2178–84.

Taruttis A, M Wildgruber, K Kosanke, N Beziere, K Licha, R Haag, M Aichler, A Walch, E Rummeny, and V Ntziachristos. 2013. Multispectral optoacoustic tomography of myocardial infarction. *Photoacoustics* 1 (1):3–8.

VanderLaan D, A Karpiouk, D Yeager, and S Emelianov. 2014. System and integrated catheter for real-time intravascular ultrasound and photoacoustic imaging. Ultrasonics Symposium (IUS), 2014 IEEE International, September 3–6, 2014.

Walsh JT. 2011. Basic interactions of light with tissue. In *Optical-Thermal Response of Laser-Irradiated Tissue*, edited by Ashley J. Welch and Martin J. C. Gemert, 13–26. Springer, the Netherlands.

Wang B, and S Emelianov. 2011. Thermal intravascular photoacoustic imaging. *Biomed Opt Express* 2 (11):3072–8.

Wang L, SL Jacques, and L Zheng. 1995. MCML—Monte Carlo modeling of light transport in multi-layered tissues. *Comput Methods Programs Biomed* 47 (2):131–46.

Wang B, A Karpiouk, D Yeager, J Amirian, S Litovsky, R Smalling, and S Emelianov. 2012a. In vivo intravascular ultrasound-guided photoacoustic imaging of lipid in plaques using an animal model of atherosclerosis. *Ultrasound Med Biol* 38 (12):2098–103.

Wang B, A Karpiouk, D Yeager, J Amirian, S Litovsky, R Smalling, and S Emelianov. 2012b. Intravascular photoacoustic imaging of lipid in atherosclerotic plaques in the presence of luminal blood. *Opt Lett* 37 (7):1244–6.

Wang B, JL Su, AB Karpiouk, KV Sokolov, RW Smalling, and SY Emelianov. 2010a. Intravascular photoacoustic imaging. *IEEE J Select Top Quantum Electron* 16 (3):588–99.

Wang B, JL Su, J Amirian, SH Litovsky, R Smalling, and S Emelianov. 2010b. Detection of lipid in atherosclerotic vessels using ultrasound-guided spectroscopic intravascular photoacoustic imaging. *Opt Express* 18 (5):4889–97.

Wang P, P Wang, H Wang, and J Cheng. 2012c. Mapping lipid and collagen by multispectral photoacoustic imaging of chemical bond vibration. *J Biomed Opt* 17 (9):96010–1.

Wang B, E Yantsen, T Larson, AB Karpiouk, S Sethuraman, JL Su, K Sokolov, and SY Emelianov. 2009. Plasmonic intravascular photoacoustic imaging for detection of macrophages in atherosclerotic plaques. *Nano Lett* 9 (6):2212–7.

Wei W, X Li, Q Zhou, KK Shung, and Z Chen. 2011. Integrated ultrasound and photoacoustic probe for co-registered intravascular imaging. *J Biomed Opt* 16 (10):106001.

Wilson K, K Homan, and S Emelianov. 2012. Biomedical photoacoustics beyond thermal expansion using triggered nanodroplet vaporization for contrast-enhanced imaging. *Nat Commun* 3:618.

Xia Y, Y Xiong, B Lim, and SE Skrabalak. 2009. Shape-controlled synthesis of metal nanocrystals: Simple chemistry meets complex physics? *Angew Chem Int Ed* 48 (1):60–103.

Yang X, SE Skrabalak, ZY Li, Y Xia, and LV Wang. 2007. Photoacoustic tomography of a rat cerebral cortex in vivo with au nanocages as an optical contrast agent. *Nano Lett* 7 (12):3798–802.

Yeager D, YS Chen, S Litovsky, and S Emelianov. 2013. Intravascular photoacoustics for image-guidance and temperature monitoring during plasmonic photothermal therapy of atherosclerotic plaques: A feasibility study. *Theranostics* 4 (1):36–46.

Yeager D, A Karpiouk, B Wang, J Amirian, K Sokolov, R Smalling, and S Emelianov. 2012. Intravascular photoacoustic imaging of exogenously labeled atherosclerotic plaque through luminal blood. *J Biomed Opt* 17 (10):106016.

Yin J, X Li, J Jing, J Li, D Mukai, S Mahon, A Edris, K Hoang, KK Shung, M Brenner, J Narula, Q Zhou, and Z Chen. 2011. Novel combined miniature optical coherence tomography ultrasound probe for in vivo intravascular imaging. *J Biomed Opt* 16 (6):060505.

Yoon SJ, S Mallidi, JM Tam, JO Tam, A Murthy, KP Johnston, KV Sokolov, and SY Emelianov. 2010. Utility of biodegradable plasmonic nanoclusters in photoacoustic imaging. *Opt Lett* 35 (22):3751–3.

Zhang E, J Laufer, and P Beard. 2008. Backward-mode multiwavelength photoacoustic scanner using a planar Fabry-Perot polymer film ultrasound sensor for high-resolution three-dimensional imaging of biological tissues. *Appl Opt* 47 (4):561–77.

Hybrid intravascular imaging in the study of atherosclerosis

CHRISTOS V. BOURANTAS, JAVIER ESCANED, CARLOS A.M. CAMPOS, HECTOR M. GARCIA-GARCIA, AND PATRICK W. SERRUYS

9.1 INTRODUCTION

Accurate evaluation of coronary artery morphology and quantification of luminal stenosis is crucial for the assessment of the extent and severity of coronary artery disease and for evaluation of atherosclerotic disease progression. Contrast coronary angiography has been the traditional method for the visualization of coronary anatomy and geometry and the quantification of luminal stenosis. The advent of quantitative coronary angiography (QCA) in 1978 by Reiber et al. allowed for a more objective and reproducible estimation of the severity of a luminal stenosis and permitted the use of this modality in numerous studies that evaluated the effect of interventional and pharmaceutical treatments (Berry et al. 2007; Burton et al. 2003; Reiber et al. 1978; Syvanne, Nieminen, and Frick 1994).

A significant limitation of QCA is the fact that the estimation of luminal stenosis relied on two-dimensional (2-D) angiographic images, which provide inaccurate estimations in case of foreshortening or overlapping. In addition, it is unable to provide information about the composition of the plaque, which, as it has been shown in numerous histology-based studies, provides useful prognostic information (Davies and Thomas 1985; Falk 1983). The advent of 3-D QCA that incorporated information from different angiographic projections in order to extract the coronary geometry and reconstruct the 3-D luminal anatomy managed to address the first limitation of QCA and allowed a more accurate assessment of the significance of coronary obstruction, although even the 3-D approach was incapable of providing information about the composition of the plaque (Bourantas et al. 2009; Yong et al. 2011).

In 1972, Bom et al. developed the first intravascular ultrasound (IVUS) catheter that permitted imaging of luminal pathology, and 16 years later, Yock et al. presented a catheter with a smaller diameter that was suitable for intracoronary imaging (Bom, Lancee, and Van Egmond 1972; Yock, Johnson, and Linker 1988). These advances opened new horizons in the study of atherosclerotic evolution as they allowed for the first time detailed imaging of the lumen and plaque morphology, quantification of plaque burden, and assessment of the composition of the plaque (Mintz et al. 2001). In more recent years, IVUS imaging has been extensively used to study the efficacy of new interventional and pharmaceutical treatments. In parallel, an effort was made to design different types of intravascular imaging catheters that would provide high-resolution imaging of the lumen, identification of plaque characteristics associated with increased vulnerability and precise detection of plaque inflammation (Nissen et al. 2004; Tardif et al. 2007). Today, several invasive imaging modalities are available. Some of these imaging systems have already been used in clinical practice such as angioscopy, thermography, optical coherence tomography (OCT), and near-infrared spectroscopy (NIRS), as discussed in Chapters 8 and 17 of this book, while other intracoronary imaging catheters remain restricted to preclinical applications, such as intravascular magnetic resonance imaging, Raman spectroscopy, photoacoustic imaging, near-infrared fluorescence imaging (NIFR), and time resolved fluorescence spectroscopy (TFRS), as demonstrated in Chapters 8 and 17 of this book as well (Figure 9.1). Each of these modalities has unique strengths but also weaknesses that have not permitted complete visualization of the vessel wall pathology. Therefore, an effort was made to design new catheters and software that would combine different intravascular techniques permitting comprehensive imaging of vessel pathophysiology.

The aim of this chapter is to review the new developments in the field. In the following sections, we focus on the presentation of the available intravascular imaging modalities, highlight the limitations of each technique, and then describe the newer hybrid imaging approaches. The advantages, limitations, and potential value of each imaging system or hybrid approach in the study of plaque progression are also reviewed.

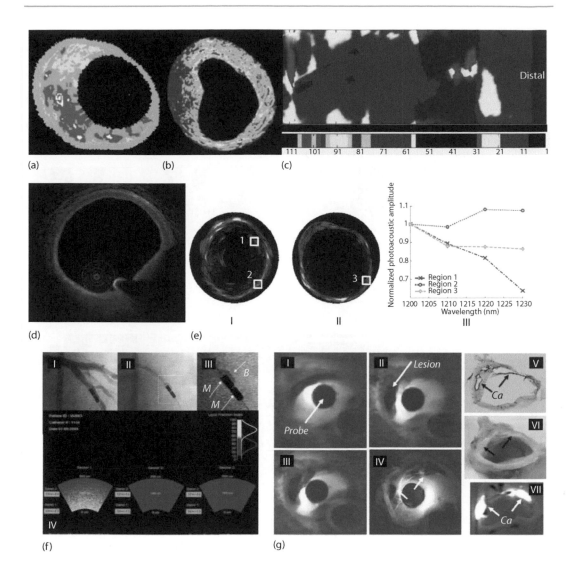

Figure 9.1 Intravascular imaging techniques developed for assessing plaque composition. (a) Virtual histology IVUS. (b) Integrated backscatter IVUS. (c) Output of an NIRS catheter. The spread out plot indicates the probability for the presence of a lipid core (yellow corresponds to high probability and red to low), while the gray areas correspond to segments with poor NIRS signal. The block chemogram provides a summary of the raw data and is illustrated on the bottom of the panel. (d) Optical coherence tomographic image portraying a lipid-rich plaque. (e) intravascular photoacoustic (IVPA) images of a diseased (I) and a normal (II) aorta; different tissues have different photoacoustic fingerprints [i.e., lipid tissue (1), normal vessel wall (2), and media-adventitia (3)], which allow identification of atheroma's composition (III). (f) X-ray images showing an intravascular magnetic spectroscopy catheter (I–III) (M indicates the probe, and B, the balloon used to obstruct the vessel and prevent flow); (IV) the output of this examination is color coded displayed, with the yellow corresponding to lipid tissue, and the blue, to nonlipid tissue. (g) Output of intravascular magnetic resonance imaging (I–IV); images were obtained from an atherosclerotic iliac artery of a cadaveric animal model. The dark areas at 9 o'clock (II, III) and 12 o'clock (IV) show the presence of calcific tissue, which was confirmed in histology by Van Kossa staining (V–VI) and by microcomputed tomography (VII). (*Continued*)

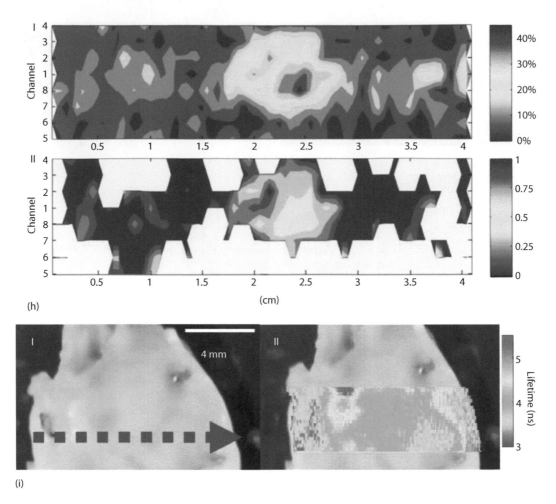

(h)

(i)

Figure 9.1 (Continued) (h) Color-coded display of the data acquired by an intravascular Raman spectroscopy catheter. The first panel shows the total cholesterol distribution along the studied vessel (in the y axis, the number of the sensors used to scan the vessel) with the yellow-red color representing increased cholesterol, whereas the second panel provides information about the presence of the nonesterified cholesterol which was measured when the total cholesterol was >5%. (i) An atherosclerotic plaque of a carotid artery imaged using a fluorescence lifetime imaging apparatus, which provides time resolved fluorescence spectroscopy (TRFS) imaging of relatively large surfaces. The final output gives information about the biochemical composition of the superficial plaque and is color coded displayed (the red corresponds to fibrotic plaque, the yellow to fibro-lipid and the cyan to normal endothelium) (II). (From Bourantas CV et al., *J Invasive Cardiol*, 25, 24A–26A, 2013. With permission.)

9.2 CURRENTLY AVAILABLE INTRAVASCULAR IMAGING MODALITIES: ADVANTAGES AND LIMITATIONS IN THE STUDY OF ATHEROSCLEROSIS

9.2.1 INTRAVASCULAR ULTRASOUND

IVUS imaging is the most widely used intravascular imaging technique. IVUS requires the advancement of a catheter, with an imaging probe at the tip, to the more distal coronary artery. The probe is able to emit ultrasound signal perpendicular to the axis of the catheter, at a frequency of 20–70 MHz, and receive the reflected signal from the vessel wall and adjacent tissues. These signals are then processed to generate cross-sectional images of the vessel. In these images, the operator is able to identify the lumen and the outer vessel wall, including the plaque burden and acquire information about the composition of the plaque (Di Mario et al. 1992; Mintz et al. 2001; Thieme et al. 1996). Several histology-based studies have shown that grayscale IVUS has a limited capability to differentiate plaque components (Hiro et al. 1996; Kubo et al. 2007; Ohtani et al. 2006). This pitfall has been, at least partially, addressed by the radiofrequency analysis of the backscatter signal (RF-IVUS), which has been extensively used to assess plaque phenotype and study atherosclerotic evolution (Figure 9.1a and b) (Amano et al. 2011; Calvert et al. 2011; Mehta et al. 2007; Stone et al. 2011). In the recently published PROSPECT and VIVA studies, RF-IVUS was used to detect plaque features associated with increased vulnerability and future events. These studies demonstrated that a thin cap fibroatheroma phenotype, a plaque burden >70%, and a minimum lumen area <4 mm^2 were independent predictors of future culprit lesions and events (Calvert et al. 2011; Stone et al. 2011). Although there are data about the ability of RF-IVUS to detect plaque characteristics that are related with poor outcomes, recent reports casted doubts about the reliability for characterizing plaques with a complex morphology, or the plaque behind calcific tissues (Sales et al. 2010; Thim et al. 2010). Other significant limitations of IVUS are associated with the increased noise and artifacts seen in some images that make the identification of specific plaque features difficult. IVUS also suffers from relatively poor resolution that does not permit accurate assessment of the thickness of the fibrous cap over necrotic cores and imaging of microfeatures (i.e., microcalcifications, macrophages, etc.) that have been related with increased vulnerability. Finally, IVUS gives no information about the 3-D geometry of the vessel and the exact distribution of the plaque on the vessel.

9.2.2 ANGIOSCOPY

Angioscopy was the first intracoronary imaging technique introduced in the middle of the 1980s. This approach involves advancing a catheter with illumination fibers on its tip into the coronary artery for direct visualization of the luminal surface. Imaging required an occlusion balloon to be inflated proximally to the catheter so as to impede flow and eliminate blood from the imaging field. The angioscopic images can be used to evaluate stent coverage; identify the presence of thrombus, plaque rupture, or dissections; and provide information about the composition of the superficial plaque (Kotani et al. 2006; Kubo et al. 2007; Uchida 2011). It has been speculated that the color of the plaque in the angioscopic images provides information regarding plaque vulnerability, with plaques appearing yellow having the histological characteristics of a vulnerable plaque (a thin fibrous cap covering a lipid-rich core) and those plaques appearing white being more stable (Ohtani et al. 2006; Thieme et al. 1996; Uchida et al. 1995). Advances in coronary angioscopy and the implementation of NIFR and microscopy have enabled more detailed assessment of the plaque and identification of lipid-rich plaques in deeper layers (Uchida et al. 2010). Unfortunately, angioscopy cannot provide information about more distal plaques, the luminal morphology, and cannot give quantitative data regarding the plaque burden, the stent, or the outer vessel wall. These limitations, as well as the fact that this approach requires proximal balloon inflation that can cause vessel trauma, have restricted the application of angioscopy in clinical practice and research.

9.2.3 THERMOGRAPHY

Intravascular thermography has been used in the past to detect the presence of vessel wall inflammation. This modality relies on the measurement of the heat released by activated inflamed tissues that are located in unstable plaques. Several reports have shown that coronary plaque temperature is higher in patients having plaques with positive remodeling and in unstable plaques compared to stable lesions (Toutouzas et al. 2007). In addition, it has been shown that thermography can detect more accurately the culprit lesion than IVUS (which can detect the culprit plaque in only 30%–40% of the cases) in patients suffering from acute coronary syndromes (Takumi et al. 2007). However, the fact that this technique cannot visualize the lumen and the plaque and that it requires a proximal balloon occlusion to prevent flow for more accurate assessment of vessel wall temperature has limited the application of thermography in research arena (Rzeszutko et al. 2006).

9.2.4 NEAR-INFRARED SPECTROSCOPY

NIRS takes advantage of the fact that different organic molecules absorb and scatter the near-infrared light at different degrees and various wavelengths. The processing of the reflected signal provides information about the chemical composition of different tissues and appears to permit reliable detection of the lipid component. The output of the NIRS analysis is called chemogram and is a 2-D plot that utilizes a color-coded map to provide information about the probability of the presence of a lipid component into the vessel wall, as shown in Figure 9.1c (red color indicates high probability, and yellow, low probability).

The reliability of this technique has been evaluated in histology-based *ex vivo* imaging studies of human tissue specimens, while the SPECTRAL study was the first report that demonstrated the feasibility of a NIRS catheter in clinical setting (Gardner et al. 2008; Waxman et al. 2009). Recently, NIRS has been used to assess changes in the composition of the plaque and the prognostic implications of plaque morphology. In the YELLOW trial, NIRS was implemented to assess the short-term effect of intensive medical treatment with rosuvastatin on the burden and composition of the plaque in obstructive (Kini et al. 2013). In addition, in the European Collaborative Project on Inflammation and Vascular Wall Remodeling in Atherosclerosis (ATHEROREMO)–NIRS Substudy, NIRS was used to examine the prognostic implications of the presence of lipid-rich atherosclerotic plaques (Bourantas et al. 2013a).

A limitation of NIRS is that this approach is able to detect superficial large lipid-rich plaques (cap thickness <450 μm, core circumferential extent >60°, plaque thickness >200 μm). On the other hand, in contrast to RF-IVUS, NIRS is able to detect more accurately lipid cores behind calcific deposits. Other pitfalls of NIRS relate to the inability of this modality to provide information about coronary geometry or to provide visualization of the lumen and the plaque.

9.2.5 OPTICAL COHERENCE TOMOGRAPHY

OCT is the optical analog of IVUS that involves the advancement of an optical fiber within the coronary arteries (Figure 9.1d) for morphological imaging. The fiber emits light perpendicular to the catheter axis and is able to receive and analyze the reflected signal, allowing visualization of intracoronary features with significantly higher axial resolution than IVUS (10–20 μm vs. 100–150 μm). Several histology-based studies have demonstrated that OCT allows accurate characterization of the composition of the superficial plaque, and today, OCT is regarded as the gold standard for the assessment of plaque phenotype (Kawasaki et al. 2006; Yabushita et al. 2002). In addition, OCT is able to portray details that are unseen by other imaging techniques, such as the presence of macrophages, neovascularizations, microcalcifications, and cholesterol crystals, and can accurately estimate the thickness of the fibrous cap. In addition, OCT can identify the presence of vessel erosion and differentiate the type of thrombus (Tearney et al. 2012).

However, OCT has a limited penetration depth (1–2 mm), and therefore, this approach is unable to visualize the entire plaque, particularly in heavily diseased segments, or quantify total plaque burden. Other limitations of OCT relate to the fact that the emitted light cannot penetrate the lipid tissue and that OCT has limited capability to differentiate the lipid from the calcific tissue when this is deeply embedded in the atheroma (Manfrini et al. 2006).

9.2.6 INTRAVASCULAR MAGNETIC RESONANCE SPECTROSCOPY

Intravascular magnetic resonance spectroscopy requires the advancement of a catheter with a magnetic resonance probe on the tip (Figure 9.1f). The transducer is able to generate a magnetic field that allows detection of the lipid component at the superficial (depth 0–100 μm) and the deeper (depth 100–250 μm) plaque in a radial sector of 60°. Imaging involves the inflation of a balloon that is located at the proximal end of the probe and is used to stabilize the probe against the vessel wall and prevent distal blood flow. The designed catheter has a large diameter (5.2F), and requires 51 seconds to acquire a single image. Furthermore, intravascular magnetic resonance spectroscopy does not allow visualization of the lumen, vessel wall, or plaque morphology and is unable to detect other tissues apart from the lipid. The proposed modality has been used in small-scale feasibility studies, but the previously mentioned limitations have impeded broad application in the clinical and research arena (Regar et al. 2006; Schneiderman et al. 2005).

9.3 FUTURE TRENDS IN INTRAVASCULAR IMAGING

9.3.1 INTRAVASCULAR MAGNETIC RESONANCE IMAGING

Intravascular magnetic resonance imaging appears able to overcome some of the limitations of the available invasive modalities, as it has an increased penetration that permits visualization of the entire vessel wall beyond calcific and lipid tissue (Figure 9.1g). The feasibility of this technique was under question due to concerns about the safety of this modality (increased heating—produced during imaging), the quality of the obtained images (increased noise), and the time that was necessary for image acquisition and processing. However, with recent advances in intravascular magnetic resonance technology, these limitations have been overcome. In a recent study, Sathyanarayana et al. developed an intravascular magnetic resonance catheter that has a diameter of 9F, which was able to acquire cross-sectional images of the vessel wall with resolution up to 300 μm and at a frame rate of 2 frame/s (Sathyanarayana et al. 2010). Additional effort will be needed to further miniaturize the intravascular coil and expedite image acquisition before this modality finds applications in the clinical setting.

9.3.2 PHOTOACOUSTIC IMAGING

Intravascular photoacoustic (IVPA) imaging takes advantage of the frequency of the sound that is produced during the thermal expansion of different irradiated tissues. Several studies have demonstrated that IVPA can differentiate normal from atherosclerotic vessels, detect the presence of neovascularization, provide information about the composition of the plaque, and visualize stent morphology (Figure 9.1e) (Su et al. 2010; Wang et al. 2010). Moreover, the use of gold nanoparticles that have high absorption coefficients has enabled the visualization of macrophages associated with vessel wall inflammation and increased vulnerability (Wang et al. 2009). Although experimental studies have shown that IVPA may be useful in the clinical arena, this modality has currently significant limitations, including suboptimal image quality, long acquisition times required for intravascular imaging, and limited evidence regarding the safety of this technique. The advantages and limitations of this technology are described in great detail in Chapter 8 of this book.

9.3.3 RAMAN SPECTROSCOPY

Raman spectroscopy relies on the spectral analysis of the Raman scattering of a tissue after illumination with a laser beam. The Raman effect involves the scattering of light when it passes through molecules. This process changes the energy of the photons and modifies the frequency of the backscattered light. The Raman spectra is unique for each molecule and thus it can be used to identify the chemical composition of tissues (Bruggink et al. 2010). The feasibility of this technique has been tested *in vitro* and has been shown to be capable of providing a detailed analysis of the composition of plaque, detection of elastin, collagen, calcium, and esterified and nonesterified cholesterol (Figure 9.1h) (Motz et al. 2006; Romer et al. 1998, 2000; Salenius et al. 1998; van De Poll et al. 2001). However, this modality has not been effectively applied in humans because of the difficulty in acquiring a good-quality signal. High-wave number Raman spectroscopy seems to overcome this limitation, while recent advances in catheter design have enabled the construction of a catheter for intravascular imaging (Brennan et al. 2008; Chau et al. 2008).

9.3.4 NEAR-INFRARED FLUORESCENCE IMAGING

NIRF imaging is a promising modality that has been recently introduced for the study of atherosclerosis. This approach relies on the injection of activatable NIRF agents that bind different molecular targets within the plaque, which are associated with increased inflammatory activity and neovascularization (Jaffer, Libby, and Weissleder 2009). For *in vivo* NIRF imaging, an intravascular catheter has been constructed that can detect the presence of these molecular markers and provide information about plaque vulnerability. NIRF imaging catheters have been extensively tested in experimental models (Jaffer et al. 2008, 2011). However, further research is required to prove the safety of these markers before considering the use of NIRF imaging in humans.

9.3.5 TIME-RESOLVED FLUORESCENCE SPECTROSCOPY

TRFS assesses the time that is required to resolve the fluorescence that is emitted after molecules are excited by light. Numerous experimental studies have provided evidence that TRFS can characterize the composition of plaque (Figure 9.1i) (Phipps et al. 2011). Also, studies have shown that this technique can detect the presence of macrophages (Marcu et al. 2005, 2009). This modality has significant limitations, however, including poor tissue penetration (250 μm) that necessitates the catheter to be in contact with the vessel wall and a narrow field of view that does not permit imaging of the entire circumference of the vessel wall (Marcu 2010; Phipps et al. 2011). Other significant drawbacks of TRFS are its inability to visualize the luminal morphology and the vessel wall and the fact that it cannot give any information about the composition of the entire plaque.

9.3.6 INTRAVASCULAR SCINTILLATION PROBES

Intravascular scintillation (IVS) was introduced in the beginning of the century to detect vascular inflammation. Several microcatheters have been designed with a diameter of 1 mm that were tested in experimental studies which showed that these prototypes are able to detect radioactive tracers such as ^{18}F-FDG, creating hope that they may have a role in detecting high-risk inflamed plaques (Hosokawa et al. 2006; Lederman et al. 2001; Mukai et al. 2004). Nevertheless, these modalities have significant limitations, including the increased time needed for imaging, the necessity to inject particles emitted radio-pharmaceuticals, the potential need for blood removal, and their inability to assess plaque morphology. Another significant limitation of this technique is the fact that the radioactive tracer that has been mainly used in validation studies was ^{18}F-FDG, which it is uptaken not only by the inflamed vessel wall but also by the myocardium. Efforts were made to develop combined OCT-IVS catheters; however, the limitations of the IVS and the increased size of the first prototypes (3 mm) have not enabled their use in the clinical arena (Piao et al. 2005).

9.3.7 INTEGRATION OF INTRAVASCULAR IMAGING TECHNOLOGIES

As summarized previously, there are numerous intravascular imaging modalities for the study of atherosclerosis. Each of these modalities has significant advantages, as well as weaknesses, that prevent complete characterization of plaque pathology. Taking the best from each imaging modality and integrating complementary data provided by different imaging approaches would provide an obvious solution. This fusion can be achieved either through the design of new dual-probe catheters that will allow simultaneous multicoronary imaging (e.g., combination of IVUS and NIRS) or through the development of efficient methodologies that will permit offline integration of the data acquired by different imaging techniques (i.e., coronary angiography and IVUS, IVUS and computed tomographic coronary angiography [CTCA], etc). Simultaneous multimodality imaging would seem to be a superior approach, overcoming the problem of accurate coregistration and reducing processing time, and therefore is currently the most frequently used approach for hybrid intravascular imaging (Bourantas et al. 2013b).

9.4 AVAILABLE SOFTWARE AND METHODOLOGIES FOR THE CO-REGISTRATION OF INTRAVASCULAR IMAGING DATA

9.4.1 FUSION OF INTRAVASCULAR ULTRASOUND AND X-RAY ANGIOGRAPHY

The first data fusion technique for IVUS and x-ray angiography was proposed by Klein et al. in 1992. The proposed approach included the extraction of the 3-D luminal centerline from two angiographic projections and the placement of the IVUS borders onto the 3-D luminal backbone (Klein et al. 1992). Although this technique had significant limitations, as it was unable to estimate the orientation of the IVUS contours onto the 3-D centerline, it opened new horizons in coronary representation. A few years later, Lengyel et al. utilized side branches to determine the absolute orientation of the IVUS contours, but this approach made substantial approximations in the extraction of the luminal centerline, and thus, it did not have clinical applications (Lengyel, Greenberg, and Pop 1995). Shekhar et al. and Subramanian et al. proposed more accurate methodologies for the fusion of IVUS and x-ray data, but both approaches required multiple angiographic images and thus increased screening time and radiation dose, and thus, they had a limited value in the research arena (Shekhar et al. 1996; Subramanian et al. 2000).

The first clinically applicable method was proposed by Wahle (Wahle et al. 1998). This approach used the position of the guidewire in the lumen to approximate the trajectory of the IVUS catheter and required only two sets of orthogonal angiographic images to extract the IVUS catheter path. The detected IVUS borders were then placed perpendicularly onto the extracted path and their relative orientation was estimated using the sequential triangulation algorithm (Prause et al. 1997). Finally, an efficient methodology was implemented to define the absolute orientation of the IVUS borders, which relied on the comparison of the IVUS catheter and of the luminal silhouette in the angiographic projections with the projected 3-D path and 3-D model onto these images. The only drawback of this approach was the approximations that were made in the extraction of the catheter path. More specifically, the extraction of the catheter path was based on the identification of corresponding points in the two angiographic projections, and then "normal" lines were drawn from these points that were intersected in 3-D points that defined the 3-D catheter path. However, quite often, these lines did not intersect, and in that case, the 3-D point was approximated in the middle of the distance between the two lines. These rough estimations of position were likely to distort the morphology of the 3-D path.

Slager et al. proposed an approach that followed similar steps to reconstruct the coronary anatomy, but they implemented a different methodology for the extraction of the catheter path (Slager et al. 2000). According to this technique, a 3-D circular segment was used to define the 3-D trajectory of the IVUS catheter; this segment was stepwise adapted in 3-D space until the computed biplane projections matched with the silhouette of the catheter in both angiographic images. Limitations of this approach relate to the inability to accurately reconstruct segments that appeared foreshortened in the angiographic images, while its validation showed an increased error in distance between the silhouette of the path in the x-ray images and the projected 3-D path of up to 1.6 mm.

Bourantas et al. (2005) introduced a more accurate catheter path extraction methodology, which used a cubic B-spline to approximate the IVUS trajectory in each of the two projections. Both B-splines were extruded parallel to their plane, forming two surfaces; the intersection of the two surfaces was a 3-D curve that corresponded to the 3-D trajectory of the IVUS catheter. Validation of the proposed methodology *in vivo* and *in vitro* showed that this technique overcomes many of the common problems, such as the presence of foreshortening and it was able to reconstruct arteries with a complex geometry (S-shaped arteries), as well as segments where the catheter path was not visible over the entire course (Bourantas et al. 2005).

Although the more recent techniques appear reliable and are able to provide geometrically correct models, they did not have applications in the clinical arena. The two user-friendly systems that are available today and incorporate a visualization module that provides comprehensive imaging of the final objects have failed to expand the applications of this hybrid technique (Figure 9.2) (Bourantas et al. 2008; Wahle, Olszewski, and Sonka 2004). This could be attributed to the limited added value of the information provided by the 3-D models with regard to treatment planning and to the fact that coronary reconstruction is a time-consuming process, requiring the implementation of tedious protocols during image acquisition. To address the later pitfall,

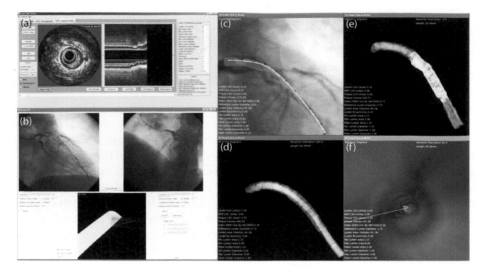

Figure 9.2 Methodology implemented to reconstruct the coronary anatomy using the ANGIOCARE software. (a) Segmentation of the intravascular ultrasound sequence; (b) detection of the catheter path in two angiographic images and extraction of its 3-D geometry; (c) back-projection of the reconstructed model onto the angiographic image; (d) 3-D visualization of the obtained model; the outer vessel wall is portrayed in a transparent fashion allowing imaging of the distribution of the plaque; (e) the user can interact with the final object, select a segment of interest, and acquire quantitative data regarding its length, the luminal, outer vessel wall, and the plaque dimensions that are displayed in the left–bottom side of the screen; (f) virtual endoscopy and assessment of the vessel anatomy from inside. (From Bourantas CV et al., *J Am Coll Cardiol*, 61, 1369–1378, 2013. With permission.)

we have recently introduced a simplified reconstruction methodology that is similar to the one proposed by Lengyel et al. but implements a more sophisticated methodology for the accurate extraction of the catheter centerline. The proposed approach requires only two angiographic projections and an IVUS sequence to reconstruct coronary anatomy and therefore can be used even for the retrospective reconstruction of data that have been acquired during a conventional IVUS examination (Bourantas et al. 2013c).

The fusion of IVUS with the angiographic data has been used over the last years in research to study the association between the local hemodynamic forces and the atherosclerotic evolution. Several *in vivo* studies have shown that low or oscillating shear stresses are associated with atherosclerotic disease progression in native and stented segments, while the PREDICTION study has provided robust evidence that the local hemodynamic patterns are independent predictors of future cardiovascular events (Papafaklis et al. 2010; Stone et al. 2012; Wentzel et al. 2001).

9.4.2 FUSION OF INTRAVASCULAR ULTRASOUND AND COMPUTED TOMOGRAPHY

van der Giessen et al. were the first who introduced a methodology for the integration of IVUS and CTCA data (van der Giessen et al. 2010). The proposed method used CTCA to extract the luminal centerline and anatomical landmarks seen in both IVUS and CTCA to identify the position and the orientation of the IVUS images onto the 3-D centerline. This method was validated in 35 coronaries from 23 patients. Three-dimensional reconstruction was feasible in 31 cases (89%), while in four arteries, there were insufficient anatomical landmarks.

Limitations of the proposed reconstruction methodology are related to the inability to reconstruct segments without side branches, as at least two landmarks are necessary to identify correspondence between CTCA and IVUS. In addition, this process was laborious and time consuming and required extensive human interaction in all the reconstruction steps. Finally, although the authors tested the reproducibility of the blood flow measurements, they did not examine the accuracy of the methodology to reconstruct the coronary anatomy using a reference gold standard methodology.

Similar coregistration approaches have been proposed recently by Bougers et al. and Voros et al. for the fusion of IVUS and CTCA (Boogers et al. 2012; Voros et al. 2011). Although all these techniques can be implemented only in patients who have CTCA, coronary angiography, and IVUS examination, they have attracted attention since they allow us to appreciate the limitations and capabilities of noninvasive imaging for assessing plaque burden and composition. Today, it is evident that CTCA provides reliable identification of the plaque and accurate evaluation of the luminal, outer vessel wall, and plaque dimensions and is able to assess with a moderate accuracy the necrotic core, the fibrofatty, the fibrous, and the calcific burden (Boogers et al. 2012; Papadopoulou et al. 2011; Voros et al. 2011).

9.4.3 COMBINED INTRAVASCULAR ULTRASOUND AND NEAR-INFRARED SPECTROSCOPY IMAGING

Several studies that compared IVUS and NIRS showed that there was a significant discrepancy in the estimations of the two techniques with regard to the lipid component (Brugaletta et al. 2011; Pu et al. 2012). This can be attributed to the different qualities and limitations of the two modalities. A new technology has been recently introduced named TVC (MC 7 system, InfraRedx, Burlington, Massachusetts), which incorporates both an NIRS light source and an IVUS probe on the catheter tip, permitting simultaneous NIRS and IVUS imaging. The catheter has already been used *in vivo* to compare plaque characteristics in males and females and examine plaque morphology in culprit and nonculprit lesions (Bharadwaj et al. 2016). Three studies—two in patients admitted with a ST-elevation myocardial infarction and one in patients with an non-ST-elevation myocardial infarction—showed that culprit lesions have specific morphological characteristics, i.e., an increased plaque burden and lipid component that allow their differentiation from the nonculprit lesions with a high accuracy (Madder et al. 2013, 2015, 2016). Nevertheless, there are no data about the predictive value of combined NIRS-IVUS to identify lesions that are likely to progress and cause cardiovascular events. To answer these questions, two studies, the PROSPECT II (NCT02171065) and the Lipid Rich Plaque (NCT02033694), have commenced and are expected to provide robust data about the value of hybrid-intravascular imaging in detecting high-risk plaques.

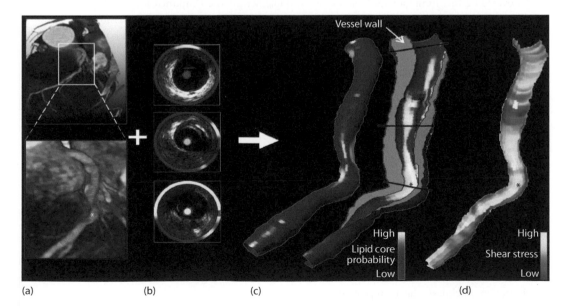

Figure 9.3 Fusion of CTCA (a), IVUS and NIRS (b). The obtained 3-D reconstructed artery is shown in Panel (c) and allows evaluation of vessel geometry and of the distribution of the plaque and identification of the location of the lipid-rich plaques. (d) Shear stress distribution onto the luminal surface. (From Wentzel JJ et al., *Circ Cardiovasc Imaging*, 3, e6–e7, 2010. With permission.)

Moreover, the output of this combined NIRS-IVUS catheter has been merged with CTCA data. This fusion has provided comprehensive 3-D models that allow assessment of the vessel wall geometry and of the distribution of the plaque and identification of the lipid-rich plaques on the vessel wall. In addition, these 3-D objects can be used for blood flow simulation and evaluation of the association between shear stress and plaque composition (Figure 9.3) (Wentzel et al. 2010).

9.4.4 FUSION OF OPTICAL COHERENCE TOMOGRAPHY AND CORONARY ANGIOGRAPHY

The first methodology for the fusion of OCT and x-ray data was proposed in 2011 by Tu et al. (2011). This approach was able to coregister the 3-D QCA and the OCT data in real time, facilitating application in the clinical setting. A limitation of this method relates to the inability to estimate the correct orientation of the OCT borders onto the 3-D model. In a more recent publication, this research group utilized the origin of side branches to determine the orientation of the OCT images, but again, they neglected the relative axial twist of the frames that occurs during the pullback of the catheter in tortuous vessels (Tu et al. 2013).

To overcome this drawback, Bourantas et al. (2008, 2012) implemented the reconstruction methodology previously proposed for the fusion of IVUS and x-ray data. This approach can be used only for the integration of the OCT data obtained at a low pull-back speed (i.e., using an M2 or M3 OCT system, LightLab Imaging, Westford, Massachusetts) and includes four steps: (1) the extraction of the OCT catheter path from two end-diastolic angiographic images, (2) the segmentation of the end-diastolic OCT frames, (3) the placement of the detected borders onto the 3-D path and the estimation of their relative axial twist with the use of the sequential triangulation algorithm, and (4) the determination of the absolute orientation of the first OCT frame using anatomical landmarks (i.e., side branches) that are visible in both OCT and x-ray angiography.

This reconstruction approach was implemented to identify the location of a ruptured plaque and demonstrate that the endothelial shear stress was increased at the ruptured site, confirming speculations and findings of previous reports (Fukumoto et al. 2008). However, this methodology did not have broad applications in the research arena since it required a short interval between the end-diastolic OCT frames. Therefore, this approach could not be implemented for the fusion of frequency domain (FD)-OCT with the x-ray data.

To overcome this limitation, Athanasiou et al. implemented the simplified methodology proposed by Bourantas et al. (2013) for the integration of IVUS and x-ray angiography data (Athanasiou et al. 2012). To validate the performance of this approach, they compared the models and the shear stress values obtained by the integration of IVUS and x-ray images with those obtained by the fusion of FD-OCT and x-ray angiography. These results demonstrated that the fusion of FD-OCT with the x-ray data provides geometrically correct reconstructions and allows detection of the segments that are exposed to a low athero-promoting shear stress environment with a high accuracy (Papafaklis et al. 2013).

This technique is anticipated to provide new insights into the mechanisms of coronary atherosclerosis, as it will allow us to examine the association between the local hemodynamic patterns and plaque characteristics that are unseen by IVUS (i.e., macrophages, neo-vessels, etc.). In addition, we will be able to detect with high accuracy the location of the ruptured plaques, assess the hemodynamic milieu in these segments, and appreciate the effect of the local hemodynamic forces on plaque destabilization and rupture (Cheng et al. 2005; Ohura et al. 2003; Stone et al. 2003, 2012). Moreover, as we have recently demonstrated, the OCT-based reconstruction of stented or scaffolded segments appear capable to provide a detailed evaluation of the local hemodynamic environment and permit identification of the flow disturbances caused by the protruded struts, allowing a more accurate evaluation of the effect of shear stress patterns on stent thrombosis and restenosis (Figure 9.4) (Bourantas et al. 2014). This reconstruction approach is likely to be useful in the future to examine *in vivo* the hemodynamic implications of different stent designs, thus allowing optimization of the configuration, thickness, and arrangement of the stent struts.

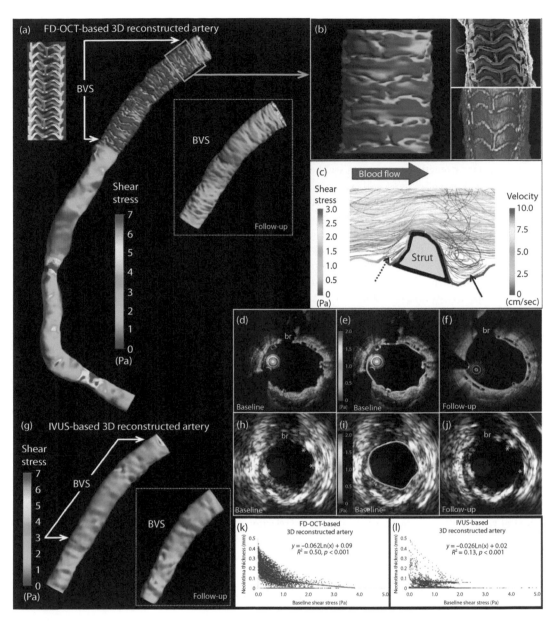

Figure 9.4 3-D reconstruction of the luminal surface of a coronary artery implanted with an Absorb bioresorbable vascular scaffold (shown at the top–left side of the figure). The model was obtained from the fusion of OCT and angiographic data (a). The shear stress distribution is portrayed in a color-coded map (the red color indicates high shear stress, and the blue, low shear stress). (b) Magnification of the proximal segment of the implanted scaffold. The struts' arrangement is apparent, which resembles the arrangement noted in a volume rendering 3-D OCT reconstruction and in an electron microscopy image acquired in a porcine model 2 weeks post device implantation. (c) Longitudinal section of the 3-D model. The blood flow streamlines with the velocities shown in a color-coded display demonstrate recirculation zones proximally and distally to the protruded struts where the measured shear stress is low, while on the top of the struts, high shear stress is noted. The high resolution of OCT allows more accurate evaluation of the lumen, identification of the protruded struts, and assessment of the neointima at 6-month follow-up (d, e, f) comparing to the IVUS, which cannot visualize details and thus the luminal borders are smother (h, i, j) and the 3-D reconstruction of the scaffolded segment has a different morphology (g). These differences appear to affect the shear stress distribution (e, i). The correlation between shear stress and neointima is higher in the OCT-based reconstruction at 6-month follow-up (r = 0.57) compared to the correlation reported in the IVUS-based model (r = 0.14) (k, l). (From Papafaklis MI et al., *EuroIntervention*, 9, 890, 2013. With permission.)

9.5 FUTURE DEVELOPMENTS IN INTRAVASCULAR HYBRID IMAGING

9.5.1 FUSION OF INTRAVASCULAR ULTRASOUND AND OPTICAL COHERENCE TOMOGRAPHY IMAGING

A hybrid IVUS-OCT imaging system is anticipated to overcome the limitations of each of the two modalities when applied independently and provide complete representation of coronary anatomy. IVUS and OCT have complementary strengths as IVUS can "see" behind lipid tissue and has an increased penetration that allows visualization of the entire vessel wall, while OCT can "see" behind calcium and has high image resolution that allows visualization of details that are not portrayed in IVUS images. Several studies provided proofs of this concept showing the superiority of the combined IVUS-OCT imaging over IVUS or OCT alone in the study of atherosclerosis (Diletti et al. 2011; Goderie et al. 2010; Gonzalo et al. 2009; Sawada et al. 2008). Hence, an effort has been made over the last 5 years to develop dual-probe catheters that would be able to provide hybrid IVUS-OCT imaging. The first catheters designed by Yin et al. and Yang et al. had the OCT and the IVUS transducer positioned side by side. This arrangement resulted in an increased outer diameter of 2.4 mm and 2.8 mm, respectively (Yang et al. 2010; Yin et al. 2010). Li et al. placed the OCT probe inside a centric hole of the IVUS transducer, but again, the size of this prototype was too big to allow intracoronary imaging (diameter of 2.5 mm) (Li et al. 2010). To overcome this limitation Yin et al. suggested a sequential arrangement of the OCT and the IVUS catheter. This modification reduced the size of the catheter to 0.69 mm, allowing *in vivo* imaging, but it failed to resolve other limitations of the first designs such as the low image acquisition rate, the moderate image quality, and the low penetration depth of this OCT signal (Yin et al. 2011). Recently, Li et al. introduced a 4F, hybrid catheter that had a penetration depth of 1 mm and provided coplanar IVUS and OCT imaging at a rate of 5 frames/s. Validation of the revision in 11 human coronaries obtained from seven autopsy cases demonstrated the potential of this hybrid imaging approach in the study of atherosclerosis. Recently, the same research group introduced an updated revision that has a smaller diameter of 3F, allows more accurate coregistration of the IVUS and OCT, and permits hybrid image acquisition at a rate of 27 frames/s (Figure 9.5).

A first-in-man study of a hybrid IVUS-OCT device is anticipated with interest as this hybrid imaging approach is expected to be useful not only in research but also in the clinical setting. This hybrid catheter can potentially optimize outcomes following percutaneous coronary interventions. IVUS would allow assessment of vessel wall dimensions, a necessity for the selection of the appropriate stent size, while OCT would provide a detailed evaluation of the final results, identification of the presence of stent underexpansion, struts malapposition, thrombus, or edge dissection and thus guide further treatment.

9.5.2 COMBINED INTRAVASCULAR ULTRASOUND AND INTRAVASCULAR PHOTOACOUSTIC IMAGING

Combined IVUS-IVPA imaging is anticipated to provide simultaneously information about the lumen and vessel wall morphology, the composition of the plaque, and the presence of inflammation. Kapriouk et al. were the first who combined a commercially available rotational IVUS catheter with a custom-designed fiber-based optical system that was able to deliver IVPA laser pulses (Karpiouk, Wang, and Emelianov 2010). Limitations of this prototype were the increased time required for hybrid imaging (approximately 25 s), the large diameter of the catheter, and the low image quality. The same year, Hsieh et al. developed a catheter that had an outer diameter of 3 mm, which combined a phased array IVUS probe and a multimode fiber with a cone-shaped mirror for optical illumination (Hsieh et al. 2010). Recently, Jensen et al. introduced a miniaturized catheter comprising an angle-polished optical fiber adjacent to a 30-MHz ultrasound transducer with a diameter of only 1.25 mm suitable for intracoronary IVUS-IVPA imaging (Jansen et al. 2010). The proposed device was tested in animal models; however, the suboptimal image quality and the increased time that is required for intravascular imaging, as well as concerns with regards the safety of IVPA, hamper its application in humans.

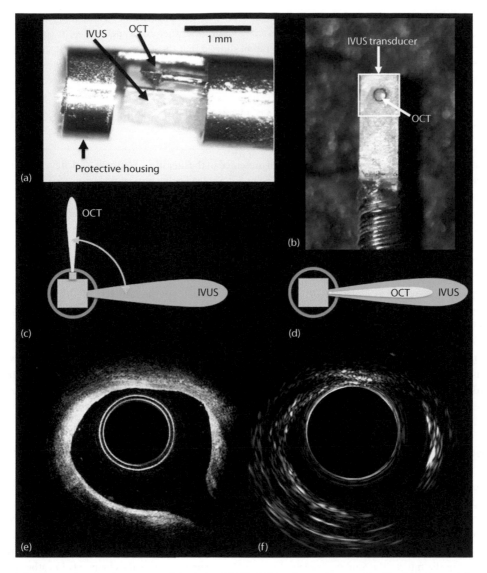

Figure 9.5 First and second revisions of the hybrid IVUS-OCT catheter designed by Harduar, Li, and Courtney. The first catheter, which has a 4F diameter, incorporates a protective housing (a) and there is a 90° offset between the IVUS and the OCT probe (c). In the new revision, the OCT catheter is integrated to the IVUS probe, the device has a smaller diameter (3F), (b) and the two transducers have a collinear alignment (d), which provides accurate coregistration of the OCT (e) and IVUS (f) images. (From Nikas D et al., *Curr Cardiovasc Imaging Rep*, 411–420, 2013. With permission.)

9.5.3 FUSION OF OPTICAL COHERENCE TOMOGRAPHY AND NEAR-INFRARED FLUORESCENCE SPECTROSCOPY

Although OCT allows reliable identification of the presence of macrophages on the vessel wall, it cannot discriminate the activated from the nonactivated macrophages and give direct information about vessel wall inflammation. Recently, Ryu et al. introduced a dual modality catheter that allows simultaneous OCT and NIRF imaging (Ryu et al. 2008). The probe features a double-clad fiber that has a single-mode core, which can transmit and receive OCT light, and a multimode light-guiding inner cladding that can transmit the NIRF excitation and receive and process the emitted fluorescence light. The coregistration of the OCT and NIRF was performed with the use of side-viewing ball-lens located at the distal end of the fiber. The diameter of the catheter is not different from a typical OCT device. Validation of the prototype *ex vivo* in coronary artery specimens from cadavers and *in vivo* in a living rabbit showed that the proposed design allows detailed and comprehensive imaging of vessel wall morphology and pathophysiology (Yoo et al. 2011). However, the safety of NIRF imaging has not been proven yet and thus further research is needed toward this direction.

9.5.4 FUSION OF INTRAVASCULAR ULTRASOUND AND TIME RESOLVED FLUORESCENCE SPECTROSCOPY

Several combined IVUS-TRFS catheters have been presented in the literature. Stephens et al. developed a hybrid catheter that featured an IVUS mechanical rotating probe, a side viewing optical fiber, and a steering wire that was connected to the distal end of the hybrid catheter and can be pushed forward to steer the device toward the luminal surface of the region of interest. The feasibility of the proposed design was examined *in vitro*; however, the large diameter of the device (5.4F) and the fact that the TRFS signal has a poor penetration and thus it requires the catheter to be pushed onto the vessel wall have not allowed its *in vivo* implementation (Stephens et al. 2009). Bec et al. introduced an updated IVUS-TRFS revision that was able to overcome the limitations of the prototype of Stephens but had a rather large-diameter (7F) (Bec et al. 2014). A miniaturized version of this prototype was presented by Ma et al. (2014). In this design (diameter of the catheter: 3.5F) the optic fiber and IVUS probe were placed in parallel in a large oval-shaped shaft (largest diameter of 5F). The feasibility of this prototype has been examined *ex vivo* in the coronary arteries of 16 patients. It was shown that combined IVUS-TRSFS imaging was able to differentiate plaque phenotypes with a high sensitivity and specificity (89%, 99%) than standalone TRFS (70%, 98%) or IVUS (45%, 94%) (Fatakdawala et al. 2015). However, the first *in vivo* applications of this device showed inaccuracies in the coregistration of the IVUS and TRFS data that were attributed to cardiac motion (Bec et al. 2015). Currently, Marcu's group is working toward the design of a new catheter that enables reliable IVUS-TRFS coregistration. This new catheter is currently tested in swine coronaries.

9.6 FUTURE PERSPECTIVE IN HYBRID INTRAVASCULAR IMAGING

Although hybrid intravascular imaging techniques carry a great future potential, these hybrid approaches have limited applications in the study of atherosclerosis. This should can be attributed partly to the fact that the existing software that has been developed for the fusion of different imaging techniques is not user-friendly and is available only in a few research laboratories. In addition, most of the hybrid catheters that have been designed have significant limitations that do not permit their applications in the clinical arena. However, things are likely to change in the future. The miniaturization of the medical devices, the new developments in catheter design, and the advances in image and signal processing are expected to overcome the current limitations and allow the construction of hybrid catheters that would be easy to use. The development of user-friendly platforms will likely permit fast and reliable fusion of different imaging techniques in the future. Whether these advances will change our clinical practice and replace standalone intravascular imaging with hybrid techniques is something that is difficult to predict, as it depends not only on the innovations in hybrid imaging but also on the evolution of standalone intravascular imaging modalities; for example, in OCT, the penetration of the signal has been increased in the second-generation systems, while polarized OCT imaging has allowed the accurate detection of the fibrous tissue. In IVUS, ACIST Kodama (ACIST Kodama Medical Systems Inc, Eden Prairie, Minnesota) has recently introduced a high definition catheter with improved image quality.

It should also be acknowledged that there is a trend toward noninvasive imaging over the last years, which has considerably reduced the applications of intravascular techniques in the study of atherosclerosis. However, we believe that hybrid imaging will still play an important role. Recently, the Lipid-Rich Plaque and the PROSPECT II studies have commenced and aim to utilize dual NIRS-IVUS three-vessel imaging to identify plaque characteristics that are associated with future cardiovascular events. In addition, a small-scale serial intravascular imaging study, which incorporates IVUS and OCT imaging (i.e., the Integrated

Biomarkers Imaging Study 4), is currently underway and is expected to provide additional information about the atherosclerotic evolution and the mechanisms that are involved in this process.

9.7 CONCLUSIONS

Hybrid intravascular imaging has emerged over the last years to address the limitations of intravascular imaging techniques, and has broadened our knowledge about plaque development. There is no doubt that most of the available hybrid approaches have failed to progress and have limited applications in research arena. Further effort is anticipated over the upcoming years in the development of new imaging approaches with optimization of catheter design and advancement of reconstruction and co-registration software that are anticipate to overcome existing limitations and to broaden the applications of the hybrid imaging in the study of atherosclerosis.

REFERENCES

Amano T, T Matsubara, T Uetani, M Kato, B Kato, T Yoshida, K Harada, S Kumagai, A Kunimura, Y Shinbo, H Ishii, and T Murohara. 2011. Lipid-rich plaques predict non-target-lesion ischemic events in patients undergoing percutaneous coronary intervention. *Circ J* 75 (1):157–66.

Athanasiou LS, CV Bourantas, PK Siogkas, AI Sakellarios, TP Exarchos, KK Naka, MI Papafaklis, LK Michalis, F Prati, and DI Fotiadis. 2012. 3D reconstruction of coronary arteries using frequency domain optical coherence tomography images and biplane angiography. *Conf Proc IEEE Eng Med Biol Soc* 2012:2647–50.

Bec J, D Ma, DR Yankelevich, D Gorpas, WT Ferrier, J Southard, and L Marcu. 2015. In-vivo validation of fluorescence lifetime imaging (FLIm) of coronary arteries in swine. *SPIE Proc Photonic Therapeutics and Diagnostics XI*, 9303.

Bec J, DM Ma, DR Yankelevich, J Liu, WT Ferrier, J Southard, and L Marcu. 2014. Multispectral fluorescence lifetime imaging system for intravascular diagnostics with ultrasound guidance: *In vivo* validation in swine arteries. *J Biophoton* 7 (5):281–5.

Berry C, PL L'Allier, J Gregoire, J Lesperance, S Levesque, R Ibrahim, and JC Tardif. 2007. Comparison of intravascular ultrasound and quantitative coronary angiography for the assessment of coronary artery disease progression. *Circulation* 115 (14):1851–7.

Bharadwaj AS, Y Vengrenyuk, T Yoshimura, U Baber, C Hasan, J Narula, SK Sharma, and AS Kini. 2016. Multimodality intravascular imaging to evaluate sex differences in plaque morphology in stable CAD. *JACC Cardiovasc Imaging* 9 (4):400–7.

Bom N, CT Lancee, and FC Van Egmond. 1972. An ultrasonic intracardiac scanner. *Ultrasonics* 10 (2):72–6.

Boogers MJ, A Broersen, JE van Velzen, FR de Graaf, HM El-Naggar, PH Kitslaar, J Dijkstra, V Delgado, E Boersma, A de Roos, JD Schuijf, MJ Schalij, JH Reiber, JJ Bax, and JW Jukema. 2012. Automated quantification of coronary plaque with computed tomography: Comparison with intravascular ultrasound using a dedicated registration algorithm for fusion-based quantification. *Eur Heart J* 33 (8):1007–16.

Bourantas CV, HM Garcia-Garcia, R Diletti, AM Carlos, S Garg, Y Zhang, and PW Serruys. 2013a. Long term consequences of lipid core plaques. *J Invasive Cardiol* 25:24A–26A.

Bourantas CV, HM Garcia-Garcia, KK Naka, A Sakellarios, L Athanasiou, DI Fotiadis, LK Michalis, and PW Serruys. 2013b. Hybrid intravascular imaging: Current applications and prospective potential in the study of coronary atherosclerosis. *J Am Coll Cardiol* 61 (13):1369–78.

Bourantas CV, FG Kalatzis, MI Papafaklis, DI Fotiadis, AC Tweddel, IC Kourtis, CS Katsouras, and LK Michalis. 2008. ANGIOCARE: An automated system for fast three-dimensional coronary reconstruction by integrating angiographic and intracoronary ultrasound data. *Catheter Cardiovasc Interv* 72 (2):166–75.

Bourantas CV, IC Kourtis, ME Plissiti, DI Fotiadis, CS Katsouras, MI Papafaklis, and LK Michalis. 2005. A method for 3D reconstruction of coronary arteries using biplane angiography and intravascular ultrasound images. *Comput Med Imaging Graph* 29 (8):597–606.

Bourantas CV, MI Papafaklis, L Athanasiou, FG Kalatzis, KK Naka, PK Siogkas, S Takahashi, S Saito, DI Fotiadis, CL Feldman, PH Stone, and LK Michalis. 2013c. A new methodology for accurate 3-dimensional coronary artery reconstruction using routine intravascular ultrasound and angiographic data: Implications for widespread assessment of endothelial shear stress in humans. *EuroIntervention* 9 (5):582–93.

Bourantas CV, MI Papafaklis, KK Naka, VD Tsakanikas, DN Lysitsas, FM Alamgir, DI Fotiadis, and LK Michalis. 2012. Fusion of optical coherence tomography and coronary angiography—*In vivo* assessment of shear stress in plaque rupture. *Int J Cardiol* 155 (2):e24–6.

Bourantas CV, AC Tweddel, MI Papafaklis, PS Karvelis, DI Fotiadis, CS Katsouras, and LK Michalis. 2009. Comparison of quantitative coronary angiography with intracoronary ultrasound. Can quantitative coronary angiography accurately estimate the severity of a luminal stenosis? *Angiology* 60 (2):169–79.

Bourantas CV, MI Papafaklis, HM Garcia-Garcia, V Farooq, R Diletti, T Muramatsu, Y Zhang, FG Kalatzis, KK Naka, DI Fotiadis, Y Onuma, LK Michalis, and PW Serruys. 2014. Short- and long-term implications of a bioresorbable vascular scaffold implantation on the local endothelial shear stress patterns. *JACC Cardiovasc Interv* 7 (1):100–1.

Bourantas CV, MI Papafaklis, A Kotsia, V Farooq, R Diletti, T Muramatsu, J Gomez-Lara, F Kalatzis, KK Naka, DI Fotiadis, C Dorange, R Rapoza, HM Garcia Garcia, Y Onuma, LK Michalis, and PW Serruys. 2014. Implications of the endothelial shear stress patterns on neointimal proliferation following drug-eluting bioresorbable vascular scaffolds Implantation: An optical coherence tomography study. *JACC Cardiovasc Interv* 7 (3):315–24.

Brennan JF, 3rd, J Nazemi, J Motz, and S Ramcharitar. 2008. The vPredict optical catheter system: Intravascular Raman spectroscopy. *EuroIntervention* 3 (5):635–8.

Brugaletta S, HM Garcia-Garcia, PW Serruys, S de Boer, J Ligthart, J Gomez-Lara, K Witberg, R Diletti, J Wykrzykowska, RJ van Geuns, C Schultz, E Regar, HJ Duckers, N van Mieghem, P de Jaegere, SP Madden, JE Muller, AF van der Steen, WJ van der Giessen, and E Boersma. 2011. NIRS and IVUS for characterization of atherosclerosis in patients undergoing coronary angiography. *JACC Cardiovasc Imaging* 4 (6):647–55.

Bruggink JL, R Meerwaldt, GM van Dam, JD Lefrandt, RH Slart, RA Tio, AJ Smit, and CJ Zeebregts. 2010. Spectroscopy to improve identification of vulnerable plaques in cardiovascular disease. *Int J Cardiovasc Imaging* 26 (1):111–9.

Burton JR, KK Teo, CE Buller, S Plante, D Catellier, W Tymchak, D Taylor, V Dzavik, and TJ Montague. 2003. Effects of long term cholesterol lowering on coronary atherosclerosis in patient risk factor subgroups: The Simvastatin/Enalapril Coronary Atherosclerosis Trial (SCAT). *Can J Cardiol* 19 (5):487–91.

Calvert PA, DR Obaid, M O'Sullivan, LM Shapiro, D McNab, CG Densem, PM Schofield, D Braganza, SC Clarke, KK Ray, NE West, and MR Bennett. 2011. Association between IVUS findings and adverse outcomes in patients with coronary artery disease: The VIVA (VH-IVUS in Vulnerable Atherosclerosis) Study. *JACC Cardiovasc Imaging* 4 (8):894–901.

Chau AH, JT Motz, JA Gardecki, S Waxman, BE Bouma, and GJ Tearney. 2008. Fingerprint and high-wavenumber Raman spectroscopy in a human-swine coronary xenograft *in vivo*. *J Biomed Opt* 13 (4):040501.

Cheng C, R van Haperen, M de Waard, LC van Damme, D Tempel, L Hanemaaijer, GW van Cappellen, J Bos, CJ Slager, DJ Duncker, AF van der Steen, R de Crom, and R Krams. 2005. Shear stress affects the intracellular distribution of eNOS: Direct demonstration by a novel *in vivo* technique. *Blood* 106 (12):3691–8.

Davies MJ, and AC Thomas. 1985. Plaque fissuring—The cause of acute myocardial infarction, sudden ischaemic death, and crescendo angina. *Br Heart J* 53 (4):363–73.

Di Mario C, SH The, S Madretsma, RJ van Suylen, RA Wilson, N Bom, PW Serruys, EJ Gussenhoven, and JR Roelandt. 1992. Detection and characterization of vascular lesions by intravascular ultrasound: An in vitro study correlated with histology. *J Am Soc Echocardiogr* 5 (2):135–46.

Diletti R, HM Garcia-Garcia, J Gomez-Lara, S Brugaletta, JJ Wykrzykowska, N van Ditzhuijzen, RJ van Geuns, E Regar, G Ambrosio, and PW Serruys. 2011. Assessment of coronary atherosclerosis progression and regression at bifurcations using combined IVUS and OCT. *JACC Cardiovasc Imaging* 4 (7):774–80.

Falk E. 1983. Plaque rupture with severe pre-existing stenosis precipitating coronary thrombosis. Characteristics of coronary atherosclerotic plaques underlying fatal occlusive thrombi. *Br Heart J* 50 (2):127–34.

Fatakdawala H, D Gorpas, JW Bishop, J Bec, D Ma, JA Southard, KB Margulies, and L Marcu. 2015. Fluorescence lifetime imaging combined with conventional intravascular ultrasound for enhanced assessment of atherosclerotic plaques: An ex vivo study in human coronary arteries. *J Cardiovasc Transl Res* 8 (4):253–63.

Fukumoto Y, T Hiro, T Fujii, G Hashimoto, T Fujimura, J Yamada, T Okamura, and M Matsuzaki. 2008. Localized elevation of shear stress is related to coronary plaque rupture: A 3-dimensional intravascular ultrasound study with in-vivo color mapping of shear stress distribution. *J Am Coll Cardiol* 51 (6):645–50.

Gardner CM, H Tan, EL Hull, JB Lisauskas, ST Sum, TM Meese, C Jiang, SP Madden, JD Caplan, AP Burke, R Virmani, J Goldstein, and JE Muller. 2008. Detection of lipid core coronary plaques in autopsy specimens with a novel catheter-based near-infrared spectroscopy system. *JACC Cardiovasc Imaging* 1 (5):638–48.

Goderie TP, G van Soest, HM Garcia-Garcia, N Gonzalo, S Koljenovic, GJ van Leenders, F Mastik, E Regar, JW Oosterhuis, PW Serruys, and AF van der Steen. 2010. Combined optical coherence tomography and intravascular ultrasound radio frequency data analysis for plaque characterization. Classification accuracy of human coronary plaques in vitro. *Int J Cardiovasc Imaging* 26 (8):843–50.

Gonzalo N, HM Garcia-Garcia, E Regar, P Barlis, J Wentzel, Y Onuma, J Ligthart, and PW Serruys. 2009. *In vivo* assessment of high-risk coronary plaques at bifurcations with combined intravascular ultrasound and optical coherence tomography. *JACC Cardiovasc Imaging* 2 (4):473–82.

Hiro T, CY Leung, RJ Russo, I Moussa, H Karimi, AR Farvid, and JM Tobis. 1996. Variability in tissue characterization of atherosclerotic plaque by intravascular ultrasound: A comparison of four intravascular ultrasound systems. *Am J Card Imaging* 10 (4):209–18.

Hosokawa R, N. Kambara, M Ohba, T Mukai, M Ogawa, H Motomura, N Kume, H Saji, T Kita, and R Nohara. 2006. A catheter-based intravascular radiation detector of vulnerable plaques. *J Nucl Med* 47 (5):863–7.

Hsieh BY, SL Chen, T Ling, LJ Guo, and PC Li. 2010. Integrated intravascular ultrasound and photoacoustic imaging scan head. *Opt Lett* 35 (17):2892–4.

Jaffer FA, MA Calfon, A Rosenthal, G Mallas, RN Razansky, A Mauskapf, R Weissleder, P Libby, and V Ntziachristos. 2011. Two-dimensional intravascular near-infrared fluorescence molecular imaging of inflammation in atherosclerosis and stent-induced vascular injury. *J Am Coll Cardiol* 57 (25):2516–26.

Jaffer FA, P Libby, and R Weissleder 2009. Optical and multimodality molecular imaging: Insights into atherosclerosis. *Arterioscler Thromb Vasc Biol* 29 (7):1017–24.

Jaffer FA, C Vinegoni, MC John, E Aikawa, HK Gold, AV Finn, V Ntziachristos, P Libby, and R Weissleder. 2008. Real-time catheter molecular sensing of inflammation in proteolytically active atherosclerosis. *Circulation* 118 (18):1802–9.

Jansen K, G Springeling, C Lancee, R Berskens, F Mastik, and AF van der Steen. 2010. An intravascular photoacoustic imaging catheter. *International Ultrasonics Symposium (IUS), 2010 IEEE* 378–81.

Karpiouk AB, B Wang, and SY Emelianov. 2010. Development of a catheter for combined intravascular ultrasound and photoacoustic imaging. *Rev Sci Instrum* 81 (1):014901.

Kawasaki M, BE Bouma J Bressner, SL Houser, SK Nadkarni, BD MacNeill, IK Jang, H Fujiwara, and GJ Tearney. 2006. Diagnostic accuracy of optical coherence tomography and integrated backscatter intravascular ultrasound images for tissue characterization of human coronary plaques. *J Am Coll Cardiol* 48 (1):81–8.

Kini AS, U Baber, JC Kovacic, A Limaye, ZA Ali, J Sweeny, A Maehara, R Mehran, G Dangas, GS Mintz, V Fuster, J Narula, SK Sharma, and PR Moreno. 2013. Changes in plaque lipid content after short-term, intensive versus standard statin therapy: The YELLOW Trial. *J Am Coll Cardiol* 62 (1):21–9.

Klein HM, RW Gunther, M Verlande, W Schneider, D Vorwerk, J Kelch, and M Hamm. 1992. 3D-surface reconstruction of intravascular ultrasound images using personal computer hardware and a motorized catheter control. *Cardiovasc Intervent Radiol* 15 (2):97–101.

Kotani J, M Awata, S Nanto, M Uematsu, F Oshima, H Minamiguchi, GS Mintz, and S Nagata. 2006. Incomplete neointimal coverage of sirolimus-eluting stents: Angioscopic findings. *J Am Coll Cardiol* 47 (10):2108–11.

Kubo T, T Imanishi, S Takarada, A Kuroi, S Ueno, T Yamano, T Tanimoto, Y Matsuo, T Masho, H Kitabata, K Tsuda, Y Tomobuchi, and T Akasaka. 2007. Assessment of culprit lesion morphology in acute myocardial infarction: Ability of optical coherence tomography compared with intravascular ultrasound and coronary angioscopy. *J Am Coll Cardiol* 50 (10):933–9.

Lederman RJ, RR Raylman, SJ Fisher, PV Kison, H San, EG Nabel, and RL Wahl. 2001. Detection of atherosclerosis using a novel positron-sensitive probe and 18-fluorodeoxyglucose (FDG). *Nucl Med Commun* 22 (7):747–53.

Lengyel J, DP Greenberg, and R Pop. 1995. Time-dependent three-dimensional intravascular ultrasound. In: Cook, R., editor. *Proceedings of the SIGGRAPH 95 Conference on Comp Graphics USA: ACM* 457–64.

Li X, J Yin, C Hu, Q Zhou, KK Shung, and Z Chen. 2010. High-resolution coregistered intravascular imaging with integrated ultrasound and optical coherence tomography probe. *Appl Phys Lett* 97 (13):133702.

Ma D, J Bec, DR Yankelevich, D Gorpas, H Fatakdawala, and L Marcu. 2014. Rotational multispectral fluorescence lifetime imaging and intravascular ultrasound: Bimodal system for intravascular applications. *J Biomed Opt* 19 (6):066004.

Madder RD, JA Goldstein, SP Madden, R Puri, K Wolski, M Hendricks, ST Sum, A Kini, S Sharma, D Rizik, ES Brilakis, KA Shunk, J Petersen, G Weisz, R Virmani, SJ Nicholls, A Maehara, GS Mintz, GW Stone, and JE Muller. 2013. Detection by near-infrared spectroscopy of large lipid core plaques at culprit sites in patients with acute ST-segment elevation myocardial infarction. *JACC Cardiovasc Interv* 6 (8):838–46.

Madder RD, M Husaini, AT Davis, S Van Oosterhout, J Harnek, M Gotberg, and D Erlinge. 2015. Detection by near-infrared spectroscopy of large lipid cores at culprit sites in patients with non-ST-segment elevation myocardial infarction and unstable angina. *Catheter Cardiovasc Interv* 86 (6):1014–21.

Madder RD, R Puri, JE Muller, J Harnek, M Gotberg, S VanOosterhout, M Chi, D Wohns, R McNamara, K Wolski, S Madden, S Sidharta, J Andrews, SJ Nicholls, and D Erlinge. 2016. Confirmation of the intracoronary near-infrared spectroscopy threshold of lipid-rich plaques that underlie st-segment-elevation myocardial infarction. *Arterioscler Thromb Vasc Biol* 36 (5):1010–5.

Manfrini O, E Mont, O Leone, E Arbustini, V Eusebi, R Virmani, and R Bugiardini. 2006. Sources of error and interpretation of plaque morphology by optical coherence tomography. *Am J Cardiol* 98 (2):156–9.

Marcu L. 2010. Fluorescence lifetime in cardiovascular diagnostics. *J Biomed Opt* 15 (1):011106.

Marcu L, Q Fang, JA Jo, T Papaioannou, A Dorafshar, T Reil, JH Qiao, JD Baker, JA Freischlag, and MC Fishbein. 2005. *In vivo* detection of macrophages in a rabbit atherosclerotic model by time-resolved laser-induced fluorescence spectroscopy. *Atherosclerosis* 181 (2):295–303.

Marcu L, JA Jo, Q Fang, T Papaioannou, T Reil, JH Qiao, JD Baker, JA Freischlag, and MC Fishbein 2009. Detection of rupture-prone atherosclerotic plaques by time-resolved laser-induced fluorescence spectroscopy. *Atherosclerosis* 204 (1):156–64.

Mehta SK, JR McCrary, AD Frutkin, WJ Dolla, and SP Marso. 2007. Intravascular ultrasound radiofrequency analysis of coronary atherosclerosis: An emerging technology for the assessment of vulnerable plaque. *Eur Heart J* 28 (11):1283–8.

Mintz GS, SE Nissen, WD Anderson, SR Bailey, R Erbel, PJ Fitzgerald, FJ Pinto, K Rosenfield, RJ Siegel, EM Tuzcu, and PG Yock. 2001. American College of Cardiology Clinical Expert Consensus document on standards for acquisition, measurement and reporting of intravascular ultrasound studies (IVUS). A report of the American College of Cardiology Task Force on Clinical Expert Consensus Documents. *J Am Coll Cardiol* 37 (5):1478–92.

Motz JT, M Fitzmaurice, A Miller, SJ Gandhi, AS Haka, LH Galindo, RR Dasari, JR Kramer, and MS Feld. 2006. *In vivo* Raman spectral pathology of human atherosclerosis and vulnerable plaque. *J Biomed Opt* 11 (2):021003.

Mukai T, R Nohara, M Ogawa, S Ishino, N Kambara, K Kataoka, T Kanoi, K Saito, H Motomura, J Konishi, and H Saji. 2004. A catheter-based radiation detector for endovascular detection of atheromatous plaques. *Eur J Nucl Med Mol Imaging* 31 (9):1299–303.

Nikas D, CV Bourantas, A Sakellarios, A Ramos, KK Naka, LK Michalis, and PW Serruys. 2013. New developments in hybrid optical coherence tomographic imaging: Current status and potential implications in clinical practice and research. *Curr Cardiovasc Imaging Rep* (6):411–20.

Nissen SE, EM Tuzcu, P Libby, PD Thompson, M Ghali, D Garza, L Berman, H Shi, E Buebendorf, and EJ Topol. 2004. Effect of antihypertensive agents on cardiovascular events in patients with coronary disease and normal blood pressure: The CAMELOT study: A randomized controlled trial. *JAMA* 292 (18):2217–25.

Ohtani T, Y Ueda, I Mizote, J Oyabu, K Okada, A Hirayama, and K Kodama. 2006. Number of yellow plaques detected in a coronary artery is associated with future risk of acute coronary syndrome: Detection of vulnerable patients by angioscopy. *J Am Coll Cardiol* 47 (11):2194–200.

Ohura N, K Yamamoto, S Ichioka, T Sokabe, H Nakatsuka, A Baba, M Shibata, T Nakatsuka, K Harii, Y Wada, T Kohro, T Kodama, and J Ando. 2003. Global analysis of shear stress-responsive genes in vascular endothelial cells. *J Atheroscler Thromb* 10 (5):304–13.

Papadopoulou SL, LA Neefjes, M Schaap, HL Li, E Capuano, AG van der Giessen, JC Schuurbiers, FJ Gijsen, AS Dharampal, K Nieman, RJ van Geuns, NR Mollet, and PJ de Feyter. 2011. Detection and quantification of coronary atherosclerotic plaque by 64-slice multidetector CT: A systematic head-to-head comparison with intravascular ultrasound. *Atherosclerosis* 219 (1):163–70.

Papafaklis MI, CV Bourantas, PE Theodorakis, CS Katsouras, KK Naka, DI Fotiadis, and LK Michalis. 2010. The effect of shear stress on neointimal response following sirolimus- and paclitaxel-eluting stent implantation compared with bare-metal stents in humans. *JACC Cardiovasc Interv* 3 (11):1181–9.

Papafaklis MI, CV Bourantas, T Yonetsu, K Kato, A Kotsia, AU Coskun, H Jia, AP Antoniadis, R Vergallo, M Tsuda, DI Fotiadis, CL Feldman, PH Stone, I-K Jang, and LK Michalis. 2013. Geometrically accurate three-dimensional coronary artery reconstruction using frequency-domain optical coherence tomography and angiographic data: New opportunities for *in vivo* endothelial shear stress assessment. *JACC Cardiovasc Interv* 6 (2S):S34.

Phipps J, Y Sun, R Saroufeem, N Hatami, MC Fishbein, and L Marcu. 2011. Fluorescence lifetime imaging for the characterization of the biochemical composition of atherosclerotic plaques. *J Biomed Opt* 16 (9):096018.

Piao D, MM Sadeghi, J Zhang, Y Chen, AJ Sinusas, and Q Zhu. 2005. Hybrid positron detection and optical coherence tomography system: Design, calibration, and experimental validation with rabbit atherosclerotic models. *J Biomed Opt* 10 (4):44010.

Prause GP, SC DeJong, CR McKay, and M Sonka. 1997. Towards a geometrically correct 3-D reconstruction of tortuous coronary arteries based on biplane angiography and intravascular ultrasound. *Int J Card Imaging* 13 (6):451–62.

Pu J, GS Mintz, ES Brilakis, S Banerjee, AR Abdel-Karim, B Maini, S Biro, JB Lee, GW Stone, G Weisz, and A Maehara. 2012. *In vivo* characterization of coronary plaques: Novel findings from comparing greyscale and virtual histology intravascular ultrasound and near-infrared spectroscopy. *Eur Heart J* 33 (3):372–83.

Regar E, B Hennen, E Grube, D Halon, RL Wilensky, R Virmani, J Schneiderman, S Sax, H Friedmann, PW Serruys, and W Wijns. 2006. First-In-Man application of a miniature self-contained intra-coronary magnetic resonance probe. A multi-centre safety and feasibility trial. *EuroIntervention* 2 (1):77–83.

Reiber JHC, F Booman, H Tan, CJ Slager, JC Schuurbiers, and JJ Gerbrands. 1978. A cardiac image analysis system. Objective quantitative processing of angiocardiograms. *IEEE Comp Cardiol* 239–2.

Romer TJ, JF Brennan, 3rd, M Fitzmaurice, ML Feldstein, G Deinum, JL Myles, JR Kramer, RS Lees, and MS Feld. 1998. Histopathology of human coronary atherosclerosis by quantifying its chemical composition with Raman spectroscopy. *Circulation* 97 (9):878–85.

Romer TJ, JF Brennan, 3rd, GJ Puppels, AH Zwinderman, SG van Duinen, A van der Laarse, AF van der Steen, NA Bom, and AV Bruschke. 2000. Intravascular ultrasound combined with Raman spectroscopy to localize and quantify cholesterol and calcium salts in atherosclerotic coronary arteries. *Arterioscler Thromb Vasc Biol* 20 (2):478–83.

Ryu SY, HY Choi, J Na, ES Choi, and BH Lee. 2008. Combined system of optical coherence tomography and fluorescence spectroscopy based on double-cladding fiber. *Opt Lett* 33 (20):2347–9.

Rzeszutko L, J Legutko, GL Kaluza, M Wizimirski, A Richter, M Chyrchel, G Heba, JS Dubiel, and D Dudek. 2006. Assessment of culprit plaque temperature by intracoronary thermography appears inconclusive in patients with acute coronary syndromes. *Arterioscler Thromb Vasc Biol* 26 (8):1889–94.

Salenius JP, JF Brennan, 3rd, A Miller, Y Wang, T Aretz, B Sacks, RR Dasari, and MS Feld. 1998. Biochemical composition of human peripheral arteries examined with near-infrared Raman spectroscopy. *J Vasc Surg* 27 (4):710–9.

Sales FJ, BA Falcao, JL Falcao, EE Ribeiro, MA Perin, PE Horta, AG Spadaro, JA Ambrose, EE Martinez, SS Furuie, and PA Lemos. 2010. Evaluation of plaque composition by intravascular ultrasound "virtual histology": The impact of dense calcium on the measurement of necrotic tissue. *EuroIntervention* 6 (3):394–9.

Sathyanarayana S, M Schar, DL Kraitchman, and PA Bottomley. 2010. Towards real-time intravascular endo-scopic magnetic resonance imaging. *JACC Cardiovasc Imaging* 3 (11):1158–65.

Sawada T, J Shite, HM Garcia-Garcia, T Shinke, S Watanabe, H Otake, D Matsumoto, Y Tanino, D Ogasawara, H Kawamori, H Kato, N Miyoshi, M Yokoyama, P W Serruys, and K Hirata. 2008. Feasibility of com-bined use of intravascular ultrasound radiofrequency data analysis and optical coherence tomography for detecting thin-cap fibroatheroma. *Eur Heart J* 29 (9):1136–46.

Schneiderman J, RL Wilensky, A Weiss, E Samouha, L Muchnik, M Chen-Zion, M Ilovitch, E Golan, A Blank, M Flugelman, Y Rozenman, and R Virmani. 2005. Diagnosis of thin-cap fibroatheromas by a self-con-tained intravascular magnetic resonance imaging probe in ex vivo human aortas and in situ coronary arteries. *J Am Coll Cardiol* 45 (12):1961–9.

Shekhar R, RM Cothren, DG Vince, and JF Cornhill. 1996. Fusion of intravascular ultrasound and biplane angiography for three-dimensional reconstruction of coronary arteries. *In: Proc Comp in Cardiology; 1996; Indianapolis Indiana: IEEE* 5–8.

Slager CJ, JJ Wentzel, JC Schuurbiers, JA Oomen, J Kloet, R Krams, C von Birgelen, WJ van der Giessen, PW Serruys, and PJ de Feyter. 2000. True 3-dimensional reconstruction of coronary arteries in patients by fusion of angiography and IVUS (ANGUS) and its quantitative validation. *Circulation* 102 (5):511–6.

Stephens DN, J Park, Y Sun, T Papaioannou, and L Marcu. 2009. Intraluminal fluorescence spectroscopy catheter with ultrasound guidance. *J Biomed Opt* 14 (3):030505.

Stone GW, A Maehara, AJ Lansky, B de Bruyne, E Cristea, GS Mintz, R Mehran, J McPherson, N Farhat, SP Marso, H Parise, B Templin, R White, Z Zhang, and PW Serruys. 2011. A prospective natural-history study of coronary atherosclerosis. *N Engl J Med* 364 (3):226–35.

Stone PH, AU Coskun, S Kinlay, ME Clark, M Sonka, A Wahle, OJ Ilegbusi, Y Yeghiazarians, JJ Popma, J Orav, RE Kuntz, and CL Feldman. 2003. Effect of endothelial shear stress on the progression of coro-nary artery disease, vascular remodeling, and in-stent restenosis in humans: *In vivo* 6-month follow-up study. *Circulation* 108 (4):438–44.

Stone PH, S Saito, S Takahashi, Y Makita, S Nakamura, T Kawasaki, A Takahashi, T Katsuki, A Namiki, A Hirohata, T Matsumura, S Yamazaki, H Yokoi, S Tanaka, S Otsuji, F Yoshimachi, J Honye, D Harwood, M Reitman, AU Coskun, MI Papafaklis, and CL Feldman. 2012. Prediction of progression of coronary artery disease and clinical outcomes using vascular profiling of endothelial shear stress and arterial plaque characteristics: The PREDICTION Study. *Circulation* 126 (2):172–81.

Su JL, B Wang, KE Wilson, CL Bayer, YS Chen, S Kim, KA Homan, and SY Emelianov. 2010. Advances in clinical and biomedical applications of photoacoustic imaging. *Expert Opin Med Diagn* 4 (6):497–510.

Subramanian KR, MJ Thubrikar, B Fowler, MT Mostafavi, and MW Funk. 2000. Accurate 3D reconstruction of complex blood vessel geometries from intravascular ultrasound images: In vitro study. *J Med Eng Technol* 24 (4):131–40.

Syvanne M, MS Nieminen, and MH Frick. 1994. Accuracy and precision of quantitative arteriography in the evaluation of coronary artery disease after coronary bypass surgery. A validation study. *Int J Card Imaging* 10 (4):243–52.

Takumi T, S Lee, S Hamasaki, K Toyonaga, D Kanda, K Kusumoto, H Toda, T Takenaka, M Miyata, R Anan, Y Otsuji, and C Tei. 2007. Limitation of angiography to identify the culprit plaque in acute myocardial infarction with coronary total occlusion utility of coronary plaque temperature measurement to identify the culprit plaque. *J Am Coll Cardiol* 50 (23):2197–203.

Tardif JC, J Gregoire, PL L'Allier, R Ibrahim, J Lesperance, TM Heinonen, S Kouz, C Berry, R Basser, MA Lavoie, MC Guertin, and J Rodes-Cabau. 2007. Effects of reconstituted high-density lipoprotein infusions on coronary atherosclerosis: A randomized controlled trial. *JAMA* 297 (15):1675–82.

Tearney GJ, E Regar, T Akasaka, T Adriaenssens, P Barlis, HG Bezerra, B Bouma, N Bruining, JM Cho, S Chowdhary, MA Costa, R de Silva, J Dijkstra, C Di Mario, D Dudek, E Falk, MD Feldman, P Fitzgerald, HM Garcia-Garcia, N Gonzalo, JF Granada, G Guagliumi, NR Holm, Y Honda, F Ikeno, M Kawasaki, J Kochman, L Koltowski, T Kubo, T Kume, H Kyono, CC Lam, G Lamouche, DP Lee, MB Leon, A Maehara, O Manfrini, GS Mintz, K Mizuno, MA Morel, S Nadkarni, H Okura, H Otake, A Pietrasik, F Prati, L Raber, MD Radu, J Rieber, M Riga, A Rollins, M Rosenberg, V Sirbu, PW Serruys, K Shimada, T Shinke, J Shite, E Siegel, S Sonoda, M Suter, S Takarada, A Tanaka, M Terashima, T Thim, S Uemura, GJ Ughi, HM van Beusekom, AF van der Steen, GA van Es, G van Soest, R Virmani, S Waxman, NJ Weissman, and G Weisz. 2012. Consensus standards for acquisition, measurement, and reporting of intravascular optical coherence tomography studies: A report from the International Working Group for Intravascular Optical Coherence Tomography Standardization and Validation. *J Am Coll Cardiol* 59 (12):1058–72.

Thieme T, KD Wernecke, R Meyer, E Brandenstein, D Habedank, A Hinz, SB Felix, G Baumann, and FX Kleber. 1996. Angioscopic evaluation of atherosclerotic plaques: Validation by histomorphologic analysis and association with stable and unstable coronary syndromes. *J Am Coll Cardiol* 28 (1):1–6.

Thim T, MK Hagensen, D Wallace-Bradley, JF Granada, GL Kaluza, L Drouet, WP Paaske, HE Botker, and E Falk. 2010. Unreliable assessment of necrotic core by virtual histology intravascular ultrasound in porcine coronary artery disease. *Circ Cardiovasc Imaging* 3 (4):384–91.

Toutouzas K, A Synetos, E Stefanadi, S Vaina, V Markou, M Vavuranakis, E Tsiamis, D Tousoulis, and C Stefanadis. 2007. Correlation between morphologic characteristics and local temperature differences in culprit lesions of patients with symptomatic coronary artery disease. *J Am Coll Cardiol* 49 (23):2264–71.

Tu S, NR Holm, G Koning, Z Huang, and JH Reiber. 2011. Fusion of 3D QCA and IVUS/OCT. *Int J Cardiovasc Imaging* 27 (2):197–207.

Tu S, SA Pyxaras, Y Li, E Barbato, JH Reiber, and W Wijns. 2013. *In vivo* flow simulation at coronary bifurcation reconstructed by fusion of 3-dimensional x-ray angiography and optical coherence tomography. *Circ Cardiovasc Interv* 6 (2):e15–7.

Uchida Y 2011. Recent advances in coronary angioscopy. *J Cardiol* 57 (1):18–30.

Uchida Y, F Nakamura, T Tomaru, T Morita, T Oshima, T Sasaki, S Morizuki, and J Hirose. 1995. Prediction of acute coronary syndromes by percutaneous coronary angioscopy in patients with stable angina. *Am Heart J* 130 (2):195–203.

Uchida Y, Y Sugiyama, T Tomaru, S Kawai, R Kanamaru, and E Shimoyama. 2010. Two-dimensional visualization of cholesterol and cholesteryl esters within human coronary plaques by near-infrared fluorescence angioscopy. *Clin Cardiol* 33 (12):775–82.

van De Poll SW, TJ Romer, OL Volger, DJ Delsing, TC Bakker Schut, HM Princen, LM Havekes, JW Jukema, A van Der Laarse, and GJ Puppels. 2001. Raman spectroscopic evaluation of the effects of diet and lipid-lowering therapy on atherosclerotic plaque development in mice. *Arterioscler Thromb Vasc Biol* 21 (10):1630–5.

van der Giessen AG, M Schaap, FJ Gijsen, HC Groen, T van Walsum, NR Mollet, J Dijkstra, FN van de Vosse, W Niessen, PJ de Feyter, A F van der Steen, and JJ Wentzel. 2010. 3D fusion of intravascular ultrasound and coronary computed tomography for in-vivo wall shear stress analysis: A feasibility study. *Int J Cardiovasc Imaging* 26 (7):781–96.

Voros S, S Rinehart, Z Qian, G Vazquez, H Anderson, L Murrieta, C Wilmer, H Carlson, K Taylor, W Ballard, D Karmpaliotis, A Kalynych, and C Brown, 3rd. 2011. Prospective validation of standardized, 3-dimensional, quantitative coronary computed tomographic plaque measurements using radiofrequency backscatter intravascular ultrasound as reference standard in intermediate coronary arterial lesions: Results from the ATLANTA (Assessment of Tissue Characteristics, Lesion Morphology, and Hemodynamics by Angiography with Fractional Flow Reserve, Intravascular Ultrasound and Virtual Histology, and Noninvasive Computed Tomography in Atherosclerotic Plaques) I study. *JACC Cardiovasc Interv* 4 (2):198–208.

Wahle A, ME Olszewski, and M Sonka. 2004. Interactive virtual endoscopy in coronary arteries based on multimodality fusion. *IEEE Trans Med Imaging* 23 (11):1391–403.

Wahle A, GPM Prause, SC DeJong, and M Sonka. 1998. 3-D fusion of biplane angiography and intravascular ultrasound for accurate visualization and volumetry. *Med Image Comput Comput Assist Interv Miccai '98* 1496:146–55.

Wang B, JL Su, J Amirian, SH Litovsky, R Smalling, and S Emelianov. 2010. Detection of lipid in atherosclerotic vessels using ultrasound-guided spectroscopic intravascular photoacoustic imaging. *Opt Express* 18 (5):4889–97.

Wang B, E Yantsen, T Larson, AB Karpiouk, S Sethuraman, JL Su, K Sokolov, and SY Emelianov. 2009. Plasmonic intravascular photoacoustic imaging for detection of macrophages in atherosclerotic plaques. *Nano Lett* 9 (6):2212–7.

Waxman S, SR Dixon, P L'Allier, JW Moses, JL Petersen, D Cutlip, JC Tardif, RW Nesto, JE Muller, MJ Hendricks, ST Sum, CM Gardner, JA Goldstein, GW Stone, and MW Krucoff. 2009. *In vivo* validation of a catheter-based near-infrared spectroscopy system for detection of lipid core coronary plaques: Initial results of the SPECTACL study. *JACC Cardiovasc Imaging* 2 (7):858–68.

Wentzel JJ, R Krams, JC Schuurbiers, JA Oomen, J Kloet, WJ van Der Giessen, PW Serruys, and CJ Slager. 2001. Relationship between neointimal thickness and shear stress after Wallstent implantation in human coronary arteries. *Circulation* 103 (13):1740–5.

Wentzel JJ, AG van der Giessen, S Garg, C Schultz, F Mastik, FJH Gijsen, PW Serruys, AFW van der Steen, and E Regar. 2010. *In Vivo* 3D distribution of lipid-core plaque in human coronary artery as assessed by fusion of near infrared spectroscopy-intravascular ultrasound and multislice computed tomography scan. *Circ Cardiovasc Imaging* 3 (6):E6–7.

Yabushita H, BE Bouma, SL Houser, HT Aretz, IK Jang, KH Schlendorf, CR Kauffman, M Shishkov, DH Kang, EF Halpern, and GJ Tearney. 2002. Characterization of human atherosclerosis by optical coherence tomography. *Circulation* 106 (13):1640–5.

Yang HC, J Yin, C Hu, J Cannata, Q Zhou, J Zhang, Z Chen, and KK Shung. 2010. A dual-modality probe utilizing intravascular ultrasound and optical coherence tomography for intravascular imaging applications. *IEEE Trans Ultrason Ferroelectr Freq Control* 57 (12):2839–43.

Yin J, X Li, J Jing, J Li, D Mukai, S Mahon, A Edris, K Hoang, KK Shung, M Brenner, J Narula, Q Zhou, and Z Chen. 2011. Novel combined miniature optical coherence tomography ultrasound probe for *in vivo* intravascular imaging. *J Biomed Opt* 16 (6):060505.

Yin J, HC Yang, X Li, J Zhang, Q Zhou, C Hu, KK Shung, and Z Chen. 2010. Integrated intravascular optical coherence tomography ultrasound imaging system. *J Biomed Opt* 15 (1):010512.

Yock PG, EL Johnson, and DT Linker. 1988. Intravascular ultrasound: Development and clinical potential. *Am J Card Imaging* 2 (3):185–93.

Yong AS, AC Ng, D Brieger, HC Lowe, MK Ng, and L Kritharides. 2011. Three-dimensional and two-dimensional quantitative coronary angiography, and their prediction of reduced fractional flow reserve. *Eur Heart J* 32 (3):345–53.

Yoo H, JW Kim, M Shishkov, E Namati, T Morse, R Shubochkin, JR McCarthy, V Ntziachristos, BE Bouma, FA Jaffer, and GJ Tearney. 2011. Intra-arterial catheter for simultaneous microstructural and molecular imaging *in vivo*. *Nat Med* 17 (12):1680–4.

MULTIMODALITY PROBES FOR HYBRID IMAGING

Preclinical evaluation of multimodality probes

YINGLI FU AND DARA L. KRAITCHMAN

10.1 INTRODUCTION

Cardiovascular disease remains the number one cause of death in the developed countries. Medical imaging, e.g., magnetic resonance imaging (MRI), x-ray fluoroscopy, computed tomography (CT), ultrasound, positron emission tomography (PET), single photon emission tomography (SPECT), and optical imaging, plays an important role in understanding the mechanism of cardiovascular disease and, in some instances, diagnosing and tracking cardiovascular disease progression. The advances of cardiovascular imaging are mainly driven by the fast development of highly sensitive and specific imaging probes, even at the molecular level, and the imaging systems that provide superior spatial and temporal resolution for these probes *in vitro* and *in vivo*. In general, these imaging probes for cardiovascular imaging can be classified into two categories: (1) probes with single imaging detectability and (2) multimodality imaging probes that enable multiple *in vivo* imaging visualization (e.g., detectable by optical, MRI, and PET simultaneously). Some of these imaging probes may contain a therapeutic component that enables concomitant targeted therapy and *in vivo* imaging (Cyrus et al. 2008). Ideally, multimodality imaging probes should take advantage of complementary imaging modalities to provide anatomical, functional, and metabolic information with high sensitivity and spatial resolution and enable both noninvasive and invasive imaging, thereby providing comprehensive information of cardiovascular processes for diagnostic and therapeutic interventions. This could be accomplished by employing multiple imaging probes or a single multifunctional probe that possesses multiple imaging visibilities. One classic example of the later is the development of the first triple fusion reporter (TFR) gene probe that enables fluorescence imaging, bioluminescent imaging (BLI), and PET imaging in the same living subject (Ray et al. 2004). Since then, a plethora of other innovative imaging probes have been developed and applied to improve the understanding of disease progression or cell fate in the case of cell therapies (Nahrendorf et al. 2008; Fu et al. 2013; Kedziorek et al. 2013). This application of the multimodality imaging probes heavily relies on the development of imaging hardware and software as well.

In this chapter, we will describe the current development of multimodality imaging probes with emphasis on those that show promise for clinical translations. The advantages and disadvantages of these probes will be highlighted and seminal preclinical evaluations in the context of cardiovascular disease models will be discussed.

10.2 MRI PROBES

The high spatial resolution of MRI, together with its ability to generate three-dimensional (3-D) anatomical information and the lack of ionized radiation, makes it attractive for preclinical and clinical cardiovascular application. MRI detects the net magnetic moment of a collection of nuclei in a strong magnetic field after a radiofrequency pulse. In biological systems, MRI is essentially an image of the protons presented in water and fat as described in Part I of this book. Tissue contrast in MRI is achieved by the difference in proton density or intrinsic spin–spin (T1) and spin–lattice (T2) relaxation times. However, the intrinsic contrast provided by the water T1 and T2 and changes in their values caused by tissue pathology are often too limited to enable a sensitive and specific diagnosis. Therefore, contrast materials, called MRI probes, are increasingly added exogenously to generate appreciable magnetic resonance (MR) signals. These probes are designed to locally modify the magnetic properties of nearby water protons, creating either hyperintense (T1-weighted) or hypointense (T2- and T2*-weighted) MR signal contrast. In general, MRI probes fall into three classes: paramagnetic, superparamagnetic, and chemical exchange saturation transfer (CEST).

10.2.1 PARAMAGNETIC MRI PROBES

The most commonly used paramagnetic MRI probes are gadolinium (Gd)-based chelate agents. At physiologically low concentrations, paramagnetic MRI probes shorten the T1 relaxation time of nearby water protons, leading to hyperintense signals on T1-weighted imaging. Chemically, the Gd compounds are encapsulated with multidentate ligands to ensure the safety with respect to metal loss. If the protective chelation complex is disrupted or lost, then the highly toxic metal ions will be released. Additionally, due to the intrinsic low moment of Gd, the linkage of multiple Ga chelates with carriers, such as nanoparticles, peptides, and protein/liposome assemblies, is often required to increase the payload of the probe, leading to improved imaging sensitivity and MRI signal amplification (Loai et al. 2012; Paulis et al. 2012).

The first clinically developed Gd-based chelate agent is Gd diethylenetriamine pentaacetate complex (Gd-DTPA, Magnevist, Bayer HealthCare Pharmaceuticals) (Aime and Caravan 2009). Due to its low toxicity and high thermodynamic and kinetic stability, Gd-DTPA was approved by the U.S. Food and Drug Administration (FDA) for use in humans in 1988. Since then, other Gd-based chelate agents have been also developed. Examples include Gd-DTPA-bis-methylamide (BMA) (Omniscan, GE Healthcare), Gd-hydroxypropyl (HP)-tetraazacyclododecane-triacetic acid (DO3A) (ProHance, Bracco Diagnostics), Gd-DTPA-bis-methoxyethylamide (BMEA) (Optimark, Covidien Pharmaceuticals), Gd-ethoxybenzyl (EOB)-DTPA (Eovist, Bayer HealthCare Pharmaceuticals), and Gd-benzyloxypropionyl tetraacetate (BOPTA) (MultiHance, Bracco Diagnostics). Compared with Gd-DTPA, which is highly osmolar, Gd-BOPTA and Gd-EOB-DTAP are low in osmolarity and therefore are better tolerated by the host in particular for liver imaging. These Gd-chelated compounds are widely used as an extracellular blood pool contrast agents for T1-weighted imaging to enhance signal in the vessels for MR angiography, dynamic perfusion assessment in the heart, and viability assessment of myocardium in delayed contrast-enhanced imaging (Gerber et al. 2008;

Azene et al. 2014). The clearance route for Gd-based MRI probes is mainly through kidneys, with the exception of Eovist and MultiHance, which are partially eliminated through the liver. The biological elimination half-life in patients with normal renal function is ~1.5 h, while it could be as long as >30 h in patients with advanced renal impairment (Thomsen et al. 2006). Transmetallation is likely to occur when Gd-chelated agents present in the body for such a long period of time, which may contribute to the development of nephrogenic systemic fibrosis (Morcos 2008). The commercially available linear, nonionic gadodiamide is thermodynamically instable. Therefore, it carries excess chelates to ensure the absence of free Gd^{3+} in the pharmaceutical solutions over their shelf lives (Morcos 2008).

From a therapeutic stand point, Gd-based MRI probes are rarely used in cardiac cell labeling primarily due to the toxicity concerns and the low sensitivity for MRI once the probes become intracellular (Bulte and Kraitchman 2004a). In preclinical investigations, these probes are often used in higher concentration and are required to be conjugated with a carrier to improve their permeability to cell membrane. For example, a Gd-based, Cy3-labeled Gadofluorine M contrast agent used for embryonic stem cell (ESC)-derived cardiac progenitor cell tracking in the myocardium was designed to have a hydrophilic tail to enhance internalization into the cells (Adler et al. 2009). Gadofluorine M did not adversely affect cell viability *in vitro* and transplanted cells could be imaged *in vivo* 2 weeks post injection in both infarcted and normal mice and could be imaged both with MRI and fluorescent imaging (Adler et al. 2009).

Recently, a number of protein-, antibody-, or nanoparticles-based paramagnetic molecular probes have also been developed for targeted MRI in disease diagnosis and therapy. Apoptosis of cardiomyocytes plays a critical role in ischemic heart disease; thus, modulation of apoptosis may provide a valuable tool for targeted cardiac imaging and therapeutic interventions. Hiller et al. developed an annexin-conjugated Gd-loaded small liposome probe to image apoptosis in an isolated perfused rat heart (Hiller et al. 2006). In a similar study, Briley-Saebo et al. demonstrated the ability of an antibody-conjugated Gd compound to specifically delineate atherosclerotic plaques in mouse using MRI (Briley-Saebo et al. 2008). In another study, a theranostic MRI probe using $\alpha_v\beta_3$-targeting, rampamycin-containing, Gd-labeled, perfluorocarbon (PFC) nanoparticle was investigated for in balloon-injured rabbits as a means to detect and prevent areas of restenosis (Cyrus et al. 2008). With MR signal, enhancement from $\alpha_v\beta_3$-targeted paramagnetic nanoparticles on T1-weighted, black blood MRIs of injured vascular segments was demonstrated. In addition, the rampamycin-containing targeted nanoparticle was able to inhibit plaque formation and stenosis based on MR angiograms compared to sham injections or targeted nanoparticles without rampamycin (Cyrus et al. 2008).

In addition to ${}^1H$ MRI probes, nonproton MRI probes, such as ${}^{19}F$ in PFCs, have also demonstrated high potential for cardiovascular imaging and cardiac stem cell tracking (Partlow et al. 2007; Kraitchman and Bulte 2009). Because the background signal from fluorine is negligible in the body, a high sensitivity to exogenously introduced ${}^{19}F$ agents can be achieved. Upon systemic administration, PFC nanoparticles are preferentially phagocytosed by circulating monocytes or macrophages. Thus, ${}^{19}F$ MRI signal mainly reflects macrophage infiltration or inflammation (Stoll et al. 2012). Consequently, these novel imaging probes can be utilized to monitor immune cell responses in myocardial infarction and rejection of donor organs after transplantation (Neubauer et al. 2007; Flogel et al. 2008). In a murine acute myocardial ischemia model, ${}^{19}F$ MRI revealed a time-dependent infiltration of injected biochemically inert nanoemulsions of PFCs at the border zone of infarcted areas; histology demonstrated colocalization of PFCs with monocytes/macrophages (Flogel et al. 2008). When PFCs are utilized, there is the potential for multimodality imaging, such as ultrasound in combination with ${}^{19}F$ MRI. The disadvantages of nonproton MR probes are the low abundance relative to water that makes detection challenging and the need for specialized MRI hardware to detect nonproton signals.

10.2.2 SUPERPARAMAGNETIC MRI PROBES

Superparamagnetic MRI probes include many species, such as cobalt, iron platinum, and iron oxide. Among those, superparamagnetic iron oxide (SPIO)-based agents are the most widely used MRI probes for biomedical imaging, in particular for cell tracking. Unlike Gd-based contrast agents, SPIOs cause a substantial signal loss or hypointense signal in the vicinity of iron oxide particles on T2*-weighted MRI irrespective of whether SPIOs are internalized into the cells or not. Therefore, the sensitivity of imaging SPIO-labeled cells is much higher than paramagnetic probes (Azene et al. 2014). SPIOs are generally coated with dextran or carboxydextran to improve their biocompatibility. Two SPIOs, low-molecular-weight dextran coated ferumoxide (Feridex, Berlex Laboratories, and Endorem, Guerbet) and carboxydextrane-coated ferucarbotran (Resovist, Bayer Healthcare), were approved for MRI of liver tumors (Ros et al. 1995) and thus showed tremendous promise for clinical translation. However, economic factors resulted in these agents ultimately being commercially abandoned.

After intravenous injections, SPIOs are incorporated into macrophages via endocytosis. Therefore, the uptake of SPIOs by phagocytic monocytes and macrophages provides a valuable tool to monitor the involvement of macrophages in inflammatory processes, such as vulnerable plaque development in carotid artery (Chan et al. 2014). In one study, Sosnovik et al. demonstrated the accumulation of long circulating SPIOs in the infarcted myocardium due to the uptake by infiltrating macrophages on T2*-weighted MRI (Sosnovik et al. 2007). This study also showed high correlation between the amount of injected iron oxide probes and image contrast generated within the myocardium. Conjugating the probe with a near-infrared (NIR) fluorophore also provided additional benefit to image infiltrating macrophages/monocytes *in vivo* with NIR fluorescence tomography (Sosnovik et al. 2007). Recently, radiolabeled iron oxide nanoparticles have been found to significantly accumulate in the heart of apoE$^{-/-}$ mice compared with that of healthy control animals, suggesting that they may be useful to detect macrophages in the atherosclerosis plaques of coronary arteries (de Barros et al. 2014).

The ability to label nonphagocytic cells in culture using derivatized SPIOs, followed by transplantation or transfusion in living subjects, has enabled the monitoring of cellular biodistribution *in vivo* including cell migration and trafficking during cellular therapeutic interventions. The sensitivity for detecting SPIO-labeled cells mainly depends on magnetic field strength, the concentration of intracellular iron, and cell numbers. It has shown that as few as 20 labeled cells per 1000 cells in a voxel can be detected by MRI at 1.5T due to the "blooming" effect; i.e., the artifact created by SPIO-labeled cells is much larger than the volume occupied by the cells (Arbab et al. 2003; Zhang 2004) At higher magnetic field, *in vivo* single cell detection can be achieved (Shapiro et al. 2006). One advantage of SPIO labeling is that iron oxides can be integrated into the body and recycled into the native iron pool should labeled cells die. However, the signal void created by the labeled cells is often difficult to differentiate from endogenous sources of iron, such as hemorrhage and susceptibility artifacts, which also cause hypointensities on T2*-weighted MRI (Azene et al. 2014). Additionally, debate remains as to whether the hypointensities at a later time point represent transplanted cells, engrafted cells, lost iron particles from cells, or macrophages with iron uptake. Thus, efforts on developing positive contrast techniques to track the susceptibility of off-resonance artifacts created by iron-labeled cells have been made. Many of these techniques require either specific pulse sequences (e.g., spin echo or gradient echo) or postprocessing methods (e.g., inversion recovery with on-resonance water suppression [IRON], sweep imaging with Fourier transformation [SWIFT], positive contrast with alternating repetition time steady-state free precession [PARTS]) (Stuber et al. 2007; Cukur et al. 2010; Eibofner et al. 2010; Zhou et al. 2010). The positive contrast SPIO imaging technique called inversion recovery with on-resonance water suppression (IRON) was developed that saturates the water and fat peaks so that the off-resonance protons in close proximity to the SPIO-labeled cells are enhanced (Stuber et al. 2007). This technique has been used for detection of SPIO-labeled stem cells in a rabbit model of peripheral arterial disease (Figure 10.1) (Kraitchman and Bulte 2008).

Although no clinical trials using SPIO-labeled cells have been initiated for cardiac repair, many preclinical studies on MR-based tracking of SPIO-labeled stem cells in the heart have been performed in varied animal models to address the questions regarding optimal cell delivery route, timing, dosage, cell type, and retention (Kraitchman et al. 2003; Ebert et al. 2007; Zhou et al. 2010). In one study, mouse ESCs were labeled with SPIO prior to transplantation into mice, and hypointensities in ischemic myocardium were observed 4 weeks after delivery, suggesting the successful incorporation of labeled ESCs within infarcted myocardium (Ebert et al. 2007). A similar study done by Amado et al. demonstrated substantial retention of SPIO-labeled bone marrow-derived stromal cells in infarcted myocardium at 8 weeks (Amado et al. 2005). More recently, Drey et al. used micron-sized SPIOs to label mesenchymal stem cells (MSCs) and showed the feasibility of *in vivo* tracking of as few as 10^5 labeled MSCs in infarcted murine heart 4 weeks after intramyocardial injection (Drey et al. 2013). Using large-animal models, our group has successfully demonstrated the detection of SPIO-labeled MSCs in infarcted pigs, dogs, and in critical limb ischemia rabbit (Kraitchman et al. 2003; Bulte and Kraitchman 2004b; Kraitchman and Bulte 2008). In reperfused myocardial infarcted pigs (Figure 10.2),

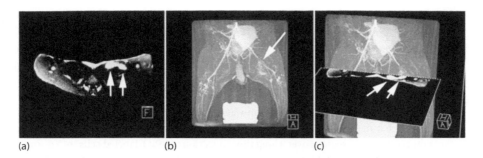

(a)   (b)   (c)

Figure 10.1 Positive contrast detection of SPIO-labeled MSCs in a rabbit model of peripheral arterial disease using inversion-recovery with on-resonance water suppression (IRON). (a) An axial positive contrast imaging with IRON shows two injection sites (arrows) as bright hyperintensities. (b) A maximum intensity projection of a 3-D T2-prepared MR angiogram shows the region of superficial femoral artery occlusion (arrow) in a rabbit 24 h after occlusion. (c) Fusion of the positive contrast images (a) and MR angiogram (b) reveals the location of SPIO-labeled MSCs relative to collateral neovasculature. (Adapted from Kraitchman DL, Bulte, JW, *Basic Res Cardiol*, 103, 105–113, 2008. With permission.)

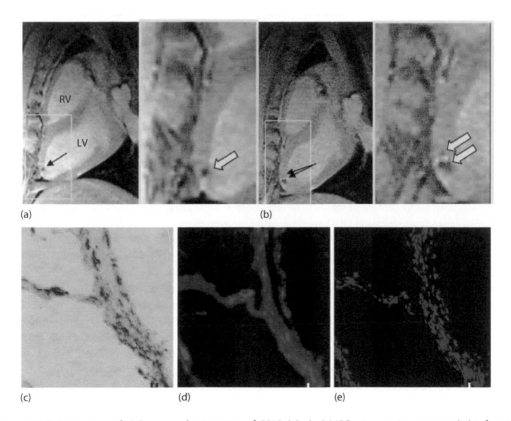

Figure 10.2 Detection of delivery and migration of SPIO-labeled MSCs in a swine myocardial infarction model. (a, b) Long-axis MR images show hypointense lesions (arrows) caused by MSCs acquired within 24 h (a) and 1 week (b) of injection with the inset on the bright demonstrating expansion of the lesion over 1 week. (c–e) Intracellular iron as detected by diamino-benzidine Prussian Blue staining (c) matches colabeling of MSCs with the fluorescent dyes Dil (d) and DAPI (e) on adjacent histological sections at 24 h after injection in another animal, indicating that the SPIOs are still contained within the MSCs. RV: right ventricle; LV: left ventricle. (Adapted from Kraitchman, DL et al., *Circulation*, 107, 2290–2293, 2003. With permission.)

MSCs were colabeled with SPIOs (ferumoxides) by "magnetofection" and DiI (I) prior to x-ray fluoroscopic-guided transmyocardial injection. The detection of SPIO-labeled MSCs on MRI was accomplished and confirmed by fluorescence microscopy postmortem (Kraitchman et al. 2003). Interestingly, approximately 30% of the injections of SPIO-labeled cells delivered under x-ray fluoroscopic guidance were not successful as confirmed by the lack of visualization of labeled cells on MRI of the heart, highlighting the power of cellular labeling to determine the success of delivery (Kraitchman et al. 2003). Subsequently, the migration of SPIO-labeled stem cells in the peri-infarcted myocardium was noted over 8 weeks in a reperfused dog infarction model (Bulte and Kraitchman 2004b). While these studies confirmed the presence of SPIO-labeled cells in the infarcted and peri-infarcted regions over a long time, injection of SPIO-labeled cells into normal myocardium was no longer detected at 4 weeks postdelivery (Soto et al. 2006). Another potential multimodality use of SPIOs is to bind antibodies to the iron oxide nanoparticles to enable cell selection and sorting followed by noninvasive imaging (Verma et al. 2015).

Despite the promise of iron oxides in cardiac stem cell tracking, clinical translations have been hampered by the recent removal of clinical formulation of SPIOs from the market for economic reasons, and regulatory hurdles with the addition of an investigational new device for delivery on an MRI platform that is not familiar to interventional cardiologists. Nevertheless, many investigators continue to explore cell labeling strategies and applications with an off-label use of FDA-approved ultrasmall SPIO, ferumoxytol (Feraheme, AMAG Pharmaceuticals).

10.2.3 CEST PROBES

MR probes that utilize the properties of CEST are a novel class of contrast agents that are designed to contain a narrow band of off-resonance protons that exchange with the protons in tissue water. When the saturated CEST protons exchange with tissue water protons, the on-resonance water signal drops, leading to decreased signal intensity in the location where CEST contrast agents present. CEST probes are considered as "switchable" contrast agents as they can be turned on or off depending on the specific saturation pulses applied. A variety of CEST probes have been developed for medical applications. Agents with amide (-NH), amine (-NH$_2$), and hydroxyl (-OH) groups are particularly suitable for producing diamagnetic CEST contrast (Yang et al. 2013). One of the advantages of CEST probes is the possibility of generating families of CEST probes based on the different resonance frequencies of the exchangeable protons, so that "multiple colors" can be created to enable simultaneous detection of different targets after image postprocessing (McMahon et al. 2008). Based on this concept, an MR reporter gene that overexpresses lysine-rich protein has been developed as an endogenous CEST probe and shown to be detectable in the brain (Gilad et al. 2007). In particular, non-radioactive fluorodeoxyglucose (FDG) has exchangeable protons that can be used for CEST-MRI as well as ^{19}F MRI (Rivlin et al. 2013). Despite recent advances in CEST probe design, the sensitivity of CEST probes is low in general. One mechanism to increase sensitivity is to use a paramagnetic CEST probe (Evbuomwan et al. 2012). For example, a fluorescent label can be bound to a europium complex to yield a dual-modality probe for optical and CEST-MRI, respectively (Ali et al. 2012). Because CEST imaging requires paired images obtained with and without radiofrequency irradiation, motion can be problematic and most CEST-MRI has been performed in the brain. However, two recent studies have looked at CEST-MRI in the heart. Vandsburger et al. developed a steady-state CEST-MRI sequence for examining fibrosis in mouse myocardial infarction using a PARACEST contrast agent (Vandsburger et al. 2015). Haris et al. have used CEST-MRI in the heart with exchangeable amine protons from creatine used in the creatine kinase reaction to provide energy to the heart and shown the potential for this technique to be more sensitive than MR spectroscopy for examining myocardial infarction in sheep and swine (Haris et al. 2014).

10.3 X-RAY PROBES

Radiographic iodinated contrast agents are perhaps the most commonly prescribed drugs in the history of modern medicine (Singh and Daftary 2008). Intravenously delivered iodinated contrast has been utilized extensively in x-ray-based imaging, including x-ray fluoroscopy and CT, to visualize vascular structures like the arteries and veins (e.g., CT angiography) in the heart and periphery. The recent development of iodinated nanoparticles, N1177, has made it feasible to identify ruptured vs. nonruptured atherosclerotic plaques in rabbits (Van Herck et al. 2010). New probes, such as PEGylated, low-generation dendrimer-entrapped gold nanoparticles, have recently emerged and have been tested for cardiovascular imaging (Liu et al. 2014). X-ray probes together with x-ray imaging modalities provide high spatial resolution and allow real-time interactivity. However, most x-ray probes are highly toxic when used intracellularly even at low concentrations, making them unsuitable for cardiac cell labeling and tracking. In addition, the lack of soft tissue visualization and concern about ionizing radiation also limit their cardiac application. However, iodinated contrast agents for vascular imaging are suitable for anatomical visualization in combination with radionuclide probes for molecular imaging.

Recently, our group has developed x-ray-visible microcapsule formulations that allow the use of high payload of x-ray probes without cell toxicity (Barnett et al. 2006, 2011; Kedziorek et al. 2012; Fu et al. 2014). Because these x-ray contrast probes are retained in the microcapsule rather than intracellularly, high concentration of such probes can now be utilized to enable serial noninvasive tracking of encapsulated cells using conventional clinical x-ray equipment. Alginate microcapsules with addition of barium sulfate allowed the confirmation of MSC delivery success using conventional x-ray fluoroscopy and improved the retention of allogeneic MSCs in a rabbit model of peripheral arterial disease (Figure 10.3) (Kedziorek et al. 2012). However, the large size of the microcapsules (~300–500 μm) may prevent direct injection of encapsulated cells into the coronary arteries or myocardium of the heart primarily due to embolization concern or induction of conduction abnormalities. Since the cells are trapped within the microcapsules, direct incorporation of the cells is also unlikely. Presumably, these techniques would be better suited for deposition of cells outside of the heart, where the encapsulated cells may improve cardiac function via paracrine mechanism, i.e., encapsulated cells release cytokines or growth factors to enhance angiogenesis and recruit native stem cells to the heart and differentiate into cardiomyocytes. Based on this concept, Fu et al. have recently demonstrated the feasibility and safety of delivering x-ray-visible microencapsulated hMSCs into the pericardial space in an immunocompetent swine model (Fu et al. 2014). One multimodality imaging method is to use x-ray imaging to enable high temporal resolution of the heart for interventional techniques in combination with high spatial resolution of anatomical detail from MRI. Using this real-time x-ray imaging fused with segmented myocardial borders from 3-D whole-heart MRI to enhance visualization of coronary vasculature and the myocardial wall, precise intrapericardial deposition of barium sulfate-containing microencapsulated hMSCs was achieved (Figure 10.4) (Fu et al. 2014). Contrast agent impregnation is not limited to radiopaque contrast agents. Indeed, using perfluoro-octyl bromide (PFOB), a variety of imaging techniques can be performed singly or in a combined fashion ranging from ultrasound (based on PFCs), ^{19}F MRI, or x-ray imaging (based on bromine radiopacity) (Barnett et al. 2010). Thus, microencapsulation in combination with contrast agents may provide a method to monitor the delivery success and track engraftment using a well-accepted x-ray fluoroscopic imaging platform commonly used in cardiovascular application or in combination with ultrasound or MRI.

10.4 RADIONUCLIDE PROBES

Radionuclide imaging, i.e., PET and SPECT, has the highest sensitivity (PET: 10^{-11} to 10^{-12} mol/L; SPECT: 10^{-10} to 10^{-11} mol/L) among all currently used imaging modalities with the ability to quantify radioisotope levels (Massoud and Gambhir 2003). Radionuclide probes have been routinely used to assess cardiac metabolic function, viability, contractile function, as well as to noninvasively monitor cell fate (Kendziorra et al. 2008; Castellani et al. 2010). PET imaging probes are labeled with positron emitting radionuclides (e.g., ^{18}F, ^{13}N, and ^{11}C), whereas SPECT probes are labeled with γ-emitting radionuclides (e.g., ^{111}In, ^{99m}Tc, and ^{125}I), as mentioned in Part I of this book. However, the high sensitivity to the radioisotopes also means that anatomical localization cannot be obtained without fusion with alternate imaging modalities, such as MRI or CT.

In clinical diagnosis and preclinical investigations, ^{18}F and ^{11}C are two most widely used PET radiotracers because of their availability, chemical characteristics, and nuclear properties. Currently, ^{11}C radionuclide

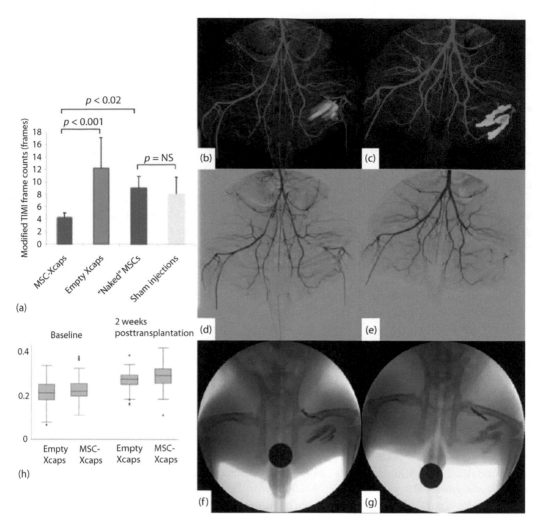

Figure 10.3 X-ray visible microcapsule for MSC delivery in peripheral arterial disease (PAD) rabbits. (a) A bar graph of the average modified thrombolysis in myocardial infarction (TIMI) frame count, as a measure of collateral vessel development, demonstrates a significant improvement in distal filling only in the PAD rabbits that received microencapsulated MSCs (*$p < 0.001$ empty microcapsules vs. MSC-Xcaps; p = NS naked MSCs vs. sham injections). (b–g) Representative digital subtraction angiogram (DSA, red) obtained during peak contrast opacification performed at 2 weeks post injection of MSCs-Xcaps (b) and empty microcapsules (c) with an overlay of microcapsules injections (green) obtained from mask image of DSA. The small collateral vessels are somewhat obscured by the Xcap radiopacity. However, the increased collateralization can be appreciated in the MSC-Xcap-treated animal DSA (d) relative to the Xcap-treated animal (e). Native mask digital radiographs demonstrate the location of the MSC-Xcaps (f) and empty Xcaps (g) in the same animals. (h) Box-whisker plot shows the difference between left and right distal deep femoral artery diameters at baseline and 2 weeks after superficial femoral artery occlusion in treated (MSC-Xcaps) and untreated animals (Empty Xcaps). (Adapted from Kedziorek DA et al., *Stem Cells*, 30, 1286–1296, 2012. With permission.)

probes are mainly synthesized to image cardiac metabolism, such as fatty acid metabolism (Coggan et al. 2009). However, ^{11}C has a relatively short half-life of 20 min, making it difficult to synthesize and image within a short time window. In contrast, the half-life of ^{18}F is approximately 110 min, allowing time-consuming multistep radiosyntheses and long imaging window. Its low β+-energy (0.64 MeV) provides a short positron linear range in tissue, leading to high-resolution PET images. Since the first evaluation in 1978, ^{18}F-FDG has been routinely used for myocardial viability assessment (Segall 2002). Recently, the application of ^{18}F-FDG has expanded to image atherosclerotic plaque inflammation (Blomberg et al. 2013) and label stem cells for

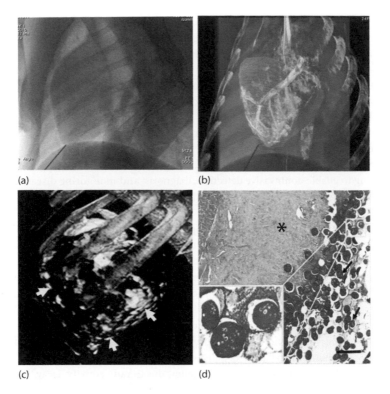

Figure 10.4 X-ray fusion with MRI (XFM)-guided intrapericardial delivery of x-ray visible human MSC BaCaps in swine. (a) Fluoroscopically guided pericardial puncture shows the lack of visualization of coronary vasculature and myocardial borders. (b) Image obtained with XFM (gray scale indicates the x-ray portion of the image, and color indicates MR imaging) of the pig heart shows enhanced coronary vasculature and ventricular boundaries that may enable more precise targeting of stem cell therapeutics. (c) C-arm CT image of the heart obtained 1 week after BaCaps delivery shows the distribution of the BaCaps (arrows). (d) Hematoxylin–eosin staining of the heart 1 week after XFM-guided delivery of BaCaps with hMSCs (arrows). Insert shows viable hMSCs, with clear nuclear morphology and absence of a foreign-body reaction. (Adapted from Fu Y et al., *Radiology*, 131424, 2014. With permission.)

tracking after transplantation (Wolfs et al. 2013). As an alternative, Tahara and colleagues developed a new radionuclide probe for imaging atherosclerotic plaque using ^{18}F-labeled mannose (FDM), an isomer of glucose whose receptors are expressed on a subset of macrophages in high risk plaques (Tahara et al. 2014). This study demonstrated that ^{18}F-FDM uptake was proportional to the plaque macrophage population in a rabbit model (Tahara et al. 2014). ^{18}F has also been used to label other biomolecules for targeted cardiovascular imaging. For instance, an ^{18}F-labeled glycosylated $\alpha_v\beta_3$ integrin antagonist (^{18}F-galaco-RGD) was synthesized, and its feasibility of targeting $\alpha_v\beta_3$ integrin expression was demonstrated on PET with focal ^{18}F-galaco-RGD uptake after coronary occlusion and reperfusion in rats (Higuchi et al. 2008; Sherif et al. 2012). In addition, high levels of ^{18}F-galaco-RGD uptake in the perfusion defect area early after myocardial infarction were associated with the absence of significant left ventricular remodeling after 12 weeks of follow-up (Sherif et al. 2012). A variety of other ^{18}F-labeled biomolecules, such as ^{18}F-annexvin V (Murakami et al. 2004) and ^{18}F-FBzBMS (Higuchi et al. 2013), have been developed for targeting apoptosis and endothelin subtype-A receptor in the infarcted myocardium. In ischemic animals, the accumulation of ^{18}F-annexin V in the infarcted area was three times higher than that in the noninfarcted area and was correlated to histological detection (Murakami et al. 2004).

Radionuclide probes with γ emitters, e.g., ^{111}In, ^{99m}Tc, and ^{125}I, are often used for SPECT imaging. One of the advantages of these radiotracers is their longer half-life, higher stability, and improved labeling efficiency as compared to PET radiotracers. Direct cell labeling with ^{111}In oxine ($t_{1/2} \approx 2.8$ days) initially developed for lymphocyte labeling was translated for cardiac stem cell imaging. In a canine myocardial infarction model, the trafficking of ^{111}In oxine and SPIO-labeled MSCs could be monitored by clinical SPECT/CT up to 7 days, while MRI with a lower sensitivity failed to detect the dual radiolabeled/SPIO-labeled MSCs (Figure 10.5) (Kraitchman et al. 2005). Subsequently, many other investigators have demonstrated the varied retention of radiolabeled stem cells in the heart after intravenous, intramyocardial, intracoronary, or interstitial retrograde coronary venous delivery (Aicher et al. 2003; Brenner et al. 2004; Hou et al. 2005; Zhou et al. 2005; Tran et al. 2006; Blackwood et al. 2009; Wisenberg et al. 2009; Lyngbaek et al. 2010). The minimal detection limits of cells with direct radiotracer labeling range from 2,900 cells to 25,000 cells depending on the choice of radiotracer and cell type (Jin et al. 2005).

In addition to direct labeling with radioisotopes, new radionuclide nanoparticles are being developed for targeted imaging to enable noninvasive detection, diagnosis, and monitoring disease progression. These probes will share the high sensitivity of traditional radiotracers but also have a high specificity due to surface chemistry conjugation and will likely be readily clinically translatable. In one study, de Baross et al. synthesized ^{125}I-labeled dextran-coated iron oxide nanoparticles to detect macrophages in the atherosclerosis plaques of coronary arteries (de Barros et al. 2014). A biodistribution study showed significant accumulation of the probe in the heart of apoE$^{-/-}$ mice (de Barros et al. 2014). In another study, Li et al. reported the use of ^{111}In-labeled liposome nanoparticles with surface conjugation with antibodies against the low-density lipoprotein receptor LOX1. SPECT imaging displayed a "hot spot" signal in atherosclerotic plaques in apoE$^{-/-}$ mice (Li et al. 2010).

Hurdles for development of radionuclide probes include the relatively low spatial resolution of PET/SPECT imaging, the need for a generator/cyclotron to produce radionuclide, the potential for radiation damage to the cells (Jin et al. 2005; Gholamrezanezhad et al. 2009; Gildehaus et al. 2011), leakage of radiotracers over the time course (Aicher et al. 2003), and short imaging window due to radioactivity decay.

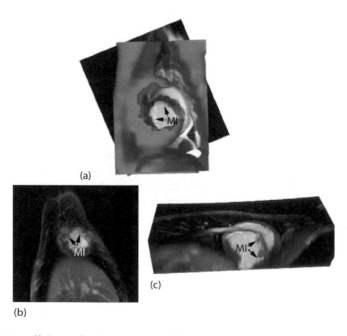

Figure 10.5 Dynamic trafficking of indium-111 oxine-labeled allogeneic MSCs to myocardial infarction (MI) dogs. MSCs were colabeled with indium-111 oxine and SPIOs. (a) Short-axis view of alignment of CT (gold) with MRI (gray scale) and SPECT (red) shows focal uptake in the septal region of the MI in a representative dog. (b) Focal uptake on SPECT (red) in another animal demonstrates localization of the MSCs to the infarcted myocardium in the short-axis (b) and long-axis (c) views. SPECT, owing to the higher sensitivity, was able to detect approximately 8000 labeled MSCs/g of tissue, whereas MRI was unable to detect the SPIO-labeled MSCs. (Adapted from Kraitchman DL et al., *Circulation*, 112, 1451–1461, 2005. With permission.)

10.5 ULTRASOUND PROBES

Echocardiography or ultrasound imaging has been recognized as a powerful imaging tool for cardiac structure and function evaluations. Due to its noninvasiveness, low cost, lack of ionizing radiation, and portability, ultrasound imaging is the most commonly used imaging modality in clinical practice and preclinical investigations. Ultrasound probes for molecular imaging share the common property of "acoustic activity" in the appropriate energy and frequency of an ultrasound field, i.e., causing different acoustic impedance between tissues. The majority of ultrasound probes are gas-filled microbubbles surrounded by a lipid, protein, or polymer shell, which increase echogenicity. The customary size of microbubbles ranges from 1 μm to 7 μm, preferably around 3 μm (Schutt et al. 2003). The stability of microbubbles is dependent on the type of gas in the microbubbles and the nature of shell composition. In general, low-solubility gases, e.g. PFCs, increase the stability and circulation time of microbubbles *in vivo* (Schutt et al. 2003). The present PFC-based microbubble ultrasound probes were primarily designed to remain within the vascular space to image the blood pool. Therefore, for cardiovascular imaging, free circulated microbubbles have been used to enhance left ventricular endocardial border opacification and evaluate the response of therapies, including stem cells, on perfusion and function in ischemic cardiomyopathy (Nanda et al. 2003; Inaba et al. 2014). Subsequently, a variety of targeted ultrasound probes were developed using monoclonal antibodies (Yan et al. 2014), peptides (Hyvelin et al. 2014; Leng et al. 2014), or proteins as targeting moieties to facilitate microbubble adhesion to endothelial targets. Using a biodegradable polymer microbubble bearing a short synthetic peptide with specific human E-selectin affinity, Leng et al. demonstrated persistent ultrasound contrast enhancement of the ischemic region of the heart in rats 4 hours after transient coronary occlusion (Leng et al. 2014). Similar results were reported in reperfused transient ischemic rat hearts, where high late-phase enhancement within the ischemic area was correlated with the expression of E-selectin 24 hours after reperfusion (Hyvelin et al. 2014). Thus, targeting ultrasound probes could be useful for clinical myocardial ischemic memory imaging to identify acute coronary syndromes.

In addition, ultrasound-visible microbubbles have been adapted as vehicles for delivering genes, proteins, or drugs to the target tissues by selective ultrasound targeted microbubble destruction (UTMD) (Bekeredjian et al. 2003; Fujii et al. 2011; Ling et al. 2013; Yan et al. 2014). Using matrix metalloproteinase (MMP) 2 antibody-conjugated cationic microbubbles carrying *Timp3* plasmids (inhibitor of MMP 2 and MMP 9), Yan et al. demonstrated significantly higher accumulation of microbubbles in the infarcted region. Upon UTMD, microbubble delivery of *Timp3* gene significantly increased TIMP3 protein levels in the infarct scar and border zone at 3 days post-UTMD, which led to smaller and thicker infarcts and improved cardiac function (Yan et al. 2014).

More recently, ultrasound probes have also been employed in stem cell labeling and tracking. Direct labeling of bone marrow-derived MSCs by double-layer polymeric microbubbles could be detected *in vivo* by ultrasound for up to 7 days in the mouse thigh (Fu et al. 2011). In a similar vein, neural progenitor cells have been efficiently labeled with cationic microbubbles, allowing a clinical ultrasound system to detect single cell *in vitro* at 7 MHz (Cui et al. 2013). Interestingly, the labeled cells could be detected in the left ventricle after intravenous injection and were still visible in the liver 5 days after delivery. This study indicated that microbubble stability was improved with internalization as free microbubbles only last for a few minutes *in vivo* (Cui et al. 2013), These microbubbles could also be used to carry payloads such as genetic material for cell transfection with the added benefit that the genetic material would only be released when exposed to ultrasonic engineering, allowing selective targeting of cell expression. Another method is to create targeted ultrasound probes that will bind specific cell surface markers for tracking progenitor or stem cells. In one study, endothelial progenitor cells (EPCs) were genetically modified to express the mouse H-2Kk protein and transplanted subcutaneously in rats. Contrast-enhanced ultrasound demonstrated *in vivo* detection of EPCs with liquid microbubbles conjugated with anti-H-2Kk antibodies (Kuliszewski et al. 2009).

Besides gas-filled microbubbles, gold nanoparticles have also been used as ultrasound probes for longitudinal tracking of stem cells. Gold nanoparticle-labeled MSCs immobilized in PEGylated fibrin gel were able to be detected by ultrasound-guided photoacoustic imaging in the lower limb of the rats over 1 week

(Nam et al. 2012). Unlike direct labeling with microbubbles, where detection of microbubbles ceases once the labels are degraded or destructed, the gold nanoparticles may still remain and can be imaged even if the transplanted cells are destroyed.

Although ultrasound probes have relatively high sensitivity, several challenges must be addressed before their wide application in cardiac imaging and cell tracking can be fully realized. These include the poor spatial resolution (submillimeter) of ultrasound imaging, low stability of ultrasound contrast agents, and large size of the microbubbles, which prevents the internalization of large amount of probes, leading to agent dilution or loss with cell division. While many ultrasound targeted agents can readily move from the vascular space to the extravascular space, the translocation from the perivascular space to deep within tissue and specific targets has remained challenging.

10.6 OPTICAL PROBES

Optical imaging probes, including fluorescence probes and bioluminescence probes, are photon-emission probes that provide high sensitivity for cell tracking, with detectability of 10^{-9} to 10^{-12} mol/L and 10^{-15} to 10^{-17} mol/L, respectively (Zhang and Wu 2007; Fu and Kraitchman 2010; Ransohoff and Wu 2010). Because optical imaging probes do not generate tissue contrast, they are often used in combination with high-sensitivity anatomical imaging techniques, such as CT and MRI, to provide structural information. A low-cost alternative for anatomical localization in combination with fluorescent or bioluminescent optical imaging systems is 2-D digital light images. These systems are well suited for small-animal preclinical imaging studies. However, the low-energy photon attenuation restricts tissue penetration, making visualization of deep structure (e.g., blood vessels and the heart) difficult when using external detectors in larger animals. Thus, clinical translation of small-animal techniques requires more invasive internal detectors for large animal or clinical studies.

Despite the aforementioned limitations, optical probes are useful for preclinical investigations to provide insight regarding disease progression, therapeutic response, and cell fate (Waldeck et al. 2008). NIR fluorophores conjugated with biomolecules, e.g., antibodies, proteins, peptides, and functionalized nanoparticles, could be used for imaging cardiac angiogenesis and inflammation. In a recent study, single-walled carbon nanotubes (SWNTs) with intrinsic NIR photoluminescence were developed for NIR imaging and thermal ablation of vascular macrophages (Kosuge et al. 2012). The uptake of Cy5.5-conjugated SWNTs by macrophages was demonstrated both in cell culture and in ligated murine carotid arteries on fluorescence and NIR imaging. Simultaneous laser light exposure to the ligated left carotid arteries induced apoptosis in the neointima and adventitia of the arteries, which colocalized with macrophages detected microscopically (Kosuge et al. 2012). In another study, multipotent progenitor cells (MPCs), including mesenchymal stromal cells, bone marrow mononuclear cells, and peripheral blood mononuclear cells, were labeled with an NIR fluorophore, I-786, and subsequently transplanted in a swine model of myocardial infarction (Ly et al. 2009). *In vivo* NIR imaging demonstrated that MPC distribution and retention immediately after intracoronary delivery varied depending on cell type (Ly et al. 2009). This study showed that cell retention in the myocardium is dependent on the cell population, which could potentially impact the clinical efficacy of cardiac cell therapy.

Quantum dots (QDs) are another class of fluorescence probes that have been employed for vulnerable plaque detection in atherosclerotic lesions and stem cell labeling and tracking in the cardiovascular system. These nanosized inorganic semiconducting probes have unique optical advantages over organic fluorescence dyes, including narrower band emission and broader band excitation with a high quantum yield, exceptional photostability, and resistance to chemical and metabolic degradation (Medintz et al. 2005; Cesar 2014). In addition, QDs could be easily functionalized with versatile chemistry modification strategies to enable targeted detection. In an *in vitro* study, Marrache and Dhar developed a synthetic high-density lipoprotein nanoparticle bearing diagnostically active QDs for optical detection of macrophage apoptosis in vulnerable plaques (Marrache and Dhar 2013). QDs have also been used for stem cell tracking in the heart. QD-labeled human MSCs can be unambiguously detected *in vivo* and in postmortem histological sections at least 8 weeks after delivery (Rosen et al. 2007). Though QDs alleviate some of the problems associated with organic dyes (e.g., low absorbance and photobleaching), concerns about the toxicity of QDs, which contain heavy metals, have primarily limited these probes to the preclinical arena.

10.7 REPORTER GENE/PROBES

Unlike the aforementioned multimodality imaging probes, which are essentially exogenously added contrast agents, the reporter gene/probes require genetic engineering of the cells with appropriate plasmids or viral vectors to induce the cells to express a specific protein, receptor, or enzyme that can be detected either directly by imaging (endogenous probes) or by the introduction of a reporter probe. To prevent dilution of the reporter upon cell division, a stable transfection using viral promoter (e.g., lentivirus and adenovirus) is often required to integrate the reporter gene into host chromatin. Reporter gene/probes were initially developed for postmortem analysis of the tissue. Perhaps the earliest example of reporter gene/probes is green/red fluorescence protein, which can be directly detected by fluorescence imaging for histological analysis. Since reporter gene products are expressed only by living cells, false-positive detection is less likely than direct labeling techniques. As such, reporter gene/probes are extremely useful for cardiac cell-based therapies to assess the status of cell survival, migration, and fate following transplantation.

10.7.1 MRI REPORTER GENE/PROBE

Several MR reporters have been developed, including beta-galactosidase (Louie et al. 2000), iron storage proteins (e.g., ferritin, transferring, and transferrin receptor) (Moore et al. 2001; Genove et al. 2005; Deans et al. 2006; Pawelczyk et al. 2006; Liu et al. 2009; Naumova et al. 2010), and artificial proteins (e.g., lysine-rich protein) (Gilad et al. 2007). In 2000, Louie et al. prepared an MR reporter probe for cellular imaging where the cells were transfected to express beta-galactosidase that can enzymatically cleave the blocking group on chelated paramagnetic ions, leading to increased signal intensity on T1-weighted images (Louie et al. 2000). Recently, overexpression of transgenic human ferritin receptor and ferritin heavy chain subunit has been performed in various cells, including tumor cells (Moore et al. 2001), neural stem cells (Pawelczyk et al. 2006), and ESCs (Liu et al. 2009), such that signal amplification can be realized by accumulating more irons within the cells. Preclinical studies have demonstrated the feasibility of overexpressing mouse skeletal myoblasts with an MR reporter, ferritin. These transgenic cells were successfully detected by MRI *in vitro* and *in vivo* after transplantation into the infarcted mouse heart (Naumova et al. 2010). This technique was also used to image injected cardiac progenitor cells in an infarcted rat heart (Campan et al. 2011). Follow-up studies demonstrated iron uptake up to 4 weeks after transplantation on T2*-weighted MRI (Campan et al. 2011). Because ferritin is a native protein responsive for iron storage, its overexpression is not expected to lead to iron toxicity. Similarly, an endogenous MRI reporter probe based on lysine-rich protein has been developed for CEST-MRI in the brain (Gilad et al. 2007). Although it has the potential to create multicolor images of different exchangeable proton residues, its cardiac application could be extremely challenging because of cardiac motion and special CEST imaging procedures.

10.7.2 PET/SPECT REPORTER GENE/PROBE

Frequently used PET/SPECT reporter gene constructs include transporter-based sodium-iodide symporter for SPECT imaging (Miyagawa et al. 2005; Lee et al. 2008), receptor-based dopamine type 2 receptor (Sun et al. 2001; Yaghoubi et al. 2001), and the most widely used enzyme-based herpes simplex virus type 1 thymidine kinase (HSV1-tk) or its mutant form HSV1-sr39tk (Gambhir et al. 2000; Cao et al. 2006) for PET imaging.

The reporter probes for imaging thymidine kinase reporter genes are radiolabeled pyrimidine nucleoside analogues and acycloguanosine, such as 9-[4-[^{18}F] fluoro-3-(hydroxymethyl)butyl]guanine (^{18}F-FHBG) and ^{123}I- or ^{124}I-5-iodo-2′-fluoro-1-beta-D-arabinofuranosyluracil (^{123}I-/^{124}I-FIAU). After injection, the reporter probe can be detected by PET imaging after it is phosphorylated by *HSV1-tk/HSV1-sr39tk* and trapped inside the cells. One of the first applications of *HSV1-sr39tk* reporter gene demonstrated the feasibility of tracking gene expression in rat myocardium quantitatively using [^{18}F]-FHBG PET imaging (Wu et al. 2002). Subsequently, the detection limit of adenoviral titers was found to be as low as 1×10^7 plaque-forming units. Serial microPET studies demonstrated that myocardial [^{18}F]-FHBG accumulation peaked on days 3 to 5 and was no longer identified on days 10 to 17 (Inubushi et al. 2003). Drastic signal loss from late time point was also observed on microPET imaging in a different study where embryonic cardiomyoblasts expressing *HSV1-sr39tk* reporter gene were transplanted in rat myocardium (Wu et al. 2003). This signal reduction was likely attributed to acute donor cell death from immune rejection, inflammation, viral toxicity, ischemic environment, or apoptosis. Using a large animal model of myocardial infarction, Gyongyosi et al. demonstrated the first successful translation of PET imaging of HSV1-tk reporter gene to track cardiac stem cell biodistribution after intramyocardial injection using electromechanical mapping guidance (Figure 10.6) (Gyongyosi et al. 2008). Focal ^{18}F-FHBG tracer uptake in the anterior myocardial wall was observed in two injection sites 8 hours after autologous MSC transplantation (Gyongyosi et al. 2008). Enzyme-based PET reporter gene has the advantage of signal amplification. Thus, a very low level of reporter gene expression or small number of transplanted cells can often be detected using radionuclide imaging. The major limitations include potential immune response elicitation to the foreign reporter gene product, limited reporter probe trapping due to rate limited probe transport into the cells, and silencing of the reporter gene over time leading to inability to detect the transplanted cells (Luker et al. 2002). Although radionuclide reporter genes and probes have a

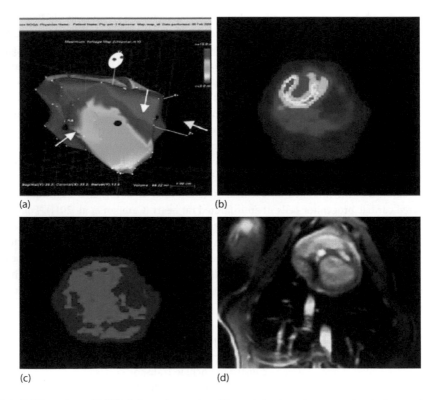

(a) (b)

(c) (d)

Figure 10.6 PET imaging of MSC delivery in a swine MI model using electromechanical mapping guidance. (a) Endocardial mapping of a pig heart 16 days after MI. MSCs transfected with a truncated thymidine kinase reporter gene were intramyocardially injected into the border zone of the infarction (white arrows), and unlabeled MSCs were delivered into noninfarcted posterior wall (yellow arrow). (b) ^{13}N-ammonia PET with transmission scan of the pig heart shows perfusion defect in the anterior wall and apex 16 days after MI. (c) The locations of two injection sites of reporter gene transfected MSCs are demonstrated by ^{18}F-FHBG PET image of the pig heart 8 hours after injection. Unlabeled MSCs could not be detected. (d) Registration of ^{18}F-FHBG PET (hot scale) with MRI (gray scale) demonstrating tracer uptake only at MSCs injection sites. (Adapted from Gyongyosi M et al., *Circ Cardiovasc Imaging*, 1, 94–103, 2008. With permission.)

high sensitivity to a small number of cells, anatomical information is lacking. Thus, CT or MRI is needed to provide localization of probe activity (Judenhofer et al. 2008; Cherry 2009).

10.7.3 OPTICAL REPORTER GENE PROBES

Fluorescent proteins, e.g., green fluorescent protein (GFP) and red fluorescent protein (RFP), are among the earliest and well-established optical reporter probes that have been widely used primarily for *in vitro* gene expression identification and postmortem histological verification. GFP derived from jellyfish *Aequorea victoria* has been used to identify the presence of transplanted bone marrow- and adipose tissue-derived MSCs in the infarcted mouse myocardium (van der Bogt et al. 2009). However, fluorescent reporter proteins have inherent limitations, i.e., significantly high autofluorescence background and scattered photon attenuation. Although fluorescence techniques, such as fluorescence-mediated molecular tomography, which permits tomographic reconstruction, improved the detection depth up to 1 mm (Graves et al. 2003), it is unlikely that such limited penetration depth will be enough to allow *in vivo* cardiac imaging in large animals or man.

Another type of optical reporter is bioluminescence reporter, such as firefly luciferase from *Photinus pyralis* and renilla luciferase from *Renilla reniformis*. Exogenous expression of a luciferase enzyme, followed by systemic delivery of its substrate (e.g., D-luciferin), forms the foundation of *in vivo* BLI. Due to the lack of background signal in living subjects, bioluminescent reporter has extremely high imaging sensitivity (10^{-15} to 10^{-17} mol/L), making it suitable for cell tracking and gene therapy monitoring in small-animal models (Massoud and Gambhir 2003). The application of bioluminescence reporter together with BLI has been demonstrated in rats and mice for tracking the survival, proliferation, and cardiac-specific differentiation of ESCs and biodistribution of induced pluripotent stem cells (iPSCs) in the infarcted myocardium (Li et al. 2008; Martens et al. 2014). In the latter case, early massive iPSC loss from the injection site and pulmonary accumulation were noted on BLI, suggesting that tissue engineering approaches for cardiac stem cell delivery may be necessary in order to limit cell distribution and improve cell retention within the myocardium. As with other optical imaging probes, bioluminescent reporter probe also suffers from limited light penetration and requires another imaging method for anatomical localization. In addition, it requires the injection of a large amount of potentially immunogenic substrates, which makes the clinical translation unlikely.

Although reporter gene/probe approach can directly report the viability of labeled cells, safety concerns due to genetic alteration remain. The other primary inherent problems with reporter probes are whether genetic expression or uptake of reporter probes affects cell function and whether a small number of cells can generate sufficient reporter gene products to enable visualization.

10.8 MULTIMODALITY PROBES

The purpose of developing multimodality probes is to take advantage of the strength from each imaging modality to provide high-sensitivity and superior anatomical details of the target. By integrating individual strength of different modality probes, multimodality imaging probes offer a powerful tool to enhance the assessment of critical pathophysiological processes and cell therapeutic efficacy. One of the classic examples of such a probe is the TFR that consists of truncated thymidine kinase for PET imaging, firefly luciferase for BLI, and monomeric RFP for fluorescence imaging. In a murine myocardial infarction model, early survival, proliferation, and migration of ESCs transfected with a lentiviral vector carrying the TFR gene were revealed on both BLI and PET images (Cao et al. 2006). While reporter gene imaging provides a way to determine cell fate in infarcted or normal subjects, delivery of reporter probe (e.g., luciferin, ^{18}F-FHBG) by systemic injection could be costly for large animals and may be hindered in ischemic tissues. To address these issues, Kedziorek et al. recently coupled reporter gene imaging with x-ray-visible microencapsulation techniques to allow targeted reporter probe delivery in rabbits (Kedziorek et al. 2013). The group transfected rabbit MSCs with a TFR gene that enabled cell viability assessment by BLI. TFR-labeled MSCs were then encapsulated in a PFOB-containing alginate microcapsule that allowed x-ray-guided cell delivery into the hind limb of a rabbit. Since the injection sites could be easily visualized on CT images and targeted, small amounts of the reporter probe (i.e., luciferin) could be administered directly to the transplantation site for *in vivo* viability assessment (Figure 10.7). The fluorine moiety of PFOB could be used for ^{19}F MRI. Subsequently, Fu and coworkers demonstrated the feasibility of *in vivo* PET-MRI tracking of PFOB microencapsulated TFR-labeled human MSCs in rabbits using a high-resolution clinical brain PET (Figure 10.8) (Fu et al. 2013). Follow-up BLI demonstrated high cell survival 2 weeks after delivery, which may be attributed to microcapsule immune protection or enhanced oxygen tension provided by PFCs.

Additionally, a number of investigators have synthesized nanoparticle-based probes for multimodality imaging in biological subjects or tumor models. In an interesting study, Nahrendorf et al. developed a tri-reporter nanoparticle probe for PET/MRI/fluorescent imaging and detection of vulnerable plaque in apoE$^{-/-}$ mice (Nahrendorf et al. 2008). The probe consisted of a dextran-coated iron oxide nanoparticle core for T2-weighted MRI of macrophages, the nuclear tracer ^{64}Cu conjugated on the dextran coat via DTPA for PET imaging, and the NIR fluorochrome VT680 for fluorescence imaging. Thus, this probe provided a highly sensitive tool to quantitatively assess the macrophage burden in atherosclerosic lesions and allowed rigorous probe validation by fluorescence-based techniques on the cellular and molecular level. Moreover, hybrid imaging probes may be useful not only for disease detection but also for therapeutic intervention. In the scenario of PET/optical imaging probe, the PET isotope-labeled macrophage-targeted nanocarrier could be used to localize the vulnerable plaques, and the fluorophores attached to the same nanocarrier could guide local delivery of therapeutic agents with a fluorescence-sensing intravascular catheter (Yoo et al. 2011).

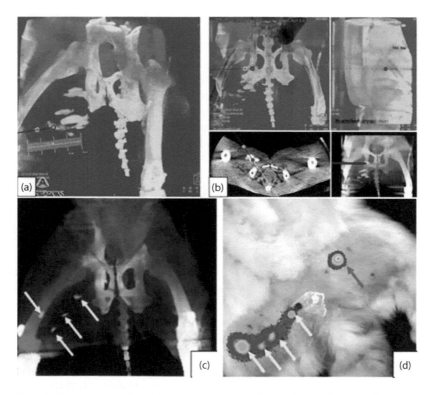

Figure 10.7 Targeted reporter probe delivery under C-arm CT guidance. (a) X-ray fluoroscopic overlay in oblique projection on the C-arm CT in preparation for needle targeting to PFOB microcapsules injection sites. Orange circle indicates the skin entry point; blue circle shows the target point. (b) Planning of the needle entry to the target point in coronal (top left), sagittal (top right), axial (bottom left), and multiplanar reformat (bottom right). (c) C-arm CT image demonstrates the visualization of PFOB microcapsule injections (yellow arrows) in the right hind leg. (d) In the same rabbit, targeted luciferin injections 24 hours post transplantation reveal viable PFOB-microencapsulated MSCs in the right thigh (yellow arrows) that correspond to C-arm CT (c) while nontargeted injections of luciferin into the left thigh only shows one visible injection site (blue arrow). (Adapted from Kedziorek DA et al., *Theranostics*, 3, 916–926, 2013. With permission.)

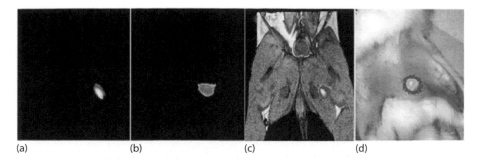

Figure 10.8 PET-MRI tracking of human MSCs using PFOB microcapsules and triple fusion (TF) reporter gene labeling. (a) ^{19}F MR image of PFOB Caps containing TF-hMSCs in the rabbit thigh. (b) PET image of PFOB Caps in the same rabbit. (c) Fusion of ^{19}F MR image (blue) and PET image (red) with anatomical ^{1}H MR shows the concordance "hot spot" and the location of PFOB Caps injection site. (d) Bioluminescence imaging of the rabbit reveals highly viable encapsulated TF-hMSCs 2 weeks after delivery. (Adapted from Fu Y et al., *J Cardiovasc Magn Reson*, 15, M1, 2013. With permission.)

10.9 SUMMARY

The development of multimodality imaging probes relevant to cardiovascular imaging has been advanced significantly over the past two decades. There have been a large number of preclinical investigations that have exploited single or multimodality imaging probes to aid in pathophysiological assessment, therapeutic intervention, and optimization in cardiac cell type, dosing, and delivery timing and route. Each imaging modality probe possesses its own unique attributes and limitations that may set the specific challenges. Due to the relatively low sensitivity of MRI, MR-based imaging probes are often designed to either target highly expressed biomarkers or with a nanocarrier to increase the sensitivity. While radionuclide imaging probes offer the highest sensitivity, the limited resolution of radionuclide imaging and radiation exposure need to be considered for cardiovascular imaging. In the clinical and preclinical setting, PET and SPECT images are often acquired in conjunction with high-resolution anatomic CT or MR images to provide anatomic colocalization of the target. The future trend will be toward increased employment of multimodality imaging probes or fusion of multiple imaging modalities, such as PET, CT, or MRI, with x-ray fluoroscopic imaging to enable real-time interactivity with high sensitivity to the target and superior anatomical information for better diagnosis and therapy for cardiovascular diseases.

REFERENCES

Adler ED, Bystrup A, Briley-Saebo KC, Mani V, Young W, Giovanonne S, Altman P, Kattman SJ, Frank JA, Weinmann HJ, Keller GM and Fayad ZA. 2009. *In vivo* detection of embryonic stem cell-derived cardiovascular progenitor cells using Cy3-labeled gadofluorine M in murine myocardium. *JACC Cardiovasc Imaging* 2(9): 1114–1122.

Aicher A, Brenner W, Zuhayra M, Badorff C, Massoudi S, Assmus B, Eckey T, Henze E, Zeiher AM and Dimmeler S. 2003. Assessment of the tissue distribution of transplanted human endothelial progenitor cells by radioactive labeling. *Circulation* 107(16): 2134–2139.

Aime S and Caravan P. 2009. Biodistribution of gadolinium-based contrast agents, including gadolinium deposition. *J Magn Reson Imaging* 30(6): 1259–1267.

Ali MM, Bhuiyan MP, Janic B, Varma NR, Mikkelsen T, Ewing JR, Knight RA, Pagel MD and Arbab AS. 2012. A nano-sized PARACEST-fluorescence imaging contrast agent facilitates and validates *in vivo* CEST MRI detection of glioma. *Nanomedicine (Lond)* 7(12): 1827–1837.

Amado LC, Saliaris AP, Schuleri KH, St John M, Xie JS, Cattaneo S, Durand DJ, Fitton T, Kuang JQ, Stewart G, Lehrke S, Baumgartner WW, Martin BJ, Heldman AW and Hare JM. 2005. Cardiac repair with intramyocardial injection of allogeneic mesenchymal stem cells after myocardial infarction. *Proc Natl Acad Sci U S A* 102(32): 11474–11479.

Arbab AS, Bashaw LA, Miller BR, Jordan EK, Bulte JW and Frank JA. 2003. Intracytoplasmic tagging of cells with ferumoxides and transfection agent for cellular magnetic resonance imaging after cell transplantation: Methods and techniques. *Transplantation* 76(7): 1123–1130.

Azene N, Fu Y, Maurer J and Kraitchman DL. 2014. Tracking of stem cells *in vivo* for cardiovascular applications. *J Cardiovasc Magn Reson* 16(1): 7.

Barnett BP, Arepally A, Stuber M, Arifin DR, Kraitchman DL and Bulte JW. 2011. Synthesis of magnetic resonance-, x-ray- and ultrasound-visible alginate microcapsules for immunoisolation and noninvasive imaging of cellular therapeutics. *Nat Protoc* 6(8): 1142–1151.

Barnett BP, Kraitchman DL, Lauzon C, Magee CA, Walczak P, Gilson WD, Arepally A and Bulte JW. 2006. Radiopaque alginate microcapsules for x-ray visualization and immunoprotection of cellular therapeutics. *Mol Pharm* 3(5): 531–538.

Barnett BP, Ruiz-Cabello J, Hota P, Liddell R, Walczak P, Howland V, Chacko VP, Kraitchman DL, Arepally A and Bulte JW. 2010. Fluorocapsules for improved function, immunoprotection, and visualization of cellular therapeutics with MR, US, and CT imaging. *Radiology* 258(1): 182–191.

Bekeredjian R, Chen S, Frenkel PA, Grayburn PA and Shohet RV. 2003. Ultrasound-targeted microbubble destruction can repeatedly direct highly specific plasmid expression to the heart. *Circulation* 108(8): 1022–1026.

Blackwood KJ, Lewden B, Wells RG, Sykes J, Stodilka RZ, Wisenberg G and Prato FS. 2009. *In vivo* SPECT quantification of transplanted cell survival after engraftment using (111)In-tropolone in infarcted canine myocardium. *J Nucl Med* 50(6): 927–935.

Blomberg BA, Akers SR, Saboury B, Mehta NN, Cheng G, Torigian DA, Lim E, Del Bello C, Werner TJ and Alavi A. 2013. Delayed time-point 18F-FDG PET CT imaging enhances assessment of atherosclerotic plaque inflammation. *Nucl Med Commun* 34(9): 860–867.

Brenner W, Aicher A, Eckey T, Massoudi S, Zuhayra M, Koehl U, Heeschen C, Kampen WU, Zeiher AM, Dimmeler S and Henze E. 2004. 111In-labeled CD34+ hematopoietic progenitor cells in a rat myocardial infarction model. *J Nucl Med* 45(3): 512–518.

Briley-Saebo KC, Shaw PX, Mulder WJ, Choi SH, Vucic E, Aguinaldo JG, Witztum JL, Fuster V, Tsimikas S and Fayad ZA. 2008. Targeted molecular probes for imaging atherosclerotic lesions with magnetic resonance using antibodies that recognize oxidation-specific epitopes. *Circulation* 117(25): 3206–3215.

Bulte JW and Kraitchman DL. 2004a. Iron oxide MR contrast agents for molecular and cellular imaging. *NMR Biomed* 17(7): 484–499.

Bulte JW and Kraitchman DL. 2004b. Monitoring cell therapy using iron oxide MR contrast agents. *Curr Pharm Biotechnol* 5(6): 567–584.

Campan M, Lionetti V, Aquaro GD, Forini F, Matteucci M, Vannucci L, Chiuppesi F, Di Cristofano C, Faggioni M, Maioli M, Barile L, Messina E, Lombardi M, Pucci A, Pistello M and Recchia FA. 2011. Ferritin as a reporter gene for *in vivo* tracking of stem cells by 1.5-T cardiac MRI in a rat model of myocardial infarction. *Am J Physiol Heart Circ Physiol* 300(6): H2238–H2250.

Cao F, Lin S, Xie X, Ray P, Patel M, Zhang X, Drukker M, Dylla SJ, Connolly AJ, Chen X, Weissman IL, Gambhir SS and Wu JC. 2006. *In vivo* visualization of embryonic stem cell survival, proliferation, and migration after cardiac delivery. *Circulation* 113(7): 1005–1014.

Castellani M, Colombo A, Giordano R, Pusineri E, Canzi C, Longari V, Piccaluga E, Palatresi S, Dellavedova L, Soligo D, Rebulla P and Gerundini P. 2010. The role of PET with 13N-ammonia and 18F-FDG in the assessment of myocardial perfusion and metabolism in patients with recent AMI and intracoronary stem cell injection. *J Nucl Med* 51(12): 1908–1916.

Cesar CL. 2014. Quantum dots as biophotonics tools. *Methods Mol Biol* 1199: 3–9.

Chan JM, Monaco C, Wylezinska-Arridge M, Tremoleda JL and Gibbs RG. 2014. Imaging of the vulnerable carotid plaque: Biological targeting of inflammation in atherosclerosis using iron oxide particles and MRI. *Eur J Vasc Endovasc Surg* 47(5): 462–469.

Cherry SR. 2009. Multimodality imaging: Beyond PET/CT and SPECT/CT. *Semin Nucl Med* 39(5): 348–353.

Coggan AR, Kisrieva-Ware Z, Dence CS, Eisenbeis P, Gropler RJ and Herrero P. 2009. Measurement of myocardial fatty acid esterification using [1-11C]palmitate and PET: Comparison with direct measurements of myocardial triglyceride synthesis. *J Nucl Cardiol* 16(4): 562–570.

Cui W, Tavri S, Benchimol MJ, Itani M, Olson ES, Zhang H, Decyk M, Ramirez RG, Barback CV, Kono Y and Mattrey RF. 2013. Neural progenitor cells labeling with microbubble contrast agent for ultrasound imaging *in vivo*. *Biomaterials* 34(21): 4926–4935.

Cukur T, Yamada M, Overall WR, Yang P and Nishimura DG. 2010. Positive contrast with alternating repetition time SSFP (PARTS): A fast imaging technique for SPIO-labeled cells. *Magn Reson Med* 63(2): 427–437.

Cyrus T, Zhang H, Allen JS, Williams TA, Hu G, Caruthers SD, Wickline SA and Lanza GM. 2008. Intramural delivery of rapamycin with alphavbeta3-targeted paramagnetic nanoparticles inhibits stenosis after balloon injury. *Arterioscler Thromb Vasc Biol* 28(5): 820–826.

de Barros AL, Chacko AM, Mikitsh JL, Al Zaki A, Salavati A, Saboury B, Tsourkas A and Alavi A. 2014. Assessment of global cardiac uptake of radiolabeled iron oxide nanoparticles in apolipoprotein-E-deficient mice: Implications for imaging cardiovascular inflammation. *Mol Imaging Biol* 16(3): 330–339.

Deans AE, Wadghiri YZ, Bernas LM, Yu X, Rutt BK and Turnbull DH. 2006. Cellular MRI contrast via coexpression of transferrin receptor and ferritin. *Magn Reson Med* 56(1): 51–59.

Drey F, Choi YH, Neef K, Ewert B, Tenbrock A, Treskes P, Bovenschulte H, Liakopoulos OJ, Brenkmann M, Stamm C, Wittwer T and Wahlers T. 2013. Noninvasive *in vivo* tracking of mesenchymal stem cells and evaluation of cell therapeutic effects in a murine model using a clinical 3.0 T MRI. *Cell Transplant* 22(11): 1971–1980.

Ebert SN, Taylor DG, Nguyen HL, Kodack DP, Beyers RJ, Xu Y, Yang Z and French BA. 2007. Noninvasive tracking of cardiac embryonic stem cells *in vivo* using magnetic resonance imaging techniques. *Stem Cells* 25(11): 2936–2944.

Eibofner F, Steidle G, Kehlbach R, Bantleon R and Schick F. 2010. Positive contrast imaging of iron oxide nanoparticles with susceptibility-weighted imaging. *Magn Reson Med* 64(4): 1027–1038.

Evbuomwan OM, Kiefer G and Sherry AD. 2012. Amphiphilic EuDOTA-tetraamide complexes form micelles with enhanced CEST sensitivity. *Eur J Inorg Chem* 2012(12): 2126–2134.

Flogel U, Ding Z, Hardung H, Jander S, Reichmann G, Jacoby C, Schubert R and Schrader J. 2008. *In vivo* monitoring of inflammation after cardiac and cerebral ischemia by fluorine magnetic resonance imaging. *Circulation* 118(2): 140–148.

Fu H, Wang J, Chen X, Leng X, Thorne S and Villanueva FS. 2011. Long term *in vivo* stem cell tracking using contrast ultrasound. *Circulation* 124: A16631.

Fu Y, Azene N, Ehtiati T, Flammang A, Gilson WD, Gabrielson K, Weiss CR, Bulte JW, Solaiyappan M, Johnston PV and Kraitchman DL. 2014. Fused x-ray and MR imaging guidance of intrapericardial delivery of microencapsulated human mesenchymal stem cells in immunocompetent swine. *Radiology* 131424.

Fu Y and Kraitchman DL. 2010. Stem cell labeling for noninvasive delivery and tracking in cardiovascular regenerative therapy. *Expert Rev Cardiovasc Ther* 8(8): 1149–1160.

Fu Y, Mease R, Chen Y, Wang G, Kedziorek D, Solaiyappan M and Kraitchman DL. 2013. PET-MRI tracking of imaging-visible microencapsulated stem cells in immunocompetent rabbits. *J Cardiovasc Magn Reson* 15(Suppl 1): M1.

Fujii H, Li SH, Wu J, Miyagi Y, Yau TM, Rakowski H, Egashira K, Guo J, Weisel RD and Li RK. 2011. Repeated and targeted transfer of angiogenic plasmids into the infarcted rat heart via ultrasound targeted microbubble destruction enhances cardiac repair. *Eur Heart J* 32(16): 2075–2084.

Gambhir SS, Bauer E, Black ME, Liang Q, Kokoris MS, Barrio JR, Iyer M, Namavari M, Phelps ME and Herschman HR. 2000. A mutant herpes simplex virus type 1 thymidine kinase reporter gene shows improved sensitivity for imaging reporter gene expression with positron emission tomography. *Proc Natl Acad Sci U S A* 97(6): 2785–2790.

Genove G, DeMarco U, Xu H, Goins WF and Ahrens ET. 2005. A new transgene reporter for *in vivo* magnetic resonance imaging. *Nat Med* 11(4): 450–454.

Gerber BL, Raman SV, Nayak K, Epstein FH, Ferreira P, Axel L and Kraitchman DL. 2008. Myocardial first-pass perfusion cardiovascular magnetic resonance: History, theory, and current state of the art. *J Cardiovasc Magn Reson* 10(1): 18.

Gholamrezanezhad A, Mirpour S, Ardekani JM, Bagheri M, Alimoghadam K, Yarmand S and Malekzadeh R. 2009. Cytotoxicity of 111In-oxine on mesenchymal stem cells: A time-dependent adverse effect. *Nucl Med Commun* 30(3): 210–216.

Gilad AA, McMahon MT, Walczak P, Winnard PT, Jr., Raman V, van Laarhoven HW, Skoglund CM, Bulte JW and van Zijl PC. 2007. Artificial reporter gene providing MRI contrast based on proton exchange. *Nat Biotechnol* 25(2): 217–219.

Gildehaus FJ, Haasters F, Drosse I, Wagner E, Zach C, Mutschler W, Cumming P, Bartenstein P and Schieker M. 2011. Impact of indium-111 oxine labelling on viability of human mesenchymal stem cells *in vitro*, and 3D cell-tracking using SPECT/CT *in vivo*. *Mol Imaging Biol* 13(6): 1204–1214.

Graves EE, Ripoll J, Weissleder R and Ntziachristos V. 2003. A submillimeter resolution fluorescence molecular imaging system for small animal imaging. *Med Phys* 30(5): 901–911.

Gyongyosi M, Blanco J, Marian T, Tron L, Petnehazy O, Petrasi Z, Hemetsberger R, Rodriguez J, Font G, Pavo IJ, Kertesz I, Balkay L, Pavo N, Posa A, Emri M, Galuska L, Kraitchman DL, Wojta J, Huber K and Glogar D. 2008. Serial noninvasive *in vivo* positron emission tomographic tracking of percutaneously intramyocardially injected autologous porcine mesenchymal stem cells modified for transgene reporter gene expression. *Circ Cardiovasc Imaging* 1(2): 94–103.

Haris M, Singh A, Cai K, Kogan F, McGarvey J, DeBrosse C, Zsido GA, Witschey WRT, Koomalsingh K, Pilla JJ, Chirinos JA, Ferrari VA, Gorman JH, Hariharan H, Gorman RC and Reddy R. 2014. A technique for *in vivo* mapping of myocardial creatine kinase metabolism. *Nat Med* 20(2): 209–214.

Higuchi T, Bengel FM, Seidl S, Watzlowik P, Kessler H, Hegenloh R, Reder S, Nekolla SG, Wester HJ and Schwaiger M. 2008. Assessment of alphavbeta3 integrin expression after myocardial infarction by positron emission tomography. *Cardiovasc Res* 78(2): 395–403.

Higuchi T, Rischpler C, Fukushima K, Isoda T, Xia J, Javadi MS, Szabo Z, Dannals RF, Mathews WB and Bengel FM. 2013. Targeting of endothelin receptors in the healthy and infarcted rat heart using the PET tracer 18F-FBzBMS. *J Nucl Med* 54(2): 277–282.

Hiller KH, Waller C, Nahrendorf M, Bauer WR and Jakob PM. 2006. Assessment of cardiovascular apoptosis in the isolated rat heart by magnetic resonance molecular imaging. *Mol Imaging* 5(2): 115–121.

Hou D, Youssef EA, Brinton TJ, Zhang P, Rogers P, Price ET, Yeung AC, Johnstone BH, Yock PG and March KL. 2005. Radiolabeled cell distribution after intramyocardial, intracoronary, and interstitial retrograde coronary venous delivery: Implications for current clinical trials. *Circulation* 112(9 Suppl): I150–I156.

Hyvelin JM, Tardy I, Bettinger T, von Wronski M, Costa M, Emmel P, Colevret D, Bussat P, Lassus A, Botteron C, Nunn A, Frinking P and Tranquart F. 2014. Ultrasound molecular imaging of transient acute myocardial ischemia with a clinically translatable P- and E-selectin targeted contrast agent: Correlation with the expression of selectins. *Invest Radiol* 49(4): 224–235.

Inaba Y, Davidson BP, Kim S, Liu YN, Packwood W, Belcik JT, Xie A and Lindner JR. 2014. Echocardiographic evaluation of the effects of stem cell therapy on perfusion and function in ischemic cardiomyopathy. *J Am Soc Echocardiogr* 27(2): 192–199.

Inubushi M, Wu JC, Gambhir SS, Sundaresan G, Satyamurthy N, Namavari M, Yee S, Barrio JR, Stout D, Chatziioannou AF, Wu L and Schelbert HR. 2003. Positron-emission tomography reporter gene expression imaging in rat myocardium. *Circulation* 107(2): 326–332.

Jin Y, Kong H, Stodilka RZ, Wells RG, Zabel P, Merrifield PA, Sykes J and Prato FS. 2005. Determining the minimum number of detectable cardiac-transplanted 111In-tropolone-labelled bone-marrow-derived mesenchymal stem cells by SPECT. *Phys Med Biol* 50(19): 4445–4455.

Judenhofer MS, Wehrl HF, Newport DF, Catana C, Siegel SB, Becker M, Thielscher A, Kneilling M, Lichy MP, Eichner M, Klingel K, Reischl G, Widmaier S, Rocken M, Nutt RE, Machulla HJ, Uludag K, Cherry SR, Claussen CD and Pichler BJ. 2008. Simultaneous PET-MRI: A new approach for functional and morphological imaging. *Nat Med* 14(4): 459–465.

Kedziorek DA, Hofmann LV, Fu Y, Gilson WD, Cosby KM, Kohl B, Barnett BP, Simons BW, Walczak P, Bulte JW, Gabrielson K and Kraitchman DL. 2012. X-ray-visible microcapsules containing mesenchymal stem cells improve hind limb perfusion in a rabbit model of peripheral arterial disease. *Stem Cells* 30(6): 1286–1296.

Kedziorek DA, Solaiyappan M, Walczak P, Ehtiati T, Fu Y, Bulte JW, Shea SM, Brost A, Wacker FK and Kraitchman DL. 2013. Using C-arm x-ray imaging to guide local reporter probe delivery for tracking stem cell engraftment. *Theranostics* 3(11): 916–926.

Kendziorra K, Barthel H, Erbs S, Emmrich F, Hambrecht R, Schuler G, Sabri O and Kluge R. 2008. Effect of progenitor cells on myocardial perfusion and metabolism in patients after recanalization of a chronically occluded coronary artery. *J Nucl Med* 49(4): 557–563.

Kosuge H, Sherlock SP, Kitagawa T, Dash R, Robinson JT, Dai H and McConnell MV. 2012. Near infrared imaging and photothermal ablation of vascular inflammation using single-walled carbon nanotubes. *J Am Heart Assoc* 1(6): e002568.

Kraitchman DL and Bulte JW. 2008. Imaging of stem cells using MRI. *Basic Res Cardiol* 103(2): 105–113.

Kraitchman DL and Bulte JW. 2009. *In vivo* imaging of stem cells and Beta cells using direct cell labeling and reporter gene methods. *Arterioscler Thromb Vasc Biol* 29(7): 1025–1030.

Kraitchman DL, Heldman AW, Atalar E, Amado LC, Martin BJ, Pittenger MF, Hare JM and Bulte JW. 2003. *In vivo* magnetic resonance imaging of mesenchymal stem cells in myocardial infarction. *Circulation* 107(18): 2290–2293.

Kraitchman DL, Tatsumi M, Gilson WD, Ishimori T, Kedziorek D, Walczak P, Segars WP, Chen HH, Fritzges D, Izbudak I, Young RG, Marcelino M, Pittenger MF, Solaiyappan M, Boston RC, Tsui BM, Wahl RL and Bulte JW. 2005. Dynamic imaging of allogeneic mesenchymal stem cells trafficking to myocardial infarction. *Circulation* 112(10): 1451–1461.

Kuliszewski MA, Fujii H, Liao C, Smith AH, Xie A, Lindner JR and Leong-Poi H. 2009. Molecular imaging of endothelial progenitor cell engraftment using contrast-enhanced ultrasound and targeted microbubbles. *Cardiovasc Res* 83(4): 653–662.

Lee Z, Dennis JE and Gerson SL. 2008. Imaging stem cell implant for cellular-based therapies. *Exp Biol Med (Maywood)* 233(8): 930–940.

Leng X, Wang J, Carson A, Chen X, Fu H, Ottoboni S, Wagner WR and Villanueva FS. 2014. Ultrasound detection of myocardial ischemic memory using an E-selectin targeting peptide amenable to human application. *Mol Imaging* 13(4): 1–9.

Li D, Patel AR, Klibanov AL, Kramer CM, Ruiz M, Kang BY, Mehta JL, Beller GA, Glover DK and Meyer CH. 2010. Molecular imaging of atherosclerotic plaques targeted to oxidized LDL receptor LOX-1 by SPECT/CT and magnetic resonance. *Circ Cardiovasc Imaging* 3(4): 464–472.

Li Z, Suzuki Y, Huang M, Cao F, Xie X, Connolly AJ, Yang PC and Wu JC. 2008. Comparison of reporter gene and iron particle labeling for tracking fate of human embryonic stem cells and differentiated endothelial cells in living subjects. *Stem Cells* 26(4): 864–873.

Ling ZY, Shu SY, Zhong SG, Luo J, Su L, Liu ZZ, Lan XB, Yuan GB, Zheng YY, Ran HT, Wang ZG and Yin YH. 2013. Ultrasound targeted microbubble destruction promotes angiogenesis and heart function by inducing myocardial microenvironment change. *Ultrasound Med Biol* 39(11): 2001–2010.

Liu H, Wang H, Xu Y, Shen M, Zhao J, Zhang G and Shi X. 2014. Synthesis of PEGylated low generation dendrimer-entrapped gold nanoparticles for CT imaging applications. *Nanoscale* 6(9): 4521–4526.

Liu J, Cheng EC, Long Jr RC, Yang SH, Wang L, Cheng PH, Yang JJ, Wu D, Mao H and Chan AW. 2009. Noninvasive monitoring of embryonic stem cells *in vivo* with MRI transgene reporter. *Tissue Eng Part C Methods* 15(4): 739–747.

Loai Y, Sakib N, Janik R, Foltz WD and Cheng HL. 2012. Human aortic endothelial cell labeling with positive contrast gadolinium oxide nanoparticles for cellular magnetic resonance imaging at 7 Tesla. *Mol Imaging* 11(2): 166–175.

Louie AY, Huber MM, Ahrens ET, Rothbacher U, Moats R, Jacobs RE, Fraser SE and Meade TJ. 2000. *In vivo* visualization of gene expression using magnetic resonance imaging. *Nat Biotechnol* 18(3): 321–325.

Luker GD, Sharma V, Pica CM, Dahlheimer JL, Li W, Ochesky J, Ryan CE, Piwnica-Worms H and Piwnica-Worms D. 2002. Noninvasive imaging of protein-protein interactions in living animals. *Proc Natl Acad Sci U S A* 99(10): 6961–6966.

Ly HQ, Hoshino K, Pomerantseva I, Kawase Y, Yoneyama R, Takewa Y, Fortier A, Gibbs-Strauss SL, Vooght C, Frangioni JV and Hajjar RJ. 2009. *In vivo* myocardial distribution of multipotent progenitor cells following intracoronary delivery in a swine model of myocardial infarction. *Eur Heart J* 30(23): 2861–2868.

Lyngbaek S, Ripa RS, Haack-Sorensen M, Cortsen A, Kragh L, Andersen CB, Jorgensen E, Kjaer A, Kastrup J and Hesse B. 2010. Serial *in vivo* imaging of the porcine heart after percutaneous, intramyocardially injected 111In-labeled human mesenchymal stromal cells. *Int J Cardiovasc Imaging* 26(3): 273–284.

Marrache S and Dhar S. 2013. Biodegradable synthetic high-density lipoprotein nanoparticles for atherosclerosis. *Proc Natl Acad Sci U S A* 110(23): 9445–9450.

Martens A, Rojas SV, Baraki H, Rathert C, Schecker N, Zweigerdt R, Schwanke K, Rojas-Hernandez S, Martin U, Saito S, Schmitto JD, Haverich A and Kutschka I. 2014. Substantial early loss of induced pluripotent stem cells following transplantation in myocardial infarction. *Artif Organs* 38(11): 978–984.

Massoud TF and Gambhir SS. 2003. Molecular imaging in living subjects: Seeing fundamental biological processes in a new light. *Genes Dev* 17(5): 545–580.

McMahon MT, Gilad AA, DeLiso MA, Berman SM, Bulte JW and van Zijl PC. 2008. New "multicolor" polypeptide diamagnetic chemical exchange saturation transfer (DIACEST) contrast agents for MRI. *Magn Reson Med* 60(4): 803–812.

Medintz IL, Uyeda HT, Goldman ER and Mattoussi H. 2005. Quantum dot bioconjugates for imaging, labelling and sensing. *Nat Mater* 4(6): 435–446.

Miyagawa M, Beyer M, Wagner B, Anton M, Spitzweg C, Gansbacher B, Schwaiger M and Bengel FM. 2005. Cardiac reporter gene imaging using the human sodium/iodide symporter gene. *Cardiovasc Res* 65(1): 195–202.

Moore A, Josephson L, Bhorade RM, Basilion JP and Weissleder R. 2001. Human transferrin receptor gene as a marker gene for MR imaging. *Radiology* 221(1): 244–250.

Morcos SK. 2008. Extracellular gadolinium contrast agents: Differences in stability. *Eur J Radiol* 66(2): 175–179.

Murakami Y, Takamatsu H, Taki J, Tatsumi M, Noda A, Ichise R, Tait JF and Nishimura S. 2004. 18F-labelled annexin V: A PET tracer for apoptosis imaging. *Eur J Nucl Med Mol Imaging* 31(4): 469–474.

Nahrendorf M, Zhang H, Hembrador S, Panizzi P, Sosnovik DE, Aikawa E, Libby P, Swirski FK and Weissleder R. 2008. Nanoparticle PET-CT imaging of macrophages in inflammatory atherosclerosis. *Circulation* 117(3): 379–387.

Nam SY, Ricles LM, Suggs LJ and Emelianov SY. 2012. *In vivo* ultrasound and photoacoustic monitoring of mesenchymal stem cells labeled with gold nanotracers. *PLoS One* 7(5): e37267.

Nanda NC, Kitzman DW, Dittrich HC and Hall G. 2003. Imagent improves endocardial border delineation, inter-reader agreement, and the accuracy of segmental wall motion assessment. *Echocardiography* 20(2): 151–161.

Naumova AV, Reinecke H, Yarnykh V, Deem J, Yuan C and Charles EM. 2010. Ferritin overexpression for noninvasive magnetic resonance imaging-based tracking of stem cells transplanted into the heart. *Mol Imaging* 9(4): 201–210.

Neubauer AM, Caruthers SD, Hockett FD, Cyrus T, Robertson JD, Allen JS, Williams TD, Fuhrhop RW, Lanza GM and Wickline SA. 2007. Fluorine cardiovascular magnetic resonance angiography *in vivo* at 1.5 T with perfluorocarbon nanoparticle contrast agents. *J Cardiovasc Magn Reson* 9(3): 565–573.

Partlow KC, Chen J, Brant JA, Neubauer AM, Meyerrose TE, Creer MH, Nolta JA, Caruthers SD, Lanza GM and Wickline SA. 2007. 19F magnetic resonance imaging for stem/progenitor cell tracking with multiple unique perfluorocarbon nanobeacons. *FASEB J* 21(8): 1647–1654.

Paulis LE, Geelen T, Kuhlmann MT, Coolen BF, Schafers M, Nicolay K and Strijkers GJ. 2012. Distribution of lipid-based nanoparticles to infarcted myocardium with potential application for MRI-monitored drug delivery. *J Control Release* 162(2): 276–285.

Pawelczyk E, Arbab AS, Pandit S, Hu E and Frank JA. 2006. Expression of transferrin receptor and ferritin following ferumoxides-protamine sulfate labeling of cells: Implications for cellular magnetic resonance imaging. *NMR Biomed* 19(5): 581–592.

Ransohoff KJ and Wu JC. 2010. Advances in cardiovascular molecular imaging for tracking stem cell therapy. *Thromb Haemost* 104(1): 13–22.

Ray P, De A, Min JJ, Tsien RY and Gambhir SS. 2004. Imaging tri-fusion multimodality reporter gene expression in living subjects. *Cancer Res* 64(4): 1323–1330.

Rivlin M, Horev J, Tsarfaty I and Navon G. 2013. Molecular imaging of tumors and metastases using chemical exchange saturation transfer (CEST) MRI. *Sci Rep* 3: 3045.

Ros PR, Freeny PC, Harms SE, Seltzer SE, Davis PL, Chan TW, Stillman AE, Muroff LR, Runge VM, and Nissenbaum MA. 1995. Hepatic MR imaging with ferumoxides: A multicenter clinical trial of the safety and efficacy in the detection of focal hepatic lesions. *Radiology* 196(2): 481–488.

Rosen AB, Kelly DJ, Schuldt AJ, Lu J, Potapova IA, Doronin SV, Robichaud KJ, Robinson RB, Rosen MR, Brink PR, Gaudette GR and Cohen IS. 2007. Finding fluorescent needles in the cardiac haystack: Tracking human mesenchymal stem cells labeled with quantum dots for quantitative *in vivo* three-dimensional fluorescence analysis. *Stem Cells* 25(8): 2128–2138.

Schutt EG, Klein DH, Mattrey RM and Riess JG. 2003. Injectable microbubbles as contrast agents for diagnostic ultrasound imaging: The key role of perfluorochemicals. *Angew Chem Int Ed Engl* 42(28): 3218–3235.

Segall G. 2002. Assessment of myocardial viability by positron emission tomography. *Nucl Med Commun* 23(4): 323–330.

Shapiro EM, Sharer K, Skrtic S and Koretsky AP. 2006. *In vivo* detection of single cells by MRI. *Magn Reson Med* 55(2): 242–249.

Sherif HM, Saraste A, Nekolla SG, Weidl E, Reder S, Tapfer A, Rudelius M, Higuchi T, Botnar RM, Wester HJ and Schwaiger M. 2012. Molecular imaging of early alphavbeta3 integrin expression predicts long-term left-ventricle remodeling after myocardial infarction in rats. *J Nucl Med* 53(2): 318–323.

Singh J and Daftary A. 2008. Iodinated contrast media and their adverse reactions. *J Nucl Med Technol* 36(2): 69–74; quiz 76–77.

Sosnovik DE, Nahrendorf M, Deliolanis N, Novikov M, Aikawa E, Josephson L, Rosenzweig A, Weissleder R and Ntziachristos V. 2007. Fluorescence tomography and magnetic resonance imaging of myocardial macrophage infiltration in infarcted myocardium *in vivo*. *Circulation* 115(11): 1384–1391.

Soto AV, Gilson WD, Kedziorek D, Fritzges D, Izbudak I, Young RG, Pittenger MF, Bulte JW and Kraitchman DL. 2006. MRI tracking of regional persistence of feridex-labeled mesenchymal stem cells in a canine myocardial infarction model. *J Cardiovasc Magn Reson* 8: 89–90.

Stoll G, Basse-Lusebrink T, Weise G and Jakob P. 2012. Visualization of inflammation using (19) F-magnetic resonance imaging and perfluorocarbons. *Wiley Interdiscip Rev Nanomed Nanobiotechnol* 4(4): 438–447.

Stuber M, Gilson WD, Schar M, Kedziorek DA, Hofmann LV, Shah S, Vonken EJ, Bulte JW and Kraitchman DL. 2007. Positive contrast visualization of iron oxide-labeled stem cells using inversion-recovery with on-resonant water suppression (IRON). *Magn Reson Med* 58(5): 1072–1077.

Sun X, Annala AJ, Yaghoubi SS, Barrio JR, Nguyen KN, Toyokuni T, Satyamurthy N, Namavari M, Phelps ME, Herschman HR and Gambhir SS. 2001. Quantitative imaging of gene induction in living animals. *Gene Ther* 8(20): 1572–1579.

Tahara N, Mukherjee J, de Haas HJ, Petrov AD, Tawakol A, Haider N, Tahara A, Constantinescu CC, Zhou J, Boersma HH, Imaizumi T, Nakano M, Finn A, Fayad Z, Virmani R, Fuster V, Bosca L and Narula J. 2014. 2-deoxy-2-[18F]fluoro-D-mannose positron emission tomography imaging in atherosclerosis. *Nat Med* 20(2): 215–219.

Thomsen HS, Morcos SK and Dawson P. 2006. Is there a causal relation between the administration of gadolinium based contrast media and the development of nephrogenic systemic fibrosis (NSF)? *Clin Radiol* 61(11): 905–906.

Tran N, Li Y, Maskali F, Antunes L, Maureira P, Laurens MH, Marie PY, Karcher G, Groubatch F, Stoltz JF and Villemot JP. 2006. Short-term heart retention and distribution of intramyocardial delivered mesenchymal cells within necrotic or intact myocardium. *Cell Transplant* 15(4): 351–358.

van der Bogt KE, Schrepfer S, Yu J, Sheikh AY, Hoyt G, Govaert JA, Velotta JB, Contag CH, Robbins RC and Wu JC. 2009. Comparison of transplantation of adipose tissue- and bone marrow-derived mesenchymal stem cells in the infarcted heart. *Transplantation* 87(5): 642–652.

Van Herck JL, De Meyer GR, Martinet W, Salgado RA, Shivalkar B, De Mondt R, Van De Ven H, Ludwig A, Van Der Veken P, Van Vaeck L, Bult H, Herman AG and Vrints CJ. 2010. Multi-slice computed tomography with N1177 identifies ruptured atherosclerotic plaques in rabbits. *Basic Res Cardiol* 105(1): 51–59.

Vandsburger M, Vandoorne K, Oren R, Leftin A, Mpofu S, Delli Castelli D, Aime S and Neeman M. 2015. Cardio-chemical exchange saturation transfer magnetic resonance imaging reveals molecular signatures of endogenous fibrosis and exogenous contrast media. *Circ Cardiovasc Imaging* 8(1).

Verma VK, Kamaraju SR, Kancherla R, Kona LK, Beevi SS, Debnath T, Usha SP, Vadapalli R, Arbab AS and Chelluri LK. 2015. Fluorescent magnetic iron oxide nanoparticles for cardiac precursor cell selection from stromal vascular fraction and optimization for magnetic resonance imaging. *Int J Nanomedicine* 10: 711–726.

Waldeck J, Hager F, Holtke C, Lanckohr C, von Wallbrunn A, Torsello G, Heindel W, Theilmeier G, Schafers M and Bremer C. 2008. Fluorescence reflectance imaging of macrophage-rich atherosclerotic plaques using an alphavbeta3 integrin-targeted fluorochrome. *J Nucl Med* 49(11): 1845–1851.

Wisenberg G, Lekx K, Zabel P, Kong H, Mann R, Zeman PR, Datta S, Culshaw CN, Merrifield P, Bureau Y, Wells G, Sykes J and Prato FS. 2009. Cell tracking and therapy evaluation of bone marrow monocytes and stromal cells using SPECT and CMR in a canine model of myocardial infarction. *J Cardiovasc Magn Reson* 11: 11.

Wolfs E, Struys T, Notelaers T, Roberts SJ, Sohni A, Bormans G, Van Laere K, Luyten FP, Gheysens O, Lambrichts I, Verfaillie CM and Deroose CM. 2013. 18F-FDG labeling of mesenchymal stem cells and multipotent adult progenitor cells for PET imaging: Effects on ultrastructure and differentiation capacity. *J Nucl Med* 54(3): 447–454.

Wu JC, Chen IY, Sundaresan G, Min JJ, De A, Qiao JH, Fishbein MC and Gambhir SS. 2003. Molecular imaging of cardiac cell transplantation in living animals using optical bioluminescence and positron emission tomography. *Circulation* 108(11): 1302–1305.

Wu JC, Inubushi M, Sundaresan G, Schelbert HR and Gambhir SS. 2002. Positron emission tomography imaging of cardiac reporter gene expression in living rats. *Circulation* 106(2): 180–183.

Yaghoubi SS, Wu L, Liang Q, Toyokuni T, Barrio JR, Namavari M, Satyamurthy N, Phelps ME, Herschman HR and Gambhir SS. 2001. Direct correlation between positron emission tomographic images of two reporter genes delivered by two distinct adenoviral vectors. *Gene Ther* 8(14): 1072–1080.

Yan P, Chen KJ, Wu J, Sun L, Sung HW, Weisel RD, Xie J and Li RK. 2014. The use of MMP2 antibody-conjugated cationic microbubble to target the ischemic myocardium, enhance Timp3 gene transfection and improve cardiac function. *Biomaterials* 35(3): 1063–1073.

Yang X, Song X, Li Y, Liu G, Ray Banerjee S, Pomper MG and McMahon MT. 2013. Salicylic acid and analogues as diaCEST MRI contrast agents with highly shifted exchangeable proton frequencies. *Angew Chem Int Ed Engl* 52(31): 8116–8119.

Yoo H, Kim JW, Shishkov M, Namati E, Morse T, Shubochkin R, McCarthy JR, Ntziachristos V, Bouma BE, Jaffer FA and Tearney GJ. 2011. Intra-arterial catheter for simultaneous microstructural and molecular imaging *in vivo*. *Nat Med* 17(12): 1680–1684.

Zhang H. 2004. Iron oxide nanoparticles-poly-L-lysine complex. In: Molecular Imaging and Contrast Agent Database. https://www.ncbi.nlm.nih.gov/books/NBK23706/.

Zhang SJ and Wu JC. 2007. Comparison of imaging techniques for tracking cardiac stem cell therapy. *J Nucl Med* 48(12): 1916–1919.

Zhou R, Idiyatullin D, Moeller S, Corum C, Zhang H, Qiao H, Zhong J and Garwood M. 2010. SWIFT detection of SPIO-labeled stem cells grafted in the myocardium. *Magn Reson Med* 63(5): 1154–1161.

Zhou R, Thomas DH, Qiao H, Bal HS, Choi SR, Alavi A, Ferrari VA, Kung HF and Acton PD. 2005. *In vivo* detection of stem cells grafted in infarcted rat myocardium. *J Nucl Med* 46(5): 816–822.

Multimodality probes for cardiovascular imaging

JAMES T. THACKERAY AND FRANK M. BENGEL

11.1 INTRODUCTION

The era of single-modality imaging systems is drawing to close, with an influx of multimodality imaging systems to the marketplace, facilitating one-stop imaging in the interrogation of anatomic and physiologic or molecular indices of disease, whether in the form of positron emission tomography (PET)/x-ray computed tomography (CT), single-photon emission CT (SPECT)/CT, or the now expanding deployment of PET/magnetic resonance (MR) (Bengel et al. 2009; Nekolla et al. 2009; Nensa et al. 2013). However, while the advances in instrumentation have been far reaching, the development of multimodality targeted imaging probes has lagged behind somewhat, due in part to specific and significant challenges in terms of synthesis and variable sensitivity between modalities necessitating complex stoichiometry in combining agents (Louie 2010).

To minimize impact on normal physiology, reduce potential toxicity, and ensure natural pharmokinetics, multimodality probes are necessarily small, ideally at a nanometer scale. The expansion of nanotechnology and microfluidics has facilitated chemistry research and development of novel multimodality imaging probes. Many early multimodality probes were limited to the combination of a fluorophore with some form of contrast agent, whether enveloped within a liposome or micelle or crosslinked to some form of nanoparticle. In more recent years, a variety of alternative designs have been synthesized and developed, combining fluorescence and optical imaging with MR, CT, PET, and SPECT labels, facilitating the efficient combination of anatomic and physiologic or molecular information from a single imaging probe (Table 11.1).

While extensive preliminary evaluations have been completed in oncology and tumor models, a wider array of cardiovascular application is only recently beginning to be explored. In particular, molecular disease states such as atherosclerosis and the evaluation of vulnerable plaque have emerged as attractive targets for the development of novel multimodality probes. Moreover, the chemical nature of the multimodality probes, i.e., as a delivery vehicle, renders them suitable for concurrent targeting of molecular therapy, as so-called theranostic agents. In this chapter, we will introduce the various platforms for multimodality imaging probes and the inherent advantages and challenges to these designs. We will then examine the cardiovascular applications in which multimodality imaging probes are desirable and the preclinical testing of select probes in these disease settings. Finally, we will discuss the concept of theranostic probes and their potential utility in cardiovascular disease.

11.2 LIPOSOMES AND MICELLES AS MULTIMODALITY CARRIERS

Perhaps the simplest approach to concerted delivery of multiple contrast agents is their incorporation within a lipid-based carrier. Liposomes are essentially hollow spheres of lipids either as a bilayer or a monolayer, comprising a lipid-rich hydrophobic shell and an aqueous core, which have been used for drug and gene delivery. Multimodality agents can be incorporated into liposomes in a variety of ways. This can include (i) the encapsulation of hydrophilic CT or MR contrast agents (iodine or gadolinium based) within the aqueous core of liposomes, (ii) lipid-linking of metal chelators (for gadolinium, gamma- or positron-emitting metals) and optical probes (e.g., rhodamine phosphotidylethanolamine), or (iii) a combination of these. Liposomes can be further functionalized by addition of targeting moieties to the lipophilic membrane, allowing for selective delivery of contrast agent and/or radioisotope to the tissue of interest.

Liposomes are readily synthesized by dispersion of lipid components in solution, resulting in self-assembly into a variety of monolayer and bilayer lipid spheres (Al-Jamal and Kostarelos 2007). The incorporation of a soluble contrast agent in the liposome is achieved by inclusion in the aqueous solution, generating a wide range of contrast agent concentration within the aqueous core. Contrast agents that have been encapsulated into the liposome core include iodine-based CT contrast agents (Zheng et al. 2007), gadolinium-based MR contrast agents (Mikhaylova et al. 2009), perfluorocarbons that are detectable by both ultrasound and MR (Janjic et al. 2008; Pisani et al. 2008), and fluorescent dyes (Louie 2010). Modification of the lipid heads with chelating agents such as 1,4,7,10-tetraazacyclododecane-1,4,7,10-tetraacetic acid (DOTA) or diethylenetriaminepentaacetic acid (DTPA), fluorophores, and targeting moieties is carried out prior to assembly (Kamaly et al. 2008; Vuu et al. 2005; Zielhuis et al. 2006), and these modified lipids are included in the solution to generate multimodal liposomes. Functional lipid linkages can include gadolinium (Gd) for MR, ^{99m}Tc, ^{111}In, or ^{64}Cu for radionuclide imaging and rhodamine phosphotidylethanolamine or fluorescein-lipid for optical imaging (Kamaly et al. 2008; Lijowski et al. 2009). Aqueous core liposomes have been generated with diameters as small as 75 nm (Zheng et al. 2007) but are typically around 150–200 nm, depending on the extent of modification.

Liposome incorporation extends the biological half-life of tracers by ~50%, resulting in longer circulation time and reduced probe clearance (Zheng et al. 2007). Moreover, further modification of the liposome bilayer can allow interaction with the membrane of target cells, facilitating the delivery of contents as contrast agents. The capacity for targeted delivery of liposomal contents has important therapeutic connotations as well.

Table 11.1 Multimodality imaging probe designs, advantages, and disadvantages

Design	Modalities	Size (nm)	Advantages	Disadvantages
Liposomes/ micelles	MR, CT, fluorescence, optical, PET, SPECT	100–200	• Well documented from pharmacologic research • Simple synthesis, self–assembly • Multiple payloads possible • Hydrophilic and hydrophobic conjugation • Intracellular targeting	• Large size, biliary clearance • Uneven stoichiometry in batch • Potential for toxicity
Nanoparticles				
QDs	Fluorescence, optical, MR, CT, PET, SPECT	5–20	• Inherent optical capacity • Crosslink modifications readily applied • Coating to modify targetry • Transition metal doping to reduce toxicity	• Fixed sensitivity of optical imaging unequal to MR or PET • Can require more complex chemistry • Potential toxicity, ROS, and free radicals
Iron oxide	MR, CT, fluorescence, optical, PET, SPECT	7–80	• Inherent CT/MR capacity • Coating modifications routine • Multiple couplings possible by encapsulation or covalent bonds • Size can be limited to ease clearance	• Toxicity due to hepatic/biliary clearance • MR signal weaker than coupled molecules • Signal for cell tracking not restricted to living cells
Silica	MR, CT, fluorescence, optical, PET, SPECT	30–200	• Flexibility for conjugation • Self-assembly synthesis • Combination with QD or FeO • Improved stoichiometric capacity for signal combination	• Relatively large particle size, hepatic/biliary clearance • Potential toxicity • Synthetic composition
Polymers	MR, optical, PET, SPECT	70–300	• Well documented from pharmacologic research • Biocompatibility (chitosan)	• Nonnatural • Relatively large size, hepatic/biliary clearance • Potential toxicity
Dendrimers	MR, optical, SPECT	5–10	• Very small size, renal clearance • Multiple branches for different functionalities • Capacity for specific stoichiometry of functionalities	• More complex chemistry • Control of concentrations for different sensitivities required

(Continued)

Table 11.1 (Continued) Multimodality imaging probe designs, advantages, and disadvantages

Design	Modalities	Size (nm)	Advantages	Disadvantages
Nanoparticles				
HDL-like nanoparticles	MR, CT, fluorescence, optical, SPECT, PET	8–11	• Biocompatible, natural components • Inherent homing capacity • Relatively simple conjugation chemistry	• Natural clearance, potential metabolism of particles • Precise stoichiometry of components necessary
Small molecule	MR, CT, fluorescence, optical, SPECT, PET	Wide range	• Very small size, renal clearance • Well defined for radiolabeling • Biocompatible, natural components	• Complex protein chemistry • Specific stoichiometry required for combinations • Nonspecific binding
Reporter genes	Fluorescence, optical, SPECT, PET	N/A	• Long-term imaging capacity • Cell tracking and differentiation • Signal limited to living cells	• Large size of fusion reporter genes complicates molecular biology • Limitation of anatomic imaging functionalities

Mitchell et al. studied the incorporation of multiple contrast agents into liposomes for multimodality imaging (Mitchell et al. 2013). Incorporation of a DOTA macrocycle chelator attached to the lipid head group with ethylene glycol (EG) spacers allowed the inclusion of Gd^{3+}, $^{64}Cu^{2+}$, $^{111}In^{3+}$, or fluorescent lipids within the liposome. Phantom studies with Gd-DOTA liposomes demonstrated 85%–90% reduction of T1 signal relative to water control, suggesting effective contrast enhancement for MR. *In vivo* administration of ^{64}Cu and ^{111}In liposomes indicated differential distribution patterns, depending on the EG spacers and polyethylene glycol (PEG) coverage, targeting to the kidney (theorized to be with smaller sizes), liver, and spleen. This study suggests the possibility of multimodality imaging by specific liposome, with the added potential for theranostic delivery of drug payload in tumors, atherosclerosis, or heart disease (Mitchell et al. 2013).

Another study aimed to generate nanomicelles that mimicked activated leukocytes by conjugating the associated antigen I domain to the lipid shell with fluorescent dyes (AlexaFluor, Thermo Fisher Scientific) and a superparamagnetic iron oxide core. *In vitro* testing in HeLa cells demonstrated specific binding to the intercellular adhesion molecule-1 (ICAM-1), inhibited by addition of missense F262S/F292G, along with T2*-weighted imaging of the cell pellet that demonstrated negative enhancement, with a detection limit as low as 1 µg in 100 µL. *In vivo* testing in mice administered lipopolysaccharide (LPS) to induce inflammation showed that the liposome-iron oxide particles were localized selectively to the LPS injection site over phosphate buffered saline (PBS)-administered control, with longer-term accumulation in the liver, suggesting a combination of ICAM-1 independent phagocytosis (20%) and selective binding in the LPS-induced inflammation regions. Accumulation at the injection site was consistently higher for LPS than for PBS up to 72 h (Chen et al. 2011), demonstrating the capacity for selective targeting of multimodality nanomicelles.

11.3 NANOPARTICLES

An alternative platform for multimodality probe design is broadly termed nanoparticles, which has proved to be a highly active area of research in the last decade. Nanomaterials provide a versatile starting product for the conjugation of multiple imaging functionalities and targeting moieties, and relatively simple synthesis methods allow for substitution of imaging probes or targeting molecules for multifunctional applications.

Essentially, nanoparticles comprise a single particle (usually spherical but have also been produced in a variety of alternative shapes), often with inherent imaging functionality, which can then be built upon to incorporate additional functionalities by chelator coupling, crosslinks, micelle shells, lipid-bonds, and/or antibody binding. The core nanoparticle starting materials include quantum dots (QDs), MR-visible iron oxide, encapsulating silica, or polymers as particles or dendrimers.

11.3.1 Quantum dots

QDs were among the earliest nanoparticle platforms, exhibiting excitation/emission profiles attractive for optical imaging and conducive to chemical modification. Commercially available CdSe/ZnS, CdTe/ZnS, and CdSeTe/CdS QDs can be readily modified by direct attachment of chelators such as DOTA or DTPA, allowing for incorporation of either Gd^{3+} for MR or ^{64}Cu for PET imaging (Cai et al. 2007, Oostendorp et al. 2008, Park et al. 2008; Schipper et al. 2007). CdSe/CdZnS QDs can also be labeled with ^{18}F-polyethylene glycolipid (Chen et al. 2008). Targeting moieties such as vascular endothelial growth factor (VEGF) can be incorporated by thiol crosslinking with N-hydroxysuccinimide ester maleimide (Chen et al. 2008). Alternatively, multimodal imaging functionality can be incorporated using a coencapsulation method, whereby QDs can be enveloped within lipid micelles, which can further incorporate magnetic nanoparticles within the core (Mulder et al. 2006; Roullier et al. 2008).

A major limitation of multimodality imaging using QDs is that different modalities have different sensitivities and require widely divergent amounts of the common probe for functional imaging. As evidenced in a study with VEGF conjugation and DOTA chelation of ^{64}Cu, PET and optical imaging were conducted with separate probe administration, as the amount of probe required for optical imaging was an order of magnitude higher than that required for PET imaging (Chen et al. 2008). This can be potentially solved by manipulating the concentration or specific activity of the radioisotope, reducing the effective ratio of labeled QDs to unlabeled QDs (Louie 2010). By contrast, the challenge is reversed for MR contrast agents, which require amplification of Gd per nanoparticle compared to near infrared imaging of QDs. This complication can be overcome in a number of ways, including the attachment of streptavidin and biotinylated Gd-DTPA to augment Gd per particle or the construction of micelles, effectively coencapsulating a magnetic nanoparticle such as iron oxide within an inner core of QDs and outer core of PEG phospholipid (Park et al. 2008).

QDs can also be doped with transition metals either to enhance optical emission parameters or to add MR functionality, as demonstrated by doping with manganese (Mn), indium (In), and cobalt (Co), among other metals. Addition of Mn to silicon (Si) nanoparticles generated a red shift in the emission and excitation wavelengths for optical imaging (Zhang et al. 2007).

The other drawback to QDs is the perceived long-term toxicity after administration. Several studies have demonstrated cytotoxicity of QDs due in part to production of reactive oxygen species and free radicals (Ipe et al. 2005; Lovric et al. 2005). While it has been generally believed that coating of the QD prevented significant degradation, accumulation, and deleterious effects, *in vitro* studies suggest that adverse toxicity may persist regardless of the QD coating.

Nevertheless, preclinical studies have established several multimodality candidate probes based around QDs. Marrache and Dhar developed nanoparticles via self-assembly of poly(lactic-co-glycolic acid) (PLGA), PLGA-b-PEG QDs, and cholesteryl oleate (CO) (Marrache and Dhar 2013). The core was surrounded by a 1,2-distearoyl-*sn*-glycero-3-phosphoethanoamine-polyethylene-glycol-carboxylic acid (DSPE-PET-COOH) embedded with cholesterol and apolipoprotein A-1 mimetic 4F peptide (sequence FAEKFKEAVKDYFAKFWD) targeted to mitochondria for biodegradation, and to thereby provide internalized high-density lipoprotein (HDL) as a cardioprotective agent. Nanoparticles of ~100 nm showed limited toxicity in RAW macrophages and mesenchymal stem cells (MSCs) *in vitro* and demonstrated effective localization to mitochondria by live cell imaging. Targeted nanoparticles homed to the mitochondria of nonapoptotic cells, whereas nontargeted nanoparticles were found in the cytoplasm. Highest concentration of cadmium (Cd) (indicative of QD homing) was found in the heart and spleen, with clearance from the plasma over 24 h to ~50% of initial levels. Cholesterol levels and circulating triglycerides were markedly reduced within 5 h of administration, persisting over the first 24 h, likely due to the binding of cholesterol to the triphenylphosphonium (TPP)-HDL-apoA-I nanoparticles, suggesting theranostic potential (Marrache and Dhar 2013). The particles also allow for additional labeling for other modalities.

11.3.2 IRON OXIDE

Iron oxide nanoparticles have been reliable contrast agents used clinically for several years, and the biocompatibility and low toxicity of iron have supported these nanoparticles as platforms for multimodality imaging probes. Iron oxide nanoparticles can be synthesized with diameters as small as 7–10 nm, reaching sizes of 100–200 nm with additional conjugated molecules. Additional functionalities can be incorporated into iron oxide nanoparticles by direct covalent conjugation or encapsulation within a luminescent shell, which can be further functionalized. The vast majority of multimodal iron oxide nanoparticles combine the inherent MR properties with a covalent linkage to fluorescence or near-infrared optical imaging molecules such as rhodamine (Banerjee and Chen 2009; Gallagher et al. 2009), Cy5.5 (Kelly et al. 2005; Pittet et al. 2006), or AlexaFluor (Gunn et al. 2008; Maxwell et al. 2008). Alternatively, radionuclide imaging functionality can be incorporated by appending DOTA or DTPA for chelation of ^{64}Cu or [111]In (Nahrendorf et al. 2008; Natarajan et al. 2008), by PEG-lipid conjugation of [18]F (Devaraj et al. 2009), or direct coupling to the coating of the iron oxide as exemplified by [124]I conjugation to Mn-coated iron oxide nanoparticles (Choi et al. 2008).

In a clinical case study, ultra-small superparamagnetic iron oxide nanoparticles were administered 4 days after ST-elevation myocardial infarction in a 50-year-old patient. On baseline scan, small foci of subendocardial dark regions were identified within the late gadolinium enhancement positive segments. After administration of nanoparticles, hypoenhancement of T2-weighted images, as well as hyperenhancement of the anterior wall, was noted, sustained over 48 h after injection and receding by 96 h, suggesting a better localization of the infarct and border zone than late gadolinium enhancement alone (Yilmaz et al. 2012). The group later reported a larger study in 14 patients, in which similar distribution patterns were reported. Flow cytometry established the accumulation of nanoparticles in inflammatory leukocytes, particularly within activated macrophages. *In vitro* assessment showed that incubation of cultured macrophages with ferucarobtran (up to 0.25 mg Fe/mL) did not alter cell morphology or viability. Taken together, the studies suggest that ultra-small superparamagnetic iron oxide nanoparticles have a superior safety profile for clinical application and may be useful in identifying scar and peri-infarct regions of the myocardium at early stages postinfarct (Yilmaz et al. 2013).

Iron oxide particles have been used to label erythrocytes for MR blood pool imaging (Rahmer et al. 2013). *In vitro* imaging showed that signal from iron oxide-loaded erythrocytes was comparable to a higher dose of iron oxide nanoparticles, achieving similar resolution with 3.3 times lower iron concentration. *In vivo* imaging in healthy mice showed a defined blood pool image that was maintained to 24 h after injection, with periodic modulation matching to the heart rate. Overlay of blood pool images with anatomical MR images from the same mouse that was gated for respiration and heart contraction resulted in a reasonable alignment of ferrous oxide (FeO) erythrocyte blood pool signal within the anatomic defined ventricles.

11.3.3 SILICA

Alternatively, nanoparticles can be coated with inert matrices such as silica, providing a stable and water soluble matrix for containment of molecular probes including Gd-DTPA, QDs, iron oxide, or optical agents rhodamine ortris(bipyridine)ruthenium(II)-chloride (Ru(bpy)) (Kim et al. 2007; Lee et al. 2006; Rieter et al. 2007a; 2007b; Santra et al. 2005; Yang et al. 2008). Synthesis of silica nanoparticles is achieved by reverse microemulsion mixing of surfactant, oil, and water to synthesize nanoparticles in multiple nanoreactions, as well as subsequent addition of silica shells by condensation with tetraethylorthosilicate. This layering process can be repeated to incorporate further QDs and functionalities within the nanoparticle, as has been demonstrated with gadolinium-containing gold flecked nanoparticles with size selection achieved by water/surfactant ratio separation (Sharma et al. 2008). Silica nanoparticles have been prepared in a variety of sizes (30–200 nm diameter).

Kumar et al. provided the initial evaluation of organically modified silica nanoparticles (20–25 nm) radiolabeled with [124]I for PET imaging and also conjugated these nanoparticles with near infrared fluorophores for optical imaging (Kumar et al. 2010). PET imaging, fluorescence, and biodistribution studies in severe combined immunodeficiency (SCID) mice demonstrated a comparable signal between the modalities with accumulation in the liver, spleen, and stomach, with lower distribution in the kidney, heart, and lung and no histopathology-defined toxicity (Kumar et al. 2010).

Woolley et al. developed a stable dye-doped silica nanoparticle functionalized with polyamidoamine (PAMAM) dendrimers to facilitate conjugation of platelet activation antibodies (Woolley et al. 2013). The final nanoparticles were 159 nm in diameter, with a molecular brightness 74 times greater than a single dye molecule. Coupled to PAMAM for platelet targeting improved mono dispersion with reduced aggregation of particles. When functionalized with CD41, a 2.5-fold increase in platelet binding was observed for nanoparticles, confirmed by scanning electron microscopy (Woolley et al. 2013).

With the larger size of silica nanoparticles, toxicity is a concern. Thakor et al. compared the toxicity of a silica-based, gold, and Raman-active organic nanoparticle (Thakor et al. 2011). Nanoparticles (9.6×10^{10}) were administered to 120 mice (60 male, 60 female) and evaluated over >2 weeks. Gold was detected in the liver and spleen as early as 5 min after injection, with a gradual decline over 2 weeks in the liver and a peak accumulation in the spleen at 24 h after injection. Terminal deoxynucleotidyl transferase (TdT) dUTP Nick-End Labeling (TUNEL) staining demonstrated no associated apoptosis. Antioxidant enzyme gene expression (superoxide dismutase, catalase, hemoxygenase) was elevated in liver tissue after nanoparticle administration. However, nanoparticles were determined to have no toxicity after administration in mice (Thakor et al. 2011), dependent in part on maintaining effective particle clearance by limiting size.

11.3.4 POLYMERS

Other polymers have also been used in the synthesis of nanoparticles, including polystyrene, polyethyleneimine, polymethacrylates, and chitosan. Building from pharmacological delivery technology, polymers can be readily engineered to encapsulate contrast agents or to crosslink with other imaging functionalities (Doiron et al. 2009; Louie 2010; Seo et al. 2008). In general, polymers have included MR and fluorescence probes, incorporating manganese doped iron oxides, fluorophores, and QDs. Chitosan is a biocompatible polymer with capacity for electrostatic interaction with cells, facilitating delivery of imaging probes (Tan and Zhang 2007; Yuan et al. 2006). Polymer nanoparticles have been generated at 70–300 nm diameter.

Liu et al. prepared a nanoparticle comprising C-type atrial natriuretic factor (C-ANF) targeting moieties and DOTA chelators coupled to a poly(methyl methacrylate) core for labeling with [64]Cu (20–22 nm) (Liu et al. 2011). In healthy mice, C-ANF-targeted nanoparticles exhibited slower clearance and extended presence in blood—56% at 1 h after injection compared to 25% for nontargeted nanoparticles, which were rapidly cleared by the liver and kidney. Nanoparticles were administered to mice following femoral artery ligation to induce hindlimb ischemia and promote angiogenesis. On day 7 after surgery, uptake of ^{64}Cu-DOTA-C-ANF nanoparticles was significantly increased (~threefold) compared to the nonischemic leg, responsive to a saturating dose of unlabeled C-ANF nanoparticles. Immunofluorescence of nanoparticles colocalized to neovessel endothelial cells and vascular smooth muscle cells (Liu et al. 2011).

Chitosan nanoparticles have been studied for the delivery of cyclosporine to an eye injury rabbit model, with ^{99m}Tc loaded concurrently to track the efficacy of therapeutic targeting. Nanoparticles were demonstrated to bind selectively to the corneal surface, as confirmed by *in vivo* SPECT imaging, and could potentially facilitate targeted drug delivery to the injury site (Yuan et al. 2006). The porous nature of chitosan and other polymers renders these nanoparticles particularly attractive for nuclear imaging approaches due to the free diffusion and delivery of radionuclides.

11.3.5 DENDRIMERS

As opposed to encapsulation of contrast agents, nanoparticles have also been engineered as aggregates of functional branches to which imaging probes and targeting moieties can be appended. These dendrimers generally comprise polyamidoamine branched chains and can be readily coupled to a variety of probes by amine chemical reactions. Due to the lack of encapsulated particles, dendrimers have been synthesized with extremely small sizes (5–10 nm in diameter), and multiple branches proffer multiple permutations of multimodality functionalities including Gd-DTPA, chelators for ^{64}Cu, ^{99m}Tc, or ^{111}In, gold, and fluorescent dyes like rhodamine, Alexafluor, or Cy5.5 (Boswell et al. 2008; Sarin et al. 2008; Talanov et al. 2006; Xu et al. 2007).

Wen et al. synthesized gadolinium loaded dendrimer-entrapped gold (Au) nanoparticles for dual CT and MR imaging (Wen et al. 2013). Amine-terminated generation-five poly-amidoamine dendrimers were modified with a gadolinium chelator and PEG monomethyl ether to bind gold, with a final average diameter of 4.6 nm. Gd-Au dendrimers demonstrated enhanced x-ray attenuation and improved enhancement of T1-weighted images in a concentration-dependent manner, though the presence of gold may have lowered the relaxivity. *In vitro* studies demonstrated maintained cell viability with an Au concentration up to 50 μM. *In vivo* imaging studies established localization of dendrimers in the heart, liver, and bladder over 45 minutes of clearance; comparable distribution was observed in MR images. The relevance of two anatomical imaging modalities is questionable, but these dendrimers may additionally provide an alternative platform for incorporation of radioligands and PET/SPECT imaging. Nevertheless, the requisite concentrations for sufficient resolution must be determined (Wen et al. 2013).

11.3.6 HDL-LIKE NANOPARTICLES

The uses of low-density liporotein (LDL), HDL, or nanoparticles designed to mimic HDL are appealing platforms for multimodality probes due to their biocompatibility, small size, and facilitative chemistry properties for encapsulation, surface conjugation, and protein binding of imaging probes and targeting moieties. Natural LDL and HDL have been labeled with Gd-chelates, fluorescence dyes, near-infrared dyes, and targeting molecules to prevent binding to native LDL and HDL receptors. Alternatively, synthetic HDL has been constructed from the core components of native HDL with the inclusion of a gold, iron oxide, or QD nanoparticle core. These HDL-like nanoparticles have been conjugated to Gd-DTPA and rhodamine, providing a combination of CT, T1-weighted MR, T2-weighted MR, and fluorescence imaging capacity. Without targeting appendages, these nanoparticles distribute in a similar fashion to native HDL, rendering them a biological and intriguing multimodality imaging probe for cardiovascular disease.

Frias et al. incorporated Gd-1,2-dimyristoyl-*sn*-glycero-3-phosphoethanolamine-DTPA (DMPE-DTPA) and 1,2-dipalmitoyl-*sn*-glyero-3-phospoethanolamine-N-(7-nitro-2,1,3-benzoxadiazol-4-yl) (NBD-DPPE) into HDL expressing apolipoprotein A-1 (Frias et al. 2004). The HDL-like contrast agent exhibited a relaxivity of 65 MHz, with an r1 value of 10.4 mM^{-1} s^{-1}. ApoE knockout mice were injected with HDL-like contrast agent and scanned at 24 h after injection. Enhancement of the abdominal aorta wall was visible immediately after injection, proportional to the degree of cellular content of the plaque. The signal was substantially reduced by 48 h, whereas without apoA-HDL, contrast was retained for a longer period, suggesting natural clearance of the nano contrast agent. Postmortem confocal microscopy demonstrated integration of HDL contrast within the laminar cells of the vessel wall (Frias et al. 2004). LDL has also been labeled with ^{125}I and ^{99m}Tc for SPECT imaging of microvasculature endothelium in atherosclerosis (Sabal et al. 2006).

Skajaa et al. encased an iron oxide core within HDL and conjugated to Cy5.5 as a nanoparticle (7–13 nm in diameter) for dual MR and optical imaging (Skajaa et al. 2011). *In vitro* testing demonstrated maintained cell viability after HDL administration *in vivo*, in both macrophages and hepatocytes over 8 h of incubation, and uptake was saturable. Presence of increasing concentrations of unlabeled HDL significantly reduced uptake in macrophages, hepatocytes, and endothelial cells; higher uptake was observed with HDL nanoparticles compared to larger PEG nanoparticles with the same core and Cy5.5 conjugation. Transmission electron microscopy established presence of FeO-HDL within atherosclerotic plaques of ApoE$^{-/-}$ mice, with wider distribution to hepatocytes, and excretion via the biliary system (Skajaa et al. 2010).

11.4 MACROMOLECULAR CARRIERS AND SMALL MOLECULE PROBES

Naturally occurring macromolecules can also be used as a delivery carrier of imaging agents. Conjugation chemistry allows the coupling of chelators, fluorophores, and targeting moieties directly to the macromolecule. Macromolecular carriers include polylysines, polyglutamic acid, dextrans, and peptides. The relative concentration of Gd-chelate conjugations per polylysine compared to fluorescence labels is essential for effective MR enhancement and tissue transport (Uzgiris et al. 2006). Dextran conjugates with Gd-DTPA and rhodamine (GRID) have proven useful as an *ex vivo* cell labeling technique and to track cells *in vivo*. Peptide carriers are similar platforms to polymers, although the loading capacity for imaging functionalities is inherently dependent on the peptide size. As such, radiolabeling of peptides for PET or SPECT is more feasible than Gd-coupling for MR. Chelation of ^{64}Cu, ^{111}In, and ^{177}Lu has been demonstrated in small peptides, with concurrent labeling with fluorescence tags (Edwards et al. 2008, 2009).

Albumin exhibits a larger proportion of free amines for chemical modification, which facilitates higher loading of Gd and greater MR contrast potential. Maleylated bovine serum albumin has been used as a dual-modality MR/PET imaging agent targeted to macrophage scavenger receptors, labeled with Gd-DOTA for MR and ^{64}Cu-DOTA for PET in a 11:1 stoichiometric ratio (Louie 2010). Alternatively, Gd-DTPA has been combined with biotin and rhodamine or fluorescein for MR and optical/fluorescence imaging. The inclusion of biotin was suggested to allow subsequent administration of avidin to neutralize nonspecific binding of albumin, as validated in a tumor model (Dafni et al. 2003).

A final category of multimodality imaging probe platforms are small molecules, which minimize the connection between different functional imaging agents but often involve complex synthesis procedures. The advantage of this approach is the minimal size of the generated probe, which typically allows for renal clearance and reduced toxicity. Perhaps the simplest approach is the direct radioiodination of fluorophores, providing nuclear imaging functionality to a fluorescence probe (Pandey et al. 2005). Alternatively, Gd has been effectively coupled to rhodamine, fluorescein, coumarin, and spiropyran via DOTA (Mishra et al. 2006; Mizukami et al. 2009; Tu and Louie 2007) and to Oregon Green via DTPA (Dirksen et al. 2004). These probes generally provide excessive fluorescence signal for microscopy due to the ratio of Gd-chelate to fluorescence probe and the necessary dose for effective MR contrast enhancement. Conversely, conjugation of DOTA to cypate has allowed multimodality imaging with ^{64}Cu, ^{177}Lu, or ^{111}In with optical imaging without excessive cypate signal due to the comparable sensitivities of the modalities (Edwards et al. 2008).

11.5 REPORTER GENES

An alternate approach to multimodality imaging is by engineering cells for endogenous expression of multiple reporter genes. Unlike the previous examples, this technique generally utilizes a single modality probe since the target cells directly express any secondary and tertiary functionality. Inclusion of fluorescent genes (e.g., green fluorescent protein [GFP], luciferase) in plasmid vectors is relatively commonplace (Che et al. 2005; Sun et al. 2009; Terrovitis et al. 2010), which enables the incorporation of receptor or thymidine kinase reporter genes for targeted nuclear imaging in combination with fluorescence in a single fusion construct. A triple-fused reporter gene has been incorporated into a recombinant adenoviral vector, including enhanced GFP, firefly luciferase, and herpes simplex virus thymidine kinase (HSV1-tk) for PET imaging in a variety of cell lines (Ponomarev et al. 2004). Labeling of bone marrow stem cells injected into the left forelimb demonstrated capacity for fluorescence, bioluminescence, and PET monitoring of cells (Pei et al. 2012). Sun et al. evaluated several combinations of bicistronic reporter genes under a common tetracycline-response element, experimenting with combinations of luciferase, β-galactosidase, dopamine D_2-receptor, and HSV1-tk genes (Sun et al. 2001). Dual PET reporter showed 98% correlation between 3-(2′-^{18}F-fluoroethyl)spiperon (^{18}F-FESP, D_2) and ^{18}F-9-[(1-fluoro-3-hydroxy-2-propoxy)methyl]guanine (^{18}F-FHBG, HSV1-tk) binding to the PET targets (Sun et al. 2001). A new generation of triple-modality reporter genes incorporated mutant of thermostable firefly luciferase, which exhibited a twofold increase in bioluminescence *in vivo*, partially countering the sensitivity inequality between modalities (Ray et al. 2007).

More recently, gene transfer of the enzyme tyrosinase, and the associated product melanin, has been proposed as a trimodal reporter for photoacoustic, MR, and PET imaging due to broad optical absorption (photoacoustic), metal chelating capacity (MR), and binding of N-(2-(diethylamino)ethyl)-^{18}F-5-fluoropicolinamide (^{18}F-P3BZA, PET) (Qin et al. 2013). *In vivo* imaging in tumor bearing mice showed distinct and sensitive photoacoustic and PET identification of transplanted cells; though sensitivity was lower, spatial resolution of MR improved precise localization of the signal.

11.6 CARDIOVASCULAR APPLICATIONS

The combination of anatomic and physiologic information from a single probe is particularly advantageous in cardiovascular disease, in which precise localization of altered molecular processes provides insight into positive and negative remodeling as well as facilitating targeted and monitored therapy. While much of the early *in vivo* characterization of nanoparticles and other multimodality imaging tracers has been conducted in oncology animal models, the expansion of targeting functionalities has facilitated a wider application in cardiovascular disease. Among multimodality probe molecular targets are programmed cell death, cellular inflammation, and angiogenesis. Moreover, multimodal probes have been used for cell labeling to monitor cell-based therapies. In recent years, a wider application of multimodality imaging probes in animal models of myocardial infarction, ischemia, allogenic transplant, and particularly in atherosclerosis has also been found (Table 11.2).

11.6.1 APOPTOSIS

As with single-modality tracers, apoptosis can be investigated by targeting annexin signaling (Bauwens et al. 2011; Laufer et al. 2009; Wang et al. 2013). PEGylated micelles were engineered to include an iron oxide core or Gd-DTPA lipids, with additional fluorescent lipid markers incorporated into the lipid bilayer and further functionalized with Annexin A5-cys to target apoptosis. Superparamagnetic micelles exhibited increased relaxivity on transmission electron microscopy as defined by high r2/r1 ratio (as a negative contrast agent), with an average diameter of ~10 nm. Functionalized liposomes showed high calcium-dependent binding to phosphatidylserine in ellipsometry measurements (Andree et al. 1990). MR imaging of apoptotic Jurkat cell pellets displayed hypointesity on T2-weighted MR and hyperintensity on T1-weighted MR, with prolonged relaxation rates confirmed. Fluorescence imaging verified localization within the apoptotic cells (Figure 11.1) (van Tilborg et al. 2006). More extensive *in vivo* studies remain in progress.

11.6.2 INFLAMMATION

Myocardial inflammation has emerged as a potential therapeutic target in the early stages after infarction, whereby modulation of the inflammatory microenvironment and leukocyte trafficking can improve endogenous wound healing. While several publications have demonstrated imaging of postinfarct inflammation by macrophage accumulation of ^{18}F-fluorodeoxyglucose (FDG), this method is limited due to natural cardiomyocyte ^{18}F-FDG uptake, which must be suppressed, and can therefore complicate the precise localization of the inflammatory signal. As such, specific multimodality agents targeted to inflammatory cells and cytokines are desirable.

In an early example, ^{111}In-IgG was incorporated into a PEGylated liposome to evaluate infection and inflammation in rats after *Staphylococcus aureus* injection into skeletal muscle. Liposomes functionalized with ^{111}In-IgG showed enhanced accumulation within the abscess over 48 h, but relatively slow blood clearance. There was no uptake in kidneys, suggesting a liver/biliary excretion of the PEGylated liposomes (Boerman et al. 1995).

Table 11.2 Clinical applications of multimodality imaging probes in cardiology

Application	Target	Probe	Findings	Ref
Cellular apoptosis	Annexin a5-cys	Micelle, iron oxide core, fluorescence molecules	• Increased relaxivity on transmission electron microscopy • Targeting of apoptotic cells in vitro • MR imaging of apoptotic cell pellets with T2 hypoenhancement and T1 hyperenhancement	van Tilborg et al. (2006)
Inflammation	S. aureus infection	Liposome, ^{111}In-IgG	• Increased accumulation in infected muscle • Liver/biliary excretion	Boerman et al. (1995)
	Myocarditis	Silica-coated iron oxide nanoparticle, rhodamine β-thiocyanate	• Biocompatible, <3 h blood half-life • T2* negative enhancement proportional to the inflammation severity • Association with macrophages confirmed by fluorescence associated cell sorting (FACS), immunohistochemistry	Moon et al. (2012)
	VCAM-1	Iron oxide nanoparticle, fluorescence	• Nanoparticle accumulation at local site of inflammation by TNFα injection (periphery) • Significant MR enhancement of aortic arch and root in ApoE−/− mice (20x), refractory to atorvostatin treatment • Association with endothelial cells by FACS, immunohistochemistry	Kelly et al. (2005), Nahrendorf et al. (2006)
Atherosclerosis	MDA2, Fab fragments, E06	Micelles, Gd or Mn contrast, rhodamine	• Circulation time increased by 12–18 h • Improved biocompatibility by substitution of Mn for Gd • Abdominal aortic wall MR enhancement by 88%–95% over 48 h in ApoE−/− mice • Colocalization with CD68+ macrophages by immunohistochemistry	Briley-Saebo et al. (2008, 2012)
	CD36+ foam cells	Micelles, Gd contrast, fluorescence	• Increased accumulation by foam cells in vitro compared to nontargeted micelles • Rapid (30–60 min) MR enhancement of aortic wall • Foam cell accumulation confirmed by immunohistochemistry	Dellinger et al. (2013)

(Continued)

Table 11.2 (Continued) Clinical applications of multimodality imaging probes in cardiology

Application	Target	Probe	Findings	Ref
Atherosclerosis	Macrophages	Liposome, iodixanol contrast, rhoadmine	• CT enhancement by iodixanol in non-pegylated liposomes in cell pellets, 100-fold increase in Hounsfield units compared to unlabeled macrophages • Enhanced ct contrast and rhodamine accumulation of aortic arch and sinus in ApoE$^{-/-}$ mice at 48 h	Bhavane et al. (2013)
	Plaque	Short chain dextrans, ^{89}Zr, fluorophore	• Blood half-life of 3.7 h, clearance by liver and biliary system • Increased PET and MR signal at aortic root and arch in ApoE$^{-/-}$ mice • Autofluorescence near infrared imaging and flow cytometry localized signal to monocytes and macrophages • Treatment with anti-inflammatory CCR2 inhibition reduced dextran accumulation in plaques	Majmudar et al. (2013)
	Macrophages	Iron oxide nanoparticle, dextran outer core, ^{64}Cu, VT680	• 7- to 20-fold increase in plaque binding compared to MR detection • Association with monocytes and macrophages by FACS and immunohistochemistry	Nahrendorf et al. (2008)
	Adiponectin	Nanoparticle or liposome, ^{125}I-adiponectin, Atto655	• Increased aortic arch targeting compared to untargeted liposomes/nanoparticles • Colocalization with CD68 macrophages	Almer et al. (2011)
	RGD angiogenesis	Nanoparticle, Cy5.5	• Selective localization to neointima of aorta in ApoE$^{-/-}$ mice • Increased in vivo optical signal at ligated carotid artery compared to free contralateral side	Kitagawa et al. (2012)

(Continued)

Table 11.2 (Continued) Clinical applications of multimodality imaging probes in cardiology

Application	Target	Probe	Findings	Ref
Atherosclerosis	HDL	HDL, gold, iron oxide, or QDs	• Uptake into atherosclerotic lesions in ApoE$^{-/-}$ elevated compared to other nanoparticles • Independent of the core contents and modality • Additional targeting to collagen peptide EP3533 greater association with extracellular matrix • Monitoring of lesion regression after high fat diet withdrawal in ApoE$^{-/-}$ mice	Chen et al. (2013), Cormode et al. (2008)
Aortic aneurysm	Monocytes, macrophages	Dextran-coated iron oxide nanoparticles, ^{18}F-PET$_3$N$_3$	• Aortic aneurysm induced by angiotensin II system treatment • Threefold increase in ^{18}F-PET signal, larger aorta diameter by MR contrast enhancement • Specificity of nanoparticle accumulation to monocytes and macrophages by FACS	Nahrendorf et al. (2011)
Cell tracking	Neural stem cells	Dextran polymer, Gd-DTPA, rhodamine	• Administration of labeled cells 3 months after middle cerebral artery occlusion • MR contrast enhancement of cell transplant visible over 14 days • Localization of cell engraftment confirmed by fluorescence histology	Brekke et al. (2007), Modo et al. (2004a, 2004b)
	MSCs	Gd$_2$O$_3$ nanoparticles, Eu, fluorescence	• Cells remained viable over 7–10 days in culture • Enhanced MR contrast, but not as extensive as with iron oxide	Shi et al. (2010)
	Embryonic stem cells	Fusion reporter gene, red fluorescence protein (RFP), firefly lucieferase (Fluc), HSV-tk, iron oxide	• Tracking of transplanted stem cells over 3 months after occlusion of middle cerebral artery and implantation • Differentiation of cells into multiple neural lineages	Daadi et al. (2013)

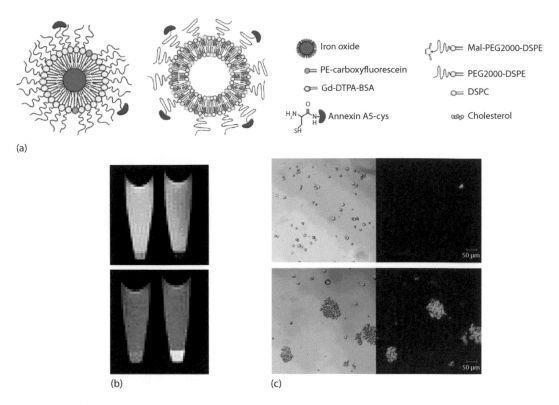

Figure 11.1 (a) Design of superparamagnetic micelles conjugated to Annexin-A5 for targeting apoptotic cells. Micelles contain a core of iron oxide or Gd-DTPA-bovine serum albumin for MR contrast enhancement. (b) Spin-echo images of cell pellets with (right) and without (left) exposure to superparamagnetic micelles; hypoenhancement on T2-weighted images (upper) and hyperenhancement on T1-weighted images (lower). (c) Confocal transmission (left) and fluorescence (right) images of apoptotic Jurkat cells exposed to micelles with (lower) and without (upper) Annexin A-5 targeting. (From van Tilborg GA et al., *Bioconjug Chem*, 17, 741–749, 2006. With permission.)

Dual-modality labeling has been demonstrated with silica-coated iron oxide nanoparticles conjugated with fluorescent rhodamine β-isothiocyanate for assessing degrees of inflammation in an experimental rat model of autoimmune myocarditis (Moon et al. 2012). Nanoparticle blood half-life was 2.8 ± 0.5 h and was found to be biocompatible. Myocarditis hearts exhibited extensive inflammation, with significant accumulation of internalized nanoparticles and flow cytometry-identified infiltrating immune cells, predominantly macrophages (80%), in the enlarged left ventricle. After administration of nanoparticles (24 h), negative enhancement on T2*-weighted MR was apparent in the left ventricle of myocarditis hearts but not control hearts. The area and degree of contrast enhancement on T2*-weighted MR were proportional to the inflammation grades assigned from hematoxylin/eosin staining, suggesting a noninvasive method for tracking the pathogenesis of myocardial inflammation.

Kelly et al. identified a specific moiety within adhesion molecule vascular cell adhesion molecule (VCAM)-1 mediated cell internalizing peptides that bound VCAM-1 and blocked leukocyte endothelial interactions (Kelly et al. 2005). They incorporated this moiety into an iron oxide-fluorescence-labeled nanoparticle to target inflammation, induced in mice by tumor necrosis factor (TNF)-α injection to the ear. Nanoparticles were imaged by confocal microscopy, revealing accumulation in the inflamed ear vessels at 4 h and 24 h after administration, while accumulation was absent in the contralateral ear. MR imaging in ApoE$^{-/-}$ mice showed enhancement of the aortic arch and root, consistent with inflammation of atherosclerotic plaque and accumulation of nanoparticles. Flow cytometry demonstrated nanoparticle binding to endothelial cells, with limited accumulation in macrophages, suggesting an alternative target for atherosclerotic imaging (Kelly et al. 2005). Subsequently, Nahrendorf et al. demonstrated 20× greater enhancement on T2-weighted MR than previously reported, confirming uptake into endothelial cells of ApoE$^{-/-}$ mice by immunohistochemistry (Nahrendorf et al. 2006). Treatment with atorvostatin reduced the accumulation of nanoparticles in the aortic root, with ~50% reduction of MR contrast enhancement and fluorescence, which occurred in parallel with markedly lower VCAM-1 expression at the aortic root. Binding of VCAM-1 is suggested to provide an early indicator of plaque inflammation and may be a functional early biomarker of plaque vulnerability (Nahrendorf et al. 2006).

11.6.3 ATHEROSCLEROSIS AND VULNERABLE PLAQUE

Perhaps the most relevant and well-reported cardiovascular application of multimodality imaging probes to date is in the field of atherosclerosis and the quantification of plaque. The combination of high-resolution MR or CT contrast with radioisotopes for PET or SPECT imaging bears potential for the identification of not only plaque and vessel anatomy but also greater insight into plaque physiology and vulnerability (Fuster et al. 2005a, 2005b). Concurrent labeling with fluorophores and tags for optical imaging provides useful correlations for localization of the probe signal. A particular focus in atherosclerosis has been the development of modified HDL and HDL-like nanoparticles, which exhibit natural homing to atherosclerotic plaque and are readily modified with multimodality functional groups.

Briley-Saebo et al. generated a series of micelles with variable targeting antibodies to evaluate pharmacokinetics and biodistribution in ApoE$^{-/-}$ mice (Briley-Saebo et al. 2008). Micelles were prepared containing: MDA2 (for malondialdehyde lysine epitopes on modified LDL or other proteins), IK17 (for Fab fragments), E06 (for native murine IgM antibody to oxidized phospholipid), IgG (control), or without targeting antibodies. Each targeting antibody extended the blood half-life by 12–18 h compared to IgG or untargeted micelles, with higher accumulation in the spleen, liver, and kidney at 48 h after injection. MR images demonstrated enhancement of abdominal aortic wall by 60%–90% at 24 h and 88%–95% at 48 h after injection, with minimal enhancement of the liver in ApoE$^{-/-}$ mice. Rhodamine in MDA2 and E06 micelles were partially colocalized to macrophage-marker CD68 within the atherosclerotic lesions. In a later study, MDA2 and IK17 micelles were conjugated with Mn as a biocompatible MR contrast agent in comparison with Gd^{3+} (Briley-Saebo et al. 2012). Relaxivities were lower with Mn enhancement, but significantly higher in ApoE$^{-/-}$ mouse blood compared with wild-type blood. There was lower accumulation of Mn contrast agent in the liver and kidney up to 168 h compared with Gd, suggesting a potential benefit of this more biocompatible contrast, with a limited tradeoff of reduced contrast enhancement. Indeed, at 48 h after injection, vessel wall enhancement with Mn-MDA2 was higher than Gd-MDA2. Colocalization of Mn MDA2 micelles with CD68 macrophages was demonstrated by confocal microscopy (Figure 11.2).

In a different study, micelle targeting could be enhanced by the addition of CD36 ligands to a Gd-labeled liposome. *In vitro* uptake assays displayed increased uptake of modified liposomes into CD36-expressing foam cells compared with nontargeted liposomes. Administration of the modified liposomes to ApoE$^{-/-}$ mice showed enhancement of the aorta wall within 30–60 min after administration and was associated with CD36 expression in atherosclerotic foam cells. Toxicity studies demonstrated minimal accumulation in the liver and kidney over initial hours after injection, and no toxicity was reported (Dellinger et al. 2013).

In vitro studies have confirmed accumulation of conjugated liposomes by macrophages. When labeled with rhodamine (103–270 nm) for fluorescence imaging and conjugated with iodixanol (93–225 nm) for *in vitro* CT imaging, liposomes exhibited substantial uptake in RAW 264.7 macrophages. After 4 h, only cells incubated with non-PEGylated 400 nm liposomes provided a coherent signal, corresponding to the highest maximal uptake at 24 h (600 Hounsfield units vs. 6 in blank pellet). ApoE$^{-/-}$ mice with some calcification present in the aorta were administered 400 nm liposomes containing both rhodamine and iodixanol. By 48 h, contrast was cleared from the circulation, and enhanced contrast was observed at the aortic arch and sinus, regions that exhibited increased macrophage levels and a moderate increase in rhodamine labeled liposome accumulation (Bhavane et al. 2013).

In an alternative approach, crosslinked short-chain dextrans (13 nm in diameter) were conjugated with desferoxamine for labeling by zirconium-89 (^{89}Zr) for PET imaging and a fluorophore for near-infrared imaging at an average ratio of 8:1. *In vivo* testing in healthy mice demonstrated a blood half-life of 3.7 h, with the highest accumulation at 48 h in the liver, spleen, and lymph nodes and lower accumulation in the heart, aorta,

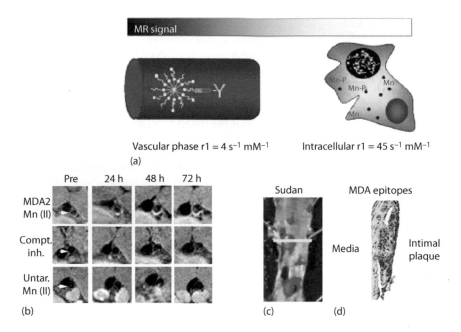

Figure 11.2 (a) Design of MDA2-antibody-targeted Mn-containing micelles to deliver Mn to intracellular compartment. (b) Enhancement of abdominal aorta wall in ApoE$^{-/-}$ mice over 72 h after administration of MDA2-targeted micelles provides clear contrast compared to competitive inhibition (Compt. Inh.) and untargeted Mn-micelles (Untar. Mn(II)). (c) Sudan IV staining shows atherosclerotic lesions in the abdominal aorta. (d) MDA epitope staining shows localization of micelles throughout the affected vessel. (From Briley-Saebo KC et al., *J Am Coll Cardiol*, 59, 616–626, 2012. With permission.)

and kidney. In ApoE$^{-/-}$ mice, significantly higher activity was found within the aortic root and arch compared to healthy mice by *in vivo* PET/MR imaging and *ex vivo* by autoradiography/oil red-O staining, suggesting dextran uptake within atherosclerotic plaques (Majmudar et al. 2013). Autofluorescence of dextran nanoparticles was concentrated in cells that stained positive for CD11b, a marker of inflammatory monocytes and macrophages, which was confirmed by flow cytometry. siRNA silencing of chemokine receptor 2 (CCR2) reduced circulating levels and homing of inflammatory cells, and lowered [89]Zr-dextran nanoparticles accumulation at the aortic arch and root compared to untreated ApoE$^{-/-}$ mice (Majmudar et al. 2013).

In an attempt to enhance homing to atherosclerotic plaques, Almer et al. coupled the globular domain of adiponectin to nanoparticles (proticles, 188–263 nm) and liposomes (100–160 nm) (Almer et al. 2011). For *in vitro* testing, adiponectin was radiolabeled with ^{125}I and particles were conjugated to Atto655 for fluorescence imaging and displayed a maximum coupling efficiency of 150 ng adiponectin/10 μg oligonucleotides. In ApoE$^{-/-}$ mice, adiponectin-coupled liposomes and proticles showed increased accumulation in aortic arch plaque compared to liposomes alone, with some colocalization to CD68-stained macrophages (Figure 11.3) (Almer et al. 2011). The relatively large size of this nanoparticle raises questions related to its natural distribution and toxicity, but the initial experiments hold promise.

Nanoparticles have also been targeted to angiogenesis within large atherosclerotic vessels. Kitagawa et al. generated a nanoparticle (12–14 nm) functionalized with an arginine-glycine-aspartate (RGD) peptide sequence and conjugated with Cy5.5 for near-infrared fluorescence imaging to evaluate lesions in

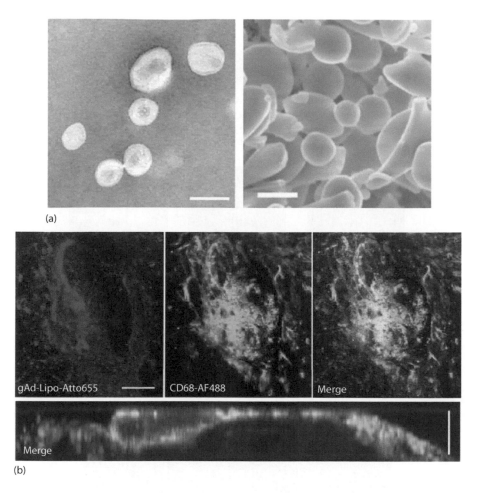

Figure 11.3 (a) Transmission electron microscopy image of liposomes (left) and encapsulated nanoparticles (proticles) targeted to adiponectin. (b) Immunostaining of ApoE$^{-/-}$ mouse aortic sections with anti CD68 for macrophages (green) colocalizes to nanoparticle-AlexaFluor targeted to the globular domain of adiponectin (red). (From Almer G et al., *Int J Nanomedicine*, 6, 1279–1290, 2011. With permission.)

the carotid arteries of healthy and diabetic mice (Kitagawa et al. 2012). The presence of RGD enhanced the optical signal *in vivo*, displaying a significant difference between the ligated and contralateral carotid arteries. Postmortem fluorescence microscopy demonstrated the presence of macrophages infiltrating the neointima, with higher Cy5.5 intensity for the RGD-conjugated nanoparticles than the nonfunctionalized particles (Kitagawa et al. 2012).

Cormode et al. examined the utility of HDL as a nanoparticle with incorporation of Au, iron oxide, or QD for CT, MR, or optical imaging, respectively (Cormode et al. 2008). *In vitro* experiments in mouse macrophages showed more rapid uptake of Au-HDL compared to Omnipaque or Au-PEG and markedly higher x-ray attenuation. Similarly, FeO-HDL was taken up by macrophages more robustly than FeO-PEG and generated further negative enhancement of T2-weighted MR. QD-HDL evoked a ninefold increase in fluorescence versus QD-PEG. *In vivo* testing in ApoE$^{-/-}$ mice showed accumulation of both Au-HDL and QD-HDL in atherosclerotic lesions of the aorta and were clearly visible by CT (Au) or fluorescence (QD). T1-weighted MR revealed contrast enhancement by both QD- and Au-HDL at 24 h after administration in the vessel wall. T2* darkening was observed with FeO-HDL (Figure 11.4) (Cormode et al. 2008).

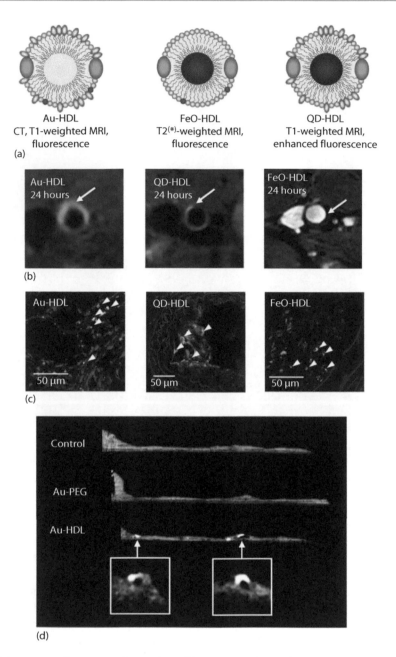

Figure 11.4 (a) Structures of nanocrystal HDLs (HDL-like nanoparticles) with Au, FeO, or QD core. (b) Enhancement of T2-weighted images with Au-HDL (left) and QD-HDL (centre) at 24 h after injection; negative enhancement of T2*-weighted images with FeO-HDL at 24 h after injection in ApoE$^{-/-}$ mice. (c) Confocal microscopy images of aortic sections show localization of HDL nanoparticles (red) in relation to macrophages (green). (d) Targeted Au-HDL shows greater hyperenhancement on *ex vivo* CT within regions of atherosclerotic plaque. (From Cormode DP et al., *Nano Lett*, 8, 3715–3723, 2008. With permission.)

HDL Gd-chelated nanoparticles have been modified by addition of collagen specific EP3533 peptides to target extracellular matrix (Chen et al. 2013). Addition of functional groups by PEG did not affect nanoparticle size (10 nm) or relaxivity (9 s^{-1} mM^{-1}). ApoE$^{-/-}$ mice were maintained on a high-cholesterol diet and after resumption of normal diet displayed regression of aortic plaque burden and a plaque compositional change from elevated macrophages and low collagen to reduced macrophages and high collagen content.

Serial imaging with EP3533 functionalized HDL nanoparticles exhibited greater association with extracellular matrix collagen as compared to a nontargeted control (Figure 11.5). Enhancement of the MR signal in the aortic wall was evident in regressing plaques by collagen-targeted HDL nanoparticles, whereas no enhancement was observed in a control group. Moreover, unmodified HDL nanoparticles demonstrated loss of enhancement in the regression group at Day 28. Fluorescence microscopy established association of nanoparticles with both collagen and CD68-positive macrophages. The degree of enhancement correlated reasonably well with the area of collagen deposition in the plaque postmortem. The study demonstrates the versatility of the HDL platform, which can be targeted to multiple facets of atherosclerosis progression and regression.

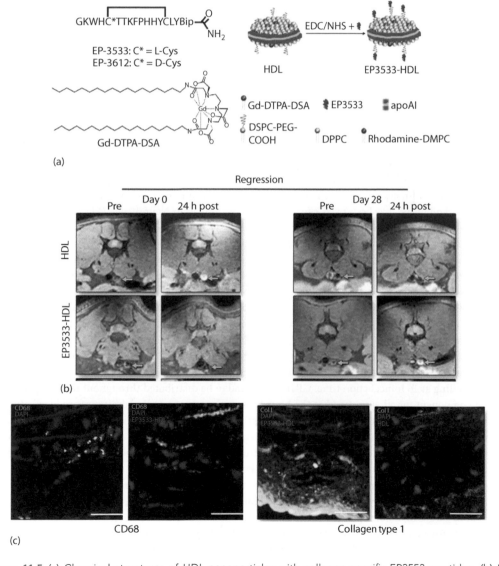

Figure 11.5 (a) Chemical structures of HDL nanoparticles with collagen-specific EP3553 peptides. (b) MR images show enhancement of the aorta at 24 h post injection on day 0. After restoration of normal chow, atherosclerotic plaque and enhancement recede. (c) Colocalization of HDL (red) and CD68 macrophages (green, left) or collagen type 1 (green, right) on fluorescence microscopy. (From Chen W et al., *JACC Cardiovasc Imaging*, 6, 373–384, 2013. With permission.)

11.6.4 Aortic aneurysm

PET- and CT-functionalized nanoparticles have been used in preclinical studies to assess severity and prognosis of aortic aneurysm. Dextran-coated iron oxide nanoparticles targeted to monocytes and macrophages were crosslinked with the near infrared fluorochrome VT680 and coupled to ^{18}F-labeled-polyethlene glycol azide (^{18}F-PEG$_3$N$_3$) via click chemistry. Aortic aneurysm was induced in ApoE$^{-/-}$ mice by angiotensin II administration. PET/CT imaging demonstrated contrast enhancement of aneurysm with a larger aorta diameter and threefold increase of ^{18}F-signal compared to wild-type mice. Scintillation imaging and fluorescence histology confirmed localization of nanoparticles to the site of aortic aneurysm. Flow cytometry showed that monocytes/macrophages comprised 90% of the inflammatory cell content of aortas from angiotensin II treated mice. The authors purported a predictive value to the nanoparticle imaging, whereby mice with higher PET signal measured at 7 days exhibited the biggest increase in aortic diameter at 28 days after systemic angiotensin II administration (Nahrendorf et al. 2011).

11.6.5 Cell tracking

Multimodality probes are also useful in labeling cells to monitor the cellular healing process. The use of multiple imaging functionalities provides a longer timeframe for visualization of cell trafficking, particularly in comparison with direct cell radiolabeling techniques, which are limited by radioactive half-life (Terrovitis et al. 2010; Zhang and Wu 2007). Moreover, the inclusion of targeting moieties can facilitate cell homing, contribute to or augment paracrine signals, and improve cell engraftment. The combination of radioisotopes or MR contrast agents with fluorophores further allows the postmortem evaluation of cell differentiation, providing additional insight into the healing process (Seo et al. 2008). As such, platforms conducive to variable stoichiometric addition are of greatest interest in cell labeling.

The so-called bifunctional GRID chelate, comprising a dextran polymer decorated with Gd-DTPA and fluorescent rhodamine, has been used effectively in stroke research. Neural cells (Maudsley Hippocampal Clone 36 [MHP36]) were cultured and split prior to incubation with GRID under proliferative conditions over 6 h. At 3 months following middle cerebral artery occlusion, mice were transplanted with 2×10^5 MHP36 cells in the contralateral hemisphere to track the migration toward the ischemic lesion within the ipsilateral hemisphere. T2-weighted MR images showed enhanced contrast of the stem cell transplant, and revealed significant transhemisphere migration of the stem cells in the stroke-induced animal, beginning from 7 days, and reaching maximum at 14 days after transplant. Localization was confirmed by postmortem fluorescence histology, demonstrating differentiation of transplanted cells by colocalization of rhodamine with the pan neuronal marker (NeuN) (Modo et al. 2004a, 2004b). In another study, Brekke et al. confirmed the viability and normal function of GRID labeled neural stem cells (Brekke et al. 2007). They demonstrated stable presence of the rhodamine fluorescence signal over 7 days after cell labeling in vitro, with maximal fluorescence and Gd loading at 16 h incubation time. GRID was found to be distributed throughout the cytoplasm, with a small component found within and at the boundaries of lysosomes. Cell viability was reduced after 24 h of incubation, with a decrease in proliferation potentially related to reactive oxygen species accumulation. However, at 16 h of incubation with GRID, cell migration was unaffected, and cells exhibited no impedance of differentiation to astrocytes, neurons, or oligodenrocytes. *In vitro* MR studies showed improved contrast with larger number of GRID labeled cells, but the authors acknowledge that the potential toxicity of the labeling procedure may limit therapeutic efficacy of these modified stem cells (Figure 11.6) (Brekke et al. 2007).

MSCs have been similarly labeled with superparamagnetic iron oxide nanoparticles to evaluate the effect of magnetic targeting following balloon occlusion injury in rabbit femoral artery. The selected particle concentrations provided 95% cell viability under starvation conditions, and showed normal MSC differentiation and growth factor secretion. Balloon occlusion of the femoral artery was followed by placement of a cylindrical magnet around the leg and administration of 10^5 allogeneic MSCs into the vessel lumen. The magnet was kept in place for 40 min to improve cell retention. *Ex vivo* MR demonstrated accumulation of magnetically labeled cells to ~50% of the vessel circumference, with no attachment in control arteries, confirmed by

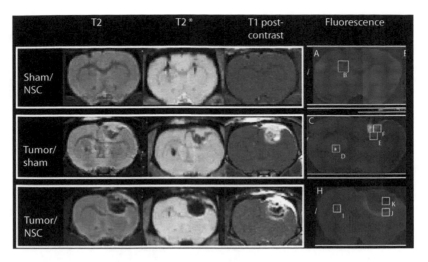

Figure 11.6 T2-, T2*-, and T1-weighted contrast-enhanced MR images of gadolinium-rhodamine-dextran-labeled neural stem cells administered to rats with glioma-induced brain tumors. Cells are visualized as hypoenhancement of T2 and T2* images and hyperenhancement of T1 images. Localization of cells was confirmed by fluorescence images of rhodamine. (From Brekke C et al., *Neuroimage*, 37, 769–782, 2007. With permission.)

confocal microscopy. At 3 weeks after the injury, the intima/media ratio was lower in magnetic cell delivery arteries compared to control suggesting reduced restenosis (Riegler et al. 2013).

Shi et al. incorporated anionic Eu^{3+} with Gd_2O_3 hybrid nanoparticles as a dual-modality contrast agent for T1-weighted MR and postmortem fluorescence microscopy and internalized this contrast agent into hMSCs for transplantation (Shi et al. 2010). Labeled cells showed maintained viability, differentiation, and activity over 7–10 days. Contrast enhancement of T1-weighted MR was evident with 50–200 cells/μL and provided a functional method for tracking of cells, with the possibility of additional functional modification for nuclear imaging. The difference in contrast was not as pronounced as with superparamagnetic iron oxide cores (Shi et al. 2010).

In a study of posttransplant immuno-rejection, a polyethylene-grafted polyethylenimine carrier was functionalized by the addition of superparamagentic iron oxide nanoparticles and CD3 single chain antibodies to target T cells (Guo et al. 2012). In addition, a therapeutic gene (DGKα) or nonfunctional plasmid was conjugated to the nanoparticle. The particles were 106 nm with a zeta-potential of +19 mV. *In vitro* studies demonstrated high selectivity for T cells, with functional transfection of primary T cells by the therapeutic gene. Transplanted heart in rats receiving nanocomplexes with null plasmid deoxyribonucleic acid (DNA) exhibited hypointense regions and endocardial dark linear signal on T2*-weighted images particularly at 3 and 10 days posttransplant with parallel near-infrared fluorescence intensity in the graft region. By contrast, rats receiving nanocarriers with DGKα plasmid DNA or nontargeted (without CD3 antibody) nanocarriers exhibited neither of these *in vivo* signals, nor did the native heart, suggesting selective imaging of T cell accumulation in the transplanted heart (Figure 11.7).

An alternative approach for cell labeling and tracking is transduction of multimodal reporter genes. Daadi et al. studied a trimodality reporter gene system in a rat model of ischemic stroke (90 min ischemia-reperfusion of middle cerebral artery) (Daadi et al. 2013). Embryonic-derived neural stem cells were engineered to express monomeric red fluorescence protein, firefly luciferase, and HSV-tk then subsequently labeled with superparamagnetic iron oxide and transplanted to the lesioned striatum. Cells were effectively tracked over 3 months by both PET and MR, with a reduction in the infarct size of treated animals with small lesions. After 3 months, postmortem immunohistochemistry demonstrated dispersal of neural stem cells throughout the lesioned parenchyma, with differentiation to neurons, astrocytes, and oligodendrocytes (Daadi et al. 2013).

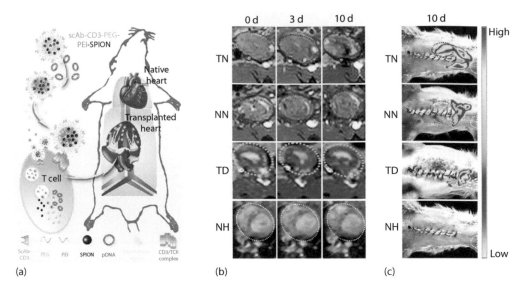

Figure 11.7 (a) Schematic of heart transplant model and administration of labeled superparamagnetic iron oxide nanoparticles targeted to T-cells by conjugation of CD3 antibody. (b) MR images at 0, 3, and 10 days posttransplant with scAbCD3-targeted (TN) and scAbCD3-targeted DGKα pDNA (TD). Signal intensity is enhanced compared to nontargeted (NN) and native heart (NH) and reduced at 10 days with therapeutic DGKα administration. (c) Near-infrared imaging shows the reduction of nanoparticle signal at 10 days in DGKα group. (From Guo et al., *ACS Nano*, 6, 10646–10657, 2012. With permission.)

11.7 THERANOSTICS

Multimodality probe platforms lend themselves to the delivery of therapeutic payloads in concert with imaging functionalities. "Theranostics" allow a single administered agent to simultaneously assess disease severity, target a specific cell or tissue, deliver treatment (pharmacologic, gene, or cell therapy), and monitor the physiologic response. The majority of early preclinical evaluation for theranostics has been conducted in cancer models, where targeted therapy delivery (e.g., to tumor cells) is clearly advantageous. The specific tissue microenvironments of a variety of cardiovascular disorders are appropriate for theranostics, including postinfarct inflammation, adverse ventricular remodeling, angiogenesis, atherosclerosis, and vulnerable plaque. As multimodality agents continue to emerge, a greater integration of therapeutic functionality bears significant translational potential for cardiovascular medicine.

Synthetic hollow nanoparticles have been used for targeted doxorubicin delivery in a cancer model, with a functional on/off capability due to pH-sensitive imidazole structure that releases doxorubicin in acidic surroundings (i.e., the intracellular endosome). Nanoparticle surface phenolic ester-PEG segments were labeled with ^{123}I to allow noninvasive SPECT imaging. Theranostic nanoparticles were cleared from the blood and distributed rapidly (within 30 min), with selective delivery to the HeLa tumors, particularly at early stages after administration (Lu et al. 2011).

Takahama et al. used a fluorescence-labeled liposome to deliver the antiarrhythmic amiodarone in an ischemia-reperfusion rat model. PEGylated liposomes incorporated amiodarone (3 mg/kg) into the lipid core and were administered prior to transient occlusion of a coronary vessel in a rat (Takahama et al. 2013). Liposomal delivery of amiodarone reduced the frequency of ventricular fibrillation during the reperfusion phase at a lower dose than free intravenous amiodarone, suggesting potential to reduce dose-dependent deleterious side effects (Takahama et al. 2013).

Schmieder et al. assessed an $\alpha_v\beta_3$ integrin targeted nanoparticle containing the anti-angiogenic agent fumagillin as a theranostic imaging probe in tumor bearing mice (Schmieder et al. 2008). *In vitro* testing demonstrated selective binding by competitive inhibition of HT1080 cell adhesion to fibronectin in the presence of the nanoparticle and was confirmed by flow cytometry. Tumor bearing athymic nude mice were imaged at 14 days after cell implantation with the nanoparticles, exhibiting 117% increase in contrast of 1.8% of the tumor volume within 2 h of injection (versus 51% enhancement of 0.2% of volume in control tumor-bearing mice). $\alpha_v\beta_3$ integrin expression was confirmed by immunohistological staining and revealed distribution of nanoparticles in the tumor periphery, restricted to vasculature and specific to endothelial cells. The inclusion of fumagillin evoked a change in the nanoparticle distribution, with smaller, contracted pockets of nanoparticles and narrower dispersion; however, no differences in tumor volume were observed, which may have reflected the relatively slow growth of the xenograft and lower angiogenesis than in other tumor cell types (Schmieder et al. 2008). Winter et al. evaluated a theranostic nanoparticle containing Gd DTPA, PEGylated vitronectin antagonist, lecithin, and therapeutic fumagillin (0.2 mol) in atherosclerotic rabbits (Winter et al. 2006). Subsequent imaging was performed using fumagillin-free nanoparticles at 1 week after treatment. MR imaging demonstrated signal enhancement in the vessel wall that was substantially reduced at 1 week among rabbits that received fumagillin-conjugated nanoparticles (60%–80%). Those rabbits with the highest signal enhancement at the time of treatment with theranostic nanoparticles also showed the greatest reduction of $\alpha_v\beta_3$ integrin expression after 1 week, coupled with a marked reduction in neovascularization evaluated by histology (Winter et al. 2006). The combination of a therapeutic agent into the nanoparticles enabled immediate regression of angiogenesis, with concurrent reductions in atherosclerotic burden. In a subsequent study, rabbits were treated with atorvastatin alone or in combination with fumagillin nanoparticles over 8 weeks. Enhancement of the aorta wall was shown to decrease at 1 week after fumagillin treatment but shows gradual recovery at 2, 3, and 4 weeks after treatment, suggesting a recession of angiogenesis. Signal enhancement was not affected by statin treatment over 8 weeks, but in combination with fumagillin was markedly reduced in the initial week following treatment, with the gradual recovery of signal ablated after a second fumagillin administration to the statin group. Antiangiogenic therapy and neovascular pruning is theorized to stabilize plaque. Noninvasive imaging of $\alpha_v\beta_3$-targeted nanoparticles showed a prolongation of plaque stability with the combination of atorvastatin and fumagillin (Figure 11.8) (Winter et al. 2008).

11.8 CLINICAL AND EXPERIMENTAL SIGNIFICANCE

The advantages of multimodality over single-modality imaging probes can be categorized to those that benefit researchers (e.g., distribution kinetics, image analysis) and those that benefit clinics and patients (e.g., increased throughput, reduced administered dose). For research, the inclusion of multiple modality functionality within a single probe provides a number of advantages. First, kinetic measurements of probe distribution can be calculated more accurately with the confluence of imaging signals from a single source. The combined information provides a truer representation of tracer kinetics, even with the addition of less-well-characterized agents. In the case of nanoparticles, the kinetics of distribution and clearance are greatly simplified and can be modified by the inclusion of targeting molecules, providing for specialized or optimized kinetics of the individual probe. Second, the inclusion of both anatomic (MR, CT, optical) and physiologic (MR, PET, SPECT) functionality within a single probe provides for more effective localization of the physiologic signal, taking advantage of the relative strengths of each imaging modality. The alignment of nuclear images with structural CT or MR is simplified by the common focal signal detected in each. Third, the inclusion of a fluorescent molecule facilitates *ex vivo* evaluation and colocalization in tissues, providing invaluable corroborating evidence for the cellular target and source of the *in vivo* imaging signal. Moreover, in cell tracking studies, the combination of imaging functionalities permits not only *ex vivo* confirmation of *in vivo* findings but also can identify cell fate and differentiation at the target site. From a clinical application perspective, the development of multimodality agents potentially benefits the patient by increasing the sensitivity and specificity of a diagnostic scan. The requisite scan time for each individual patient could also be reduced, allowing for improved patient comfort and higher patient throughput for the clinic. Reduced frequency of scanning has the ancillary benefit of reducing patient dose in the case of nuclear diagnostic scans.

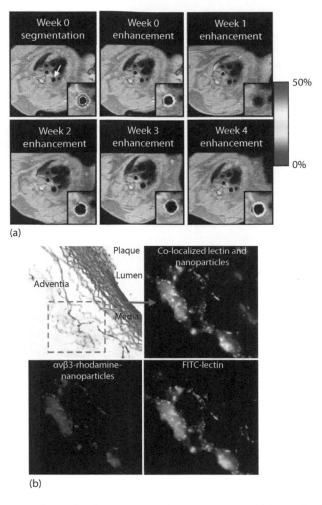

(a)

(b)

Figure 11.8 (a) MR images of signal enhancement with integrin-targeted fumagillin-loaded nanoparticles in the thoracic aorta of atherosclerotic rabbits. Color inset show degree of signal enhancement, reduced at weeks 1, 2, and 3 by antiangiogenic fumagillin. (b) Light microscopy of atherosclerotic plaque shows neovascularization (upper left); αvβ3 targeted nanoparticles expressing rhodamine (red, lower left) and fluorescein-labeled lectin targeting vascular endothelium (green, lower right) demonstrate colocalization of signal with nanoparticles (upper right). (From Winter PM et al., *JACC Cardiovasc Imaging*, 1, 624–634, 2008. With permission.)

11.9 SUMMARY

The combination of CT or MR contrast agents with radioisotopes, near-infrared imaging molecules, and fluorophores aims to take advantage of fused instrumentation, providing for the simultaneous collection of anatomic, physiologic, and targeted molecular information from the distribution and kinetics of a single molecular probe. A myriad of particle structures and configurations have been synthesized ranging from simple encapsulation within lipid carriers to more complex conjugated nanostructures to modified natural particles. As such, there is a wide range of particle sizes, specific tuning, and functional combinations available, which require further and more stringent evaluation in preclinical disease models. There remain challenges to be addressed, including the issue of biocompatibility, long-term toxicity, limiting particle size, maximizing targeting efficiency, and the variable sensitivities of different modalities necessitating stoichiometric balancing of functionalities; however, the advances in microfluidics and nanochemistry have begun to address these issues directly. Furthermore, a greater combination of imaging modalities is being investigated, moving beyond the initial scope of fluorophores coupled to MR contrast agent, CT contrast agent, or PET/SPECT radioisotopes.

Multimodality imaging probes have been evaluated in cardiovascular disease, including postinfarct apoptosis, infection, inflammation, and atherosclerosis. These probes have also been applied in labeling cells for long-term tracking of initial distribution, engraftment, and differentiation. The development of theranostics, including therapeutic agents within the design of multimodality imaging probes, has shown promise for targeted delivery of diverse treatment in cardiovascular disease, targeting angiogenesis and inflammation, and providing a simultaneous means for evaluation of disease diagnosis, prognosis, and regression to therapy.

11.10 CONCLUSIONS

Cardiovascular disease is particularly amenable to multimodality imaging approaches, wherein specific alterations of myocardial or vascular physiology can be associated with specific changes in morphology, as in postinfarct ventricular remodeling and vulnerable plaque in atherosclerosis. An array of targeting molecules has already been conjugated to nanoparticles and liposomes, allowing for a more focused imaging approaches than with contrast agent or some radionuclides alone. The development of specific dual- and triple-modality probes have inherent benefits to the research community, whereby the specificity and selectivity of the imaging probe can be more readily confirmed by postmortem fluorescence microscopy. Moreover, these particles are an attractive means to deliver targeted treatment, whether drug, gene, molecule, or cell based, and the conjugation with imaging agent(s) allow for direct monitoring of therapeutic efficacy. The combination of multiple imaging functionalities may translate to greater clinical efficiency, taking better advantage of the fused instrumentation to increase the accuracy of a single diagnostic imaging study. As the chemistry catches up with the technology, the true power and benefit of fused instrumentation can be more fully realized.

REFERENCES

Al-Jamal WT, and Kostarelos K. 2007. Liposome-nanoparticle hybrids for multimodal diagnostic and therapeutic applications. *Nanomedicine* 2:85–98.

Almer G, Wernig K, Saba-Lepek M et al. 2011. Adiponectin-coated nanoparticles for enhanced imaging of atherosclerotic plaques. *Int J Nanomed* 6:1279–1290.

Andree HA, Reutelingsperger CP, Hauptmann R et al. 1990. Binding of vascular anticoagulant alpha (vac alpha) to planar phospholipid bilayers. *J Biol Chem* 265:4923–4928.

Banerjee SS, and Chen DH. 2009. A multifunctional magnetic nanocarrier bearing fluorescent dye for targeted drug delivery by enhanced two-photon triggered release. *Nanotechnology* 20:185103.

Bauwens M, De Saint-Hubert M, Devos E et al. 2011. Site-specific 68ga-labeled annexin a5 as a PET imaging agent for apoptosis. *Nucl Med Biol* 38:381–392.

Bengel FM, Higuchi T, Javadi MS, Lautamaki R. 2009. Cardiac positron emission tomography. *J Am Coll Cardiol* 54:1–15.

Bhavane R, Badea C, Ghaghada KB et al. 2013. Dual-energy computed tomography imaging of atherosclerotic plaques in a mouse model using a liposomal-iodine nanoparticle contrast agent. *Circ Cardiovasc Imaging* 6:285–294.

Boerman OC, Storm G, Oyen WJ et al. 1995. Sterically stabilized liposomes labeled with indium-111 to image focal infection. *J Nucl Med* 36:16391644.

Boswell CA, Eck PK, Regino CA et al. 2008. Synthesis, characterization, and biological evaluation of integrin alphavbeta3-targeted PAMAM dendrimers. *Mol Pharm* 5:527–539.

Brekke C, Williams SC, Price J, Thorsen F, Modo M. 2007. Cellular multiparametric MRI of neural stem cell therapy in a rat glioma model. *Neuroimage* 37:769–782.

Briley-Saebo KC, Nguyen TH, Saeboe AM et al. 2012. In vivo detection of oxidation-specific epitopes in atherosclerotic lesions using biocompatible manganese molecular magnetic imaging probes. *J Am Coll Cardiol* 59:616–626.

Briley-Saebo KC, Shaw PX, Mulder WJ et al. 2008. Targeted molecular probes for imaging atherosclerotic lesions with magnetic resonance using antibodies that recognize oxidation-specific epitopes. *Circulation* 117:3206–3215.

Cai W, Chen K, Li ZB, Gambhir SS, Chen X. 2007. Dual-function probe for PET and near-infrared fluorescence imaging of tumor vasculature. *J Nucl Med* 48:1862–1870.

Che J, Doubrovin M, Serganova I et al. 2005. hNIS-IRES-eGFP dual reporter gene imaging. *Mol Imaging* 4:128–136.

Chen K, Li ZB, Wang H, Cai W, Chen X. 2008. Dual-modality optical and positron emission tomography imaging of vascular endothelial growth factor receptor on tumor vasculature using quantum dots. *Eur J Nucl Med Mol Imaging* 35:2235–2244.

Chen W, Cormode DP, Vengrenyuk Y et al. 2013. Collagen-specific peptide conjugated HDL nanoparticles as MRI contrast agent to evaluate compositional changes in atherosclerotic plaque regression. *JACC Cardiovasc Imaging* 6:373–384.

Chen X, Wong R, Khalidov I et al. 2011. Inflamed leukocyte-mimetic nanoparticles for molecular imaging of inflammation. *Biomaterials* 32:7651–7661.

Choi JS, Park JC, Nah H et al. 2008. A hybrid nanoparticle probe for dual-modality positron emission tomography and magnetic resonance imaging. *Angew Chem Int Ed Engl* 47:6259–6262.

Cormode DP, Skajaa T, van Schooneveld MM et al. 2008. Nanocrystal core high-density lipoproteins: A multimodality contrast agent platform. *Nano Lett* 8:3715–3723.

Daadi MM, Hu S, Klausner J et al. 2013. Imaging neural stem cell graft-induced structural repair in stroke. *Cell Transplant* 22:881–892.

Dafni H, Gilead A, Nevo N et al. 2003. Modulation of the pharmacokinetics of macromolecular contrast material by avidin chase: MRI, optical, and inductively coupled plasma mass spectrometry tracking of triply labeled albumin. *Magn Reson Med* 50:904–914.

Dellinger A, Olson J, Link K et al. 2013. Functionalization of gadolinium metallofullerenes for detecting atherosclerotic plaque lesions by cardiovascular magnetic resonance. *J Cardiovasc Magn Reson* 15:7.

Devaraj NK, Keliher EJ, Thurber GM, Nahrendorf M, Weissleder R. 2009. 18F labeled nanoparticles for in vivo PET-CT imaging. *Bioconjug Chem* 20:397–401.

Dirksen A, Langereis S, de Waal BF et al. 2004. Design and synthesis of a bimodal target-specific contrast agent for angiogenesis. *Org Lett* 6:4857–4860.

Doiron AL, Homan KA, Emelianov S, Brannon-Peppas L. 2009. Poly(lactic-co-glycolic) acid as a carrier for imaging contrast agents. *Pharm Res* 26:674–682.

Edwards WB, Akers WJ, Ye Y et al. 2009. Multimodal imaging of integrin receptor-positive tumors by bioluminescence, fluorescence, gamma scintigraphy, and single-photon emission computed tomography using a cyclic RGD peptide labeled with a near-infrared fluorescent dye and a radionuclide. *Mol Imaging* 8:101–110.

Edwards WB, Xu B, Akers W et al. 2008. Agonist-antagonist dilemma in molecular imaging: Evaluation of a monomolecular multimodal imaging agent for the somatostatin receptor. *Bioconjug Chem* 19:192–200.

Frias JC, Williams KJ, Fisher EA, Fayad ZA. 2004. Recombinant HDL-like nanoparticles: A specific contrast agent for MRI of atherosclerotic plaques. *J Am Chem Soc* 126:16316–16317.

Fuster V, Fayad ZA, Moreno PR et al. 2005a. Atherothrombosis and high-risk plaque—Part II: Approaches by noninvasive computed tomographic/magnetic resonance imaging. *J Am Coll Cardiol* 46:1209–1218.

Fuster V, Moreno PR, Fayad ZA, Corti R, Badimon JJ. 2005b. Atherothrombosis and high-risk plaque—Part I: Evolving concepts. *J Am Coll Cardiol* 46:937–954.

Gallagher JJ, Tekoriute R, O'Reilly JA et al. 2009. Bimodal magnetic-fluorescent nanostructures for biomedical applications. *J Mater Chem* 19:4081–4084.

Gunn J, Wallen H, Veiseh O et al. 2008. A multimodal targeting nanoparticle for selectively labeling t cells. *Small* 4:712–715.

Guo Y, Chen W, Wang W et al. 2012. Simultaneous diagnosis and gene therapy of immuno-rejection in rat allogeneic heart transplantation model using a T-cell-targeted theranostic nanosystem. *ACS Nano* 6:10646–10657.

Ipe BI, Lehnig M, Niemeyer CM. 2005. On the generation of free radical species from quantum dots. *Small* 1:706–709.

Janjic JM, Srinivas M, Kadayakkara DK, Ahrens ET. 2008. Self-delivering nanoemulsions for dual fluorine-19 MRI and fluorescence detection. *J Am Chem Soc* 130:2832–2841.

Kamaly N, Kalber T, Ahmad A et al. 2008. Bimodal paramagnetic and fluorescent liposomes for cellular and tumor magnetic resonance imaging. *Bioconjug Chem* 19:118–129.

Kelly KA, Allport JR, Tsourkas A et al. 2005. Detection of vascular adhesion molecule-1 expression using a novel multimodal nanoparticle. *Circ Res* 96:327–336.

Kim JS, Rieter WJ, Taylor KM, An H, Lin W. 2007. Self-assembled hybrid nanoparticles for cancer-specific multimodal imaging. *J Am Chem Soc* 129:8962–8963.

Kitagawa T, Kosuge H, Uchida M et al. 2012. RGD-conjugated human ferritin nanoparticles for imaging vascular inflammation and angiogenesis in experimental carotid and aortic disease. *Mol Imaging Biol* 14:315–324.

Kumar R, Roy I, Ohulchanskky TY et al. 2010. In vivo biodistribution and clearance studies using multimodal organically modified silica nanoparticles. *ACS Nano* 4:699–708.

Laufer EM, Winkens HM, Corsten MF et al. 2009. PET and SPECT imaging of apoptosis in vulnerable atherosclerotic plaques with radiolabeled annexin a5. *Q J Nucl Med Mol Imaging* 53:26–34.

Lee JH, Jun Y, Yeon SI, Shin JS, Cheon J. 2006. Dual-mode nanoparticle probes for high-performance magnetic resonance and fluorescence imaging of neuroblastoma. *Angew Chem Int Ed Engl* 45: 8160–8162.

Lijowski M, Caruthers S, Hu G et al. 2009. High sensitivity: High-resolution SPECT-CT/MR molecular imaging of angiogenesis in the Vx2 model. *Invest Radiol* 44:15–22.

Liu Y, Pressly ED, Abendschein DR et al. 2011. Targeting angiogenesis using a c-type atrial natriuretic factor-conjugated nanoprobe and PET. *J Nucl Med* 52:1956–1963.

Louie A. 2010. Multimodality imaging probes: Design and challenges. *Chem Rev* 110:3146–3195.

Lovric J, Cho SJ, Winnik FM, Maysinger D. 2005. Unmodified cadmium telluride quantum dots induce reactive oxygen species formation leading to multiple organelle damage and cell death. *Chem Biol* 12:1227–1234.

Lu PL, Chen YC, Ou TW et al. 2011. Multifunctional hollow nanoparticles based on graft-diblock copolymers for doxorubicin delivery. *Biomaterials* 32:2213–2221.

Majmudar MD, Keliher EJ, Heidt T et al. 2013. Monocyte-directed RNAI targeting CCR2 improves infarct healing in atherosclerosis-prone mice. *Circulation* 127:2038–2046.

Marrache S, and Dhar S. 2013. Biodegradable synthetic high-density lipoprotein nanoparticles for atherosclerosis. *Proc Natl Acad Sci U S A* 110:9445–9450.

Maxwell DJ, Bonde J, Hess DA et al. 2008. Fluorophore-conjugated iron oxide nanoparticle labeling and analysis of engrafting human hematopoietic stem cells. *Stem Cells* 26:517–524.

Mikhaylova M, Stasinopoulos I, Kato Y, Artemov D, Bhujwalla ZM. 2009. Imaging of cationic multifunctional liposome-mediated delivery of Cox-2 siRNA. *Cancer Gene Ther* 16:217–226.

Mishra A, Pfeuffer J, Mishra R et al. 2006. A new class of Gd-based DO3A-ethylamine-derived targeted contrast agents for MR and optical imaging. *Bioconjug Chem* 17:773–780.

Mitchell N, Kalber TL, Cooper MS et al. 2013. Incorporation of paramagnetic, fluorescent and PET/SPECT contrast agents into liposomes for multimodal imaging. *Biomaterials* 34:1179–1192.

Mizukami S, Takikawa R, Sugihara F, Shirakawa M, Kikuchi K. 2009. Dual-function probe to detect protease activity for fluorescence measurement and 19F MRI. *Angew Chem Int Ed Engl* 48:3641–3643.

Modo M, Mellodew K, Cash D et al. 2004a. Mapping transplanted stem cell migration after a stroke: A serial, in vivo magnetic resonance imaging study. *Neuroimage* 21:311–317.

Modo M, Roberts TJ, Sandhu JK, Williams SC. 2004b. In vivo monitoring of cellular transplants by magnetic resonance imaging and positron emission tomography. *Expert Opin Biol Ther* 4:145–155.

Moon H, Park HE, Kang J et al. 2012. Noninvasive assessment of myocardial inflammation by cardiovascular magnetic resonance in a rat model of experimental autoimmune myocarditis. *Circulation* 125:2603–2612.

Mulder WJ, Koole R, Brandwijk RJ et al. 2006. Quantum dots with a paramagnetic coating as a bimodal molecular imaging probe. *Nano Lett* 6:1–6.

Nahrendorf M, Jaffer FA, Kelly KA et al. 2006. Noninvasive vascular cell adhesion molecule-1 imaging identifies inflammatory activation of cells in atherosclerosis. *Circulation* 114:1504–1511.

Nahrendorf M, Keliher E, Marinelli B et al. 2011. Detection of macrophages in aortic aneurysms by nanoparticle positron emission tomography-computed tomography. *Arterioscler Thromb Vasc Biol* 31:750–757.

Nahrendorf M, Zhang H, Hembrador S et al. 2008. Nanoparticle PET-CT imaging of macrophages in inflammatory atherosclerosis. *Circulation* 117:379–387.

Natarajan A, Gruettner C, Ivkov R et al. 2008. Nanoferrite particle based radioimmunonanoparticles: Binding affinity and in vivo pharmacokinetics. *Bioconjug Chem* 19:1211–1218.

Nekolla SG, Martinez-Moeller A, Saraste A. 2009. PET and MRI in cardiac imaging: From validation studies to integrated applications. *Eur J Nucl Med Mol Imaging* 36 Suppl 1:S121–S130.

Nensa F, Poeppel TD, Beiderwellen K et al. 2013. Hybrid PET/MR imaging of the heart: Feasibility and initial results. *Radiology* 268:366–373.

Oostendorp M, Douma K, Hackeng TM et al. 2008. Quantitative molecular magnetic resonance imaging of tumor angiogenesis using cNGR-labeled paramagnetic quantum dots. *Cancer Res* 68:7676–7683.

Pandey SK, Gryshuk AL, Sajjad M et al. 2005. Multimodality agents for tumor imaging (PET, fluorescence) and photodynamic therapy. A possible "see and treat" approach. *J Med Chem* 48:6286–6295.

Park JH, von Maltzahn G, Ruoslahti E, Bhatia SN, Sailor MJ. 2008. Micellar hybrid nanoparticles for simultaneous magnetofluorescent imaging and drug delivery. *Angew Chem Int Ed Engl* 47:7284–7288.

Pei Z, Lan X, Cheng Z et al. 2012. A multimodality reporter gene for monitoring transplanted stem cells. *Nucl Med Biol* 39:813–820.

Pisani E, Tsapis N, Galaz B et al. 2008. Perfluorooctyl bromide polymeric capsules as dual contrast agents for ultrasonography and magnetic resonance imaging. *Adv Funct Mater* 18:2963–2971.

Pittet MJ, Swirski FK, Reynolds F, Josephson L, Weissleder R. 2006. Labeling of immune cells for in vivo imaging using magnetofluorescent nanoparticles. *Nat Protoc* 1:73–79.

Ponomarev V, Doubrovin M, Serganova I et al. 2004. A novel triple-modality reporter gene for whole-body fluorescent, bioluminescent, and nuclear noninvasive imaging. *Eur J Nucl Med Mol Imaging* 31:740–751.

Qin C, Cheng K, Chen K et al. 2013. Tyrosinase as a multifunctional reporter gene for photoacoustic/MRI/PET triple modality molecular imaging. *Sci Rep* 3:1490.

Rahmer J, Antonelli A, Sfara C et al. 2013. Nanoparticle encapsulation in red blood cells enables blood-pool magnetic particle imaging hours after injection. *Phys Med Biol* 58:3965–3977.

Ray P, Tsien R, Gambhir SS. 2007. Construction and validation of improved triple fusion reporter gene vectors for molecular imaging of living subjects. *Cancer Res* 67:3085–3093.

Riegler J, Liew A, Hynes SO et al. 2013. Superparamagnetic iron oxide nanoparticle targeting of MSCs in vascular injury. *Biomaterials* 34:1987–1994.

Rieter WJ, Kim JS, Taylor KML et al. 2007a. Hybrid silica nanoparticles for multimodal imaging. *Angew Chem Int Ed Engl* 46:3680–3682.

Rieter WJ, Taylor KM, Lin W. 2007b. Surface modification and functionalization of nanoscale metal-organic frameworks for controlled release and luminescence sensing. *J Am Chem Soc* 129:9852–9853.

Roullier V, Grasset F, Boulmedais F et al. 2008. Small bioactivated magnetic quantum dot micelles. *Chem Mater* 20:6657–6665.

Sabal G, Menzel EJ, Sinzinger H. 2006. Comparative binding of 125I- and 99mTc-labeled native and glycated low-density lipoprotein to human microvascular endothelial cells-potential for atherosclerosis imaging? *J Receptors Signal Transduct* 26:693–707.

Santra S, Bagwe RP, Dutta D et al. 2005. Synthesis and characterization of fluorescent, radio-opaque, and paramagneitc silican nanoparticles for multimodal bioimaging applications. *Adv Mater* 17:2165–2169.

Sarin H, Kanevsky AS, Wu H et al. 2008. Effective transvascular delivery of nanoparticles across the blood-brain tumor barrier into malignant glioma cells. *J Transl Med* 6:80.

Schipper ML, Cheng Z, Lee SW et al. 2007. MicroPET-based biodistribution of quantum dots in living mice. *J Nucl Med* 48:1511–1518.

Schmieder AH, Caruthers SD, Zhang H et al. 2008. Three-dimensional MR mapping of angiogenesis with alpha5beta1(alpha nu beta3)-targeted theranostic nanoparticles in the MDA-MB-435 xenograft mouse model. *Faseb J* 22:4179–4189.

Seo SB, Yang J, Lee ES et al. 2008. Nanohybrids via a polycation-based nanoemulsion method for dual-mode detection of human mesenchymal cells. *J Mater Chem* 18:4402–4407.

Sharma P, Brown SC, Bengtsson N et al. 2008. Gold-speckled multimodal nanoparticles for noninvasive bioimaging. *Chem Mater* 20:6087–6094.

Shi Z, Neoh KG, Kang ET, Shuter B, Wang SC. 2010. Bifunctional Eu(3+)-doped Gd(2)o(3) nanoparticles as a luminescent and t(1) contrast agent for stem cell labeling. *Contrast Media Mol Imaging* 5:105–111.

Skajaa T, Cormode DP, Falk E et al. 2010. High-density lipoprotein-based contrast agents for multimodal imaging of atherosclerosis. *Arterioscler Thromb Vasc Biol* 30:169–176.

Skajaa T, Cormode DP, Jarzyna PA et al. 2011. The biological properties of iron oxide core high-density lipoprotein in experimental atherosclerosis. *Biomaterials* 32:206–213.

Sun N, Lee A, Wu JC. 2009. Long term non-invasive imaging of embryonic stem cells using reporter genes. *Nat Protoc* 4:1192–1201.

Sun X, Annala AJ, Yaghoubi SS et al. 2001. Quantitative imaging of gene induction in living animals. *Gene Ther* 8:1572–1579.

Takahama H, Shigematsu H, Asai T et al. 2013. Liposomal amiodarone augments anti-arrhythmic effects and reduces hemodynamic adverse effects in an ischemia/reperfusion rat model. *Cardiovasc Drugs Ther* 27:125–132.

Talanov VS, Regino CA, Kobayashi H et al. 2006. Dendrimer-based nanoprobe for dual modality magnetic resonance and fluorescence imaging. *Nano Lett* 6:1459–1463.

Tan WB, and Zhang Y. 2007. Multi-functional chitosan nanoparticles encapsulating quantum dots and Gd-DTPA as imaging probes for bio-applications. *J Nanosci Nanotechnol* 7:2389–2393.

Terrovitis JV, Smith RR, Marban E. 2010. Assessment and optimization of cell engraftment after transplantation into the heart. *Circ Res* 106:479–494.

Thakor AS, Luong R, Paulmurugan R et al. 2011. The fate and toxicity of Raman-active silica-gold nanoparticles in mice. *Sci Transl Med* 3:79ra33.

Tu C, and Louie AY. 2007. Photochromically-controlled, reversibly-activated MRI and optical contrast agent. *Chem Commun (Camb)* 1331–1333.

Uzgiris EE, Sood A, Bove K et al. 2006. A multimodal contrast agent for preoperative MR imaging and intra-operative tumor margin delineation. *Technol Cancer Res Treat* 5:301–309.

van Tilborg GA, Mulder WJ, Deckers N et al. 2006. Annexin a5-functionalized bimodal lipid-based contrast agents for the detection of apoptosis. *Bioconjug Chem* 17:741–749.

Vuu K, Xie J, McDonald MA et al. 2005. Gadolinium-rhodamine nanoparticles for cell labeling and tracking via magnetic resonance and optical imaging. *Bioconjug Chem* 16:995–999.

Wang MW, Wang F, Zheng YJ et al. 2013. An in vivo molecular imaging probe (18)F-Annexin B1 for apoptosis detection by PET/CT: Preparation and preliminary evaluation. *Apoptosis* 18:238–247.

Wen S, Li K, Cai H et al. 2013. Multifunctional dendrimer-entrapped gold nanoparticles for dual mode CT/MR imaging applications. *Biomaterials* 34:1570–1580.

Winter PM, Caruthers SD, Zhang H et al. 2008. Antiangiogenic synergism of integrin-targeted fumagillin nanoparticles and atorvastatin in atherosclerosis. *JACC Cardiovasc Imaging* 1:624–634.

Winter PM, Neubauer AM, Caruthers SD et al. 2006. Endothelial alpha(v)beta3 integrin-targeted fumagillin nanoparticles inhibit angiogenesis in atherosclerosis. *Arterioscler Thromb Vasc Biol* 26:2103–2109.

Woolley R, Roy S, Prendergast U et al. 2013. From particle to platelet: Optimization of a stable, high brightness fluorescent nanoparticle based cell detection platform. *Nanomedicine* 9:540–549.

Xu H, Regino CA, Koyama Y et al. 2007. Preparation and preliminary evaluation of a biotin-targeted, lectin-targeted dendrimer-based probe for dual-modality magnetic resonance and fluorescence imaging. *Bioconjug Chem* 18:1474–1482.

Yang J, Lee J, Kang J et al. 2008. Magnetic sensitivity enhanced novel fluorescent magnetic silica nanoparticles for biomedical applications. *Nanotechnology* 19:075610.

Yilmaz A, Dengler MA, van der Kuip H et al. 2013. Imaging of myocardial infarction using ultrasmall superparamagnetic iron oxide nanoparticles: A human study using a multi-parametric cardiovascular magnetic resonance imaging approach. *Eur Heart J* 34:462–475.

Yilmaz A, Rosch S, Yildiz H, Klumpp S, Sechtem U. 2012. First multiparametric cardiovascular magnetic resonance study using ultrasmall superparamagnetic iron oxide nanoparticles in a patient with acute myocardial infarction: New vistas for the clinical application of ultrasmall superparamagnetic iron oxide. *Circulation* 126:1932–1934.

Yuan X, Li H, Yuan Y. 2006. Preparation of cholesterol-modified chitosan self-aggregated nanoparticles for delivery of drugs to ocular surface. *Carbohydr Polym* 65:337.

Zhang SJ, and Wu JC. 2007. Comparison of imaging techniques for tracking cardiac stem cell therapy. *J Nucl Med* 48:1916–1919.

Zhang X, Brynda M, Britt RD et al. 2007. Synthesis and characterization of manganese-doped silicon nanoparticles: Bifunctional paramagnetic-optical nanomaterial. *J Am Chem Soc* 129:10668–10669.

Zheng J, Liu J, Dunne M, Jaffray DA, Allen C. 2007. In vivo performance of a liposomal vascular contrast agent for CT and MR-based image guidance applications. *Pharm Res* 24:1193–1201.

Zielhuis SW, Seppenwoolde JH, Mateus VA et al. 2006. Lanthanide-loaded liposomes for multimodality imaging and therapy. *Cancer Biother Radiopharm* 21:520–527.

QUANTITATIVE ANALYSES AND CASE ILLUSTRATIONS OF HYBRID IMAGING

Recent developments and applications of hybrid imaging techniques

PIOTR J. SLOMKA, DANIEL S. BERMAN, AND GUIDO GERMANO

12.1 INTRODUCTION

Hybrid imaging methods, which combine complementary strengths of different modalities, allow a combination of anatomy with physiology. This approach has been very successful clinically in oncological imaging. Cardiac hybrid imaging in particular presents some unique challenges for hybrid imaging, since the heart is a moving organ. The motion of the heart is typically averaged during nuclear imaging (positron emission tomography [PET] and single-photon emission computed tomography [SPECT]), but it is captured at a specific phase by the faster anatomical modalities such as CT or MRI. This represents a challenge for cardiac imaging with hybrid devices such as PET/computed tomography (CT), SPECT/CT or PET/ magnetic resonance imaging (MRI), which are designed for oncological applications. Therefore, hybrid imaging in cardiovascular applications can sometimes be more easily accomplished with software techniques, where data are obtained from separate standalone scans. Nevertheless, cardiovascular imaging is also possible with dedicated hybrid scanners. The new instrumentation developments in hybrid imaging are covered in other chapters of this book. In this chapter, we cover recent developments in cardiovascular hybrid imaging with a focus of the software techniques such as improved image resolution of PET scanners and software registration techniques for contrast and noncontrast CT and MRI with SPECT and PET data. We also describe some recent methods for quantification of hybrid image data. In addition, we discuss various clinical and research applications of hybrid cardiovascular imaging and methods for the combination of the coronary CT angiography (CTA) with PET and SPECT for improved diagnosis of the coronary artery disease (CAD) and the combination of the nuclear data with calcium scoring. We also cover emerging applications of vascular imaging with hybrid technology and highlight some possibilities for hybrid cardiac PET/MRI applications.

12.2 RECENT TECHNIQUES

12.2.1 HIGH-RESOLUTION PET IMAGING FOR HYBRID PET/CT

One of the difficulties in hybrid imaging is the mismatch of the image resolution between modalities, with much lower resolution of PET or SPECT as compared to that of CT or magnetic resonance (MR) images. Another issue is patient motion, resulting in image blur. This hampers precise alignment of the organ structures. Several recent technical developments may allow significant increase of PET resolution and, consequently, precise fusion in hybrid imaging can be obtained.

12.2.2 ENHANCED PET RESOLUTION

A recent technical advance of hybrid PET/CT, which is important for cardiac imaging, is to use the three-dimensional (3-D) modeling of scanner-specific point spread function (PSF) maps in 3-D iterative reconstruction. Similar SPECT resolution recovery methods have also been developed. This allows achieving the apparent tomographic resolution of 2 mm and also significantly improves the image contrast (Panin et al. 2006). Most PET vendors now include this approach as an option in the reconstruction with hybrid PET/CT scanners (Jakoby et al. 2011). In particular, the combination of the time-of-flight and correction of the PSF seems to result in the best improvement of image quality (Bettinardi et al. 2011; Akamatsu et al. 2012). The CT images obtained with the CT component of the hybrid imaging are used in these corrections. Cardiac PET imaging with the correction of the PSF has been reported, demonstrating better definition of subtle defects (Le Meunier et al. 2010). An example of improved image resolution for 18F-fluorodeoxyglucose (^{18}F-FDG) imaging with the use of resolution recovery is shown in Figure 12.1 (Le Meunier et al. 2010).

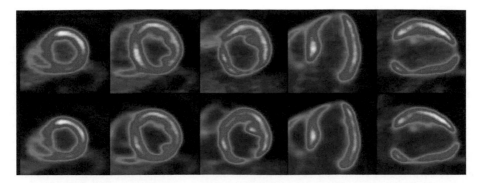

Figure 12.1 Standard iterative 3D OSEM reconstruction (top) and improved high-definition reconstruction incorporating resolution recovery (bottom) in the ^{18}F-FDG viability PET scan. (With kind permission from Springer Science+Business Media: *J. Nucl. Cardiol.*, Enhanced definition PET for cardiac imaging, 17, 2010, 424, Le Meunier L, Slomka PJ, Dey D, Ramesh A, Thomson LE, Hayes SW, Figure 5.)

12.2.3 RESPIRATORY GATING FOR HYBRID PET/CT

It may be desirable to correct for the cardiac PET image blurring and CT attenuation correction misregistration artifacts being caused by the movement of the heart during breathing by respiratory gating (Büther et al. 2009; Martinez-Möller et al. 2007). Breathing motion can be detected by a video camera, registering the movement of a marker on the patient's abdomen, or by an inductive respiration monitor with an elastic belt around the patient's chest. As described in Chapter 14 of this book, these additional signals are used as triggers to sort the list-mode data into different respiratory phases (similar to electrocardiogram [ECG] gating) before image reconstruction. Respiratory gating can be also used for the correction of the upward creep of the heart during exercise-stress imaging, as the heart rate slows down and the body relaxes over time. Dual respiratory/cardiac gating is also possible with additional hardware for respiratory gating. The principle of respiratory gating for PET scan is shown in Figure 12.2. Most clinical PET centers currently do not use respiratory gating in routine cardiac PET/CT imaging, since it adds complexity to the acquisition protocols and requires additional reconstruction time. However, this correction may be important if precise coregistration of PET and anatomical scan is required in hybrid imaging.

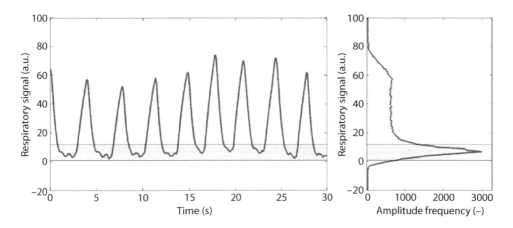

Figure 12.2 The principle of PET respiratory gating technique. The histogram (right) shows the breathing amplitudes for the entire 24-min list-mode acquisition. The optimal gating method selects the narrowest bandwidth (shaded area) containing 35% of the respiratory signal.

12.2.4 Motion-frozen techniques

Even if the resolution-recovery reconstruction methods are utilized, cardiac and respiratory motions degrade the image quality by adding unwanted blur to static perfusion images. However, if respiratory or cardiac gating is used, the count statistics in the individual gates can be significantly reduced. To avoid this, myocardial wall in ECG-gated PET or SPECT images can be tracked and all the cardiac phases can be warped into one motion-free image by motion-frozen algorithm (Slomka et al. 2004). The algorithm computes displacement vectors between each cardiac phase and utilizes those vectors to morph all the images to the end-diastolic position. The result is an image free of cardiac motion blurring but with the noise level similar to that in static summed images. We have demonstrated that this technique can improve cardiac PET resolution beyond the resolution obtained by high-definition reconstruction with resolution recovery (Le Meunier et al. 2011) and may allow better definition of subtle myocardial defects (Figure 12.3). In a preliminary development, this technique was extended to dual-gated (cardiac/respiratory) images obtained with hybrid PET/CT (Slomka et al. 2015b). Dual motion-frozen technique allows accurate superimposition of cardiac PET and CTA imaging due to the matching cardiac and respiratory phase and improved PET resolution (Figure 12.4). Although such enhancement of cardiac PET or SPECT is not routinely performed, they will likely to be considered essential in enhancing the quality of hybrid imaging. Further validation of these algorithms will be required before they are routinely applied in clinical practice.

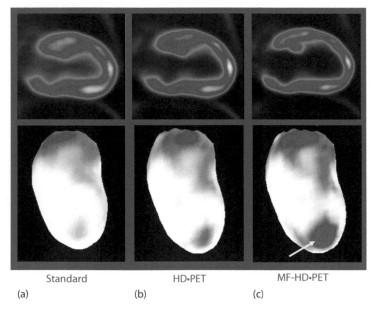

Standard	HD·PET	MF-HD·PET
(a)	(b)	(c)

Figure 12.3 Motion-frozen (MF) technique. Example of an 84-year-old patient (W = 76 kg [168 lbs], BMI = 27.1) with a history of myocardial infarction. The patient exhibits a nonreversible perfusion defect in the apical lateral region of the myocardium (left SPECT perfusion images) as well as a moderate hypokinesis in the same region of the myocardium. The figure shows an example of the increase of maximum left ventricular (LV) counts to defect contrast with cardiac MF image (Column c) in the horizontal long axis view (top row) and in the 3-D rendering (bottom row). The apical lateral metabolic defect (yellow arrow) was not seen on standard 2-D OSEM reconstruction (Column a) and not clearly on high-definition reconstruction (Column b) images. HD, high definition; OSEM, ordered subset expectation maximization. (With kind permission from Springer Science+Business Media: *J Nucl Cardiol*, Motion frozen ^{18}F-FDG cardiac PET, 18, 2011, 264, Le Meunier L, Slomka PJ, Dey D, Ramesh A, Thomson LE, Hayes SW et al. 2011. Le Meunier et al., Figure 4.)

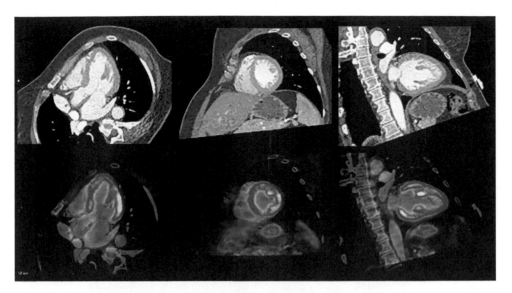

Figure 12.4 Dual-motion frozen hybrid ^{18}F-FDG/CTA scan, corrected for the cardiac and respiratory motion combined with CTA demonstrating similar wall thickness on both PET and CTA.

12.3 AUTOMATED IMAGE REGISTRATION TECHNIQUES FOR HYBRID IMAGING

12.3.1 REGISTRATION OF SPECT WITH CORONARY CTA

It has been suggested that hybrid visual analysis of SPECT/PET and coronary CTA data can synergistically improve the diagnostic value of sequential combined imaging and reduce the rate of equivocal studies (Gaemperli et al. 2007; Santana et al. 2009). The visual analysis has typically utilized manual corrections. However, the manual interactive alignment of the images from two modalities introduces subjective errors, takes a substantial amount of time, and limits the practical clinical use of such tools. Several methods for fully automatic alignment of SPECT/PET and CTA acquired on standalone modalities have been proposed to allow hybrid imaging when data are acquired on separate scanners.

Our group aimed to develop a practical tool for rapid automatic coregistration and visualization, enabling hybrid imaging of coronary CTA and myocardial perfusion SPECT (MPS) obtained from standalone scanners. To this end, we have developed a fully automated registration technique in which SPECT images are fused with standalone coronary CTA (Slomka et al. 2009). The primary difficulty in the accurate registration of myocardial perfusion imaging with coronary CTA is that disparate anatomic features appear prominent on MPS and CT images. Therefore, standard automated registration procedures (Danad et al. 2013) are prone to large errors. To provide a robust registration tool for hybrid imaging, we developed a method whereby we register presegmented MPS volumes, utilizing the left ventricular (LV) segmentation algorithm of Quantitative Gated SPECT (QGS) package (Woo et al. 2009). In addition, to match cardiac phases between MPS and coronary CTA, we performed alignment of coronary CTA reconstructed in diastolic phases with segmented "motion-frozen" slices of SPECT created from ECG-gated cardiac images. The "motion-frozen" images allow for higher image resolution of the SPECT data. We utilize geometric features from a reliable segmentation of cardiac-gated MPS volumes, where regions of myocardium and blood pool are extracted and used as an anatomical mask to deemphasize the inhomogeneities of intensity distribution caused by perfusion defects and physiological variations. A multiresolution approach is then employed to represent coarse-to-fine details of both volumes. The extracted voxels from each level are aligned using a similarity measure with a piecewise constant image model and then minimized using a gradient descent method.

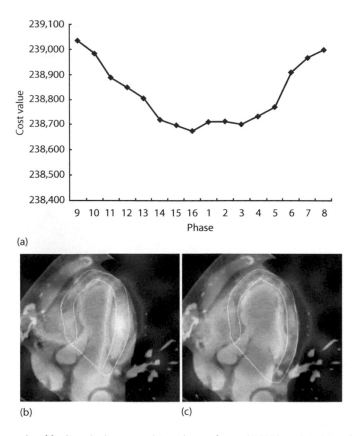

(a)

(b)   (c)

Figure 12.5 An example of finding the best matching phase of gated MPS by minimizing the registration cost functional: (a) plot of cost functional value across different phases of MPS; (b) overlaid registration result of phase 9 MPS on CT perfusion; (c) overlaid registration result of best matching phase—phase 16 with CT perfusion. Note that the best matching MPS phase is visually well correlated with CT. (This research was originally published in *Medical Physics,* Geometric feature-based multimodal image registration of contrast-enhanced cardiac CT with gated myocardial perfusion SPECT, Volume 36, 2009, 5467–5479, Woo et al., Figure 4.)

Accurate registration of the data from SPECT/PET and coronary CTA is complicated by the potential mismatch in cardiac image phase. It is possible that the best cardiac phase for the CTA is not the diastolic phase. In a further refinement of the image registration, our group has proposed an improved automated registration scheme, with correction for the cardiac phase disparities (Woo et al. 2009). We augmented the approach described previously with constrained nonlinear registration of ECG-gated MPS to adjust for the phase differences by automatic cardiac phase matching between CT and MPS. We show an example of phase matching in Figure 12.5. For phase matching, we incorporate nonlinear registration using thin-plate-spline-based warping. Results of the registration with this algorithm are shown in Figure 12.6. We demonstrated, by experimental results, that the proposed method outperformed traditional registration methods based on normalized mutual information (Radau et al. 2001). Such fully automated methods may facilitate integrated hybrid gated SPECT/PET and cardiac CT analysis for either standalone of combined scanners and may be helpful in resolving borderline cases in clinical practice, while reducing operator time and variability needed for manual adjustments.

12.3.2 CORRECTION OF SPATIAL REGISTRATION FOR HYBRID PET/CTA DATA

Latest PET/CT or SPECT/CT scanners include a high-resolution 64-slice (or higher) CT component; therefore, it is possible to combine, in one scanning session, the high-quality coronary CTA with myocardial

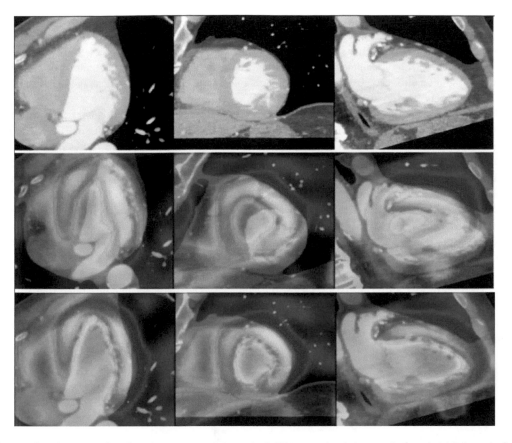

Figure 12.6 An example of registration results: original CT image (top), images before (middle), and after (bottom) registration with phase adjustments are shown. Errors were 2, 1, and 0 mm for translation and 1°, 1°, and 0° for rotation, as compared to visual alignment. (This research was originally published in *Medical Physics*, Geometric feature-based multimodal image registration of contrast-enhanced cardiac CT with gated myocardial perfusion SPECT, Volume 36, 2009, 5467–5479, Woo et al., Figure 6.)

perfusion scan (Danad et al. 2013; Di Carli, Dorbala, and Hachamovitch 2006). Nevertheless, cardiac PET and coronary CTA images obtained on a hybrid PET/CT scanner are not truly simultaneous and are typically acquired in different breathing patterns. This usually results in significant misalignments between PET and CT data and requires software registration.

We have conducted a study to verify the extent of such misalignment during typical cardiac PET/CTA protocol and validated the software registration method to correct the PET/CTA misalignment. In this study, we considered single-session hybrid PET/CT studies of rest/stress ^{13}N-ammonia PET and coronary CTA in 32 consecutive patients. All patients were studied on a whole-body 64-slice PET/CT scanner. Coronary CTA (64-slice) was performed immediately after completion of the rest-stress ^{13}N-ammonia PET protocol, without changing patient position on the imaging table. For coronary CTA, after a timing bolus, ECG-gated helical scanning was performed using ECG-based dose modulation whenever possible, during a single breath-hold (~10 sec). Retrospectively ECG-gated reconstruction of contrast-enhanced data was performed with the use of the following parameters: 0.6 mm slice thickness, 0.3 mm slice increment, 250 mm field of view, 512 × 512 matrix, and a standard cardiovascular filter setting of B30f ("medium smooth") convolution kernel. The cardiac phase with the best image quality was used for further analysis.

Automated registration of PET LV surfaces with coronary CTA volumes as described in the previous section was evaluated by comparison to expert manual alignment by two observers. Phase adjustment as described in the previous section (Woo et al. 2009) was not performed in this study. Automated coregistration was performed without any manual intervention. For validation purposes, separate expert manual alignment was performed without knowledge of the automated coregistration results. Two expert observers independently verified the alignment of coronary CTA images with emission PET data. The manual alignment parameters (three translations [x, y, z] and three rotations [xy, xz, yz]) by expert observers were recorded for the comparison with the automated coregistration. The manual alignment process took about 2–3 min for each case. The automated coregistration algorithm was applied and compared to the average visual alignment for the 64 datasets (32 stress and 32 rest).

The frequency and magnitude of the initial misalignment by automated coregistration on rest and stress studies are shown in Figure 12.7. The initial PET-coronary CTA misalignment on the hybrid PET/CT scanner was 27.2 ± 11.8 mm in x, 13.3 ± 11.5 mm in y, and 14.3 ± 9.1 mm in z-axes in rest and was 26.3 ± 10.2 mm in x, 11.1 ± 9.5 mm in y, and 11.7 ± 7.1 mm in z-axes in stress studies. These results were similar to those obtained by the visual readers. The automated coregistration processing time was about 2 seconds for each case. Overlaid PET/coronary CTA trees were superimposed on the 3-D MPS surface based on the automatic registration results between PET and coronary CTA volumes. The coronary tree did not need to be adjusted for the display in any patient, as qualitatively judged by both observers. In Figure 12.8, we show an example of 3-D visualization of coronary vessels superimposed with MPS surfaces before (Figure 12.8a) and after (Figure 12.8b) automated coregistration of PET and CTA obtained on a hybrid scanner. In Figure 12.9, we show an example of the fused images obtained by the automated coregistration of PET with coronary CTA for the same patient. This work demonstrated that software coregistration of coronary CTA and PET obtained on hybrid PET/CT scanners is still necessary due to different breathing protocols with the two different modalities, but it can be performed rapidly and automatically.

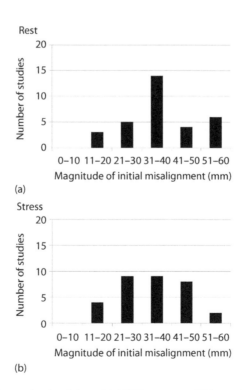

Figure 12.7 Frequency and magnitude of the initial PET-coronary CTA misalignment in rest (a) and stress (b) studies by manual registration. (With kind permission from Springer Science+Business Media: *J Nucl Cardiol*, Automatic alignment of myocardial perfusion PET and 64-slice coronary CT angiography on hybrid PET/CT, 19, 2014, 487, Nakazato R, Dey D, Alexanderson E, Meave A, Jimenez M, Romero E et al., Figure 2.)

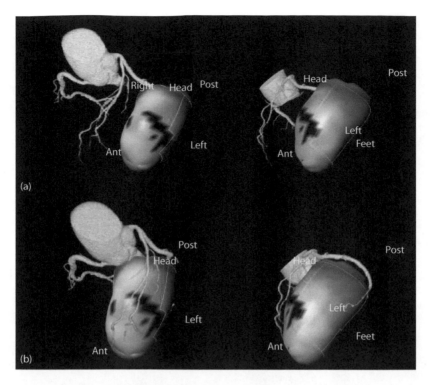

Figure 12.8 3D visualization of coronary vessels superimposed with myocardial perfusion PET ventricular surfaces, before (a) and after (b) CT misalignment correction. (With kind permission from Springer Science+Business Media: *J Nucl Cardiol*, Automatic alignment of myocardial perfusion PET and 64-slice coronary CT angiography on hybrid PET/CT, 19, 2014, 490, Nakazato R, Dey D, Alexanderson E, Meave A, Jimenez M, Romero E et al., Figure 6.)

12.3.3 REGISTRATION OF CT MAPS FOR ATTENUATION CORRECTION ON HYBRID PET/CT

Image registration also plays an important role in the CT attenuation correction on PET/CT scanners. CT-based attenuation correction for myocardial perfusion PET studies on hybrid scanners is challenging because of respiratory motion effects on the PET-CT alignment. Significant differences can occur as compared to traditional standalone PET scanners with photon transmission attenuation correction, despite manual alignment affecting PET image quality (Slomka et al. 2008). Manual alignment techniques have been suggested for the correction of such misalignments (Gould et al. 2007), but they introduce observer variability. In new developments for hybrid imaging, robust automatic registration techniques have been proposed to align CT attenuation maps with PET. Bond et al. (2008) have developed a method for cardiac CT-PET registration where the heart region is identified by multiresolution template matching and followed by cropping. They implemented both rigid and deformable methods for image alignment. In Figure 12.10, we show an example PET case and CT attenuation correction scan, which was originally misregistered due to respiratory motion (Figure 12.10a) and is corrected by the automated rigid registration method (Figure 12.10b). This registration method is now implemented by the PET/CT vendor and has been initially validated for attenuation correction using cardiac calcium scans by Zaidi et al. (2013). The deformable registration is required when calcium scans are used for attenuation correction because of mismatched respiratory and cardiac phases between PET and cardiac coronary calcium scan. Clinical validation of this fully automated CT attenuation scan/PET registration method in a larger population (n = 171) has been performed recently for the ^{82}Rb PET.

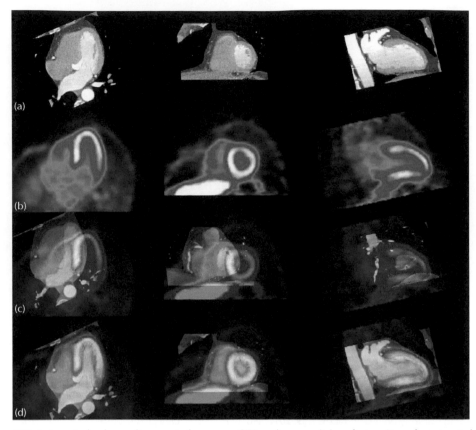

Figure 12.9 Automated volume alignment of coronary CTA and myocardial perfusion. Rows from top to bottom show in multiplanar orientations original coronary CTA images (a), original myocardial perfusion PET images (b), fused unregistered PET and coronary CTA images (c), and same images after automated volume registration (d). Subsequently, 3-D transformation parameters obtained from volume registration can be used to align associated coronary trees with PET ventricular surfaces. (With kind permission from Springer Science+Business Media: *J Nucl Cardiol*, Automatic alignment of myocardial perfusion PET and 64-slice coronary CT angiography on hybrid PET/CT, 19, 2014, 490, Nakazato R, Dey D, Alexanderson E, Meave A, Jimenez M, Romero E et al., Figure 7.)

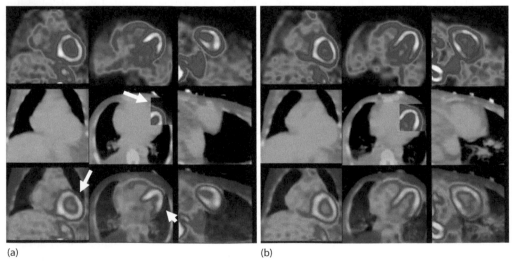

(a)   (b)

Figure 12.10 Misregistered (white arrows) results of PET/CT without software correction (a) and automatically corrected PET/CT alignment with rigid algorithm (b).

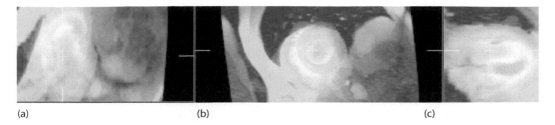

(a)   (b)   (c)

Figure 12.11 Three orthogonal slices: horizontal long axis (a), short axis (b), and vertical long axis (c) of a SPECT study registered and fused with the corresponding cardiac cine MR image. (This research was originally published in *Journal of Magnetic Resonance Imaging*, Automated image registration of gated cardiac single-photon emission computed tomography and magnetic resonance imaging, Volume 29, 2004, 283–290, Aladl et al., Figure 8.)

The authors have demonstrated that the prediction of obstructive disease on angiography is significantly improved by the automatic registration with rigid option, when compared to standard clinical manual registration of CT attenuation correction (CTAC) maps (Slomka et al. 2015a). Although these software techniques are not yet widely used on hybrid imaging equipment, they have great potential to streamline imaging with hybrid technology.

12.3.4 4-D SPECT/MR REGISTRATION

We also investigated the feasibility of the hybrid imaging by standalone gated MR images with gated SPECT acquired in separate scanning sessions. In this case, the full-gated MR as well as gated SPECT datasets are typically available and all the phases can be used for automated registration, to improve the robustness of the method. To this end, we developed a computer algorithm for the 4-D registration of SPECT with MR image data, where cardiac motion is considered as an additional dimension. In our approach, fully automatic, approximate presegmentation to cardiac MR data based on the analysis of motion was performed (Aladl et al. 2004). A novel spatial-temporal preprocessing technique was used as the initial step to eliminate time-invariant voxels on MR images. Subsequently, a 4-D registration was performed with a mutual information technique (Wells et al. 1996), which iteratively adjusted a set of six rigid body parameters to align gated SPECT volume with multiple phases of cardiac MR. In Figure 12.11, we show an example of this standalone MR/SPECT registration.

12.4 RECENT APPLICATIONS OF HYBRID IMAGING

12.4.1 IMPROVED DIAGNOSTIC ACCURACY WITH HYBRID SPECT/PET AND CORONARY CTA

12.4.1.1 VISUAL ANALYSIS OF COMBINED HYBRID IMAGING

Precise 3-D fusion of functional (PET, SPECT) and anatomical (CTA) imaging may allow an accurate assignment of epicardial lesions to the subtended ischemic myocardial territories. Consequently, more accurate detection of disease can be achieved (Danad et al. 2013; Flotats et al. 2011; Gaemperli, Bengel, and Kaufmann 2011; Namdar et al. 2005; Plass et al. 2011). It is thought that the hybrid imaging can play an important role in patients with nonconclusive SPECT or coronary CTA results (Schaap et al. 2013a). Implications for patient treatment have been studied with the conclusion that hybrid SPECT/CT imaging can lead to the similar treatment decisions as the results of invasive angiography (Schaap et al. 2013b). The potential value of hybrid PET/CT imaging was also demonstrated in patients being evaluated for the coronary bypass surgery (Plass et al. 2011). The clinical scenarios, in which such hybrid PET/CT or SPECT/CTA imaging could be applicable, are currently being investigated (Flotats et al. 2011).

The hybrid PET/CTA protocols remain complex and involve additional radiation dose, as well as injection of CT contrast, when compared to MPS alone. Recently, it has been reported that hyperemic (stress) blood flow alone (without rest scan) may be sufficient for detecting obstructive CAD (Danad et al. 2013). Therefore, combining physiologic and anatomic imaging could be potentially accomplished with PET stress-only imaging combined with coronary CTA, which could significantly reduce the overall radiation burden and study time for the patient (Kajander et al. 2009, 2010). Furthermore, prospective gating and other dose-reducing techniques allow acquisition of high quality coronary CTA with doses lower than 1 milliSievert (Achenbach et al. 2011). Thus, while a combination of stress-only PET with low-dose CTA may be feasible, it remains to be seen if such novel hybrid imaging protocols become clinically adopted. A principal obstacle to this application is that it is usually not possible to define clinically which patients would need both tests, before either an MPS or a coronary CTA study is done.

12.4.1.2 CTA-GUIDED QUANTIFICATION OF MYOCARDIAL PERFUSION

Moving one step beyond the subjective visual interpretation of the hybrid imaging data, fused CTA data could be potentially used to improve the accuracy of software quantification for MPS, rather than simply aiding visual analysis. Our group has demonstrated this principle in 35 sequential patients with 64-slice coronary CTA, MPS, and available results of the invasive coronary angiography performed within 3 months (Slomka et al. 2009). Three-dimensional coronary vessels and CTA slices were extracted and fused with quantitative MPS results mapped on the LV surfaces and SPECT vascular regions.

The hybrid quantification scheme was designed as follows. Automatically coregistered CTA images and extracted trees were used to correct the MPS contours and to adjust the standard vascular region definitions for MPS quantification. Fused images allow overlaying default vascular territory boundaries on the 3-D ventricular MPS surfaces, with color-coded perfusion information and with the coregistered volume-rendered segmented 3-D coronary tree as described previously. Vascular territories could then be adjusted segment by segment (based on a 17-segment American Heart Association model), utilizing the anatomic guidance provided by the coronary CTA. MPS contours or vascular territories could be adjusted based on the comparison with CTA. MPS perfusion analysis was performed individually for each vessel with the use of the "Group" function in quantitative perfusion SPECT (QPS) software, in which 17 segments are assigned to a vascular territory based on the perfusion defect pattern (Sharir et al. 2000; Slomka et al. 2005b). The portion of the total perfusion deficit (TPD) corresponding to a given territory was used for the automated quantification in each vessel (Slomka et al. 2005b).

Our group has demonstrated that such anatomically guided technique for adjustment of MPS contours and territories improved quantitative results as compared to SPECT imaging alone. Automated coregistration of MPS and coronary CTA had the success rate of 96% as assessed visually; the average registration errors were 4.3 ± 3.3 mm in translation and 1.5 ± 2.6 deg in rotation on stress and were 4.2 ± 3.1 mm in translation and 1.7 ± 3.2 deg in rotation on rest. MPS vascular region definition was adjusted in 17 studies and LV contours were adjusted in 11 studies using coregistered CTA images as a guide. CTA-guided SPECT analysis resulted in improved area under the receiver operator characteristics curves for the detection of right coronary artery (RCA) and left circumflex (LCX) lesions as compared to standard MPS analysis 0.84 ± 0.08 vs. 0.70 ± 0.11 for LCX ($p = 0.03$) and 0.92 ± 0.05 vs. 0.75 ± 0.09 ($p = 0.02$) for RCA (Figure 12.12). Examples of studies in which the quantification results were modified by this technique are shown in Figures 12.13 through 12.15. Fused CTA allowed for an objective adjustment of these parameters by the operator. In this work, the CTA and MPS images were acquired separately on standalone scanners and automated image registration was used to combine the data. We have subsequently applied a CTA-guided quantification technique to the data obtained on hybrid PET/CT scanners, allowing rapid integrated 3-D display and coronary CTA-guided contour and territory adjustment on PET (Nakazato et al. 2012).

12.4.2 HYBRID PET/SPECT IMAGING COMBINED WITH CALCIUM SCAN FROM CT

Integrated PET/CT or SPECT/CT hardware also enables incorporation of the coronary artery calcium (CAC) scan (Hong et al. 2002) into routine hybrid clinical imaging. The CAC scan is similar to the low-dose CT scan performed for the attenuation correction of PET, but it involves cardiac gating and breath-hold imaging. CAC scanning does not require injection of iodinated contrast and is much easier to perform than the coronary

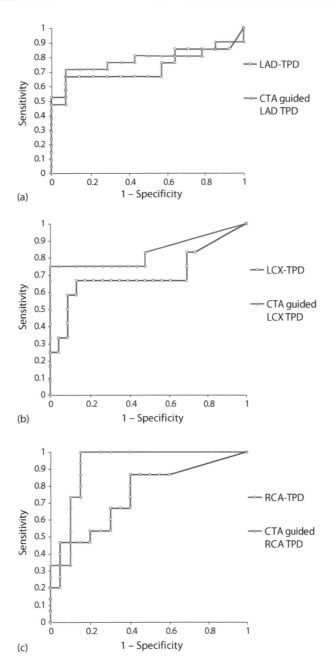

Figure 12.12 ROC curves for the disease detection in individual vessels by partial total perfusion deficit per vessel in standalone MPS (blue) and CTA-guided MPS (pink) for left anterior descending (LAD) (a), LCX (b), and RCA (c) vascular territories. (This research was originally published in *JNM*. Slomka et al. Quantitative analysis of myocardial perfusion SPECT anatomically guided by coregistered 64-slice coronary CT angiography. *J Nucl Med* 2009; 50:1621–1630. © by the Society of Nuclear Medicine and Molecular Imaging, Inc.)

CTA. It could be considered as an inexpensive add-on to PET or SPECT scan when using hybrid imaging. The CAC scan provides assessment of each patient's global coronary atherosclerotic burden. It has been shown that CAC scan adds incremental prognostic information to the nuclear scan (Schenker et al. 2008). Marked incremental prognostic value by combining CAC and SPECT MPI has recently been reported in a study of 4897 patients followed for a median of 2.5 years for cardiac events (Engbers et al. 2016). The additional

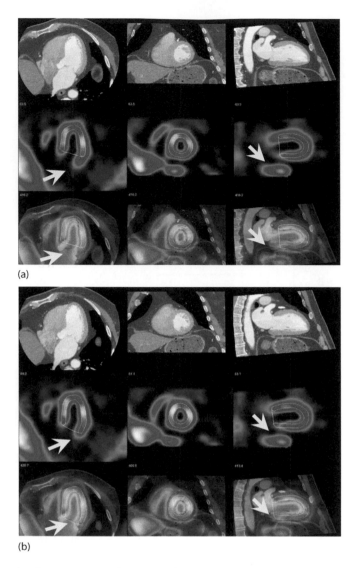

(a)

(b)

Figure 12.13 Example of MPS contour adjustment after MPS-CTA coregistration. In panel (a), co-registered CTA and MPS images are shown with MPS contours overlaid (white lines). Image fusion reveals that the valve plane is determined incorrectly (yellow arrows). The same images are shown after MPS contour adjustment (b), revealing RCA defect on MPS. Invasive angiography confirmed a ≥70% stenosis in the RCA. (This research was originally published in *JNM*. Slomka et al. Quantitative analysis of myocardial perfusion SPECT anatomically guided by coregistered 64-slice coronary CT angiography. *J Nucl Med* 2009; 50:1621–1630. © by the Society of Nuclear Medicine and Molecular Imaging, Inc.)

diagnostic value of calcium scan was also evaluated for myocardial perfusion obtained by SPECT or PET studies. A study by Schepis et al. (2007a) reported that the calcium score offered incremental diagnostic information over SPECT data for identifying patients with significant CAD and negative myocardial perfusion results. In another study utilizing hybrid PET/CT (Bybee et al. 2010), about 30% of patients demonstrated subclinical CAD detected by the presence of CAC, despite having normal myocardial perfusion. In our recent study, we have demonstrated that a combined quantitative +per-vessel score derived from the CAC and PET scan was superior in detection of obstructive disease as compared to PET imaging alone (Brodov et al. 2015). The CAC scan is also helpful to the clinician in the interpretation of MPI scans in patients with mild or equivocal perfusion defects. These studies support the concept that patients without known CAD undergoing PET/CT myocardial perfusion should perhaps undergo a same setting coronary calcification study. It has been

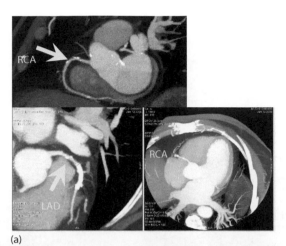

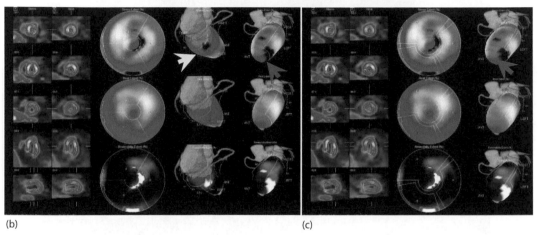

Figure 12.14 Example of MPS contour and territory adjustment based on CTA in a 72-year-old male patient. Original coronary CTA images (a) were interpreted to show a nonsignificant, <50% proximal RCA lesion and significant LAD lesion. Coregistration of CTA and MPS images was performed and indicated the need for contour adjustment (as shown in Figure 12.13). Quantification after contour adjustment reveals a 3% defect in the typical RCA territory (yellow arrow) and defect between LAD and LCX (red arrow) (b). After adjustments of the coronary territory based on the superimposed CTA coronary tree, the MPS lesion is assigned to LAD (c). Invasive angiography revealed a 50%–69% RCA lesion and a 90% LAD lesion. In this case, CTA-guided analysis allowed identification of the additional RCA lesion in MPS and reassignment of the second defect from LAD/LCX to LAD. (This research was originally published in *JNM*. Slomka et al. Quantitative analysis of myocardial perfusion SPECT anatomically guided by coregistered 64-slice coronary CT angiography. *J Nucl Med* 2009; 50:1621–1630. © by the Society of Nuclear Medicine and Molecular Imaging, Inc.)

proposed that CAC scanning should routinely be combined with either SPECT or PET MPI, with the view that adding CAC to MPI detects subclinical atherosclerosis not detected by MPI, improves risk assessment, increases diagnostic accuracy as well as physician certainty in interpretation, and leads to greater change in patient management (Rozanski, Slomka, and D 2016).

Typically, in current practice with hybrid systems, the CAC scan is acquired separately, immediately before the PET/CT scan (Di Carli, Dorbala, and Hachamovitch 2006). This regimen with separate CT CAC and attenuation scans increases the radiation dose to the patient (Kim, Einstein, and Berrington de Gonzalez 2009) and lengthens the overall scan protocol. However, it should be possible to utilize the CT calcium scans for attenuation correction and therefore eliminate the need for the CT scans on a hybrid PET/CT or SPECT/CT scanner. Vendors are developing manual or even automated tools for this purpose, as described in the previous section. In Figure 12.16, we show a screen snapshot of the software utility from one vendor

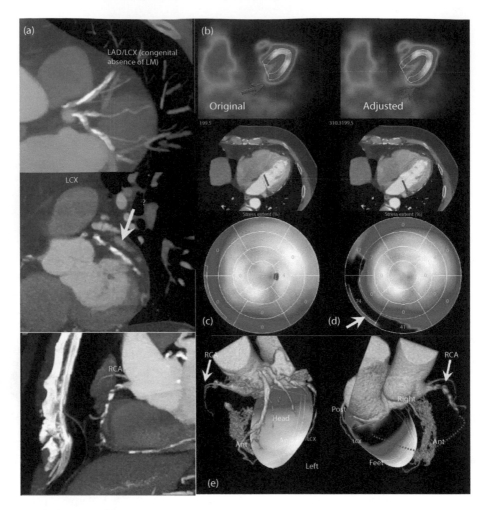

Figure 12.15 Example of MPS contour adjustment based on coregistered CTA in a 60-year-old male with congenital absence of left main artery. The calcium score was 1427. The original coronary CTA images (a) were reported as nondiagnostic in LCX and LAD arteries due to heavy calcifications in LCX potentially significant but not interpretable (yellow totally occluded). Coregistration of MPS with CTA allowed valve plane position adjustments (red arrows and lines) for MPS quantification based on the CTA (b). MPS quantitative analysis before contour adjustments (c) resulted in normal values (TPD = 1%). The repeated analysis after valve plane adjustment resulted in 7% TPD in the RCA territory (d). The fused 3D image of coronary anatomy and perfusion with corrected contours (e) illustrates the correspondence of the occluded RCA with the perfusion defect and normal perfusion in LAD and LCX region. Invasive angiography revealed 100% proximal RCA stenosis and no significant LAD or LCX disease.

demonstrating the user interface for manual realignment of calcium scan with SPECT for the purposes of attenuation correction. This technique has been successfully applied in SPECT/CT (Schepis et al. 2007b) and PET/CT hybrid imaging (Burknard et al. 2010). In both of these studies, the correlation between the myocardial perfusion results corrected for attenuation with calcium scan and with standard CT attenuation correction scan was excellent in all regions of the myocardium. Conversely, it may be also possible to estimate coronary calcium from nongated low-dose CT attenuation maps (Einstein et al. 2010; Mylonas et al. 2012), even if the quality of the CT maps used for attenuation correction is lower than that of the standard CT calcium scan. Studying 492 patients from three centers, Einstein et al. have demonstrated that coronary artery calcifications can be visually assessed from low-dose CT attenuation correction scans for PET/CT and SPECT/CT with high agreement for the Agatston score (Einstein et al. 2010). Weighted kappa was 0.89 (95%

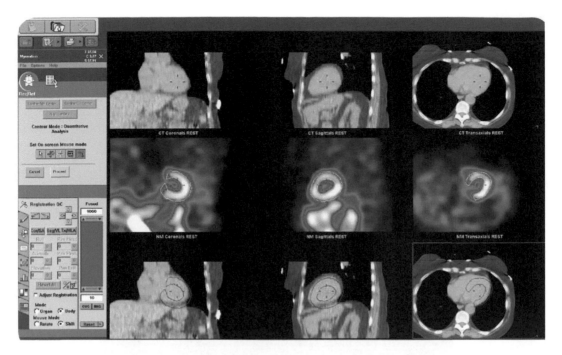

Figure 12.16 An example of CT calcium scan registered with SPECT and used for attenuation correction. Manual adjustment of registration is done by the observer by aligning the contour of the myocardium with cardiac CT. (Courtesy of Aharon Peretz, GE Healthcare.)

confidence interval [CI]: 0.88–0.91). Mylonas et al. demonstrated the quantitative assessment of calcium score from the CT attenuation maps with standard calcium score maps in 92 patients. The interclass correlation between the calcium score CT attenuation maps and calcium score CT was 0.953 (95% CI: 0.930–0.96) (Mylonas et al. 2012). Further research is warranted to standardize and simplify the hybrid cardiac PET/CT protocols, so that reliable PET or SPECT attenuation correction and calcium measurements could be obtained from a single CT scan on a hybrid scanner or even obtained using a standalone CT scan.

12.4.3 HYBRID VASCULAR IMAGING

An emerging application for the hybrid cardiovascular imaging with PET/CT is the investigation of vascular plaques. Latest PET/CT scanners are capable of combined molecular plaque imaging by PET and precise anatomical delineation of the plaque boundaries by CTA. Considering high resolution of the new hybrid PET/CT scanners especially with the new resolution-recovery techniques and motion correction techniques as described in previous sections, it should be possible to image tracer uptake in coronary lesions.

While coronary CTA can identify anatomical plaque features associated with increased risk, it does not depict the biological processes implicated in plaque rupture. Among the tracer compounds for imaging macrophage accumulation, ^{18}F-FDG is approved for human use and is used extensively for detection and staging of tumors and for imaging glucose utilization in the viable myocardium (Maddahi et al. 1994; Tillisch et al. 1986). However, the use of this tracer may be limited by myocardial uptake, which obscures the coronary arteries. Recently, in a series of studies (Dweck et al. 2012; Irkle et al. 2015; Joshi et al. 2014), it has been discovered that PET with ^{18}F-sodium-fluoride (^{18}F-NaF)—an inexpensive, widely available and Food and Drug Administration-approved tracer—identifies recently ruptured vulnerable plaque by binding to areas of active microcalcification, which is a well-established histological marker of high-risk disease. Hybrid imaging with high-resolution CTA is crucial for correct localization of the uptake in the coronary arteries both for [8]fluorodeoxyglucose (^{18}F-FDG) and for ^{18}F-NaF.

12.4.3.1 CORONARY ¹⁸F-FDG HYBRID PET/CORONARY CTA IMAGING

Few studies utilizing PET/CT hybrid equipment for coronary vascular ¹⁸F-FDG imaging have been reported to date. Increased ¹⁸F-FDG signal along the course of coronary arteries in patients with known CAD has been reported but without direct confirmation that plaques existed at the sites of increased uptake (Wykrzykowska et al. 2009). Higher ¹⁸F-FDG signal at culprit plaque sites was shown in patients with acute coronary syndrome (Rogers et al. 2010).

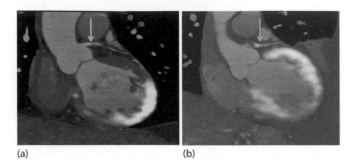

(a)   (b)

Figure 12.17 Coronary plaque imaging. Fused ¹⁸F-FDG-PET (high-definition reconstruction) and coronary CTA in two cases from our work (10). Lesion site is indicated by blue arrows. A case with AMI and coronary plaque uptake (a) and a case without uptake (b) in the left main artery.

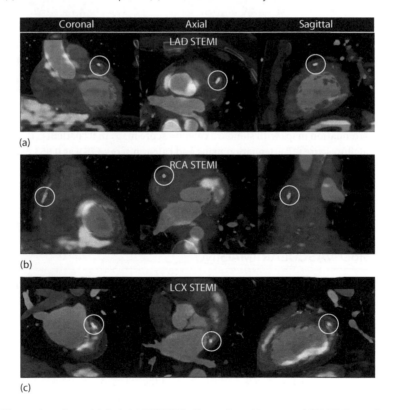

Figure 12.18 Plaque imaging with hybrid PET/CTA. Examples of increased ¹⁸F-FDG uptake at stent site in patients with acute ST-elevation myocardial infarction (STEMI) PET/CT. (a) A 54-year-old man imaged after percutaneous coronary stenting of the proximal left anterior descending artery (LAD) for STEMI. MaxTBR at the stent site was 2.1 (white circles). (b) A 49-year-old man imaged after stenting of the proximal RCA for STEMI. MaxTBR at the stent site was 2.1. (c) A 71-year-old man imaged after stenting of the mid LCX artery for STEMI. MaxTBR at the stent site was 4.0. (This research was originally published in *JNM*. Cheng et al. Coronary arterial ¹⁸F-FDG uptake by fusion of PET and coronary CT angiography at sites of percutaneous stenting for acute myocardial infarction and stable coronary artery disease. *J Nucl Med* 2012; 53:575–583. © by the Society of Nuclear Medicine and Molecular Imaging, Inc.)

Our group has studied coronary arterial ^{18}F-FDG uptake by fusion of PET and coronary CTA at sites of percutaneous stenting for acute myocardial infarction (AMI) and stable CAD (Cheng et al. 2012). CT-attenuation correction images were fused with CTA images to best overlay the 3-D position of the recently placed stent in both data sets by rigid body alignment (translation in 3-D). These registration parameters were then applied to PET data for fusion with CTA images. This method allowed robust PET/CT registration despite the lack of anatomy details on ^{18}F-FDG images.

Our study found that compared to stable disease patients, more AMI patients exhibited target to background ratio (TBR) > 2.0 at the stent site (12/20 vs. 1/7, $p = 0.04$). Two cases with and without uptake are shown in the left main artery in Figure 12.17. Further examples of increased ^{18}F-FDG uptake are shown in Figure 12.18. After adjusting for baseline demographic differences, stent-myocardium distance, and myocardial ^{18}F-FDG uptake, presentation of AMI was positively associated with the stent site TBR > 2.0 (odds ratio: 31.6, $p = 0.044$). This study demonstrated that with image fusion of ^{18}F-FDG-PET and coronary CTA increased culprit site, ^{18}F-FDG uptake could be in patients with AMI and was rarely seen in patients with stable CAD. However, this approach failed to detect increased signal at the culprit site in nearly half of AMI patients, highlighting the challenging nature of *in vivo* coronary artery plaque metabolic imaging. These results indicate detection of inflammation in the coronary arteries by hybrid PET/CTA imaging with ^{18}F-FDG remains controversial.

12.4.3.2 ^{18}F-SODIUM FLUORIDE IMAGING

Recently, it has been suggested that another PET tracer ^{18}F-sodium fluoride (^{18}F-NaF) can potentially provide new information about formation and progression of atherosclerotic plaque and vascular plaque biology (Derlin et al. 2010, 2011). ^{18}F-NaF is an established PET tracer that detects novel areas of bone formation and remodeling. It has been used in oncological PET/CT imaging and is approved for human scanning (Even-Sapir et al. 2006). Dweck et al. has utilized hybrid PET/CT scanner (Siemens Biograph mCT) to study coronary arterial uptake of ^{18}F-NaF and ^{18}F-FDG as markers of active plaque calcification and inflammation in a prospective study of 119 patients (Dweck et al. 2012). They have also utilized the high-definition PET reconstruction algorithms as described in previous sections, with resolution recovery (Panin et al. 2006) and time-of-flight corrections. Manual fusion was used to adjust the CTA to PET position. They have found that activity was higher in patients with coronary atherosclerosis versus control subjects and correlated with the calcium score. Patients with increased coronary ^{18}F-NaF activity ($n = 40$) had higher rates of prior cardiovascular events and angina. They found that quantification of coronary ^{18}F-FDG uptake was hampered by the myocardial activity and was not increased in patients with atherosclerosis versus control subjects. In a subsequent study from the same group, Joshi et al. utilized the same hybrid ^{18}F-NaF and ^{18}F-FDG PET/CTA imaging technique to identify ruptured and high-risk atherosclerotic plaques in patients with myocardial infarction ($n = 40$) and stable angina ($n = 40$) (Joshi et al. 2013). ^{18}F-NaF uptake was compared with histology in carotid endarterectomy specimens from patients with symptomatic carotid disease and with intravascular ultrasound in patients with stable angina. They have found that the highest coronary ^{18}F-NaF uptake was seen in the culprit plaques. In contrast, coronary ^{18}F-FDG uptake was commonly obscured by myocardial uptake with no differences between culprit and nonculprit plaques. They also found that marked ^{18}F-NaF uptake occurred at the site of all carotid plaque ruptures and was associated with histological evidence of active calcification, macrophage infiltration, apoptosis, and necrosis. Examples of ^{18}F-NaF uptake are shown in Figure 12.19. Additionally, in a recent study with hybrid PET/CT Dweck et al. confirmed that ^{18}F-NaF uptake identifies active tissue calcification and predicts disease progression in patients with calcific aortic stenosis (Dweck et al. 2014). In view of these findings, ^{18}F-NaF PET/CT appears to be a very promising application for hybrid PET/CT imaging for identification of vulnerable plaques in the coronary and other vessels.

Precise image fusion and image quantification will be crucial for coronary ^{18}F-NaF studies. We have developed automated methods for cardiac motion correction for ^{18}F-NaF coronary PET data (Rubeaux et al. 2016). A significant improvement in image quality (reduced noise and improved TBR) has been found despite correcting only cardiac motion. These motion correction techniques allow precise registration of anatomical

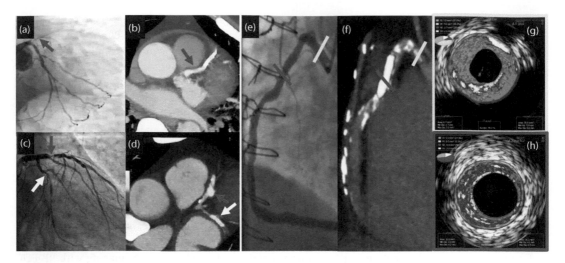

Figure 12.19 Focal ^{18}F-fluoride and ^{18}F-FDG uptake in patients with myocardial infarction and stable angina patient with acute ST-segment elevation myocardial infarction with (a) proximal occlusion (red arrow) of the left anterior descending artery on invasive coronary angiography and (b) intense focal ^{18}F-fluoride ^{18}F-NaF, tissue-to-background ratios, culprit 2.27 versus reference segment 1.09 (108% increase) uptake (yellow-red) at the site of the culprit plaque (red arrow) on the hybrid PET/CT. Patient with anterior non-ST-segment eleva-tion myocardial infarction with (c) culprit (red arrow; left anterior descending artery) and bystander nonculprit (white arrow; circumflex artery) lesions on invasive coronary angiography that were both stented during the index admission. Only the culprit lesion had increased ^{18}F-NaF uptake ^{18}F-NaF, tissue-to-background ratios, culprit 2.03 versus reference segment 1.08 (88% increase) on PET/CT (d) after percutaneous coronary interven-tion. Corresponding ^{18}F-FDG PET/CT showing no uptake either at the culprit (^{18}F-FDG, tissue-to-background ratios, culprit 1.62 versus reference segment 1.49 [9% increase]) or the bystander stented lesion. Note the intense uptake within the ascending aorta. In a patient with stable angina with previous coronary artery bypass grafting, invasive coronary angiography (e) showed nonobstructive disease in the RCA. Corresponding PET/CT scan (f) showed a region of increased ^{18}F-NaF activity (positive lesion, red line) in the mid-RCA (tissue-to-background ratio, 3.13) and a region without increased uptake in the proximal vessel (negative lesion, yellow line). Radiofrequency intravascular ultrasound shows that the ^{18}F-NaF negative plaque (g) is principally com-posed of fibrous and fibro-fatty tissue (green) with confluent calcium (white with acoustic shadow) but little evidence of necrosis. On the contrary, the ^{18}F-NaF-positive plaque (h) shows high-risk features such as a large necrotic core (red) and microcalcification (white). (Reprinted from *The Lancet*, 383, Joshi NV, Vesey AT, Williams MC, Shah AS, Calvert PA, Craighead FH et al., 18F-fluoride positron emission tomography for identification of ruptured and high-risk coronary atherosclerotic plaques: A prospective clinical trial, 705–713, Copyright 2014, with permission from Elsevier.)

lesions seen on CTA and alignment of all PET counts (not only end diastolic counts) with end diastolic CTA images. Examples from this study are shown in Figure 12.20.

To date, hybrid coronary imaging has been performed with CTA and PET obtained with one session on the same scanner. However, in the clinical application of coronary imaging, CTA may need to be performed before PET, usually on a separate scanner—with a subset of patients being selected based on their coronary CTA results to undergo follow-up with PET coronary artery plaque imaging alone, i.e., with CTA used as a gatekeeper for coronary PET. Importantly, if this sequential approach is employed, the standard coronary CTA, which is acquired in end-inspiration, will have to be fused with the free-breathing coronary PET study. Such software fusion applica-tion may be key for future clinical deployment of coronary PET imaging, with CTA used as a gatekeeper for PET.

12.5 CARDIAC APPLICATIONS FOR HYBRID PET/MRI

Recently, clinical PET/MR systems were introduced as described in another chapter of this book. The primary application for PET/MR is currently in the oncological imaging, but several hypothetical clinical cardiovas-cular applications have been suggested by the cardiovascular imaging experts (Adenaw and Salerno 2013; Quick 2014; Ratib, Nkoulou, and Schwaiger 2013; Rischpler et al. 2013). The major advantages of PET/MR

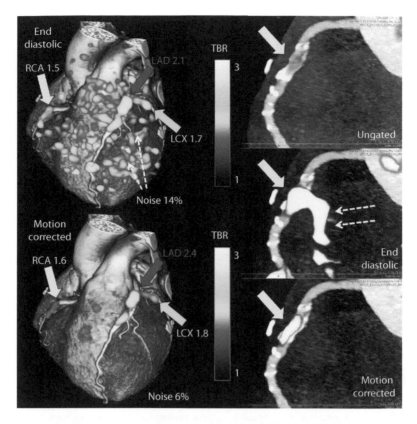

Figure 12.20 Hybrid imaging of cardiac motion-corrected ¹⁸F-NaF PET and CTA. Left: 3-D rendering: end-diastolic PET image (25% of PET counts) (left top) and motion-frozen image (left bottom). High image noise (% standard deviation) is seen in end-diastolic image (white dotted arrows). Increased 18F-NaF uptake (broad arrows) is seen in RCA (yellow), left anterior descending (LAD; red), and LCX (green). Right: 2-D multiplanar-reformatted view of RCA lesion (yellow arrows) with low lesion signal in ungated image (right top), noise in end-diastolic image (white arrows, right middle), and high lesion signal with reduced noise in motion corrected image (right bottom). (This research was originally published in *JNM*. Rubeaux et al. Motion correction of 18F-sodium fluoride PET for imaging coronary atherosclerotic plaques. *J Nucl Med* 2016;57:54–59. © by the Society of Nuclear Medicine and Molecular Imaging, Inc.)

over PET/CT in this context are lack of radiation exposure from CT and the possibility of advanced functional and molecular cardiac imaging with MR. The disadvantage is the electromagnetic field generated by MR, which may prevent imaging of patients with pacemakers, implantable cardioverter defibrillators, or other implanted devices susceptible to electromagnetic interference. Preliminary clinical case reports have been published, demonstrating potential clinical applications of simultaneous cardiac PET/MR. It has been suggested that simultaneous hybrid PET/MR of ¹³N-ammonia and late gadolinium enhancement (LGE) may allow diagnosis of myocardial hibernation (Figure 12.21) (Lau et al. 2013). Another group demonstrated the use of simultaneous PET/MR to assist in the diagnosis of the active cardiac sarcoidosis (Figure 12.22) (White et al. 2013).

It should be noted that it is possible that some of the PET/MR hybrid imaging applications can be more conveniently accomplished by standalone image and software registration as described in previous sections. For example, our group had studied the definition and size of the infarct with co-registered delayed enhancement MR and SPECT using software registration (Slomka et al. 2005a). Our study was done for cross-validation purposes; however, it could be imagined that combined stress SPECT or PET with delayed enhancement MR for viability definition obtained on standalone scanners could be performed clinically for viability imaging instead of traditional ¹⁸F-FDG viability scan. It remains to be seen if hybrid cardiac PET/MR enters clinical practice in cardiology. Nevertheless, this hybrid device is very likely to be exploited as a powerful research and cross-validation tool for new PET or MR cardiovascular imaging techniques.

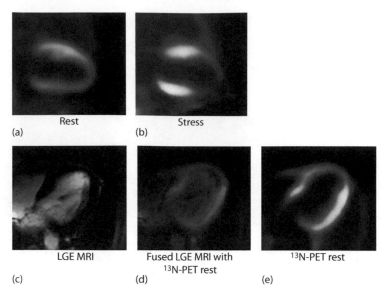

Figure 12.21 Example of cardiac PET/MR. ^{13}N-ammonia and LGE PET/MR. Two-chamber long axis views of rest (a) and stress (b) ^{13}N-ammonia PET/MR myocardial perfusion scan demonstrating a predominantly reversible anterior/anteroapical perfusion defect consistent with ischemia. (c) LGE imaging shows no delayed contrast uptake in the anterior/anteroapical wall, indicating the absence of infarction. (d) Fused PET perfusion and MR LGE images show no LGE in the area of resting hypoperfusion detected by PET (e) ^{13}N-ammonia PET demonstrating mild anterior and apical hypoperfusion at rest. (With kind permission from Springer Science+Business Media: *J Nucl Cardiol*, Demonstration of intermittent ischemia and stunning in hibernating myocardium, 20, 2014, 910, Lau J, Laforest R, Priatna A, Sharma S, Zheng J, Gropler RJ et al., Figure 2.)

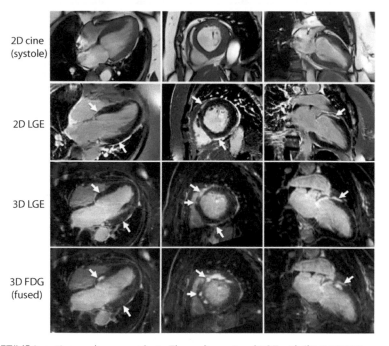

Figure 12.22 PET/MR in active cardiac sarcoidosis. Three-dimensional LGE with ^{18}F-FDG PET suggestive of active inflammation surrounding regions of established scar. Respective sequential short-axis images from the base to the mid-ventricle for T2-weighted MR, LGE MR, and ^{18}F-FDG PET imaging (fused to LGE imaging). White arrows indicate regions of established scar by LGE imaging. Far right, images at a wider field of view to show splenic nodules with enhancement on both LGE and ^{18}F-FDG imaging. (Reprinted from *Circulation*, Volume 127, Issue 22, White et al., Active cardiac sarcoidosis: first clinical experience of simultaneous positron emission tomography – magnetic resonance imaging for the diagnosis of cardiac disease, e639–e641, 2013, with permission from Wolters Kluwer Health.)

12.6 CHALLENGES AND FUTURE POSSIBILITIES

Several promising clinical and research applications for the hybrid vascular imaging have been demonstrated in this chapter. Certainly, for research purposes, the hybrid imaging is an excellent tool. It is, however, somewhat less apparent if a mainstream clinical application can be identified for cardiovascular hybrid imaging. For example, in the diagnosis of coronary disease, it is thought that only 10%–15% of patients would require both anatomical (coronary CTA) and myocardial perfusion scans (Berman et al. 2006). Performing hybrid scan on all patients is not economically feasible and would add unnecessary radiation burden to patients who would have needed only one of the scans. The main challenge is that it is not known ahead of time (before the test results of one scan are known) who should be undergoing the hybrid imaging; therefore, it becomes logistically difficult to adjudicate scans on a hybrid scanner to the appropriate patients who would benefit from it. Potentially, stress-only low-dose PET or SPECT with low-dose coronary CTA could be utilized in a streamlined dose-optimized hybrid protocol if there are patients in whom the need for both tests is clear. A potentially more practical solution could be the utilization of the selective hybrid imaging with software approach of standalone sequentially acquired data.

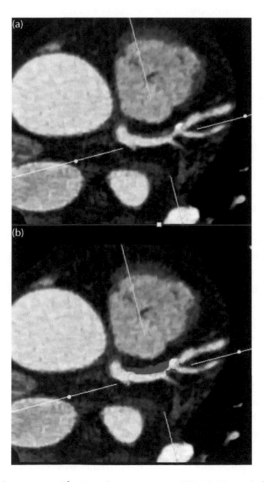

Figure 12.23 Automated plaque quantification from coronary CTA (a). Noncalcified (red) and calcified (yellow) plaque regions are detected (b). CT-defined volume of plaque could be used in the future for the partial volume correction of hybrid vascular images.

Another potential future application for hybrid imaging is the improvement of the image resolution and quantitative accuracy of the PET or SPECT scan with the use of CT or MR data, for critically demanding research applications. Such improvements could be achieved by anatomically guided partial volume effects correction for hybrid imaging. Partial volume effects correction techniques have been found effective for oncological PET/CT imaging (Marouka et al. 2013). For vascular imaging, partial volume effects in terms of bias and variability will be large for coronary plaque imaging. Coronary CTA offers abilities to delineate lipid plaque volume accurately, and therefore, it should be possible to apply partial volume effect correction methods, based on the plaque volume and location. The volume of the noncalcified lesions determined by the automated software (Rubeaux et al. 2016) from CTA could be used in the future for partial volume effects correction for PET after coregistration (Figure 12.23). Such accurate absolute measurements of PET uptake could be important for research where high accuracy may be required for measuring the effects of therapy and thus warrant the use of hybrid imaging.

Nevertheless, some clinical hybrid imaging applications may be quite feasible for the widespread clinical deployment. For example, it is possible to imagine that PET or SPECT with CT calcium scan could be routinely utilized, considering the evidence of added value of the CT calcium scan to the SPECT and PET myocardial perfusion imaging. The routine availability of the noncontrast CT scan obtained with hybrid equipment may allow additional future possibilities. For example, other cardiac variables such as pericardial fat volume can be derived via automatic software from the noncontrast CT (Dey et al. 2008). This new cardiac variable is related to presence of both myocardial ischemia and stenosis (Nazakato et al. 2012) and may offer additional value over the CT calcium score (Forouzandeh et al. 2013).

12.7 SUMMARY

We have described recent developments in hybrid cardiovascular imaging with the focus on software developments and applications. Hybrid cardiovascular imaging by PET/CT or SPECT/CT is hampered by comparatively low resolution of nuclear imaging modality, as well as cardiac and respiratory motion, which is resolved differently in CT and PET/SPECT images. Corrections for image resolution, cardiac, and respiratory motion have been proposed to improve the agreement between the two imaging components. Hybrid imaging may be performed with dedicated combined scanners or alternatively with software-enabled combination of standalone modalities. Robust automated software techniques have been developed for the registration of standalone SPECT/PET modalities with CTA, noncontrast CT, and MR. Due to mismatch in respiratory patterns, PET/CTA imaging with hybrid PET/CT scanners obtained in one scanning session still requires software registration for combined visualization and quantification.

Several reports demonstrate improved diagnostic accuracy of hybrid PET or SPECT/CTA imaging for the detection of the CAD with either dedicated hybrid scanners or standalone sequential imaging. It is also possible to improve the quantification of SPECT or PET by CTA-guided correction of the LV contours and vascular territories. A coronary calcium scan can be easily obtained with the hybrid scanners and it can provide additional diagnostic and prognostic value. A potentially noncontrast CT calcium scan could be used for attenuation correction, streamlining the clinical protocols, and reducing radiation exposure. Alternatively, calcium estimates could be obtained from CT attenuation maps. Hybrid imaging may also be used for anatomically guided molecular imaging of atherosclerotic plaque. Several reports utilize vascular PET/CTA hybrid imaging to investigate the vulnerable plaques in the aorta, carotid and coronary vessels. The most promising PET agent for this application appears to be ^{18}F-NaF. Novel applications have been suggested for hybrid PET/MR, including the diagnosis of myocardial hibernation or diagnosis of cardiac sarcoidosis. Challenges for hybrid cardiovascular imaging include the cost of the procedures, radiation dose, as well as identification of patients who would benefit from such exams. There are, however, several clinical and research applications where hybrid imaging will prove viable and potentially some of the approaches will become part of routine clinical testing.

ACKNOWLEDGMENTS

We thank Joanna Liang for the editing and proofreading assistance of this paper. Some of the research described in this review was supported in part by a grant (*Cardiac Imaging Research Initiative*) from the Adelson Medical Research Foundation at Cedars-Sinai and Grant R01HL089765 from the National Heart, Lung, and Blood Institute/National Institutes of Health (NHLBI/NIH). Its contents are solely the responsibility of the authors and do not necessarily represent the official views of the NHLBI/NIH.

REFERENCES

Achenbach S, Goroll T, Seltmann M, Pflederer T, Anders K, Ropers D et al. 2011. Detection of coronary artery stenoses by low-dose, prospectively ECG-triggered, high-pitch spiral coronary CT angiography. *JACC Cardiovasc Imaging* 44: 328–37.

Adenaw N, Salerno M. 2013. PET/MRI: Current state of the art and future potential for cardiovascular applications. *J Nucl Cardiol* 206: 976–89.

Akamatsu G, Ishikawa K, Mitsumoto K, Taniguchi T, Ohya N, Baba S et al. 2012. Improvement in PET/CT image quality with a combination of point-spread function and time-of-flight in relation to reconstruction parameters. *J Nucl Med* 5311: 1716–22.

Aladl U, Hurwitz G, Dey D, Levin D, Drangova M, Slomka P. 2004. Automated image registration of gated cardiac single-photon emission computed tomography and magnetic resonance imaging. *J Magn Reson Imaging* 193: 283–90.

Berman DS, Hachamovitch R, Shaw LJ, Friedman JD, Hayes SW, Thomson LE et al. 2006. Roles of nuclear cardiology, cardiac computed tomography, and cardiac magnetic resonance: Assessment of patients with suspected coronary artery disease. *J Nucl Med* 471: 74–82.

Bettinardi V, Presotto L, Rapisarda E, Picchio M, Gianolli L, Gilardi M. 2011. Physical performance of the new hybrid PET/CT Discovery-690. *Med Phys* 3810: 5394–411.

Bond S, Kadir T, Hamill J, Casey M, Platsch G, Burckhardt D et al. Automatic registration of cardiac PET/CT for attenuation correction. Nuclear Science Symposium Conference Record, 2008 NSS'08 IEEE; 2008 19–25 October 2008; Dresden, Germany: IEEE; 2008. p. 5512–7.

Brodov Y, Gransar H, Dey D, Shalev A, Germano G, Friedman JD et al. 2015. Combined Quantitative assessment of myocardial perfusion and coronary artery calcium score by hybrid 82Rb PET/CT improves detection of coronary artery disease. *J Nucl Med* 569: 1345–50.

Burkhard N, Herzog BA, Husmann L, Pazhenkottil AP, Burger IA, Buechel RR et al. 2010. Coronary calcium score scans for attenuation correction of quantitative PET/CT 13N-ammonia myocardial perfusion imaging. *Eur J Nucl Med Mol Imaging* 373: 517–21.

Büther F, Dawood M, Stegger L, Wübbeling F, Schäfers M, Schober O et al. 2009. List mode-driven cardiac and respiratory gating in pet. *J Nucl Med* 505: 674–81.

Bybee KA, Lee J, Markiewicz R, Longmore R, McGhie AI, O'Keefe JH et al. 2010. Diagnostic and clinical benefit of combined coronary calcium and perfusion assessment in patients undergoing PET/CT myocardial perfusion stress imaging. *J Nucl Cardiol* 172: 188–96.

Cheng VY, Slomka PJ, Le Meunier L, Tamarappoo BK, Nakazato R, Dey D et al. 2012. Coronary arterial 18F-FDG uptake by fusion of PET and coronary CT angiography at sites of percutaneous stenting for acute myocardial infarction and stable coronary artery disease. *J Nucl Med* 534: 575–83.

Danad I, Raijmakers PG, Appelman YE, Harms HJ, de Haan S, van den Oever ML et al. 2013. Hybrid imaging using quantitative H215O PET and CT-based coronary angiography for the detection of coronary artery disease. *J Nucl Med* 54: 55–63.

Derlin T, Richter U, Bannas P, Begemann P, Buchert R, Mester J et al. 2010. Feasibility of 18F-sodium fluoride PET/CT for imaging of atherosclerotic plaque. *J Nucl Med* 516: 862–5.

Derlin T, Toth Z, Papp L, Wisotzki C, Apostolova I, Habermann CR et al. 2011. Correlation of inflammation assessed by 18F-FDG PET, active mineral deposition assessed by 18F-fluoride PET, and vascular calcification in atherosclerotic plaque: A dual-tracer PET/CT study. *J Nucl Med* 527: 1020–7.

Dey D, Suzuki Y, Suzuki S, Ohba M, Slomka PJ, Polk D, Shaw LJ, Berman DS. 2008. Automated quantitation of pericardiac fat from noncontrast CT. *Invest Radiol* 43: 145–53.

Di Carli MF, Dorbala S, Hachamovitch R. 2006. Integrated cardiac PET-CT for the diagnosis and management of CAD. *J Nucl Cardiol* 132: 139–44.

Dweck MR, Chow MW, Joshi NV, Williams MC, Jones C, Fletcher AM et al. 2012. Coronary arterial 18F-sodium fluoride uptake: A novel marker of plaque biology. *J Am Coll Cardiol* 5917: 1539–48.

Dweck MR, Jenkins WS, Vesey AT, Pringle MA, Chin CW, Malley TS et al. 2014. 18F-NaF Uptake is a marker of active calcification and disease progression in patients with aortic stenosis. *Circ Cardiovasc Imaging* 7.

Einstein AJ, Johnson LL, Bokhari S, Son J, Thompson RC, Bateman TM et al. 2010. Agreement of visual estimation of coronary artery calcium from low-dose CT attenuation correction scans in hybrid PET/CT and SPECT/CT with standard Agatston score. *J Am Coll Cardiol* 5623: 1914–21.

Engbers EM, Timmer JR, Ottervanger JP, Mouden M, Knollema S, Jager PL. 2016. Prognostic Value of coronary artery calcium scoring in addition to single-photon emission computed tomographic myocardial perfusion imaging in symptomatic patients. *Circ Cardiovasc Imaging* 95.

Even-Sapir E, Metser U, Mishani E, Lievshitz G, Lerman H, Leibovitch I. 2006. The detection of bone metastases in patients with high-risk prostate cancer: 99mTc-MDP Planar bone scintigraphy, singleand multi-field-of-view SPECT, 18F-fluoride PET, and 18F-fluoride PET/CT. *J Nucl Med* 472: 287–97.

Flotats A, Knuuti J, Gutberlet M, Marcassa C, Bengel FM, Kaufmann PA et al. 2011. Hybrid cardiac imaging: SPECT/CT and PET/CT. A joint position statement by the European Association of Nuclear Medicine (EANM), the European Society of Cardiac Radiology (ESCR) and the European Council of Nuclear Cardiology (ECNC). *Eur J Nucl Med Mol Imaging* 381: 201–12.

Forouzandeh F, Chang SM, Muhyieddeen K, Zaid RR, Trevino AR, Xu J et al. 2013. Does quantifying epicardial and intrathoracic fat with noncontrast computed tomography improve risk stratification beyond calcium scoring alone? *Circ Cardiovasc Imaging* 61: 58–66.

Gaemperli O, Bengel FM, Kaufmann PA. 2011. Cardiac hybrid imaging. *Eur Heart J* 3217: 2100–8.

Gaemperli O, Schepis T, Kalff V, Namdar M, Valenta I, Stefani L et al. 2007. Validation of a new cardiac image fusion software for three-dimensional integration of myocardial perfusion SPECT and standalone 64-slice CT angiography. *Eur J Nucl Med Mol Imaging* 347: 1097–106.

Gould KL, Pan T, Loghin C, Johnson NP, Guha A, Sdringola S. 2007. Frequent diagnostic errors in cardiac PET/CT due to misregistration of CT attenuation and emission PET images: A definitive analysis of causes, consequences, and corrections. *J Nucl Med* 487: 1112–21.

Hong C, Becker CR, Schoepf UJ, Ohnesorge B, Bruening R, Reiser MF. 2002. Coronary artery calcium: Absolute quantification in nonenhanced and contrast-enhanced multi-detector row CT studies. *Radiology* 2232: 474–80.

Irkle A, Vesey AT, Lewis DY, Skepper JN, Bird JL, Dweck MR et al. 2015. Identifying active vascular microcalcification by (18)F-sodium fluoride positron emission tomography. *Nat Commun* 6: 7495.

Jakoby BW, Bercier Y, Conti M, Casey ME, Bendriem B, Townsend DW. 2011. Physical and clinical performance of the mCT time-of-flight PET/CT scanner. *Phys Med Biol* 568: 2375–89.

Joshi NV, Vesey AT, Williams MC, Shah AS, Calvert PA, Craighead FH et al. 2013. F-fluoride positron emission tomography for identification of ruptured and high-risk coronary atherosclerotic plaques: A prospective clinical trial. *Lancet* 383: 705–13.

Joshi NV, Vesey AT, Williams MC, Shah AS, Calvert PA, Craighead FH et al. 2014. 18F-fluoride positron emission tomography for identification of ruptured and high-risk coronary atherosclerotic plaques: A prospective clinical trial. *Lancet* 3839918: 705–13.

Kajander S, Joutsiniemi E, Saraste M, Pietilä M, Ukkonen H, Saraste A et al. 2010. Cardiac Positron emission tomography/computed tomography imaging accurately detects anatomically and functionally significant coronary artery disease. *Circulation* 1226: 603–13.

Kajander S, Ukkonen H, Sipilä H, Teräs M, Knuuti J. 2009. Low radiation dose imaging of myocardial perfusion and coronary angiography with a hybrid PET/CT scanner. *Clin Physiol Funct Imaging* 291: 81–8.

Kim KP, Einstein AJ, Berrington de Gonzalez A. 2009. Coronary artery calcification screening: Estimated radiation dose and cancer risk. *Arch Intern Med* 16913: 1188.

Lau J, Laforest R, Priatna A, Sharma S, Zheng J, Gropler RJ et al. 2013. Demonstration of intermittent ischemia and stunning in hibernating myocardium. *J Nucl Cardiol* 205: 908–12.

Le Meunier L, Slomka PJ, Dey D, Ramesh A, Thomson LE, Hayes SW et al. 2010. Enhanced definition PET for cardiac imaging. *J Nucl Cardiol* 173: 414–26.

Le Meunier L, Slomka PJ, Dey D, Ramesh A, Thomson LE, Hayes SW et al. 2011. Motion frozen (18) F-FDG cardiac PET. *J Nucl Cardiol* 182: 259–66.

Maddahi J, Schelbert H, Brunken R, Di Carli M. 1994. Role of thallium-201 and PET imaging in evaluation of myocardial viability and management of patients with coronary artery disease and left ventricular dysfunction. *J Nucl Med* 354: 707–15.

Martinez-Möller A, Zikic D, Botnar RM, Bundschuh RA, Howe W, Ziegler SI et al. 2007. Dual cardiac–respiratory gated PET: Implementation and results from a feasibility study. *Eur J Nucl Med Mol Imaging* 349: 1447–54.

Maruoka Y, Abe K, Baba S, Isoda T, Kitamura Y, Mizoguchi N et al. 2013. Usefulness of partial volume effect-corrected F-18 FDG PET/CT for predicting I-131 accumulation in the metastatic lymph nodes of patients with thyroid carcinoma. *Ann Nucl Med* 2710: 873–9.

Mylonas I, Kazmi M, Fuller L, Yam Y, Chen L, Beanlands RS et al. 2012. Measuring coronary artery calcification using positron emission tomography-computed tomography attenuation correction images. *Eur Heart J Cardiovasc Imaging* 139: 786–92.

Nakazato R, Dey D, Alexanderson E, Meave A, Jimenez M, Romero E et al. 2014. Automatic alignment of myocardial perfusion PET and 64-slice coronary CT angiography on hybrid PET/CT. *J Nucl Cardiol* 193: 482–91.

Nakazato R, Dey D, Cheng VY, Gransar H, Slomka PJ, Hayes SW et al. 2012. Epicardial fat volume and concurrent presence of both myocardial ischemia and obstructive coronary artery disease. *Atherosclerosis* 2212: 442–6.

Namdar M, Hany TF, Koepfli P, Siegrist PT, Burger C, Wyss CA et al. 2005. Integrated PET/CT for the assessment of coronary artery disease: A feasibility study. *J Nucl Med* 466: 930–5.

Panin VY, Kehren F, Michel C, Casey M. 2006. Fully 3-D PET reconstruction with system matrix derived from point source measurements. *IEEE Trans Med Imaging* 257: 907–21.

Plass A, Emmert MY, Gaemperli O, Alkadhi H, Kaufmann P, Falk V et al. 2011. The potential value of hybrid positron emission tomography/dual-source computed tomography imaging in coronary bypass surgery. *Heart Surg Forum* 145: E283–90.

Quick HH. 2014. Integrated PET/MR. *J Magn Reson Imaging* 392: 243–58.

Radau PE, Slomka PJ, Julin P, Svensson L, Wahlund LO. 2001. Evaluation of linear registration algorithms for brain SPECT and the errors due to hypoperfusion lesions. *Med Phys* 288: 1660–8.

Ratib O, Nkoulou R, Schwaiger M. 2013. Cardiovascular clinical applications of PET/MRI. *Clin Transl Imaging*: 1–7.

Rischpler C, Nekolla SG, Dregely I, Schwaiger M. 2013. Hybrid PET/MR imaging of the heart: Potential, initial experiences, and future prospects. *J Nucl Med* 543: 402–15.

Rogers IS, Nasir K, Figueroa AL, Cury RC, Hoffmann U, Vermylen DA et al. 2010. Feasibility of FDG imaging of the coronary arteries: Comparison between acute coronary syndrome and stable angina. *JACC Cardiovasc Imaging* 34: 388–97.

Rozanski A, Slomka P, D SB. 2016. Extending the use of coronary calcium scanning to clinical rather than just screening populations: Ready for prime time? *Circ Cardiovasc Imaging* 95.

Rubeaux M, Joshi NV, Dweck MR, Fletcher A, Motwani M, Thomson LE et al. 2016. Motion correction of 18F-NaF PET for imaging coronary atherosclerotic plaques. *J Nucl Med* 571: 54–9.

Santana CA, Garcia EV, Faber TL, Sirineni GK, Esteves FP, Sanyal R et al. 2009. Diagnostic performance of fusion of myocardial perfusion imaging (MPI) and computed tomography coronary angiography. *J Nucl Cardiol* 162: 201–11.

Schaap J, de Groot JA, Nieman K, Meijboom WB, Boekholdt SM, Post MC et al. 2013a. Hybrid myocardial perfusion SPECT/CT coronary angiography and invasive coronary angiography in patients with stable angina pectoris lead to similar treatment decisions. *Heart* 993: 188–94.

Schaap J, Kauling RM, Boekholdt SM, Nieman K, Meijboom WB, Post MC et al. 2013b. Incremental diagnostic accuracy of hybrid SPECT/CT coronary angiography in a population with an intermediate to high pre-test likelihood of coronary artery disease. *Eur Heart J Cardiovasc Imaging* 147: 642–9.

Schenker MP, Dorbala S, Hong ECT, Rybicki FJ, Hachamovitch R, Kwong RY et al. 2008. Interrelation of coronary calcification, myocardial ischemia, and outcomes in patients with intermediate likelihood of coronary artery disease. *Circulation* 11713: 1693–700.

Schepis T, Gaemperli O, Koepfli P, Namdar M, Valenta I, Scheffel H et al. 2007a. Added value of coronary artery calcium score as an adjunct to gated SPECT for the evaluation of coronary artery disease in an intermediate-risk population. *J Nucl Med* 489: 1424–30.

Schepis T, Gaemperli O, Koepfli P, Ruegg C, Burger C, Leschka S et al. 2007b. Use of coronary calcium score scans from stand-alone multislice computed tomography for attenuation correction of myocardial perfusion SPECT. *Eur J Nucl Med Mol Imaging* 341: 11–9.

Sharir T, Germano G, Waechter PB, Kavanagh PB, Areeda JS, Gerlach J et al. 2000. A new algorithm for the quantitation of myocardial perfusion SPECT. II: Validation and diagnostic yield. *J Nucl Med* 414: 720–7.

Slomka PJ, Cheng VY, Dey D, Woo J, Ramesh A, Van Kriekinge S et al. 2009. Quantitative analysis of myocardial perfusion SPECT anatomically guided by coregistered 64-slice coronary CT angiography. *J Nucl Med* 50: 1621–30.

Slomka PJ, Diaz-Zamudio M, Dey D, Motwani M, Brodov Y, Choi D et al. 2015a. Automatic registration of misaligned CT attenuation correction maps in Rb-82 PET/CT improves detection of angiographically significant coronary artery disease. *J Nucl Cardiol* 226: 1285–95.

Slomka PJ, Fieno D, Thomson L, Friedman JD, Hayes SW, Germano G et al. 2005a. Automatic detection and size quantification of infarcts by myocardial perfusion SPECT: Clinical validation by delayedenhancement MRI. *J Nucl Med* 465: 728–35.

Slomka PJ, Le Meunier L, Hayes SW, Acampa W, Oba M, Haemer GG et al. 2008. Comparison of myocardial perfusion 82Rb PET performed with CT- and transmission CT-based attenuation correction. *J Nucl Med* 4912: 1992–8.

Slomka PJ, Nishina H, Berman DS, Akincioglu C, Abidov A, Friedman JD et al. 2005b. Automated quantification of myocardial perfusion SPECT using simplified normal limits. *J Nucl Cardiol* 121: 66–77.

Slomka PJ, Nishina H, Berman DS, Kang X, Akincioglu C, Friedman JD et al. 2004. "Motion-frozen" display and quantification of myocardial perfusion. *J Nucl Med* 457: 1128–34.

Slomka PJ, Rubeaux M, Le Meunier L, Dey D, Lazewatsky JL, Pan T et al. 2015b. Dual-Gated Motion-Frozen Cardiac PET with Flurpiridaz F 18. *J Nucl Med* 56: 1876–81.

Tillisch J, Brunken R, Marshall R, Schwaiger M, Mandelkern M, Phelps M et al. 1986. Reversibility of cardiac wall-motion abnormalities predicted by positron tomography. *N Engl J Med* 31414: 884–8.

Wells WM, 3rd, Viola P, Atsumi H, Nakajima S, Kikinis R. 1996. Multi-modal volume registration by maximization of mutual information. *Med Image Anal* 11: 35–51.

White JA, Rajchl M, Butler J, Thompson RT, Prato FS, Wisenberg G. 2013. Active cardiac sarcoidosis first clinical experience of simultaneous positron emission tomography–magnetic resonance imaging for the diagnosis of cardiac disease. *Circulation* 12722: e639–41.

Woo J, Slomka PJ, Dey D, Cheng VY, Hong BW, Ramesh A et al. 2009. Geometric feature-based multimodal image registration of contrast-enhanced cardiac CT with gated myocardial perfusion SPECT. *Med Phys* 3612: 5467–79.

Wykrzykowska J LS, Williams G, Parker JA, Palmer MR, Varkey S, Kolodny G, Laham R. 2009. Imaging of inflamed and vulnerable plaque in coronary arteries with 18F-FDG PET/CT in patients with suppression of myocardial uptake using a low-carbohydrate, high-fat preparation. *J Nucl Med* 504: 563–8.

Zaidi H, Nkoulou R, Bond S, Baskin A, Schindler T, Ratib O et al. 2013. Computed tomography calcium score scan for attenuation correction of N-13 ammonia cardiac positron emission tomography: Effect of respiratory phase and registration method. *Int J Cardiovasc Imaging* 296: 1351–60.

Multimodality image fusion

MARINA PICCINELLI, JAMES R. GALT, AND ERNEST V. GARCIA

13.1 INTRODUCTION

In the last decades, the continued improvement of noninvasive imaging armamentarium (Achenbach et al. 2010; Beller 2010) and better understanding of cardiovascular diseases have driven physicians and researchers in the field to adjust the position of diagnostic imaging techniques from a *competing* role to a *synergistic* one (Achenbach et al. 2013; Blankstein and Dorbala 2010; Kramer and Narula 2010; Lee et al. 2013; van der Hoeven et al. 2012). Each modality exhibits and will most likely continue to exhibit its own strengths and weaknesses. The idea of a *one-stop-shop* (Kramer 1998) that could alone explain all aspects of myocardial disease has given way to an integrated multimodality approach according to which different modalities equally contribute to the formation of the final image eventually used for diagnosis, prognosis, and patient management.

The commonest cardiac applications for imaging multimodality frameworks deal with clinical conditions—the most prevalent being coronary artery disease (CAD) (Go et al. 2013)—for which both anatomy and function are necessary for a comprehensive assessment of the disease severity, its spatial distribution and treatment planning (Hsiao et al. 2010; Piccinelli and Garcia 2013). A number of clinical studies have been established that the degree of anatomical stenosis (measured with invasive coronary angiography [CA] or noninvasive computed tomographic angiography [CTA]) inadequately predicts the physiologic significance of obstructive CAD, particularly in intermediate lesions, resulting in many cases being referred to unnecessary invasive procedures (Achenbach et al. 2010; Blankstein and Di Carli 2010; Hacker et al. 2007; Lee et al. 2013). Relative myocardial perfusion imaging (MPI) robustly identifies the vascular territory supplied by the most severe stenosis but may underestimate true ischemia extent in patients with multivessel CAD and, given the high variability in coronary anatomy, may fail to properly single out the culprit vessel (Hachamovitch and Di Carli 2008; Underwood 2004). Additionally, after the Fractional Flow Reserve versus Angiography for Multivessel Evaluation trial demonstrated that revascularization guided by fractional flow reserve significantly improves outcomes, tests for myocardial ischemia have been incorporated into revascularization guidelines (de Bruyne et al. 2012; Lee et al. 2013; Min et al. 2012; Tonino et al. 2010).

The strength and potential of multimodality image fusion originate not only from the availability of different tests in ambiguous cases (Gaemperli et al. 2007a; Santana et al. 2009; Slomka et al. 2004), i.e., CTA-derived anatomy and MPI-derived perfusion, but also from the actual merging in the same visualization display of all the available information so that they can actually complement each other and provide an image that is more informative than any of the input singularly considered. Different types of imaging tests have been available to physicians for diagnosis for years, but they were typically interpreted side by side and fused in one's mind. Recently, studies specifically designed to address this issue have demonstrated the incremental diagnostic value of fusion frameworks with respect to side-by-side or stand-alone interpretations (Santana et al. 2009) in that the confidence in interpretation with integrated images was higher with better accuracy and less errors.

In the following sections, we will initially present two main approaches for the realization of multimodality imaging frameworks, namely, *software* versus *hardware* solution, emphasizing their advantages as well as their drawbacks. We will then consider the commonest dual-modality systems and their usage in clinical settings for image fusion.

13.2 WORKING DEFINITIONS: SOFTWARE VERSUS HARDWARE IMAGE FUSION

Generally speaking, multimodality image fusion is performed when two or more images acquired either by different devices and/or at different moments are *registered* in space and time so that colocalization of the structure and function is obtained (Giovanella and Lucignani 2010; Saraste and Knuuti 2012). This integration can be realized according to different approaches and techniques, but they all come down to two strategies: (1) *software fusion*, when images are acquired in separate scanners and dedicated image registration software is developed to perform the final integration of the volumetric datasets (Faber et al. 2004, 2011; Gaemperli et al. 2007b; Slomka and Baum 2009; Slomka et al. 2004, 2009), and (2) *hardware fusion*, when the patient is sequentially acquired by means of a hybrid device without being moved from the table (Cherry 2009; Flotats et al. 2011; Gaemperli et al. 2012; Kauffmann and Di Carli 2009; Patton et al. 2009).

13.2.1 SOFTWARE FUSION

Before the advent of hybrid devices, the only way to fuse distinct images was to make use of software tools and specifically of image registration techniques (Hutton et al. 2002; Maintz and Viergenev 1998). The increasing use of radiological images in all aspects of clinical routine has made image registration a very active field of medical image processing. Particularly, automated registration techniques have been the focus of considerable investigation (Faber et al. 2004, 2011; Gaemperli et al. 2007a, 2007b; Peifer et al. 1990; Piccinelli et al. 2014; Slomka and Baum 2009; Slomka et al. 2004, 2009). Interactive alignment may in fact still be an option in research applications and most manufacturers supply interactive software so that the user can manipulate one image with three-dimensional (3-D) rotations and translations until it matches the second image to be aligned with, but these operations are time-consuming and subjective and consequently unfeasible for clinical needs.

From the technical point of view, image registration is implemented by the optimal transformation that best aligns one of the images to the other (Hill et al. 2001; Hutton et al. 2002; Maintz and Viergenev 1998). In this optimization process, three main components have to be defined for a registration algorithm to be correctly designed: a transformation model that defines how the coordinates between the two images are related (e.g., rigid versus deformable transformations); a similarity metric that quantifies the degree of alignment between the images (e.g., mutual information, sum of squared differences (SSD), etc.); and a numerical routine that iteratively moves one of the images, evaluates the agreement between the datasets and stops when a predefined level of matching has been reached. Although a wide variety of different algorithms have been developed and proposed throughout the years, all registration techniques can be broadly grouped into two general approaches: *surface-based* and *volume-based* registration (Piccinelli and Garcia 2013). In surface-based techniques, landmark points or geometric features (edges, organ contours) are extracted—an operation commonly called segmentation—from both images and are successively aligned. Voxel-based methods directly use the original image pixel intensities to define a similarity metric and perform the spatial alignment.

Image fusion software has considerably improved in the past few years in terms of algorithm accuracy (2.0–3.0 mm) (Giovanella and Lucignani 2010), time efficiency, and the sophistication of 3-D visualization techniques and displays. The widespread acceptance of DICOM (Digital Imaging and Communication in Medicine) standard has also made this approach flexible and affordable according to the specific institution needs. First, no real limitation exists on how many images can be fused, differently from actual dual-modality hybrid systems, and the software approach remains basically device independent while not all imaging techniques can be integrated into a single apparatus. Particularly interesting, it is also the option of performing serial registration of images for follow-up studies and evaluation of treatment efficacy (Faber et al. 1991). However, the main downsides remain the difficulty in finding robust and automatically matching features between images (Piccinelli and Garcia 2013) as well as designing metrics that can robustly guide the registration process. Validation of nonlinear algorithms, which are desirable for the accurate alignment of soft tissues, is also the focus of active research (Rueckert and Aljabar 2010).

Although the inherent technical difficulties and the required specific expertise have hampered a more substantial transfer of software multimodality image fusion into clinical environments, the image registration of stand-alone acquisitions has paved the way for the development of hybrid scanners, demonstrating the clinical benefits of image integration to begin with (Slomka et al. 2009).

13.2.2 HARDWARE FUSION

In a hardware fusion approach, the radiological devices are physically integrated into a single apparatus and the images are acquired within one single imaging session. Hybrid systems such as positron emission tomography (PET)/computed tomography (CT) and single-photon emission computed tomography (SPECT)/CT represent an important step forward toward a fully integrated device, although the acquisitions are presently not truly *simultaneous*. Current clinical hybrid machinery actually enables *sequential imaging* (Cherry 2009; Patton et al. 2009). The system is designed so that the two imaging devices are latched together in a tandem configuration and the patient can be moved from one scanner to the other (Figure 13.1) while the hardware of the two devices remains separated. The system is precisely calibrated so that the coordinate transformation between the two acquired volumetric datasets is predefined and stored in the system. No image registration is in principle needed and straightforward fusion and display are performed with improved spatial registration with respect to the software approach.

In practice, even when integrated scanners are available, image registration software is still needed for a number of reasons (Flotats et al. 2011). Since acquisitions are mostly sequential, the patient may move between the scans, and specifically to cardiac applications, the heart may shift due to the effects of pharmacological stress. Additionally, imaging the breathing chest presents a problem. X-ray CT images are acquired in much less time than MPI images. A single CT slice represents a snapshot of the chest at one instant while MPI images are acquired over several minutes. As such, if standard CT procedures are followed, the two image sets will most likely not match (Goetze and Wahl 2007). Standard image registration techniques may be used to correct for this misregistration.

The development of hybrid systems that perform true simultaneous acquisitions where both modalities acquire the same volume at the same time is an active field of research. This is a particularly challenging endeavor since the different components from either modality can greatly interfere with each other, causing various types of artifacts. Promising results have been recently reached by PET/magnetic resonance (MR) and SPECT/MR prototypes, although they have not yet routinely entered clinical settings (Rischpler et al. 2013).

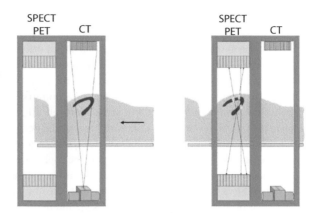

Figure 13.1 Most PET/CT and SPECT/CT scanners follow the same basic design where both scanners are integrated into a single unit with one table that carries the patient between the two scanners. Images must be acquired sequentially in both systems since the two units are separated along the table. In operation, a CT scout scan is used to determine the position of the heart for both the MPI and CT scans. The MPI and CT scans are then done in sequence (either may be first, depending on the circumstances).

13.3 PET/CT AND SPECT/CT ATTENUATION CORRECTION

While more detailed reviews of attenuation correction (AC) methods for both SPECT/CT and PET/CT can be found in other chapters of this book; herein, we briefly review the crucial role of image registration software and image realignment techniques for this important component of hybrid systems.

Attenuation correction is the most immediate exemplification of the synergistic role of multimodality imaging. Nonhomogeneous photon attenuation in the thorax is the main drawback of MPI. Attenuation artifacts can produce a decrease in both MPI sensitivity and specificity. To overcome this problem, CT-derived attenuation correction maps can be created and used to correct the original datasets greatly improving image quality (Heller et al. 2004; Kinahan et al. 2003; Machac et al. 2006) for both PET/CT and SPECT/CT. For AC purposes, the CT scan is commonly a fast low-dose acquisition that does not add much radiation burden to the patients. The crucial requirement for the correction to be performed properly remains the accurate alignment of the two volumetric datasets. As mentioned in the previous section, imaging of the heart is particularly challenging due to the inherent movement of the heart, its displacement in the cardiac cavity after pharmacological stress induction and respiratory artifacts. It has been recognized that the misregistrations between the scans pose a problem when the acquisition of the scans is sequential. This issue has been investigated (Chen et al. 2006; McCord et al. 1992), and realignment of the two scans by way of different techniques has been proposed (Fricke et al. 2004; Loghin et al. 2004; Martinex-Moller et al. 2007).

Standard practice for patients undergoing rest/stress exams is to perform two separate CT scans for AC: in resting studies, either MPI or CT can come first, while for stress studies, it is recommended that the CT scan follow the MPI acquisition to reduce misregistration artifacts. Some clinics utilize one CT scan and registration software, as shown in Figure 13.2, to ensure that the CT and MPI match throughout the imaging sequence (Nye et al. 2006).

Software approaches have been proposed also for the minimization of misregistration due to breathing artifacts. One solution is to reduce the temporal resolution of the CT to match the PET by using a very slow acquisition of a spiral CT or by averaging several CT images made over the respiratory cycle (McCord et al. 1992; Pan et al. 2005, 2006). Also, automated techniques that incorporate emission data into the CT-based attenuation map for better correction have been investigated and shown promising results (Martinex-Moller et al. 2007).

Misregistration artifacts appear to be more severe for PET than for SPECT, mainly because of the reduced spatial resolution of SPECT, but all SPECT/CT and PET/CT images require a careful quality control and, if necessary, realignment—either manually or by means of automated software—to avoid the risk in introducing artificial defects (Figure 13.3).

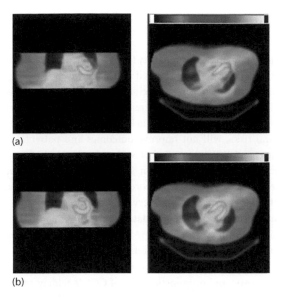

(a)

(b)

Figure 13.2 Proper registration of PET and CT is essential for accurate PET attenuation correction. In the top panels (a), the PET heart overhangs the soft-tissue into the lung on a segmented CT scan, perhaps due to movement of the patient between the two scans. The attenuation correction algorithm treated the lateral wall of the heart as if it had the same density at lung and severely under corrected for attenuation. In the bottom panels (b), the heart can be seen properly positioned after manual registration and the artefactual lateral wall defect resolves. All PET-CT images should be checked for proper alignment.

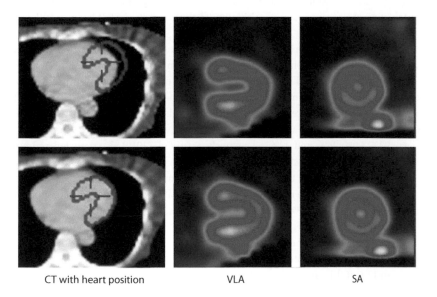

CT with heart position   VLA   SA

Figure 13.3 SPECT/CT images are acquired sequentially (first SPECT then CT or vice versa). If the patient moves between the two scans, the spatial mismatch can result in artefactual defects in the attenuation corrected images. Patient movement between the SPECT and the CT acquisitions resulted in a misalignment of the heart in the two modalities as shown in the top panels. This produced an undercorrection of the anterior wall of the heart and an artefactual anterior wall defect. Correction software was used to realign the SPECT and CT before reconstruction and the heart reconstructs as normal as shown in the bottom panels. VLA, vertical long axis; SA, short axis.

13.4 INTEGRATION OF MPI AND CTA

The most common application of image multimodality fusion in cardiovascular medicine is the assessment of CAD severity. Recent pivotal clinical trials have indeed underlined as morphological and functional data can provide complementary information regarding a specific lesion. Such comprehensive information adds to the global clinical picture and helps selecting the proper treatment. Patients will benefit from invasive procedures only if flow-limiting stenoses are present and ischemia is proven, while aggressive treatment, which does not come without periprocedural risks, will not improve outcomes and patients' symptoms (Gaemperli et al. 2012; Lee et al. 2013; Tonino et al. 2010). Integration of MPI and CTA can in fact guide a more judicious selection of patients who are scheduled for catheterization. Additionally, given that standard vascular territories do not always represent the anatomy of the patient's coronary trees, the CT-derived anatomy will improve the identification of the culprit lesion, further aiding intervention planning.

The less demanding integration of CT-derived data to MPI uses coronary calcium assessment (CCS). Technically, the inclusion is straightforward since low- and medium-quality acquisitions, namely, the same employed for AC, can be used for CCS evaluation, indeed keeping radiation dose to a minimum. CCS represents a surrogate marker for atherosclerosis burden, although low values cannot be interpreted as absence of CAD. The additional diagnostic and prognostic value of CCS is widely accepted, specifically in cases of normal SPECT or PET, where it can help uncover multivessel disease (Blankstein and Dorbala 2010; Chang et al. 2009; Rosen et al. 2009). Normal MPI associated with high CCS usually indicates low risk of events in the short term but intermediate risk in the longer term, as compared to those with little or no calcifications. In a recent study (Schenker et al. 2008), ^{82}Rb PET/CT system was used to assess the increase in event rates with increasing CCS with or without perfusion defects.

Caveats to the use of CCS and MPI resides in its inability of defining stenosis severity and detecting noncalcified plaques, which are often the culprit of acute syndromes (Hsiao et al. 2010). The clinical quality of CTA has greatly improved in the last decades, and currently, CTA represents the most promising non-invasive modality for the assessment of degree of stenosis. Additionally, with continuing improvement in spatial resolution, CT allows visualization of the vessel wall proving information on presence and composition of eccentric non-obstructive plaques, although its clinical value hasn't been completely determined yet.

The feasibility and accuracy of CAD assessment with integrated systems were first reported by Namdar et al. (2005) using hybrid PET/CT with four-slice CT to perform contrast-enhanced CTA and standard rest/stress perfusion acquisition. This and subsequent studies have confirmed the robustness of hybrid PET/CT systems in detecting CAD, resulting in improved specificity and positive predictive value (PPV) with respect to stand-alone acquisitions. In a recent study (Kajander et al. 2010) with ^{15}O-H$_2$O PET/CT (specifically 64-detector CT), the diagnostic accuracy was 98%, while sensitivity, specificity, PPV, and negative predictive value (NPV) were 93%, 99%, 96%, and 99%, respectively. Sato et al. (2010) also reported an improved performance of the ^{201}Tl SPECT/CT for detecting ≥50% stenosis on CA than stand-alone CTA: specificity increased from 80% to 92% and PPV from 69% to 85%; sensitivity remained 95%, and NPV, 97%. Rispler et al. (2007) also investigated patients with known or suspected CAD by means of SPECT/CTA with global improvement of both specificity (95%) and PPV (77%). All these studies indicate that the noninvasive detection of hemodynamically significant CAD by integrated MPI and CTA is highly accurate. Figures 13.4 and 13.5 show examples of integrated displays for the fusion data. It must be noted that to obtain a 3-D visualization like the one in Figure 13.5e, simple integrated systems are not sufficient and additional software postprocessing as described in Section 13.6 of this chapter is needed.

The incremental value of explicit registration of images acquired in separated scanners has also been addressed in numerous studies, and their results actually provided the background for the development of hybrid systems. The studies indicated that in one third of the analyzed cases, the availability of fused images improved diagnostic performance with respect to side-by-side evaluation (Gaemperli et al. 2007b; Santana et al. 2009; Slomka et al. 2009). The number of equivocal lesions decreased and the sensitivity to multivessel disease was improved.

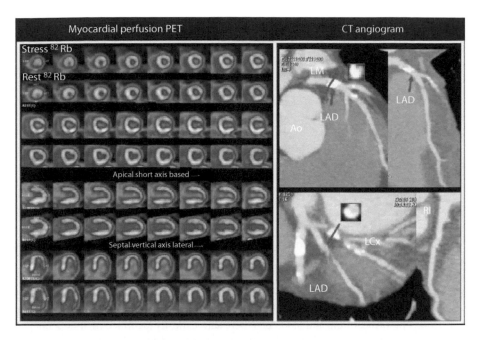

Figure 13.4 Integrated [82]Rb myocardial perfusion and CT coronary angiography images in a patient with suspected CAD. The rest and stress myocardial perfusion PET images demonstrate a small perfusion defect of severe intensity in the mid and basal lateral walls with significant but not complete reversibility. On CTA, the coronary anatomy was right-dominant (images not shown), with severe luminal narrowing of the proximal LAD and left circumflex coronary artery (LCX) (shown in cross-sectional images of the vessels in the inset). The integrated PET/CT results suggest that although the patient has extensive coronary atherosclerosis in the proximal LAD and LCX, only the LCX stenosis is hemodynamically significant. Follow-up catheter-based coronary angiography demonstrated a totally occluded LCX (filled by collaterals), with minimal disease in the proximal LAD. ANT, anterior; SEP, septal; LAT, lateral; INF, inferior; LM, left main artery; Ao, aorta; RI, ramus intermedius. (From Di Carli MF et al., *J Nucl Cardiol*, 13, 139–144, 2006. With permission.)

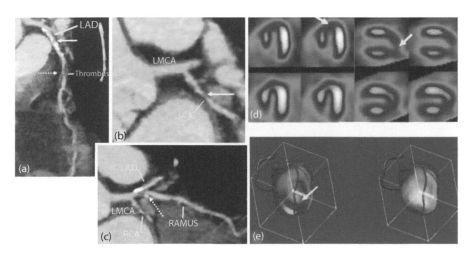

Figure 13.5 (a) CTA curved multilinear reformat of the LAD shows an irregular mostly calcified plaque causing >50% diameter stenosis (arrow). LCX (b) and RCA (c) multiplanar reformats with significant stenosis in the proximal portions of both arteries (arrows). (d) Perfusion SPECT at stress (top) and rest (bottom) with a reversible defect in the apex and antero-apical region (arrows). (e) Fused display shows overlap of arteries segmented from the CTA and the epicardium detected from the perfusion SPECT. This display graphically demonstrates that the reversible antero-apical defect is in fact a result of the LAD stenosis and that interventional therapy should be directed to the LAD lesion. RCA, right coronary artery; LCX, left circumflex artery; LMCA, left main coronary artery. (Adapted from Rispler S et al., *J Am Coll Cardiol*, 49, 1059–1067, 2007. With permission.)

From the technical point of view, in the majority of studies, image registration is performed by means of surface-based algorithms: the original images are preprocessed to extract landmarks that are eventually used for the final alignment. In the last few decades, nuclear cardiac imaging has been the focus of extensive research in order to objectively and automatically *quantify* perfusion images (left ventricle [LV] epicardial and endocardial boundaries, wall motion, ejection fraction, etc.). A number of software packages have been developed and made available to clinicians that allowed the standardization and the widespread utilization of nuclear cardiology as the most reliable tool for CAD assessment. Image processing of anatomical images on the other hand has lagged behind with regard to automated feature extraction.

The initial attempts to provide fusion of anatomy and function used biplane CA images to extract anatomy mostly with semiautomatic methods. In Faber et al. (2004), the angiographic images were manually segmented with *ad hoc* software tools to extract vessel centerlines and registered to SPECT-derived LV surface by aligning the left anterior descending (LAD) to the anterior interventricular groove; the Iterative Closest Point algorithm (Besl and McKay 1992) and a nonlinear warping to adapt the coronaries to the LV surface were successively applied and the method was validated. Figure 13.6 shows examples of the 3-D final fused models. A similar work was proposed, although both SPECT and CA acquisition parameters were used to align the models (Schindler et al. 1999).

Since the advent of CTA, researchers have been trying to make use of all the volumetric information available to perform SPECT/CTA fusion. A volume-based algorithm was developed (Slomka et al. 2009) for semiautomated registration: preprocessing of the original images was still required for the extraction of coronary tree anatomy; LV myocardium and blood pool were automatically identified in SPECT images and new volumetric dataset with masked myocardium and blood pool was created; and the CTA image was smoothed with a Gaussian filter and the images so obtained were used as inputs to the optimization algorithm. A rigid transformation was assumed and the sum of intensity differences (SSD) was used as a similarity metric. Faber and colleagues proposed in Faber et. al. (2011) a new volume-based methodology that made also use of the presence of the right ventricle (RV) to register SPECT/CTA. The resulting fused display is highly realistic, as depicted in Figure 13.7. Although the algorithm was validated using manually extracted anatomical images, preliminary results were recently presented on the complete automation of the technique (Piccinelli et al. 2014). To account for phase mismatch between SPECT and CTA, a deformable registration was also suggested

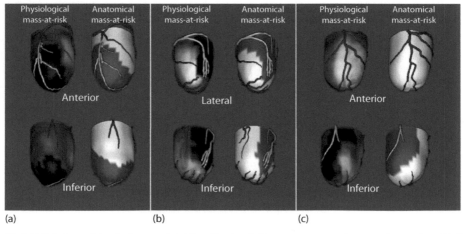

Figure 13.6 Validation of the fusion method for aligning 3-D coronary trees from angiography with 3-D epicardial surfaces from perfusion SPECT. Anatomical mass at risk (purple areas) computed from the coronary artery anatomy vs. physiological mass-at-risk (black areas) computed from perfusion quantitation. Vessels distal to the anatomic lesions are highlighted in green. Overlap of the purple and black areas is a measure of fusion accuracy, since in these patients, the perfusion abnormality was caused by the arterial blockage during balloon inflation for PTCA. (a) Example of an LAD lesion. (b) Example of an LCX lesion. (c) Example of an RCA lesion. LAD, left anterior descending artery; RCA, right coronary artery; LCX, left circumflex artery. (From Faber TL et al., *J Nucl Med*, 45, 745–753., 2004. With permission.)

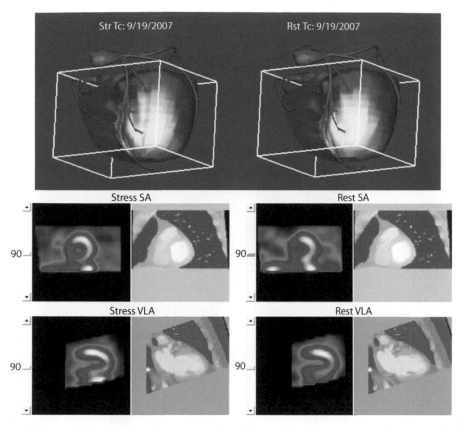

Figure 13.7 3-D and 2-D displays of anatomical and functional data as obtained with methodologies described in Faber et al. (2011). (Top) 3-D surface rendering of the RV and LV and coronary trees extracted from the CTA; RV surface is opaque, and the LV surface has mapped perfusion values from stress/rest SPECT. The user is allowed to interact with the models and rotate them. (Bottom) Short and horizontal axis slices of synchronized CTA and SPECT scans.

to select the SPECT phase that best matched the CTA and improved registration accuracy (Woo et al. 2009). In Gaemperli et al. (2007b), a different approach was described based on a volume-rendering type of display. The images are mostly manually registered, the LV epicardium is semiautomatically extracted, and the SPECT intensity is projected into the CTA-derived epicardial surface. The fusion process was highly reproducible. Figures 13.8 and 13.9 show examples of these techniques.

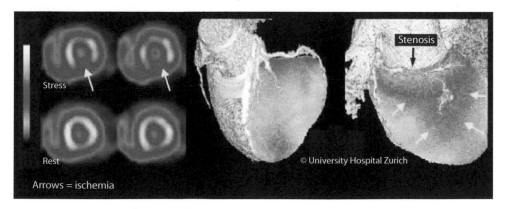

Figure 13.8 SPECT and CT fused and displayed by means of volume rendering methods. The stenosis in the LCX is seen to correspond well to the lateral wall perfusion abnormality. (Taken from Wagner HN, *J Nucl Med* 47, 13N–39N, 2006. With permission.)

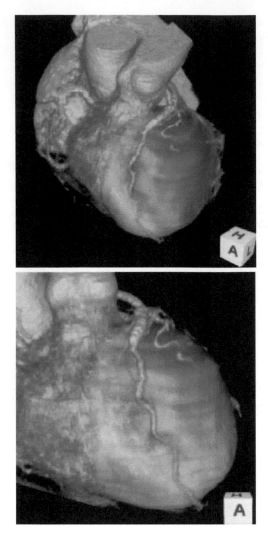

Figure 13.9 Left anterior oblique (top) and right anterior oblique (bottom) views of a 3-D fused image taken from a combined SPECT/CT scanner. The location of the left anterior descending artery (LAD) stenosis and the myocardial perfusion defect (green) can be assessed. (From Nakauro T et al., *Circulation*, 112, e47–e48, 2005. With permission.)

The accuracy of any given image registration technique can be potentially improved by adding to the algorithm design additional anatomical or functional data retrieved from either modality or by moving to more advanced transformation models, i.e., deformable versus rigid. On the other hand, these changes may affect the algorithm's performance in terms of computational power and/or time requirements and consequently limit its suitability for clinical settings (Audette et al. 2000; Rueckert and Aljabar 2010)

13.5 INTEGRATION OF MPI AND MAGNETIC RESONANCE IMAGING

Although cardiac magnetic resonance imaging (MRI) is not yet being used commonly as a clinical diagnostic procedure, it is nonetheless the gold standard for investigating cardiac anatomy and function. Its high resolution in space and time provides exquisite pictures of the moving heart. In addition, delayed enhancement images using gadolinium may be used to visualize myocardial infarcts, including those small subendocardial infarcts that are impossible to analyze with nuclear medicine. While early imaging with gadolinium may provide some measure of perfusion with MRI, nuclear medicine is still the gold standard for this application. Therefore, the registration of MRI with MPI should allow better integration of these three pieces of information: function, infarction, and perfusion. Coronary magnetic resonance angiography can be used to assess CAD but is less commonly used with respect to CTA (Hundley et al. 2010; Schuetz et al. 2010), exhibiting lower diagnostic accuracy mainly given its lower spatial resolution (Sakuma 2011).

Due to the delayed development of hybrid PET/MR or SPECT/MR, there are limited data supporting the extensive use in the clinic of these devices (Büscher et al. 2010; Gaa et al. 2004; Nekolla et al. 2009; Rischpler et al. 2013). In animal models, hybrid PET/MR was performed and functional parameters were obtained with good accuracy (Büscher et al. 2010). A pilot study has also been recently conducted comparing PET/MR and SPECT for the assessment of cardiac viability (Carballo et al. 2012). Although the authors reported good agreement between the modalities considered, they also acknowledged the need for further investigations.

The software approach consequently represents the most followed strategy for image fusion of MPI and MR images. Its feasibility and accuracy have been demonstrated in a number of studies. Registration of SPECT perfusion images with MRI of the same patient was first demonstrated by Faber et al. (1991). A surface-based method was applied: endocardial boundaries were detected in each dataset; end-diastolic (ED) and end-systolic frames were identified from volume calculations; and finally, a cost function consisting of the distance of the MRI ED surface from the SPECT ED surface plus the distance of the MRI end-systolic surface from the SPECT end-systolic surface was minimized over translation and rotation in three dimensions (Pelizzari et al. 1989). The resulting transformation was applied to the SPECT volume and interpolated in time. In clinical tests, LV and RV walls in SPECT were seen to align well with those in MR images, and areas of decreased perfusion in SPECT images were aligned with areas of decreased myocardial thickening in MR images. In the last decades, various other groups have focused on surface-based methodologies for the registration of PET or SPECT images to MR (Makela et al. 2003; Sinha et al. 1995). Crucial to surface-based methods is the extraction of anatomical landmarks from the images. In Sinha et al. (1995), LV boundaries were extracted by means of morphological and linear filtering techniques; surface fitting techniques (Pelizzari et al. 1989) were employed to find the best transformation between the two surfaces that was then used to register the original images. The method was validated by calculating the difference between user-identified landmarks in the two images after the transformation, using six MRI/PET image pairs and an average of 14 landmarks per pair. The accuracy was determined to be 1.3 ± 1.1 mm for the ED images and 1.95 ± 1.6 mm for end-systolic images. In Makela et al. (2003), deformable models were used to segment different structures in the thorax, including LV and RV, and were subsequently rigidly aligned; in Figure 13.10, a 3-D model of the heart obtained with this approach is displayed color coded with PET-FDG uptake values.

Volume-based approaches have also been explored. These methods are arduous to apply to multimodality image registration since pixel values can be significantly different and a proper similarity metric may be difficult to design. However, if the heart can be isolated from both MRI and nuclear images, then it is possible to align them using a cost function that operates on the original myocardial pixel values and attempts to reduce the differences between them. This is the approach taken by Aladl et al. (2004). When their automatic approach was compared to an interactive one performed by an expert, they found differences in translation and orientation to average ~1 pixel and ~4°, respectively, based on 20 patients.

Misko et al. (2006) described the use and value of registered SPECT and cardiac MRI. They used manual alignment tools to reorient the ED frame of the SPECT study with the ED frame of the MRI. Perfusion data were obtained from this ED SPECT slice and function was obtained from the MRI. They used delayed enhancement MRI techniques to define viability. The registered datasets were not displayed in 3-D; instead, they were divided into six segments per short axis slice, and perfusion, motion, and viability information was

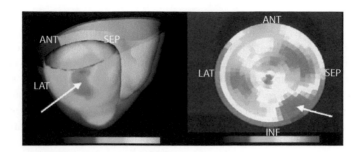

Figure 13.10 Representation of PET FDG (fluorodeoxyglucose) uptake values (left) on the heart model created from an MRI. Polar map of FDG uptake is seen in the polar map (right). (From Makela T et al., *Med Image Anal*, 7, 377–389, 2003. With permission.)

combined in 18 patients. They computed a statistic called myocardial uptake per volume, which is counts from the perfusion image normalized by the volume determined by the MRI LV boundaries. They were able to show strong relationships between reduced perfusion and delayed enhancement. Despite the good results, the authors did not address the crucial issue on whether the fusion of these modalities could aid diagnosis compared to traditional examination.

Despite the preliminary good results of the integration of PET (or SPECT) with MR (either software or hardware), the clinical relevance remains a crucial issue to further advance the development of both approaches.

13.6 QUANTITATIVE VERSUS VISUAL ASSESSMENT

Software versus hardware strategies for image fusion have different advantages and disadvantages. In most occasions, the hardware approach allows a very direct and straightforward visualization of the final *fused* image, even if images have to be double checked for misregistrations. In this case, the speed is the crucial advantage. The pivotal choice is then how the merged data are visualized for the physicians to take full advantage of the correspondence between anatomy and function. The most common approach makes use of volume-rendering techniques (Di Carli et al. 2007): the coronary arteries are selectively *visualized* from the CTA volumetric datasets and the perfusion/metabolism data are used to "*paint*" the myocardium, while the orientation of the two registered datasets is kept in synchrony. Despite being mostly qualitative, no quantitative information is retrieved from these types of display unless specific additional image preprocessing and postprocessing are applied (Rispler et al. 2007), and this visual assessment has been reported to improve diagnostic accuracy over side-by-side reading (Figures 13.4 and 13.5).

The software approach may be less immediate and needs robust and automated algorithms to enter the clinical settings but allows the extraction of quantitative data from the images being analyzed. Clinicians are eventually provided with tools for the objective interpretation and comparison of cases. Quantitative nuclear cardiology is one of the most successful examples of the potential of imaging quantification: software packages for the reading of SPECT/PET images represent today a fundamental clinical tool in the hands of cardiologists (Ficaro et al. 2007; Garcia et al. 2007; Germano et al. 2007; Liu et al. 1999; Watson 1999; Watson and Smith 2007). The same growth can potentially be envisioned for multimodality imaging devices. For example, Figure 13.11 demonstrates the implementation of the anatomical mass at risk extraction (Faber et al. 2004): given the position of a specific lesion on the coronary tree, the portion of myocardium affected by the obstruction can be estimated, visualized, and compared with perfusion data.

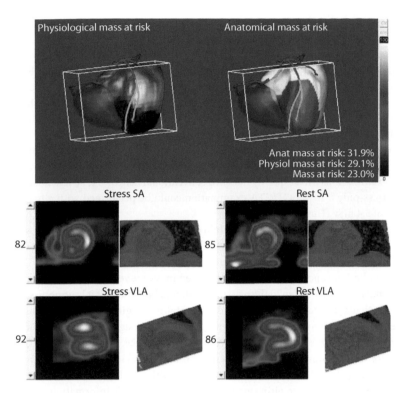

Figure 13.11 3-D and 2-D fused displays obtained with techniques proposed in Faber et al. (2011) similar to Figure 13.7. Here, the mass at risk computation (see also Figure 13.6) described in Faber et al. (2004) is applied to the more realistic myocardium anatomy and coronary tree morphology extracted from the CTA. The user can position a lesion on the coronary trees, visualize the anatomical mass at risk (right model, purple surface), and compare it to the corresponding mapped blackout perfusion data (left model). SA, short axis; VLA, vertical long axis.

Additionally, the availability of postprocessed data, such as color-coded 3-D or 4-D models, centerlines, and volumetric meshes, offers a wider range of techniques for visualization and even for direct interaction of the user with the displayed information.

13.7 RADIATION DOSE

One of the most important issues in multimodality imaging is its associated radiation burden as detailed in Chapter 19 of this book. As a general principle in diagnostic radiology the effective radiation dose given to the patient by each specific modality should be kept as low as possible and an active discussion is ongoing about the radiation given even by each stand-alone modality.

It is widely recognized that due to the short half-life of PET tracers, the radiation given by routine stress/rest PET tests is much lower than the one associated with SPECT exams, going from 7.0–9.0 mSv for SPECT to as low as 1.0–2.0 mSv (Flotats et al. 2011; Thompson and Cullom 2006) with PET. On the other hand, SPECT radiation can be significantly reduced by applying new iterative reconstruction methods and dedicated detectors and collimators optimized specifically for MPI (Esteves et al. 2009; Slomka et al. 2012). Additionally, the implementation, whenever possible, of stress-only examinations is gaining support and will further reduce the radiation exposure (Chang et al. 2010; Iskandrian et al. 2010; Pazhenkottil et al. 2010).

The effective radiation dose from cardiac CT greatly depends on the protocol implemented. Dose is minimal (~1.0 mSv) when CT scan is used to extract CCS or to perform CT-derived AC. The dose increases when clinical CTA is required and augment as the slice thickness decreases. Mean radiation values of CTA were estimated to range between 8 and 18 mSv (Hausleiter et al. 2009), but application of recently developed cardiac CT protocols, such as prospective electrocardiogram (ECG) triggering, ECG-controlled current modulation, and body mass-adapted tube voltage scan, allows a dose reduction by 60%–80%. With the latest technology on dual-source CTs using prospectively triggered high-pitch spiral acquisition, a clinical quality CTA can be performed with an associated radiation dose of 1 mSv.

Taking advantage of all these improvements for stand-alone acquisitions, dual modalities also have greatly diminished their cumulative radiation doses to values ranging from 13.0 to 5.0 mSv using ^{82}Rb PET/CT and providing CTA examinations whose quality matched conventional helical CTA (Achenbach et al. 2010; Javadi et al. 2008). A stress-only ^{99}Tc SPECT/CTA examination could be performed with a total radiation for the combined study of 5.4 mSv (Husmann et al. 2009).

As far as MR is concerned, its integration with MPI images is particularly desirable since no additional radiation is given by MR, a feature that could make PET/MR or SPECT/MR the most suitable tool for follow-up studies.

Although the radiation exposure has greatly decreased in recent years, the combined hybrid acquisition still gives additional radiation burden to the patient that should be avoided if it is unnecessary for diagnostics. The selection of patients that could benefit from dual-modality imaging is a crucial issue for clinical usage of hybrid systems, especially for CAD assessment, while it is less significant—at least from the radiation exposure point of view—with software approach. In clinical practice, an additional diagnostic test is commonly required for CAD evaluation only if the results from the previous exam are equivocal and no treatment selection can be made with confidence. Retrospective multimodality fusion of the sequentially obtained images can indeed be performed exploiting at maximum all the available information. At present, it is largely unknown what kind of patients should directly undergo such integrated examinations for clinical effectiveness and minimization of radiation dose (Flotats et al. 2011).

13.8 WORKFLOW AND COSTS

In addition to the clinical practice of adopting a sequential diagnostic approach to minimizing radiation dose, that is, to plan a second test only if the first has produced equivocal results, other primary issues such as costs and clinical workflow can influence the routine use of hybrid systems (von Schulthess and Burger 2010).

One of the crucial aspects deals with the differences in scanning times required from the two integrated devices. Nuclear cardiac images are acquired in much longer times compared to CTA that would consequently remain unused for all the duration of the functional acquisition. This can generate high costs per patient examined, particularly if CTA is a high-end model with 64 rows or more of detectors. Moreover, if used for AC only, CTA immobilization would also result in suboptimal use of an expensive device. The same consideration holds for MPI and MRI integration, although the acquisitions may be partially simultaneous.

For hybrid systems, it would also be rather difficult and expensive to integrate the rapid technological changes occurring in the radiological field, namely, top-line CT, basically creating devices that lag behind with respect to stand-alone models. Software fusion stays rather inexpensive, its only crucial component being the need for computational power, and is easily upgraded with new algorithms and approaches. Moreover, as detailed within this chapter, a software component remains necessary in hybrid systems as well to correct misregistrations and allow advance visualization techniques.

From the workflow perspective, the crucial advantage of hybrid systems is that the patient needs to be uploaded and downloaded only once instead of two times and the integrated device may fit into one room and requires only one operating team instead of two.

13.9 MULTIMODALITY IMAGE FUSION INTERPRETATION

With the spreading and availability of *fused* images, it has become critical to address the issue of competence and training in multimodality imaging techniques (Hricak et al. 2010).

The interpretation of hybrid images requires knowledge of molecular biology and metabolism as well as anatomical imaging. The continued advances in technology are now demanding a parallel expansion of our knowledge so that we are equipped to exploit the full potential offered by each technique. Radiology and nuclear cardiology are distinct imaging specialties: radiology focuses on anatomy and pathology, while nuclear cardiology focuses on biochemistry and pathophysiology.

More cross-training is indeed perceived as fundamental to facilitate the optimal use of fused techniques while retaining the necessary competence as expected by the patient. Although substantial changes in the educational and training programs would be necessary, the creation of new experts in multimodality imaging could almost be envisaged, more organ oriented than technique orientated as it is today.

13.10 CONCLUSIONS

More than a decade of research in the field of image registration to merge anatomical and functional information and the continuous technological advances in radiological imaging devices has made multimodality image fusion not only a feasible clinical instrument but also a desirable one for any radiology department and for the investigation of a wide range of cardiovascular diseases. Whether hybrid systems or retrospective image registration should be employed greatly depends on the particular needs of each institution, on their clinical workflow and costs. Regardless of the strategy selected, the full integration of image fusion approaches into the radiological practice will not only impact diagnosis, prognosis, and treatment selection procedures but also transform the profession of the radiologist himself/herself from the expert on a particular radiologic technique to the expert on the imaging of a specific organ, in our case the heart, with all available modalities.

ABBREVIATIONS

3-D, 4-D	Three-dimensional, four-dimensional
AC	Attenuation correction
CA	Coronary angiography
CAD	Coronary artery disease
CCS	Coronary calcium score
CT	Computed tomography
CTA	Computed tomographic angiography
ED, ES	End diastole, end systole
EF	Ejection fraction
FAME	Fractional flow reserve versus angiography for multivessel evaluation
FFR	Fractional flow reserve
LAD	Left anterior descending
LV, RV	Left ventricle, right ventricle
MPI	Myocardial perfusion imaging
MRI	Magnetic resonance imaging
PET	Positron emission tomography
PPV, NPV	Positive predictive value, negative predictive value
SPECT	Single-photon emission computed tomography

REFERENCES

Achenbach S, Marwan M, Ropers D et al. 2010. Coronary computed tomography angiography with a consistent dose below 1 mSv using prospectively electrocardiogram-triggered high-pitch spiral acquisition. *Eur Heart J* 31:340–46.

Achenbach S, Friedrich MG, Nagel E et al. 2013. CV imaging: What was new in 2012? *J Am Coll Cardiol Imaging* 6:714–34.

Aladl UE, Hurwitz GA, Dey D et al. 2004. Automatic image registration of gated cardiac single photon emission computed tomography and magnetic resonance imaging. *J Magn Reson Imaging* 19:283–90.

Audette MA, Ferrie FP, Peters TM. 2000. An algorithmic overview of surface registration techniques for medical imaging. *Med Image Anal* 4:201–17.

Beller GA. 2010. Recent advances and future trends in multimodality cardiac imaging. *Heart Lung Circulation* 19:193–209.

Besl PJ, McKay ND. 1992. A method for registration of 3D shapes. *IEEE Trans PAMI* 14:239–56.

Blankstein R, Di Carli F. 2010. Integration of coronary anatomy and myocardial perfusion imaging. *Nat Rev Cardiol* 7:226–36.

Blankstein R, Dorbala S. 2010. Adding calcium scoring to myocardial perfusion imaging: Does it alter physicians' therapeutic decision making? *J Nucl Cardiol* 17:168–71.

Büscher K, Judenhofer MS, Kuhlmann MT et al. 2010. Isochronous assessment of cardiac metabolism and function in mice using hybrid PET/MRI. *J Nucl Med* 51:1277–84.

Carballo D, Nkoulou R, Vincenti G et al. 2012. Value of a hybrid PET/MRI in the assessment of cardiac viability. *J Cardiovasc Magn Reson* 14:1–2.

Chang SM, Nabi F, Xu J et al. 2009. The coronary artery calcium score and stress myocardial perfusion imaging provide independent and complementary prediction of cardiac risk. *J Am Coll Cardiol* 54:1872–82.

Chang SM, Nabi F, Xu J et al. 2010. Normal stress-only versus standard stress/rest myocardial perfusion imaging. *J Am Coll Cardiol* 55:221–30.

Chen J, Caputlu-Wilson SF, Shi H et al. 2006. Automated quality control of emission-transmission misalignment for attenuation correction in myocardial perfusion imaging with SPECT-CT systems. *J Nucl Cardiol* 13:43–9.

Cherry SR. 2009. Multimodality imaging: Beyond PET/CT and SPECT/CT. *Semin Nucl Med* 39:348–53.

de Bruyne B, Pijls NHJ, Kalesan B et al. 2012. Fractional flow reserve-guided PCI versus medical therapy in stable coronary disease. *N Engl J Med* 367:991–1001.

Di Carli MF, Dorbala S, Curillova Z et al. 2007. Relationship between CT coronary angiography and stress perfusion imaging in patients with suspected ischemic heart disease assessed by integrated PET-CT imaging. *J Nucl Cardiol* 14:799–809.

Esteves FP, Raggi P, Folks RD et al. 2009. Novel solid-state-detector dedicated cardiac camera for fast myocardial perfusion imaging: Multicenter comparison with standard dual detector cameras. *J Nucl Cardiol* 16:927–34.

Faber TL, McColl RW, Opperman RM et al. 1991. Spatial and temporal registration of cardiac SPECT and MR images: Methods and evaluation. *Radiology* 179:857–61.

Faber TL, Santana CA, Garcia EV et al. 2004. Three-dimensional fusion of coronary arteries with myocardial perfusion distribution: Clinical validation. *J Nucl Med* 45:745–53.

Faber TL, Santana CA, Piccinelli M et al. 2011. Automatic alignment of myocardial perfusion images with contrast-enhanced cardiac computed tomography. *IEEE Trans Nucl Sci* 58:2296–302.

Flotats A, Knuuti J, Gutberlet M et al. 2011. Hybrid cardiac imaging: SPECT/CT and PET/CT. A joint position statement by the European Association of Nuclear Medicine (EANM), the European Society of Cardiac Radiology (ESCR) and the European Council of Nuclear Cardiology (ECNC). *Eur J Nucl Med Mol Imaging* 38:201–12.

Ficaro EP, Lee BC, Kritzman JN et al. 2007. Corridor4DM: The Michigan method for quantitative nuclear cardiology. *J Nucl Cardiol* 14:455–65.

Fricke H, Fricke E, Weise R et al. 2004. A method to remove artifacts in attenuation-corrected myocardial perfusion SPECT Introduced by misalignment between emission scan and CT-derived attenuation maps. *J Nucl Med* 45:1619–25.

Gaa J, Rummeny EJ, Seemann MD. 2004. Whole-body imaging with PET/MRI. *Eur J Med Res* 9:309–12.

Gaemperli O, Schepis T, Valenta I et al. 2007a. Cardiac image fusion from stand-alone SPECT and CT: Clinical experience. *J Nucl Med* 48:696–703.

Gaemperli O, Schepis T, Kalff V et al. 2007b. Validation of a new cardiac image fusion software for three-dimensional integration of myocardial perfusion SPECT and stand-alone 64-slice CT angiography. *Eur J Nucl Med Mol Imaging* 34:1097–106.

Gaemperli O, Saraste A, Knuuti J. 2012. Cardiac hybrid imaging. *Eur Heart Journal* 13:51–60.

Garcia EV, Faber TL, Cooke CD et al. 2007. The increasing role of quantification in nuclear cardiology: The Emory approach. *J Nucl Cardiol* 14:420–32.

Germano G, Kavanagh PB, Slomka PJ et al. 2007. Quantitation in gated perfusion SPECT imaging: The Cedars-Sinai approach. *J Nucl Cardiol* 14:433–54.

Giovanella L, Lucignani G. 2010. Hybrid versus version imaging: Are we moving forward judiciously? *Eur J Nucl Med Mol Imaging* 37:973–79.

Go AS, Mozaffarian D, Roger VL et al. 2013. Heart disease and stroke statistics—2013 update: A report from the American Heart Association. *Circulation* 127:e6–245.

Goetze S, Wahl RL. 2007. Prevalence of misregistration between SPECT and CT for attenuation-corrected myocardial perfusion SPECT. *J Nucl Cardiol* 14:200–06.

Hachamovitch R, Di Carli MF. 2008. Methods and limitations of assessing new noninvasive tests. Part II: Outcomes-based validation and reliability assessment of noninvasive testing. *Circulation* 117:2793–801.

Hacker M, Jakobs T, Hack N et al. 2007. Sixty-four slice spiral CT angiography does not predict the functional relevance of coronary artery stenoses in patients with stable angina. *Eur J Nucl Med Mol Imaging* 34:4–10.

Hausleiter J, Meyer T, Hermann F, Hadamitzky M, Krebs M, Gerber TC et al. 2009. Estimated radiation dose associated with cardiac CT angiography. *JAMA* 301:500–7.

Heller GV, Links J, Bateman TM et al. 2004. American Society of Nuclear Cardiology and Society of Nuclear Medicine joint position statement: Attenuation correction of myocardial perfusion SPECT scintigraphy. *J Nucl Cardiol* 11:229–30.

Hill DLG, Batchelor PG, Holden M et al. 2001. Medical image registration. *Phys Med Biol* 46:R1–45.

Hricak H, Choi BI, Scott AM et al. 2010. Global trends in hybrid imaging. *Radiology* 257:498–506.

Hsiao EM, Ali B, Dorbala S. 2010. Clinical role of hybrid imaging. *Curr Cardiovasc Imaging Rep* 3:324–35.

Hundley WG, Bluemke DA, Finn JP et al. 2010. ACCF/ACR/AHA/NASCI/SCMR 2010 Expert Consensus Document on Cardiovascular Magnetic Resonance. A Report of the American College of Cardiology Foundation Task Force on Expert Consensus Documents. *J Am Coll Cardiol* 55:2614–62.

Husmann L, Herzog BA, Gaemperli O, Tatsugami F, Burkhard N, Valenta I et al. 2009. Diagnostic accuracy of computed tomography coronary angiography and evaluation of stress-only single-photon emission computed tomography/computed tomography hybrid imaging: Comparison of prospective electrocardiogram-triggering vs. retrospective gating. *Eur Heart J* 30:600–7.

Hutton BF, Braun M, Thurfjell L, Lau DYH. 2002. Image registration: An essential tool for nuclear medicine. *Eur J Nucl Med* 29: 559–77.

Iskandrian AE. 2010. Stress-only myocardial perfusion imaging. *J Am Coll Cardiol* 55:231–33.

Javadi M, Mahesh M, McBride G et al. 2008. Lowering radiation dose for integrated assessment of coronary morphology and physiology: First experience with step- and-shoot CT angiography in a rubidium 82 PET-CT protocol. *J Nucl Cardiol* 15:783–90.

Kajander S, Joutsiniemi E, Saraste M et al. 2010. Cardiac positron emission tomography/computed tomography imaging accurately detects anatomically and functionally significant coronary artery disease. *Circulation* 122:603–13.

Kauffmann PA, Di Carli MF. 2009. Hybrid SPECT/CT and PET/CT imaging: The next step in noninvasive cardiac imaging. *Semin Nucl Med* 39:341–47.

Kramer CM. 1998. Integrated approach to ischemic heart disease. The one-stop shop. *Cardiol Clin* 16: 267–76.

Kramer CM, Narula J. 2010. Fusion images: More informative than the sum of individual images? *J Am Coll Cardiol Imaging* 3:985–86.

Kinahan PE, Hasegawa BH, Beyer T. 2003. X-ray-based attenuation correction for positron emission tomography/computed tomography scanners. *Semin Nucl Med* 33:166–79.

Lee AKY, Qutub MA, Aljizeeri A, Chow BJW. 2013. Integrating anatomical and functional imaging for the assessment of coronary artery disease. *Expert Rev Cardiovasc Ther* 11:1301–10.

Liu YH, Sinusa AJ, DeMan P et al. 1999. Quantification of SPECT myocardial perfusion images: Methodology and validation of Yale-CQ method. *J Nucl Cardiol* 6:190–203.

Loghin C, Sdringola S, Gould KL. 2004. Common artifacts in PET myocardial perfusion images due to attenuation-emission misregistration: Clinical significance, causes, and solutions. *J Nucl Med* 45:1029–39.

Machac J, Bacharach SL, Bateman TM et al. 2006. Quality Assurance Committee of the American Society of Nuclear Cardiology. Positron emission tomography myocardial perfusion and glucose metabolism imaging. *J Nucl Cardiol* 13:e121–51.

Martinex-Moller A, Souvatzoglou M, Navab N et al. 2007. Artifacts from misaligned CT in cardiac perfusion PET/CT studies: Frequency, effects and potential solutions. *J Nucl Med* 48:188–93.

McCord ME, Bacharach SL, Bonow RO, Dilsizian V, Cuocolo A, Freedman N. 1992. Misalignment between PET transmission and emission scans: Its effect on myocardial imaging. *J Nucl Med* 33:1209–14.

Maintz JB, Viergenev MA. 1998. A survey of medical image registration. *Med Image Anal* 2:1–36.

Makela T, Pham QC, Clarysse P et al. 2003. A 3D model-based registration approach for the PET, MR and MCG cardiac data fusion. *Med Image Anal* 7:377–89.

Min JK, Berman DS, Dunning A et al. 2012. All-cause mortality benefit of coronary revascularization vs. medical therapy in patients without known coronary artery disease undergoing coronary computed tomographic angiography: Results from CONFIRM (COronary CT Angiography EvaluatioN For Clinical Outcomes: An InteRnational Multi-center Registry). *Eur Heart J* 33:3088–97.

Misko J, Dzluk M, Skrobowska E et al. 2006. Co-registration of cardiac MRI and rest gated SPECT in the assessment of myocardial perfusion, function, and viability. *J Cardiovasc Mag Res* 8:389–97.

Nekolla SG, Martinez-Moeller A, Saraste A. 2009. PET and MRI in cardiac imaging: From validation studies to integrated applications. *Eur J Nucl Med Mol Imaging* 36:S121–30.

Namdar M, Hany TF, Koepfli P et al. 2005. Integrated PET/CT for the assessment of coronary artery disease: A feasibility study. *J Nucl Med* 46:904–05.

Nye J, Votaw J, Esteves F. 2006. Attenuation correction of cardiac PET/CT with gated, fast, and slow acquisition of the transmission map. *J Nucl Med* 47:132P.

Pan T, Mawlawi O, Nehmeh SA, Erdi YE, Luo D, Liu HH, Castillo R, Mohan R, Liao Z, Macapinlac HA. 2005. Attenuation correction of PET images with respiration-averaged CT images in PET/CT. *J Nucl Med* 46:1481–7.

Pan T, Mawlawi O, Luo D, Liu HH, Chi PC, Mar MV, Gladish G, Truong M, Erasmus J Jr, Liao Z, Macapinlac HA. 2006. Attenuation correction of PET cardiac data with low-dose average CT in PET/CT. *Med Phys* 33:3931–38.

Patton JA, Townsend DW, Hutton BF. 2009. Hybrid imaging technology: From dreams and vision to clinical devices. *Semin Nucl Med* 39:247–63.

Pazhenkottil AP, Herzog BA, Husmann L et al. 2010. Non-invasive assessment of coronary artery disease with CT coronary angiography and SPECT: A novel dose-saving fast-track algorithm. *Eur J Nucl Med Mol Imaging* 37:522–7.

Peifer JW, Ezquerra NF, Cooke CD et al. 1990. Visualization of multimodality cardiac imagery. *IEEE Trans Biomed Eng* 37:744–56.

Pelizzari CA, Chen GTY, Spelbring DR et al. 1989. Accurate three dimensional registration of CT, PET and/or MR images of the brain. *J Comput Assist Tomogr* 13:20–6.

Piccinelli M, Garcia E. 2013. Multimodality image fusion for diagnosing coronary artery disease. *J Biomed Res* 27:439–51.

Piccinelli M, Faber TL, Arepalli CD et al. 2014. Automatic detection of left and right ventricles from CTA enables efficient alignment of anatomy with myocardial perfusion data. *J Nucl Cardiol* 21:96–108.

Rispler S, Keidar Z, Ghersin E et al. 2007. Integrated single-photon emission computed tomography and computed tomography coronary angiography for the assessment of hemodynamically significant coronary artery lesions. *J Am Coll Cardiol* 49:1059–67.

Rischpler C, Nekolla SG, Dregely I et al. 2013. Hybrid PET/MR imaging of the heart: Potential, initial experiences, and future prospects. *J Nucl Med* 54:402–15.

Rosen BD, Fernandes V, McClelland RL et al. 2009. Relationship between baseline coronary calcium score and demonstration of coronary artery stenoses during follow-up MESA (Multi-Ethnic Study of Atherosclerosis). *J Am Coll Cardiol Imaging* 2:1175–83.

Rueckert D, Aljabar P. 2010. Nonrigid registration of medical images: Theory, methods and applications. *IEEE Signal Process Mag* 7:113–9.

Sakuma H. 2011. Coronary CT versus MR angiography: The role of MR angiography. *Radiology* 258:340–9.

Santana CA, Garcia EV, Faber TL et al. 2009. Diagnostic performance of fusion of myocardial perfusion imaging (MPI) and computed tomography coronary angiography. *J Nucl Cardiol* 16:201–11.

Saraste A, Knuuti J. 2012. Cardiac PET, CT, and MR: What are the advantages of hybrid imaging? *Curr Cardiol Rep* 14:24–31.

Sato A, Nozato T, Hikita H et al. 2010. Incremental value of combining 64-slice computed tomography angiography with stress nuclear myocardial perfusion imaging to improve noninvasive detection of coronary artery disease. *J Nucl Cardiol* 17:19–26.

Schenker MP, Dorbala S, Hong EC et al. 2008. Interrelation of coronary calcification, myocardial ischemia, and outcomes in patients with intermediate likelihood of coronary artery disease: A combined positron emission tomography/computed tomography study. *Circulation* 117:1693–700.

Schindler TH, Magosaki N, Jeserich M et al. 1999. Fusion imaging: Combined visualization of 3D reconstructed coronary artery tree and 3D myocardial scintigraphic image in coronary artery disease. *Int J Card Imaging* 15:357–68.

Schuetz GM, Zacharopoulou NM, Schlattmann P et al. 2010. Meta-analysis: Noninvasive coronary angiography using computed tomography versus magnetic resonance imaging. *Ann Intern Med* 152:167–77.

Sinha S, Sinha U, Czernin J et al. 1995. Noninvasive assessment of myocardial perfusion and metabolism: Feasibility of registering gated MR and PET images. *AJR Am J Roentgenol* 164:301–7.

Slomka PJ. 2004. Software approach to merging molecular with anatomic information. *J Nucl Med* 45(S):36–45.

Slomka PJ, Baum RP. 2009. Multimodality image registration with software: State of the art. *Eur J Nucl Med Mol Imaging* 1(S):44–55.

Slomka PJ, Cheng VY, Dey D et al. 2009. Quantitative analysis of myocardial perfusion SPECT anatomically guided by co-registered 64-slice coronary CT angiography. *J Nucl Med* 50:1621–30.

Slomka PJ, Dey D, Duvall WL et al. 2012. Advances in nuclear cardiac instrumentation with a view towards reduced radiation exposure. *Curr Cardiol Rep* 14:208–16.

Thompson RC, Cullom SJ. 2006. Issues regarding radiation dosage of cardiac nuclear and radiography procedures. *J Nucl Cardiol* 13:19–23.

Tonino PA, Fearon WF, De Bruyne B et al. 2010. Angiographic versus functional severity of coronary artery stenoses in the FAME study fractional flow reserve versus angiography in multivessel evaluation. *J Am Coll Cardiol* 55:2816–21.

Underwood SR. 2004. Myocardial perfusion scintigraphy: The evidence. *Eur J Nucl Med Mol Imaging* 31:261–91.

van der Hoeven, Schalij MJ, Delgrado V. 2012. Multimodality imaging in interventional cardiology. *Nat Rev Cardiol* 9:333–46.

von Schulthess GF, Burger C. 2010. Integrating imaging modalities: What makes sense from a workflow perspective? *Eur J Nucl Med Mol Imaging* 37:980–90.

Watson DD. 1999. Quantitative SPECT techniques. *Semin Nucl Med* 29:192–203.

Watson D, Smith WH. 2007. The role of quantification in clinical nuclear cardiology: The University of Virginia approach. *J Nucl Cardiol* 14:466–82.

Woo J, Slomka PJ, Dey D et al. 2009. Geometric feature-based multimodal image registration of contrast-enhanced cardiac CT with gated myocardial perfusion SPECT. *Med Phys* 36:5467–79.

Quantitative cardiac SPECT/CT

CHI LIU, P. HENDRIK PRETORIUS, AND GRANT T. GULLBERG

The principles and instrumentation of single photon emission computerized tomography (CT)/x-ray CT (single-photon emission CT [SPECT]) have been described in detail in Chapter 1 of this book. In the present chapter, we particularly focus on approaches to improving the accuracy of cardiac SPECT quantification with the corporation of hybrid SPECT/CT systems.

14.1 CORRECTIONS FOR ATTENUATION, SCATTER, COLLIMATOR-DETECTOR RESPONSES, PARTIAL VOLUME EFFECTS, AND PROJECTION TRUNCATION

14.1.1 IMAGE FORMATION

Image formation for SPECT is illustrated in Figure 14.1, depicting a dual-detector gamma camera and a source distribution f(x, y) at angle θ with respect to the x axis of Detector 1. Simultaneously, an image is acquired on Detector 2 at θ = 90°. The detectors (or heads) of the gamma camera are equipped with parallel-hole collimators, which in the ideal case only accept photons emitted from f(x, y) in the direction parallel to the collimator septa, as represented by the solid lines t_1' and t_2' in Figure 14.1. The source distribution in Figure 14.1 is a coarse cartoon rendition of a slice through the four-dimensional (4-D) nonuniform rational B-spline (NURBS) cardiac-torso (NCAT) phantom (Segars and Tsui 2002) with a CT-simulated heart (Erlandsson et al. 2012). If (t, s) is a second coordinate system rotated by an angle θ with respect to (x, y), as described in Chapter 1 of this book, t and s can be expressed in terms of x and y, and the rotation angle θ as (Boas 1983)

$$t = x \cos \theta + y \sin \theta$$
$$s = -x \sin \theta + y \cos \theta \qquad (14.1)$$

The ray sum (p(θ, t′)) is the line integral over f(t, s) with respect to s for t = t′, where t́ is bounded by t_1' and t_2', or

$$p(\theta, t') = \int_{-\infty}^{\infty} (t', s) \, ds = \int_{-\infty}^{\infty} \int_{-\infty}^{\infty} f_\theta(t, s) \delta(t - t') \, dt \, ds = \int_{-\infty}^{\infty} \int_{-\infty}^{\infty} f(x, y) \delta(x \cos\theta + y \sin\theta - t') \, dx \, dy \qquad (14.2)$$

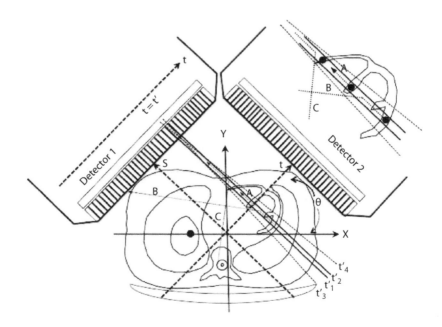

Figure 14.1 A cartoon depicting a slice through a patient placed supine in a dual-detector scintillation camera undergoing a myocardial perfusion scan. An enlargement of the heart is placed in the right upper corner for better visualization and the three black 'dots' represent radioactive spheres used to obtain the results shown in Figure 14.2. Note that the drawing is not always to scale.

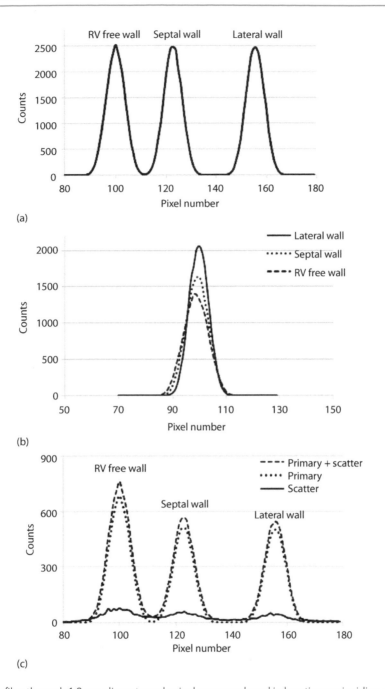

Figure 14.2 Profiles through 1.2-cm-diameter spherical sources placed in locations coinciding with the lateral wall, septal wall, and right ventricle (RV) free wall. The top two profiles (a, b) were simulated with no attenuation medium present, showing similar resolution with spherical sources at the same distance from the collimator surface (a) and deteriorating resolution by increasing distances (b), respectively. With the attenuation simulated by NCAT, the degrading effects caused by the varying distances from the body's edge are noticeable. Note that counts in the profiles without (a, b) and with (c) the attenuation are different, while the resolution (FWHM) of the sources shown in (c) are similar. The head 2 location in Figure 14.1 was used to generate the profiles in (a and c), while (b) was generated with head 2 in the left lateral oblique location to show the resolution differences most favorably.

where δ is the delta function and used when the bounds t_1' and t_2' are infinitely close. The ray sums over all t's covering the 2-D extent of the collimator (x, z) constitute a projection representing a 2-D radionuclide distribution at angle θ. A set of projections can be acquired by rotating the camera heads about the object using a step-and-shoot or a continuous mode for the data collection. Projections can immediately be framed into 2-D matrices during the collection or stored as a list of events, with an event either being the position of a detected photon or other needed information such as regular time stamps and the location of the detectors. When the 2-D projections are reordered in the z direction and stacked according to θ, they form matrices called the sinograms. These matrices are so designated because each location traces out a shape of the sine function (if acquired through 360°) of which the amplitude is dependent on the distance of the object from the center-of-rotation and the phase is determined by the angular location.

14.1.2 Image degradation

There are several physical factors contributing to the ideal image formation process and degrading the 2-D radionuclide distribution at each angle. These factors originate either (1) from inside the patient body such as body, heart, and respiratory motion, radionuclide decay, and fast-changing radionuclide distributions, (2) from the path along with the patient body to the detector such as attenuation and Compton scatter of photons, or (3) during the detection process of photons such as attenuation, Compton scatter, resolution loss/partial volume effects, and resolving time losses. While radionuclide decay and resolving time losses (the time taken to detect a single photon when a large amount of the radionuclide is present) are well understood, cardiac motion and consequent blood pool volume changes as well as fast-changing radionuclide distributions such as a first pass through the heart of a radionuclide bolus injection garnered to measure cardiac function are three other confounding factors to be discussed separately in this chapter. In this section, we focus on the photon attenuation, Compton scatter, resolution loss, and partial volume effects attributable to SPECT image degradation.

Attenuation is the removal of photons that were created and initially traveled in the direction of the detector as depicted by A and B in Figure 14.1. The photon emitting at A from the septal wall of the left ventricle (LV) is absorbed within the right ventricle by the photo electric effect, while the photon at B originating from the lateral wall of the LV and initially traveling in the direction of the detector undergoes Compton scatter, changes direction, escapes from the patient body, and is ultimately stopped by the collimator. It is also possible for Photon B to undergo multiple interactions and be absorbed in the body with the same effect. The transmission fraction (TF(t', s', θ)) of photons escaping from the patient body and being detected by the detector can be described mathematically by

$$\text{TF}(t',s',\theta) = \exp\left(-\int_{s'}^{\infty} \mu(t',s)\,ds\right) \tag{14.3}$$

with μ(t', s) being the linear attenuation coefficients at locations (t', s). The impact of attenuation on imaging inevitably reduces "good" photon detection and causes underestimation in radionuclide image quantitation.

The detection of Compton scattered photons has the opposite effect than attenuation, increasing "bad" photon detection while causing overestimation in the image quantification. When tracing the path travelled by Photon C (Figure 14.1) originating in the spine, it is noted that the Compton scatter event directs the photon into the same path originally travelled by Photons A and B, after which it escaped from the patient body to be detected possibly at t = t'. A necessary condition for the scattered photon to be detected is that the photon energy after the Compton scatter at the scattering angle be greater than or equal to the lower boundary of the selected energy window. Classical scattering, where no energy change occurs, is not distinguishable from the photons detected in the photopeak energy window and is only made up of a small percentage of the interactions with the energies used for SPECT imaging.

While photon attenuation and scatter depend on both the patient body composition and the characteristics of the detector system, spatial resolution and the resulted partial volume effects are mainly due to the latter.

As mentioned in Chapter 1 of this book, two independent resolution components can be identified in the absence of attenuation and scatter, namely, (1) intrinsic resolution due to the scintillation crystal thickness, number of photomultiplier tubes, and subsequent electronics and (2) the spatial varying geometrical acceptance of photons through the collimator influenced by hole size, hole length, and distance from the collimator. In addition to the photons emitted from an area bounded by t'_1 and t'_2 as shown in Figure 14.1, the collimator hole also accepts photons from a much larger area bounded by t'_3 and t'_4, giving rise to the distance-dependent resolution. The point spread function (PSF) characterizing the system spatial resolution can be approximated by the convolution of two Gaussian functions by

$$\sigma_s(d) = \left(\sigma_c^2(d) + \sigma_i^2\right)^{1/2} \qquad (14.4)$$

where σ_i represents the standard deviation of intrinsic resolution and σ_c is given by a linear function of distance as

$$\sigma_c(d) = \sigma_0 + \sigma_d\, d \qquad (14.5)$$

where d denotes the distance of an object from the collimator surface and σ_0 is the standard deviation of the collimator resolution at the surface. The image degradation wrought by the distance-dependent resolution is also referred to as partial volume effects which can be defined as the reduction in apparent radioactivity when the object imaged partially occupy the area inscribed in the geometrical acceptance angle of a collimator hole (Huang et al. 1979; Hutton and Osiecki 1998). According to Kessler et al. (1984), all objects smaller than 2–3 times the full-width-half-maximum (FWHM) of the system resolution are prone to the partial volume effects. As such, the LV thickness ranging from 8 mm to 12 mm imaged by a commonly used SPECT camera of 10–15 mm of spatial resolution is affected by the partial volume effects in its entirety.

Figure 14.2 summarizes the degradation effects discussed thus far using three spherical sources (1.2 cm diameter) in the geometry given in Figure 14.1 (enlargement). Projections were generated using the SIMIND Monte Carlo code (Ljungberg and Strand 1989). Profiles through two projections are used to demonstrate (1) distance-dependent resolution (Figure 14.2a and b) and the effects of attenuation and Compton scatter (Figure 14.2c). The two projections were selected with the three spherical sources at the same distance from the collimator surface (Figure 14.2a and c) and aligned as drawn in Figure 14.1. Results of the spheres in Figure 14.2b were generated in separate simulations as to distinguish them from each other.

14.1.3 IMAGE RESTORATION

Image restoration encompasses all efforts to correct for factors affecting the integrity of the original object (e.g., the heart) under investigation. During the course of this section, correction methods for attenuation, Compton scatter, and distance-dependent spatial resolution are discussed with an emphasis of the current start-of-the art in addition to the methods described in Chapter 1 of this book.

14.1.3.1 ATTENUATION COMPENSATION

Compensation for attenuation presupposes information regarding the composition of the patient's body that a photon has to traverse. Early practical attempts to create attenuation maps, other than importing high-resolution images from CT (Fleming 1989), concentrated on measuring the transmission of photons through the patient's body using a Gd-153 line source in a static (Celler et al. 1998; Gullberg et al. 1998; Tung et al. 1992) or scanning (Tan et al. 1993) geometry while acquiring emission projections. A review by King et al. (1995) has elaborated on this topic. One manufacturer deviated from Gd-153 and used scanning Ba-133 point sources in sequence with the emission in an offset fan beam geometry employing ~350 keV photons to penetrate the parallel-hole collimators (Gagnon et al. 1999). The state-of-the-art in obtaining patient-specific attenuation maps was introduced by Blankespoor et al. (1996) with an x-ray CT and scintillation camera utilizing the same patient bed in sequential SPECT/CT acquisitions. Since then, all major manufacturers embraced this SPECT/CT technology with (1) a combination of high-resolution diagnostic CT and dual-headed SPECT with the same patient bed, (2) a low-cost four-slice CT using the same gantry as the SPECT, (3) a low-cost cone-beam CT with an axial coverage of ~14 cm on the same gantry as the SPECT, and (4) a low-cost, low-dose CT combined with a small footprint three-headed SPECT system utilizing a rotating chair. In all the systems mentioned earlier, appropriate scaling is necessary for attenuation coefficients at the emission photon's energy. This scaling procedure becomes more complicated when the radionuclide to be corrected emits photons with more than a single energy (Pretorius et al. 1997).

The first attenuation compensation technique in SPECT introduced clinically was developed by Chang (1978) and used with filtered backprojection (FBP). Chang's method is implemented after SPECT reconstruction to compensate for the attenuation by multiplying each point in the patient body by the reciprocal of the TF (Equation 14.3), averaged over the entire acquisition angles, from the point to the edge of the body. Therefore, the correction factor in Chang's method is calculated by

$$C(x',y') = \left[\frac{1}{M} \sum_{i=1}^{M} TF(t',s',\theta_i) \right]^{-1}$$

(14.6)

where M is the total number of projection angles θ_i, TF is calculated by Equation 14.3, and the conversion of x' and y' to t' and s' for each θ_i is executed by Equation 14.1. An iterative step can be added where the attenuation compensated slices are forward projected (simulating the acquisition process) and subtracted from the acquired projections. The difference or error projections calculated by the subtraction are then reconstructed using FBP, corrected for attenuation using the same TF previously, and finally added to the slices originally corrected. Multiple iterations are possible, but noise may increasingly arise. Since the implementation of Chang's method, attenuation models (Gullberg et al. 1985) for projection and backprojection processes in fast iterative reconstruction techniques, such as ordered-subset expectation maximization (Hudson and Larkin 1994) and the rescaled block iterative algorithm (Byrne 1998), have became available at a time when the speed of affordable computers increased dramatically. These iterative reconstruction algorithms were derived based on the maximum likelihood expectation maximization (MLEM) technique (Chornoboy et al. 1990; Erlandsson et al. 2012; Lange and Carson 1984; Shepp and Vardi 1982). The iterative MLEM algorithm can be formulated as

$$f_j^{new} = \frac{f_j^{old}}{\sum_i h_{ij}} \sum_i \left[h_{ij} \frac{g_i}{\sum_k h_{ik} f_k^{old}} \right]$$

(14.7)

with h the system transfer matrix, g is the measured projections, while f^{new} and f^{old} represent the current and previous estimates, respectively. The attenuation factor is accounted for in the h matrix earlier.

14.1.3.2 SCATTER ESTIMATION AND COMPENSATION

Various scatter estimation methods have been proposed since classic scatter subtraction was introduced (Jaszczak et al. 1984). An extensive discussion is beyond the scope of this chapter but can be found in recent (Hutton et al. 2011) and earlier (Buvat et al. 1994) reviews. The state-of-the-art equipment mentioned in the previous section endeavors to measure 'good geometry' attenuation values, or attenuation not influenced by scattered x-rays. Therefore, the first 'scatter compensation method' still employed clinically when scatter is not independently estimated is to lower the 'good geometry' attenuation values to account for the excess counts due to scatter. Clearly, this approach is only a crude approximation of photon scatter because the true scatter depends on the composition of the patient body as well as the radionuclide distribution. Two methods of scatter estimation and compensation currently available in most of clinical SPECT systems are the triple energy window (TEW) (King et al. 1997; Ogawa et al. 1991, 1994) and effective scatter source estimation (ESSE) (Bai et al. 2000; Frey and Tsui 1997; Kadrmas et al. 1998;

Zeng et al. 1999) methods. The TEW approach employs two additional energy windows on each side of the energy window straddling the photopeak to calculate a triangular scatter estimate in the absence of down scatter from higher-energy photons. In the original implementation, very small widths of these two energy windows (Ogawa et al. 1991, 1994) even including primary photons were chosen. Needless to say, these scatter estimates are noisy and in need of smoothing before subtraction. One of the solutions was to use wider energy windows further away from the photopeak energy window to ensure better noise characteristics and less primary photon contamination. However, such a strategy tends to underestimate the scatter. The TEW is a projection-based scatter estimation method, whereas ESSE is a reconstruction-based method. With ESSE, the estimated primary distribution is blurred into an effective scatter source distribution. The effective scatter source is formed by taking into account the probability that a photon emitted at a given location will reach the scattering site where the photon that has undergone a scattering interaction is detected and finally will interact in the crystal, producing an event detectable within the photopeak energy window. The attenuated projector handles the probability of a photon being absorbed before the last scattering site. Approximations being made include the assumption of spatial invariance to precalculated blurring kernels by Monte Carlo simulation and the truncation of a Taylor series expansion of the exponential describing the probability of attenuation of the photon from the site of emission to the site of last scattering (Frey and Tsui 1997; Kadrmas et al. 1998). With use of Monte Carlo precalculated kernels, the path of the photon from emission to last scattering interaction before detection can include scattering interactions up to any order desired. However, these intermediate scatterings are assumed to occur in a uniform medium. An alternative method formulates the effective scatter source distribution by incrementally blurring and attenuating each layer of the patient forward toward the detector (Bai et al. 2000; Zeng et al. 1999). A highly desirable and accurate method of scatter compensation being made possible by the improvements of computer speed and the accuracy of attenuation maps is Monte Carlo estimated scatter (Beekman et al. 2002; de Jong et al. 2001; Kamphuis et al. 1998). In this method, the attenuation map provides information about the density of a patient's body and the reconstructed radionuclide distribution is used as the "true" source distribution. To speed up the calculation of scatter in the projection/ backprojection operations in each iteration of the reconstruction algorithm, it was demonstrated early on that one could model scatter in the forward projection operation and ignore it in the backprojection operation (Welch and Gullberg 1997). The group of Beekman et al. (Beekman et al. 2002; Kamphuis et al. 1998) identified the idea of modeling the scatter only in the forward projection but not in the backprojection operation as a dual matrix method. Further efficiency was obtained by combining stochastic photon transport of the interactions within the patient with an analytical model of the detection by the camera, a technique called convolution-based forced detection (de Jong et al. 2001). With TEW, scatter estimates can be subtracted before reconstruction or incorporated in the forward projector during reconstruction, while ESSE- and Monte Carlo-based estimations are better suited for incorporating the scatter in the forward projector. Limitations of ESSE- and Monte Carlo-based methods are the lack of information about out-of-field-of-view radioactivity.

14.1.3.3 DISTANCE-DEPENDENT RESOLUTION AND PARTIAL VOLUME COMPENSATION

Compensation techniques for the degradation of SPECT resolution caused by the partial volume effects can be divided into three categories: (1) image enhancement techniques, (2) image-domain correction techniques employing anatomical knowledge, and (3) projection-based techniques (Erlandsson et al. 2012).

The first category enlists the emission data alone and in some cases the addition of structural knowledge during or post reconstruction. During iterative reconstruction, a model of the PSF incorporated into the transfer matrix for forward and back projection facilitates the distance-dependent resolution compensation. The ordered subset (Byrne 1998; Hudson and Larkin 1994) methods improved the efficiency of iterative reconstruction and made the MLEM technique clinically feasible. The Gaussian diffusion method (McCarthy and Miller 1991), in which spatial resolution correction is implemented incrementally as small Gaussians, can further improve the computing efficiency. Theoretically, an exact radioactive distribution is possible; however, the loss of high-frequency information during the acquisition process limits the resolution recovery. Striking results can be obtained by adding anatomical priors during the iterative reconstruction to enhance edges and control noise; however, quantitative accuracy is compromised. Postreconstruction methods in which the image is deconvolved with the system PSF have also been proposed to enhance the image resolution. Restoration filtering such as the Wiener filter falls in this class.

It is obvious that the loss of high-frequency information during data acquisition limits full recovery of the true object. The next category of image domain correcting techniques is to use information from another modality, such as CT or magnetic resonance imaging, to identify and segment structures to correct for the partial volume as well as to identify and segment background structures that may have effects (such as spill-in or cross-talk) on the structures to be corrected. The partial volume corrected structure a(x) can be calculated by

$$a(x) = \frac{1}{r_i(x)}\left(b(x) - \sum_{j \neq i} A_j \phi_{ij}(x)\right) \quad x \in roi_i \quad i = 1, \dots M \qquad (14.8)$$

where $r_i(x)$ is the recovery coefficient, $b(x)$ is the uncorrected structure, A_j is activity concentration from background structures, and ϕ_{ij} is the cross-talk factor. By stepping through i, all regions of interest (roi_i) can be corrected for partial volume. Several methods can be employed to obtain recovery coefficients (Erlandsson et al. 2012).

In the third category of projection-based techniques, the Poisson model of noise is exploited. Huesman (1984) and Carson (1986) developed analytical and iterative algorithms, respectively, to determine region of interest (ROI) mean values from projection data without having to reconstruct the full data set. Other alternative methods have also been developed (Formiconi 1993) and evaluated (Vanzi et al. 2007). However, as in the case of image domain methods, segmentation of the entire image is necessary. Moore et al. (2012) simplified the method by segmenting only two to four tissue types and then iteratively maximizing the log-likelihood to estimate the mean activities of each tissue type within each volume of interest (VOI). Background activity outside the VOI is estimated by reprojecting all the data not traversing through the VOI.

14.1.3.4 PARTIAL VOLUME COMPENSATION FOR LOW-DOSE DEDICATED CARDIAC SPECT

In current SPECT/CT applications, dose reduction is gaining interest. With less injected dose, the image noise would be increased. Investigation of partial volume compensation (PVC) methods for low-count data is particularly important for applications of respiratory and/or cardiac gating and dynamic SPECT studies with tracer kinetic modeling, as PVC methods typically amplify image noise.

In a study for different noise levels, several anatomical-based PVC methods were included for performance comparison (Liu et al. 2015). These methods include pGTM (Du et al. 2005), perturbation geometry transfer matrix (pGTM) followed by the Muller-Gartner method (MGM) (Da Silva et al. 2001; Matsuda et al. 2003), and an iterative MGM approach (Liu et al. 2015) both with and without pGTM as the initial estimation. Almost all PVC approaches provided equivalent corrected results for noise-free and low-noise data. However, for low count data with high image noise, pGTM followed by MGM approach was very sensitive to noise and led to image artifacts and substantial quantification errors in the myocardium regions. In contrast, iterative PVC methods provided consistent mean images with low bias across all count levels, with slightly more artifacts with lower count levels, indicating the robustness of these iterative PVC approaches for high noise data.

Although the iterative PVC approaches can generate accurate quantification for the entire heart, these approaches amplify image noise on a voxel-by-voxel basis, which may result in a heterogeneous tracer uptake pattern in the myocardium due to noise and thus limit the capability of regional quantification. Chan et al. proposed a PVC approach that can simultaneously correct for the partial volume effect and reduce image noise (SPR) (Chan et al. 2016b). To suppress noise in SPECT images, a segmentation-free anatomical-based maximum *a posteriori* (MAP) reconstruction method that incorporates contrast CT into the reconstruction as *a priori* knowledge was used to encourage local smoothness while preventing contamination from neighboring anatomical regions (Bowsher et al. 2004; Chan et al. 2009; Vunckx et al. 2012). The same reconstruction is also used to reconstruct the anatomical template to obtain a correction factor map. The final image is obtained by multiplying this correction factor map to the original SPECT reconstructions without PVC. The key contribution of noise reduction is achieved by using anatomical-based reconstruction for both SPECT and the template images. Dog studies demonstrated that this SPR approach can simultaneously correct for partial volume effect (PVE) effectively and reduce image noise. Even with only ~10% of total counts, the SPR approach could still maintain high image quality with low noise and sharpen organ boundaries, while the MGM approach (Da Silva et al. 2001; Matsuda et al. 2003) substantially amplified image noise.

14.1.4 CORRECTION OF PROJECTION TRUNCATION FOR DEDICATED CARDIAC SPECT SYSTEM

Recently, new dedicated cardiac SPECT systems have emerged (Garcia et al. 2011; Slomka et al. 2009). These systems typically use small detectors and constrain the entire field of view (FOV) to focus just on the heart. By using multiple detectors to image the heart simultaneously, these designs can provide higher sensitivity as compared to parallel-hole SPECT. However, the dedicated cardiac SPECT systems typically have limited FOV that covers only the heart, causing a major concern of projection truncation. This effect resulted from activities outside of the FOV, such as background activity from other organs, imaged by some but not all detectors (Clackdoyle et al. 2004; Lalush and Tsui 2000; Manglos et al. 1993). Reconstruction of truncated projections could cause severe artifacts and quantitative errors (Funk et al. 2006; Lalush and Tsui 2000; Zeniya et al. 2007; Zeng and Gullberg 1990).

Several studies have investigated the impact of projection truncation on SPECT and proposed methods to compensate for the errors caused by the projection truncation. Previous studies have suggested that iterative reconstruction could provide more stable solution and potentially recover a larger volume with accurate reconstruction than analytic reconstruction methods can (Clackdoyle et al. 2004; Zhang and Zeng 2007). For a parallel-hole SPECT camera with 180° acquisition, Xiao et al. (2010) investigated the impact of truncations in both transmission and emission projections on attenuation correction and model-based scatter corrections. Their results showed that the impact of truncations from attenuation and scatter corrections on single-isotope SPECT is insignificant and should not affect clinical diagnosis. However, the truncation could have a larger impact on dual isotope acquisitions. They proposed a method by extending the projections in axial direction effectively to compensate for the truncation error. In Sabondjian et al. (2009), the authors demonstrated that the truncation in projections can cause artifacts in the reconstructed images, leading to over 100% error in quantification and reduction of the myocardial perfusion defect size by 17%. It has been reported (Branderhorst et al. 2012) that underestimation of the myocardial tracer activity was observed in a stationary small animal multipinhole SPECT. Zeniya et al. simulated a single pinhole collimator that rotated 360° for data acquisition and they demonstrated that projection truncation resulted in artifacts on the edge of the FOV and produced up to 15% over-estimation in a simulation study (Zeniya et al. 2006, 2007). By using a larger matrix for image reconstruction, the overestimation was reduced to 2%–3.5% using iterative reconstruction algorithms.

Image reconstruction from truncated projections fall into two classes: the FOV is (1) partially or (2) completely contained within the object (the interior problem). Natterer pointed out that the interior problem is not uniquely solvable (Natterer 2001). Noo et al. (2004) introduced the concept of differentiated backprojection and showed that some truncation problems are solvable for some special truncation geometries of Class (1) mentioned previously. More recent studies indicate that the solution to the interior problem is unique if the value of a tiny subregion in the ROI is known (Defrise et al. 2006; Kudo et al. 2008; Ye et al. 2007; Yu and Wang 2009). These developments have led to significant progress in solving the truncation problem for x-ray CT.

The problem in emission tomography is somewhat more complicated because of the use of SPECT/CT to solve the attenuation problem. Truncation for both modalities presents the challenge of solving a double interior problem in which both the transmission and emission projection data are truncated. Since the truncated data do not provide enough information for a unique reconstruction, some prior information must be known. An algorithm that corrected for both transmission and emission truncation was developed

(Zeng and Gullberg 2010). This algorithm required the information about a small subregion inside the FOV in both transmission and emission scans. Studies have also been performed for the cases in which this prior information is not known. By studying the null space, it was determined that the bias in the reconstruction was more sensitive to the detector sampling resolution than the number of projection views (Zeng and Gullberg 2012). However, this study shows that reconstructing truncated projections from a few projections resulted in significant image artifacts in addition to reconstructed biases. Later work (Mao and Zeng 2013) also showed that by modifying the MLEM reconstruction algorithm, these artifacts could be significantly reduced when sampling with a few projection angles. For a dedicated cardiac SPECT/CT system in which the CT projections are not truncated, a study suggested that the application of a reconstruction matrix support that matches with the body contour obtained from CT provides the most accurate reconstruction, whereas a smaller image matrix can cause artifacts and overestimation while an arbitrarily large image matrix could lead to an underestimation of counts (Chan et al. 2016a). These studies suggested that for hybrid SPECT/CT systems, the anatomical information obtained from CT could be used to improve the quantitative accuracy of dedicated cardiac SPECT with limited FOV.

14.2 CORRECTION FOR MOTION

14.2.1 MOTION CORRECTION FOR PARALLEL-HOLE SPECT

Patient motion has been recognized as a major cause for the presence of image artifacts and degradation of quantitative accuracy. Many groups have set out to develop comprehensive motion detection, estimation, and correction strategies to improve image fidelity and eliminate costly repeat acquisitions (Arata et al. 1995; Bai et al. 2009; Beach et al. 2004; Botvinick et al. 1993; Bruyant et al. 2005; Cooper et al. 1992; Dey et al. 2010; Eisner et al. 1987, 1988, 1993; Feng et al. 2006a; Fleming 1984; Friedman et al. 1989; Geckle et al. 1988; Germano et al. 1993; Huang and Yu, 1992; King et al. 2013; Kovalski et al. 2007; Leslie et al. 1997; Matsumoto et al. 2001; McNamara et al. 2008, 2009; Mukherjee et al. 2009, 2010; O'Connor et al. 1998; Pretorius et al. 2011). Most of the work on myocardial perfusion SPECT focused on garnering information regarding the consistency of the projections or sinograms to detect the motion and subsequently estimate and apply corrections in the projection or image space (Arata et al. 1995; Bai et al. 2009; Eisner et al. 1987, 1988, 1993; Geckle et al. 1988; Huang and Yu 1992; Leslie et al. 1997; Matsumoto et al. 2001; O'Connor et al. 1998). The most frequently used methods calculate projection-by-projection cross-correlations of the perfusion data (Eisner et al. 1987, 1988, 1993; Leslie et al. 1997; O'Connor et al. 1998) or using a fiducial marker placed on the patient (Leslie et al. 1997) to calculate the center of the heart using diverging squares (Eisner et al. 1988; O'Connor et al. 1998). Other approaches implement some projection/reprojection algorithm and determine the best match between the original and reprojected projections (Arata et al. 1995; Bai et al. 2009; Matsumoto et al. 2001). More recently, a stereo tracking scheme using infrared or near-infrared cameras in combination with retro-reflective markers to monitor external body and respiratory motion for correcting cardiac perfusion SPECT was introduced (Beach et al. 2004; Bruyant et al. 2005; Dey et al. 2010; Feng et al. 2006a; King et al. 2013; McNamara et al. 2008, 2009, Mukherjee et al. 2009, 2010; Pretorius et al. 2011). Respiratory gating using other means was also tested (Kovalski et al. 2007). While patient body motion occurs with varying frequency and amplitude or neither of these, the duration of a perfusion SPECT acquisition allowing for respiratory blurring always causes image degradation. The severity of this image degradation obviously depends on the amplitude and frequency of the breathing cycle. The rest of this section will be dedicated to respiratory motion monitoring, respiratory motion estimation, and respiratory motion correction, while only body and respiratory motion compensation will be illustrated. All the methods described in this subsection can only be done on gamma cameras capable of list mode data acquisitions.

Respiratory gating using an aligned signal from a respiratory elastic strap (Kovalski et al. 2007) has been shown to improve the uniformity of image intensity in the inferior and anterior regions of the LV. With this method, list mode data are reframed into four to five respiratory amplitude-binned projection sets and separately reconstructed after normalization for the irregularity of the respiratory signal. Sets of short axis slices are generated and the first respiratory bin was used as a reference for comparison to determine the displacement vectors. These displacement vectors are in turn reprojected to the transaxial frame of reference, after which the x and y components are used to shift the projections accordingly taking the rotation geometry of the detectors into account. As a result, a final reconstruction yields respiratory corrected slices.

Instead of using a resistive strap to monitor respiration, a system referred to as visual tracking system (VTS) using retroreflective markers in conjunction with near-infrared cameras can also be used (Dey et al. 2010; King et al. 2013; McNamara et al. 2008; Mukherjee et al. 2009, 2010; Pretorius et al. 2011). Although one retroreflective marker is sufficient to monitor the respiration, the implementation described endeavors to track both body and respiratory motions and uses up to seven of the markers (Figure 14.3) placed on the chest and abdomen. For respiratory motion monitoring and subsequent estimation, the accuracy of the alignment between the monitoring devices (resistive strap or VTS) and the list mode data stream are crucial. In the case of the VTS, three different methods have been devised, namely, (1) introducing a continuous pulse sequence at a known time point after the start of the VTS into the list mode data stream using either the electrocardiogram (ECG)-triggered pulse input or a second trigger pulse input if available, (2) introducing a uniquely designed pulse sequence together with the ECG-triggered pulse (O'Connor et al. 2013) at a known time point from the start of the VTS, or (3) aligning the observed VTS axial motion of the patient bed between the CT and SPECT portions of an acquisition with the information about the axial motion of the bed being recorded in the list mode data stream. To prepare for reframing into several projection sets, list mode data are divided into a sequence of 100-ms frames aligned with the VTS. As with the resistive strap method, the list mode data are binned into several respiratory frames acquired at each projection angle using the anterior–posterior (AP) signal obtained from an abdominal retroreflective marker. This reframing procedure differs from the resistive strap method in that a fixed number of respiratory frames are generated (seven or nine), allowing for some of the projection sets to have blank projections depending on the irregularity of the respiratory amplitude. Instead of the first or end expiration projection set, the projection set centrally located (four or five depending on the number of respiratory frames) is used as a reference and is judged to be better aligned with a free breathing CT or transmission map. To minimize reconstruction differences between the reference projection set and the remaining projection sets, a reference projection set is reconstructed for each of the remaining projection sets after scaling to conform to the number of projections with counts and the count level of the specific projection set under consideration. After reconstruction with the compensation for rigid body motion (Feng et al. 2006a) and estimation of the respiratory motion using an intensity-based method (Dey et al. 2010), a final reconstruction is done combining compensation for rigid-body and respiratory motion by sequentially considering each respiratory frame at an angle before stepping to the next projection. Each of the subprojections at an angle was weighted with its fractional contribution to the full projection.

A patient example is given in Figure 14.4 to demonstrate the incremental improvement achieved with the respiratory motion compensation added to the state-of-the-art attenuation, scatter, and distance-dependent resolution compensation. Results without state-of-the-art compensation are also demonstrated in Figure 14.4.

Figure 14.3 An example of the placement of the retroreflective markers on a patient's chest and abdomen.

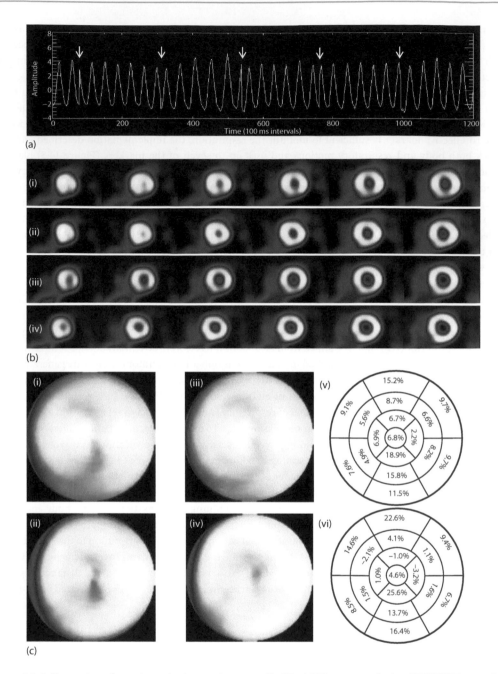

Figure 14.4 Illustration of a patient who has undergone a Tc-99m MIBI stress perfusion SPECT/CT scan with an abdominal AP signal of ~6 mm. (a) An estimated axial respiratory amplitude of 1.5 cm. The arrows show the transitions between projections. (b) SPECT images incrementally improved with (i) no compensation, (ii) compensation for respiratory motion only, (iii) compensation for attenuation, distance-dependent resolution, and Compton scatter (TEW), and (iv) compensation for attenuation, distance-dependent resolution, Compton scatter (TEW), and respiratory motion. Note the change in the shape of the LV cavity from elongated to circular with respiratory motion compensation and the change in the thickness of the walls when resolution compensation is done (compare (i) and (ii) with (iii) and (iv)). (c) Incremental changes in polar maps generated from the short axis slices shown in (b) with (c, i) and (c, ii) corresponding to (b, i) and (b, ii) with their % differences in counts shown in (c, v), while (c, iii) and (c, iv) correspond to (b, iii) and (b, iv) with the % differences in counts between them given in (c, vi). The % differences in counts are arranged in 17-segment polar map format. (Courtesy of University of Massachusetts Medical School, Worcester, Massachusetts.)

14.2.2 MOTION CORRECTION FOR DEDICATED CARDIAC SPECT

Conventional SPECT systems with parallel-hole collimators typically provide a spatial resolution of >10 mm in FWHM. Previous studies have shown that patients' diaphragm motion amplitude is about 11 mm on average, which is similar to the spatial resolution of parallel-hole SPECT systems (Liu et al. 2009). Recent dedicated cardiac SPECT systems, such as GE Discovery 530c/570c systems with pinhole collimators and cadmium zinc telluride (CZT) detectors, could achieve high spatial resolution of ~7 mm in the center of field-of-view. For these dedicated scanners, the motion amplitude is 2–3 times greater than the spatial resolution, making respiratory motion a more severe problem. Without correcting for the motion, the advantages of high spatial resolution are not fully taken for these dedicated systems.

As mentioned, conventional SPECT systems use parallel-hole collimators and slowly rotating detectors that acquire data sequentially from multiple angles for 3-D SPECT image reconstruction. Therefore, when the respiratory motion of a patient is irregular with amplitude variability and baseline shift, projection data from all angular views are not available for all respiratory phases, causing limited-angle reconstruction artifacts. These complex motion artifacts have been shown to limit the accuracy of motion estimation and thus have impacted respiratory motion correction algorithms (Dey et al. 2010). The dedicated cardiac SPECT systems with stationary data acquisition could intrinsically avoid this limitation and potentially further improve the image quality and quantitative accuracy of motion-corrected SPECT images.

Typical respiratory and cardiac motion correction methods utilize a gating technique, which uses only a small portion of all detected counts, and thus, the image noise level in the motion-corrected gated images is high, potentially compromising the quantitative accuracy and detectability of myocardial perfusion defects. Other motion correction approaches using all detected events without increasing image noise typically involves nonrigid registration of gated data, which are also noisy and could affect the registration accuracy, subsequently affecting the motion correction. Most of the dedicated cardiac SPECT systems are collimated only to the heart and can provide higher count sensitivity by several folds. Such high sensitivity could reduce the image noise in each gated image. Therefore, high-sensitivity scanners could effectively reduce motion through gating with minimal amplification of image noise, as compared to parallel-hole SPECT. In general, respiratory gating employs five to eight respiratory gates and each gated image contains only 10%–20% of all detected events. With dual respiratory and cardiac gating with typically eight cardiac gates, the counts in each dual-gated image could be further reduced to less than 2%–3%. In this dual-gating scenario, the dedicated cardiac SPECT systems with high sensitivity would be particularly helpful for noise reduction.

A commercially available external motion monitoring system can integrate both respiratory and cardiac triggering signals into the list mode data acquired using a dedicated cardiac SPECT system (GE Discovery NM 530c and NM/CT 570c, GE Healthcare) (Chan et al. 2014). Respiratory motion correction was achieved by extracting the list mode data only during the end-expiration phases. As previous studies have shown, most patients spend longer time during the end-expiration quiescent periods (Liu et al. 2009), which take about 30%–50% breathing time and are relatively "motion-free," and the rebinned SPECT data during end-expiration phases could provide a better tradeoff between motion correction and noise amplification. Figure 14.5 illustrates the different effects of end-expiration gating (five gates), cardiac gating (eight gates), and dual respiratory-cardiac gating in a patient study. Compared to the images without respiratory motion correction, the end-expiration gating (with ~40% of the counts) provides better definition of the perfusion defect in the inferior wall. By compensating for both respiratory and cardiac motions, the reconstructed images yielded a superior spatial resolution with better definition of the defect and even fine structures, such as the papillary muscle, although with a small increase in noise due to the fact that only 5% counts were used in this case. The application of respiratory motion correction in general improved the image quality and potentially quantitative accuracy.

The high-sensitivity cardiac SPECT can also potentially minimize body motion. However, a previous study of 100 patients showed that 10% of the patients still had significant body motion more than 12 mm after 4 minutes of acquisition using this type of dedicated cardiac cameras (Kim et al. 2010). Patient body motion significantly increased between 6 and 7 minutes of image acquisition. This study suggested that

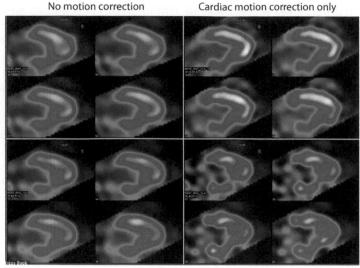

No motion correction	Cardiac motion correction only

Respiratory motion correction only	Dual respiratory-cardiac gating

Figure 14.5 An illustration of patient cardiac SPECT images obtained from a CZT dedicated cardiac SPECT using pinhole collimators without motion correction and with various combinations of respiratory and cardiac motion correction.

acquisition time shorter than 4–6 minutes, using faster dedicated cardiac SPECT systems, may avoid a significant increase in patient motion.

In summary, the features of high sensitivity, high resolution, and stationary acquisition geometry of dedicated cardiac SPECT systems provide new opportunities to correct for respiratory, cardiac, and body motions and may subsequently improve image quality and quantitative accuracy.

14.3 DYNAMIC SPECT QUANTIFICATION

14.3.1 HARDWARE, PHARMACEUTICAL, AND MODELING DEVELOPMENTS

Dynamic SPECT (Figure 14.6) is an imaging procedure that uses radiotracers to obtain images that reflect fundamental biophysiologic functions of perfusion, metabolism, and neural integrity in body organs by quantifying the temporal changes of radionuclide concentrations. In SPECT perfusion imaging, the critical aspect is to understand the biochemistry involved that limits the extraction of the tracer from the blood because of size, hydrophilicity properties, and energy requirements for transport across the capillary wall. These properties determine the rate of extraction of the tracer from the blood. Models developed early on by Krogh (1919), Crone (1963), and Renkin (1959) help us understand and quantify the rate of extraction between tracers as a function of blood flow. These models are also preliminary for more complex spatially distributed models of exchange between capillary and extravascular tissue (Bassingthwaighte 1974). However, accurate quantitation in dynamic SPECT requires development that combines the physics of the image detection process and the complexity of physical and biochemical processes in tissue into models of the time varying distribution between compartments (Figure 14.6).

The development of SPECT systems with sufficient temporal resolution to extract kinetic compartment model parameters has had a history of progress and regression. Early on, SPECT systems were designed to obtain dynamic tomographic data as shown in brain studies as early as 1963 (Kuhl and Edwards 1963) and cardiac studies in the 1990s (Stewart et al. 1990). Using a three-detector SPECT system (Lim et al. 1985), Budinger et al. (1991) demonstrated that a reliable blood input function could be obtained from 5-sec sampling intervals of reconstructed dynamic SPECT data and this sampling regime would be sufficient for

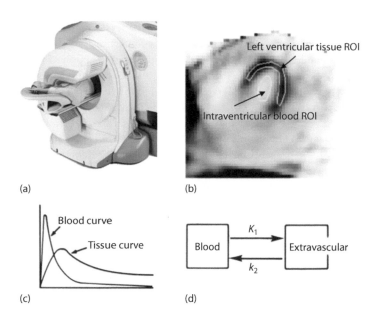

Figure 14.6 Concept of dynamic cardiac SPECT imaging. (a) Projections are acquired with a stationary ring of cadmium zinc telluride detectors (Discovery 570c, GE Healthcare) providing a consistent set of projections from time frame to time frame (b) A reconstructed slice of the sum of dynamic frames illustrating intraventricular blood and left ventricular tissue regions of interest (ROIs). (c) Illustration of TACs estimated from blood and tissue regions of interest. Note, the tissue curve will contain activity from both blood and extravascular tissue. (d) These curves are fit to a one-compartment model to obtain kinetic rate parameters for the wash-in (K_1) and wash-out (k_2). ([a, b] Courtesy of GE Healthcare. With permission.)

compartment model-based analysis of the data. The first results of a compartment model-based analysis of dynamic cardiac SPECT data were obtained using a three-detector PRISM system (Picker, Cleveland, Ohio). The use of imaging systems with sufficient temporal resolution to extract kinetic compartment model parameters using conventional methods of estimating kinetic parameters from time-activity curves (TACs) generated from dynamic reconstructed ROIs demonstrated dynamic SPECT's effectiveness for extraction of physiologic values of cardiac perfusion (Smith et al. 1994) and cardiac metabolism (Gullberg et al. 1999) in animals and similarly in humans (Chiao et al. 1994a).

The original three-detector PRISM SPECT system had the mechanical stability and control hardware to allow acquisition of 360° of data in 5 to 10 sec. The major advantage of a system of multiple-gamma cameras is the improvement in sensitivity afforded by better coverage of the solid angle; however, the arrangement of the detectors did not necessarily minimize attenuation of posterior views for nuclear cardiac imaging. Another advantage is that multiple-gamma cameras reduce the angular range of motion necessary to acquire a complete data set. The disadvantage is the cost. As development of SPECT cameras moved to hybrid SPECT/CT systems with large-FOV dual-gamma detectors, these systems were able to perform continuous rotation to complete 360° tomographic views because of the slip ring (electromechanical device that allows transmission of electrical signals from a stationary to a rotating structure) technology necessary for the CT but were not able to acquire tomographic views as rapidly as the triple-headed systems because of the large FOV cameras requiring a reduced orbit speed. However, more recently dedicated cardiac scanners (Buechel et al. 2010; Erlandsson et al. 2009; Patton et al. 2007) have emerged with new solid state detector technology, but to reduce costs, the rotating SPECT system no longer operates on the same gantry as the CT with the slip ring design; instead, the design has reverted to detector orbits with alternating reversals of the detectors. Today, the instruments available in most clinics do not have good temporal resolution and therefore present major challenges in acquiring and modeling dynamic data. However, the introduction of new dedicated cardiac SPECT cameras is changing this somewhat, but in general, to make systems cost effective, the manufacturers sell typically general purpose systems that are not optimal for dynamic capability.

Coordination of available radiotracers that have excellent flow versus extraction properties with SPECT systems with sufficient temporal resolution to extract kinetic compartment model parameters has been very irregular. SPECT tracers that have been Food and Drug Administration approved for myocardial perfusion are ^{201}Tl (thallium), ^{99m}Tc-sesstamibi, ^{99m}Tc-tetrofosmin, and ^{99m}Tc-teboroxime. ^{201}Tl, a potassium analog, has a high transcapillary extraction fraction and is rapidly cleared from arterial blood and taken up into myocardial cells by active transport (Iida et al. 2003). ^{99m}Tc-sestamibi, with a biological half-life in the myocardium of 6 hours (Maublant et al. 1992), is more widely used to image myocardial perfusion in clinical studies and in some cases is used with dual ^{201}Tl in rest(^{201}Tl)/stress(^{99m}Tc-sestamibi) imaging protocols. However, in most cases, ^{201}Tl has given way to ^{99m}Tc agents such as ^{99m}Tc-sestamibi and ^{99m}Tc-tetrofosmin because of concern for radiation dose. ^{99m}Tc-teboroxime (Leppo and Meerdink 1989), a stable, neutral, and lipid soluble tracer, has the fastest clearance time (a 10–15-min half-life in the myocardium) of the tracers and is 90% extracted in the first pass. It is highly linear with the flow. Research indicates that teboroxime has a lower degree of sensitivity to cellular metabolic impairment (Maublant et al. 1993) when compared to sestamibi or ^{201}Tl. It has been shown that dynamic imaging of ^{99m}Tc-teboroxime with compartment modeling provides a measure of response to coronary vasodilation (Chiao et al. 1994a) and is a better measure of flow (better contrast) than that can be obtained from static imaging of ^{201}Tl or, for that matter, static imaging of ^{99m}Tc-teboroxime (Di Bella et al. 2001). However, in evaluating the clinical potential of ^{99m}Tc-teboroxime, methods that took advantage of its fast wash-in and wash-out kinetics were not used, so ^{99m}Tc-teboroxime is no longer produced and not available nowadays for clinical use. This is unfortunate considering the advent of the new dedicated cardiac scanners. ^{123}I-iodorotenone, with an even higher extraction fraction (Marshall et al. 2001), is another potential cardiac perfusion agent, but intellectual property issues make this less desirable for the industry.

The conventional method used to analyze dynamic SPECT and positron emission tomography (PET) data, in which (1) ROIs are specified in a sequence of reconstructed images, (2) TACs are generated from the data values in these ROIs (Gullberg et al. 2004), and (3) then these curves are input to a nonlinear estimation program (Huesman et al. 1995), is not optimal. A better method than the conventional method to produce estimates of kinetic parameters with lower variance and bias is to model the dynamic process in a voxel as a complete tomographic inverse problem (Reutter et al. 2005). Mathematically, the time varying distribution between compartments is represented by a system of differential equations with linear coefficients corresponding to kinetic rate parameters of wash-in and wash-out from the compartments. These rate parameters are either (1) (semidirectly) estimated by modeling the TACs using spatiotemporal basis functions where the coefficients are first estimated directly from the dynamically acquired projections and then the compartment model parameters are estimated from the estimated blood and tissue TACs or (2) (directly) estimated from the projection data.

14.3.2 ESTIMATING TIME ACTIVITY CURVES DIRECTLY FROM PROJECTION DATA

The estimation of TACs directly from projections is more difficult for SPECT than PET because the tracer concentration distribution changes during the rotation of the detectors (Links et al. 1991). No constraints are placed on the dynamics of the TAC curves, but a model of the projected ROI is required. Spatiotemporal basis functions are the basic building blocks for modeling the variation of the reconstructed spatial distribution as a function of time. The *spatially* nonuniform activity concentration within a particular volume can be modeled by selecting an appropriate set of spatial basis functions defined within the volume, such as splines, point clouds of tetrahedral elements, blobs, various types of polynomial expansions, or indicator functions corresponding to voxels. Similarly, *time-varying activity concentrations* within the volumes can be modeled by selecting a set of temporal basis functions that are capable of representing typical time variations and have desired smoothness properties. These may be splines (Figure 14.7), Gaussians, or factor analysis of dynamic structures (FADS) as elaborated further in the chapter curves that fit compartment models, polynomial expansions, or several other possible spectral decompositions.

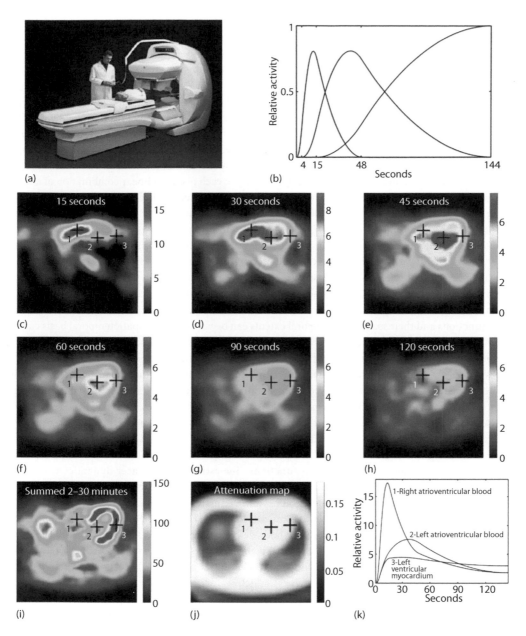

Figure 14.7 Estimating dynamic frames and time activity curves directly from projection data. Reconstructed images and TACs for a human dynamic Tc-99m-sestamibi cardiac SPECT study (36 sec for 180° rotation, stress, and bolus injection): (a) Image of a GE VG3 Millennum Hawkeye SPECT/CT dual-head scanner used to collect the data. (b) Quadratic B-splines used to reconstruct the dynamic frames The fully 4D reconstructed dynamic images are sampled at (c) 15 sec, (d) 30 sec, (e) 45 sec, (f) 60 sec, (g) 90 sec, and (h) 120 sec. (i) The image formed by summing from 2 to 30 minutes. (j) Attenuation map for the same slice obtained from the x-ray CT mounted on the same gantry as the two gamma detectors. (k) The TACs sampled at the spatial positions indicated by the black crosses. ([a] Courtesy of GE Healthcare. With permission; [b] from Reutter BW et al., *IEEE Trans Med Imaging*, 19, 434–450, 2000.)

Splines provide an efficient and accurate method of composing continuous smooth functions from fitted data of discrete samples that do not exhibit oscillatory behavior characteristic of high-degree polynomial interpolation. The direct least squares estimation of a nonuniform temporal distribution of B-splines was used in Figure 14.7 to obtain from dynamically acquired SPECT projection data dynamic frames of cardiac activity and smooth TACs that capture the relatively fast rise and fall of the tracer in the right and left ventricular blood chambers, as well as uptake and retention of the tracer in the left ventricular myocardium (Reutter et al. 2000, 2002, 2004). Recent work using splines include comparison of results of blood pool TACs using slow (Winant et al. 2012) and fast (Alhassen et al. 2014) tracer infusions. FADS is a technique to extract factors from a series of dynamic data that are physiological, i.e., blood TAC or a particular TAC for the uptake and wash-out of a particular tissue region. Factors are temporal functions (TACs) with elements that are non-negative. In the factor model, it is assumed that the activity in each pixel is a linear combination of a few factors. For cardiac studies, the estimation of factors was originally accomplished from a dynamic sequence of *reconstructed* SPECT images (Sitek et al. 2002a, 2002b). FADS was later applied to extract TACs *directly from projections* of the uptake and wash-out of ^{99m}Tc-MAG3 from the kidneys (Sitek et al. 2001) and for extracting the blood input function in rats from dynamic SPECT data acquired with slow camera rotation (Hu et al. 2008). A more general parameterization of the kinetics is the tomographic estimation of characteristic curves for each voxel. The dSPECT method (Bauschke et al. 1999; Celler et al. 2001; Farncombe et al. 1999, 2001; Humphries 2011) assumes pixel base characteristic functions as the temporal basis functions and fits TACs for each voxel directly from projection measurements acquired from single or multiple rotations. The number of basis functions and their spatial and temporal extents can be varied so that spatiotemporal basis can optimally model the spatial and temporal content of the data with the fewest number of basis functions. The goal is to parameterize spatially nonuniform activity within segmented volumes and to include smooth temporal changes within the volumes, thereby continuously modeling anatomy throughout time rather than through the use of small image voxels at discrete time segments. This provides a multiresolution structure that reduces the size of the parameter space, is ideal for estimating the parameters from projection data, and reduces noise.

Up until this point, we have ignored any organ motion caused by cardiac beating or respiration. Modeling kinetics in the moving myocardium due to periodic deformation and respiratory motion requires extending the 4-D tensor product of spatial and temporal basis functions to 5-D and 6-D tensor products using additional products of basis functions. The estimation of tracer dynamics from projections that model the change of the concentration of the tracer and the deformation of the organ as a function of time is referred to as 5-D tomography, and estimation that also includes the motion of the heart due to respiration is referred to as 6-D tomography. The 5-D and 6-D tomographies have all of the challenges of modeling both rigid and nonrigid body deformation simultaneously with the modeling of a time varying tracer concentration from tomographic projections (Gravier et al. 2007). Previous works to reconstruct 5-D data include using the dSPECT method (Feng et al. 2006b), a level set method (a numerical technique for tracking interfaces and shapes) to delineate boundaries of the heart while simultaneously estimating the time variation of the tracer concentration in the organ (Shi and Karl 2004), and splines for 5-D modeling of gated cardiac PET data (Verhaeghe et al. 2007) and of gated cardiac SPECT data (Shrestha et al. 2012). Work of 6-D tomography was also presented (Shrestha et al. 2013) where the tensor product is formed using Gaussian basis functions to represent motions of the heart due to deformation and respiration and using splines to model changes in the intensity of the signal caused by uptake and wash-out of the radiotracer. It is assumed that the data are acquired dynamically with cardiac and respiratory phase information, either by list mode acquisition or by binning into gated dynamic frames over the cardiac and respiratory cycles.

Compartmental modeling involves the solution to systems of differential equations that express the time-varying distribution between compartments. The solution to these equations involves expressions of exponentials. Therefore, it is very natural to parameterize the kinetics of the radiopharmaceutical distribution in the tissue as a spectrum of exponential basis functions in time (Cunningham and Jones 1993). Instead of estimating the kinetic parameters of models, one estimates the coefficients of a predefined spectrum of exponentially decaying factors. It is assumed that the decay constants for the exponentials span some reasonable range of physiological values and those are predefined with sufficient sampling to adequately represent the kinetics of the tracer. Efforts have been made to calculate these coefficients directly from projections (Hebber et al. 1997; Hicks et al. 1989; Maltz 2001; Matthews et al. 1997). A significant limitation of exponential spectral methods is that the solution obtained may not be unique due to the large amount of parameter redundancy, which is characteristic of sums of decaying exponential terms (Maltz 2000). It has been shown that using an exponential basis with a nonuniform resolution reconstruction grid leads to a further reduction in the dimension of the problem of direct dynamic emission computed tomography reconstruction from inconsistent projections (Maltz 2001, 2002). The solution coefficients, which must be nonnegative, can be determined by either using the MLEM algorithm or using a least squares technique with nonnegative constraints, which first determines spectral components for each projection sample and then the spectral components for each voxel are reconstructed from the projections of the spectral components (Meikle et al. 1998).

Other temporal basis functions that have been investigated for dynamic PET studies could be used in the spatiotemporal representation of dynamic SPECT data. These include some parametric decomposition (Matthews et al. 1997) using single value decomposition to determine the temporal basis functions or spectral mixture models (Hudson and Walsh 1997). Others include the use of splines (Asma and Leahy 2006; Nichols et al. 2002) or wavelets (Verhaeghe et al. 2008) for the temporal basis functions. Another choice for the temporal basis function is the top-hat function (also known as an indicator function or characteristic function for a set X that is 1 if x is an element of X and 0 otherwise) where the temporal function is composed of a sum of top-hat functions defined on independent distinct time frames (Reader et al. 2006). This 'basis function method' has been used where the nonlinear terms in the model fitting are discretized over a range of likely basis functions that encompass the possible nonlinear terms (Hong and Fryer 2010; Lodge et al. 2000; Watabe et al. 2005). In addition, a dictionary of possible basis functions has been proposed (Gunn et al. 2002). Performance comparisons among some of these methods have also been performed (Boellaard et al. 2005; Dai et al. 2011).

14.3.3 ESTIMATING KINETIC PARAMETERS DIRECTLY FROM THE PROJECTION DATA

Another approach to obtaining estimates of kinetic parameters with lower variance and bias is estimating the parameters of the kinetic model directly from projection measurements. It is assumed that the relation between the ROI and the parameters of the compartmental model and the dynamically acquired projections can be expressed in some maximum likelihood or Bayesian formulation. Chiao et al. (1994b, 1994c) estimated ROI kinetic parameters for a one-compartment model and boundary parameters for the regions from simulated transaxial PET measurements, showing that biases of estimates are reduced by including ROI specification inaccuracy. In Zeng et al. (1995), the parameters of a one-compartment model were fitted directly from the projection measurements, resulting in the elimination of the biases in parameters present when estimated from the TACs generated from the reconstructions. Later, Huesman et al. (1998) used hard boundary constraints and demonstrated bias improvement by estimating kinetic parameters for a one-compartment myocardial perfusion model directly from cone-beam SPECT projections. The compartment was used for the simulated myocardium tissue while the blood input function was assumed to be known.

For the estimation of the blood TAC and kinetic model parameters directly from projections, the organs need to be segmented to properly model the spatial and temporal distributions of the radiopharmaceutical throughout the projected FOV by identifying the blood input and formulating appropriate models for each tissue type. The segmentation step (separation of blood and organ of interest) is critical in reducing bias and can be performed prior to the estimation of the blood input function and kinetic parameters or could be part of the optimization procedure in the segmentation step. If the segmented ROI is larger than the desired organ of interest, there could be cross-talk from other organs. If the other organ has lower counts, then the TAC for the organ of interest will be too low and vice versa. The effect on bias would be less if the segmented organ is contained within the desired organ of interest. For semidirect methods where TACs are first estimated for blood and tissues and then the kinetic compartment model parameters are estimated from these TACs, the segmentation of ROIs is not essential.

Imposing a compartmental relationship (defined by a first-order differential equation) between blood and tissue activities into the spatiotemporal model will provide more temporal regularization than is provided by splines alone. In Reutter et al. (2005), a maximum likelihood approach (sum of squares) was used to estimate the kinetic model parameters directly from projections. Splines were used to model the blood input function and a physiological one-compartment model was used to model changes in activity for tissue volumes. A 4-D MAP approach with a compartmental-model-based temporal prior that constrains each voxel's behavior in time to conform to a compartmental model was developed for dynamic cardiac SPECT application (Kadrmas and Gullberg 2001). In this approach, no *a priori* limitations on kinetic parameters were applied, but the parameter estimates evolved as the algorithm iterated to a solution. The method yielded TACs with reduced noise variations and provided wash-in parameter estimates with better accuracy and lower statistical uncertainty than those obtained from conventional dynamic processing followed by compartmental modeling. The algorithm removed the bias in wash-in parameter estimates that can be caused by inconsistent projections from sampling schedules as slow as 60 sec per time frame; however, no improvement in wash-out parameter estimates was observed. Here, a one-compartment model was used as a Bayesian prior to constrain the TACs for each voxel, whereas in Reutter et al. (2005) a maximum likelihood approach was used to estimate directly from the projection coefficients of splines for the blood input function and kinetic rate parameters of a physiological one-compartment model that modeled changes in activity for tissue volumes. The Bayesian reconstruction framework is also well suited for accepting spatially regularizing priors, and such priors could readily be applied to either reconstructed voxel intensities or kinetic parameter estimates as desired.

14.3.4 DYNAMIC SPECT USING DEDICATED CARDIAC SPECT

As aforementioned, dynamic SPECT imaging using conventional SPECT scanners has challenges. The slow gantry rotation does not provide complete angular projection data needed for reconstructing accurate dynamic frames within a short period. The low sensitivity of conventional SPECT could also prevent fast sampling of short dynamic frames due to noise amplification. Recent high-sensitivity dedicated cardiac SPECT systems, in particular those with stationary or nearly stationary data acquisitions, provide new opportunities for dynamic SPECT imaging.

A recent study (Ben-Haim et al. 2013) evaluated the feasibility of quantifying a retention index, calculated as the ratio of stress and rest K_1 values, to report myocardial perfusion reserve using a solid-state dedicated cardiac camera (D-SPECT; Spectrum Dynamics). High-temporal-resolution data were obtained to calculate the input functions. The blood pool and myocardium can be clearly defined in dynamic images, potentially due to the high sensitivity of the scanner. Using a one-tissue compartmental model, 95 sets of rest/vasodilator stress patient data showed that the retention index was lower in patients with perfusion defects and in regions supplied by obstructed coronary arteries. These results suggested the feasibility of quantifying myocardial blood flow and flow reserve using dedicated solid-state SPECT cameras.

Using a stationary dedicated cardiac SPECT system, Wells et al. achieved absolute quantification of myocardial blood flow using both thallium-201 (^{201}Tl) and ^{99m}Tc-labeled radiotracers in pigs both at rest and during stress, with validation against microspheres (Wells et al. 2014; Liu and Sinusas 2014). This pilot preclinical study demonstrated the feasibility of dynamic SPECT imaging for quantification of absolute myocardial flow.

In addition to quantifying the myocardial blood flow, dynamic SPECT may play an important role in quantifying sympathetic activity. Sympathetic nuclear cardiac imaging with norepinephrine analog meta-iodobenzylguanidine (mIBG) labeled by I-123 is the most widely used imaging approach for studying cardiac sympathetic hyperactivity and has been demonstrated as a powerful predictor of survival to stratify heart failure patients (Agostini et al. 2008a, 2008b; Bax et al. 2008; Carrio et al. 2010; Chirumamilla and Travin 2011; Flotats and Carrio 2004; Jacobson et al. 2009; Matsuo et al. 1996; Nakata et al. 1998; Perrone-Filardi et al. 2011; Raffel and Wieland 2010; Strauss et al. 2010; Yamada et al. 2003; Wakabayashi et al. 2001). Relative mIBG activity is usually assessed semiquantitatively by calculating the heart-to-mediastinum ratio (HMR) of the mIBG uptake in the heart and mediastinum from 2-D planar images (Ji and Travin 2010), from which mIBG wash-out rate can also be calculated. Physiological parameters such as K_i and k_3 (tracer retention) as well as k_2 (wash-out rate) obtained from dynamic SPECT may be more sensitive and quantitatively accurate than the HMR and wash-out rate mentioned previously in predicting patient survival. One preliminary study performed at Yale University investigated I^{123}-mIBG patients with ischemic cardiomyopathy post implantation of a defibrillator as a predictor for ventricular arrhythmias using a stationary dedicated SPECT/CT system (GE Discovery 570c). As a patient example shown in Figure 14.8, both one-tissue (1T) and two-tissue (2T) compartmental models were used to fit the 4-hour time activity curves, and the 2T model clearly provided more accurate fitting of the data. These data suggest that tracer kinetic modeling on dynamic I^{123}-mIBG SPECT may be feasible for parameter estimation using the 2T model, whereas future studies are needed to explore the role of dynamic SPECT imaging for I^{123}-mIBG.

To accurately estimate the parameters through kinetic modeling of a dynamic 4-D SPECT dataset, the prerequisite is to obtain an absolute quantitative reconstruction of each 3-D SPECT image. It is unclear whether tracer kinetic modeling will provide clinical value if the primary data are not quantitative. One study performed at Yale University investigated the impact of scatter correction and attenuation correction for dynamic SPECT imaging of a pig model using Tc^{99m}-Tetrofosmin and a stationary dedicated SPECT/CT system (GE Discovery 570c). Rebinned dynamic projections were reconstructed with and without CT-based attenuation correction and with scatter correction using the TEW approach. Image-based input function was obtained from ROI quantification of the blood pool. Time activity curves obtained from the ROI of the entire myocardium were used as tissue curve. A one-tissue (two compartments) model was used to fit data to obtain K_1 and k_2. As shown in Table 14.1, scatter correction resulted in similar K_1 values but lower k_2 values, as compared to parameters obtained without scatter correction. Additional attenuation correction led to slightly higher K_1 and lower k_2. These preliminary results may suggest that quantitative corrections of attenuation and scatter have an impact on parameter estimation on dynamic SPECT imaging. More accurate parameter quantifications can be expected with additional data corrections such as partial volume and motion corrections.

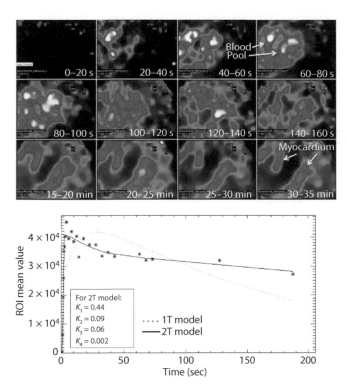

Figure 14.8 An illustration of patient dynamic I^{123}-mIBG transaxial SPECT images (top panel) and curve fitting of myocardial uptake using 1T and 2T compartmental models (bottom panel).

Table 14.1 Kinetic parameter estimations without and with corrections of scatter and attenuation in a pig dynamic SPECT study using Tc^{99m}-tetrofosmin

Scatter correction	No	Yes	Yes
Attenuation correction	No	No	Yes
K_1	0.196	0.198	0.233
k_2	0.093	0.039	0.025

14.3.5 SUMMARY OF DYNAMIC SPECT

The combination of gantry motion and the time-varying nature of the radionuclide distribution being imaged results in inconsistent projection data sets. The estimation of kinetic parameters from reconstructed TACs results in biases. The inconsistency and biases can be greatly improved with the new dedicated cardiac SPECT systems described herein. Estimating the kinetic parameters directly from projections acquired using both rotating and dedicated cardiac SPECT systems may also reduce the biases. A simulation study was performed for data acquired from a rotating SPECT system. A direct method in which the kinetic parameters for each region were simultaneously estimated directly from the projections (Reutter et al. 1998) was compared with a semidirect method where B-splines for blood and tissue TACs were estimated directly from projections using the same projection model (Reutter et al. 2000) and spline estimated TACs. This simulation study showed that the direct approach resulted in the best parameter estimation, although the results were only slightly better than those calculated by the semidirect method while significantly better than those of the conventional method. Furthermore, other simulations (Reutter et al. 2005) showed that using B-splines to model the input function and directly estimating the compartmental model parameters and blood input function can improve quantitation of dynamic SPECT by reducing the standard deviation of uptake and wash-out parameters for the septal and lateral defects by 17% to 41%, respectively, as compared to the semidirect methods that estimate time activity curves first and then fit compartmental models to the curves. We emphasize that the estimation of the time activity curves and kinetic model parameters directly derived from the projections reduces bias and improves parameter efficiencies for both rotating SPECT and dedicated cardiac SPECT systems. Ultimately, the correct combination of camera and radiotracer is critical for obtaining accurate and sensitive measures of biological function along with the correct modeling of the physics of the image detection process.

REFERENCES

Agostini D, Verberne HJ, Burchert W, Knuuti J, Povinec P, Sambuceti G, Unlu M, Estorch M, Banerjee G & Jacobson AF. 2008a. I-123-mIBG myocardial imaging for assessment of risk for a major cardiac event in heart failure patients: Insights from a retrospective European multicenter study. *Eur J Nucl Med Mol Imaging*, 35, 535–546.

Agostini D, Verberne HJ, Hamon M, Jacobson, AF & Manrique A. 2008b. Cardiac 123I-MIBG scintigraphy in heart failure. *Q J Nucl Med Mol Imaging*, 52, 369–377.

Alhassen F, Nguyen N, Bains S, Gould R, Seo Y, Bacharach S, Song X, Shao L, Gullberg G & Aparci C 2014. Myocardial blood flow measurement with a conventional dual-head SPECT/CT with spatiotemporal iterative reconstructions—A clinical feasibility study. *Am J Nucl Med Mol Imaging*, 4, 53–59.

Arata L, Pretorius PH & King MA. 1995. Correction of organ motion in SPECT using re-projection data. *IEEE Nucl Sci Symp and Med Imaging Con*, 3, 1456–1460.

Asma E & Leahy RM. 2006. Mean and covariance properties of dynamic PET reconstruction from list mode data. *IEEE Trans on Med Imag*, 25, 42–54.

Bai C, Maddahi J, Kindem J, Conwell R, Gurley M & Old R. 2009. Development and evaluation of a new fully automatic motion detection and correction technique in cardiac SPECT imaging. *J Nucl Cardiol*, 16, 580–589.

Bai CY, Zeng GSL & Gullberg GT. 2000. A slice-by-slice blurring model and kernel evaluation using the Klein-Nishina formula for 3D scatter compensation in parallel and converging beam SPECT. *Phys Med Biol,* 45, 1275–1307.

Bassingthwaighte JB. 1974. A concurrent flow model for extraction during transcapillary passage. *Circ Res,* 35, 483–503.

Bauschke HH, Noll D, Celler A & Borwein, JM. 1999. An EM algorithm for dynamic SPECT. *IEEE Trans Med Imaging,* 18, 252–261.

Bax JJ, Kraft O, Buxton AE, Fjeld JG, Parizek P, Agostini D, Knuuti J, Flotats A, Arrighi J, Muxi A, Alibelli MJ, Banerjee G & Jacobson AF 2008. I-123-mIBG scintigraphy to predict inducibility of ventricular arrhythmias on cardiac electrophysiology testing a prospective multicenter pilot study. *Circ Cardiovasc Imaging,* 1, 131–140.

Beach RD, Pretorius PH, Boening G, Bruyant, PP, Feng B, Fulton RR, Gennert MA, Nadella S & King MA. 2004. Feasibility of stereo-infrared tracking to monitor patient motion during cardiac SPECT imaging. *IEEE Trans Nucl Sci,* 51, 2693–2698.

Beekman FJ, De jong HWAM & Van Geloven S. 2002. Efficient fully 3-D iterative SPECT reconstruction with Monte Carlo-based scatter compensation. *IEEE Trans Med Imaging,* 21, 867–877.

Ben-Haim S, Murthy VL, Breault C, Allie R, Sitek A, Roth N, Fantony J, Moore SC, Park MA, Kijewski M, Haroon A, Slomka P, Erlandsson K, Baavour R, Zilberstien Y, Bomanji, J & Di Carli, M F. 2013. Quantification of myocardial perfusion reserve using dynamic SPECT imaging in humans: A feasibility study. *J Nucl Med,* 54, 873–879.

Blankespoor SC, Wu X, Kalki K, Brown JK, Tang HR, Cann CE & Hasegawa BH. 1996. Attenuation correction of SPECT using x-ray CT on an emission-transmission CT system: Myocardial perfusion assessment. *IEEE Trans Nucl Sci,* 43, 2263–2274.

Boas M. 1983. *Mathematical Methods in the Physical Sciences,* New York, John Wiley & Sons.

Boellaard R, Knaapen P, Rijbroek A, Luurtsema GJ & Lammertsma AA. 2005. Evaluation of basis function and linear least squares methods for generating parametric blood flow images using 15O-water and positron emission tomography. *Mol Imag Biol,* 7, 273–285.

Botvinick EH, Zhu YY, O'Connell WJ & Dae MW. 1993. A quantitative assessment of patient motion and its effect on myocardial perfusion SPECT images. *J Nucl Med,* 34, 303–310.

Bowsher JE, Hong Y, Hedlund LW, Turkington TG, Akabani G, Badea A, Kurylo WC, Wheeler CT, Cofer GP, Dewhirst MW & Johnson GA Utilizing MRI information to estimate F18-FDG distributions in rat flank tumors. IEEE Nuclear Science Symposium and Medical Imaging Conference Record, October 16–22, 2004. 2488–2492 Vol. 4.

Branderhorst W, Have FVD, Vastenhouw B, Viergever MA & Beekman FJ. 2012. Murine cardiac images obtained with focusing pinhole SPECT are barely influenced by extra-cardiac activity. *Phys Med Biol,* 57, 717.

Bruyant PP, Gennert MA, Speckert GC, Beach RD, Morgenstern JD, Kumar N, Nadella S & King MA. 2005. A robust visual tracking system for patient motion detection in SPECT: Hardware solutions. *IEEE Trans Nucl Sci,* 52, 1288–1294.

Budinger TF, Araujo L, Ranger N, Coxson P, Klein G, Huesman RH & Alavi A. 1991. Dynamic SPECT feasibility studies. *J Nucl Med,* 32, 955 [abstract].

Buechel RR, Herzog BA, Husmann L, Burger IA, Pazhenkottil AP, Treyer V, Valenta I, Von Schulthess P, Nkoulou R, Wyss CA & Kaufmann PA. 2010. Ultrafast nuclear myocardial perfusion imaging on a new gamma camera with semiconductor detector technique: First clinical validation. *Eur J Nucl Med Mol Imaging,* 37, 773–778.

Buvat I, Benali H, Todd-Pokropek A & Di Paola R. 1994. Scatter correction in scintigraphy: The state of the art. *Eur J Nucl Med,* 21, 675–694.

Byrne CL. 1998. Accelerating the EMML algorithm and related iterative algorithms by rescaled block-iterative methods. *IEEE Trans Image Process,* 7, 100–109.

Carrio I, Cowie MR, Yamazaki J, Udelson J & Camici PG. 2010. Cardiac sympathetic imaging with mIBG in heart failure. *JACC Cardiovasc Imaging,* 3, 92–100.

Carson RE. 1986. A maximum-likelihood method for region-of-interest evaluation in emission tomography. *J Comput Assist Tomogr,* 10, 654–663.

Celler A, Farncombe TH, Bever CA, Noll D, Maeght J, Harrop R & Lyster D. 2001. Performance of the dynamic single photon emission computed tomography (dSPECT) method for decreasing or increasing activity changes. *Phys Med Biol,* 45, 3525–3543.

Celler A, Sitek A, Stoub E, Hawman P, Harrop R & Lyster D. 1998. Multiple line source array for SPECT transmission scans: Simulation, phantom and patient studies. *J Nucl Med,* 39, 2183–2189.

Chan C, Dey J, Grobshtein Y, Wu J, Liu YH, Lampert R, Sinusas AJ & Liu C. 2016a. The impact of system matrix dimension on small FOV SPECT reconstruction with truncated projections. *Med Phys,* 43, 213.

Chan C, Fulton R, Feng D & Meikle S. 2009. Regularized image reconstruction with an anatomically adaptive prior for positron emission tomography. *Phys Med Biol,* 54, 7379–7400.

Chan C, Harris M, Le M, Biondi J, Grobshtein Y, Liu YH, Sinusas AJ & Liu C. 2014. End-expiration respiratory gating for a high-resolution stationary cardiac SPECT system. *Phys Med Biol,* 59, 6267–6287.

Chan C, Liu H, Grobshtein Y, Stacy MR, Sinusas AJ & Liu C. 2016b. Noise suppressed partial volume correction for cardiac SPECT/CT. *Med Phys,* 43, 5225.

Chang LT. 1978. Method for attenuation correction in radionuclide computed tomography. *IEEE Trans Nucl Sci,* 25, 638–643.

Chiao P-C, Ficaro EP, Dayanikli F, Rogers WL & Schwaiger M. 1994a. Compartmental analysis of technetium-99m-teboroxime kinetics employing fast dynamic SPECT at rest and stress. *J Nucl Med,* 35, 1265–1273.

Chiao P-C, Rogers WL, Clinthorne NH, Fessler JA & Hero AO. 1994b. Model-based estimation for dynamic cardiac studies using ECT. *IEEE Trans Med Imaging,* 13, 217–226.

Chiao P-C, Rogers WL, Fessler JA & Hero AO. 1994c. Model-based estimation with boundary side information or boundary regularization. *IEEE Trans Med Imaging,* 13, 227–234.

Chirumamilla A & Travin MI. 2011. Cardiac applications of 123I-mIBG imaging. *Semin Nucl Med,* 41, 374–387.

Chornoboy ES, Chen CJ, Miller MI, Miller TR & Snyder DL. 1990. An evaluation of maximum-likelihood reconstruction for SPECT. *IEEE Trans Med Imaging,* 9, 99–110.

Clackdoyle R, Noo F, Junyu G & Roberts JA. 2004. Quantitative reconstruction from truncated projections in classical tomography. *IEEE Trans Nucl Sci,* 51, 2570–2578.

Cooper JA, Neumann PH & McCandless BK. 1992. Effect of patient motion on tomographic myocardial perfusion imaging. *J Nucl Med,* 33, 1566–1571.

Crone C. 1963. Permeability of capillaries in various organs as determined by use of the "indicator diffusion" method. *Acta Physiol Scand,* 58, 292–305.

Cunningham VJ & Jones T. 1993. Spectral analysis of dynamic PET studies. *J Cereb Blood Flow Metab,* 13, 15–23.

Da silva AJ, Tang HR, Wong KH, Wu MC, Dae MW & Hasegawa BH. 2001. Absolute quantification of regional myocardial uptake of 99mTc-Sestamibi with SPECT: Experimental validation in a porcine model. *J Nucl Med,* 42, 772–779.

Dai X, Chen Z & Tian J. 2011. Performance evaluation of kinetic parameter estimation methods in dynamic FDG-PET studies. *Nucl Med Commun,* 32, 4–16.

De Jong HWAM, Slijpen ETP & Beekman FJ. 2001. Acceleration of Monte Carlo SPECT simulation using convolution-based forced detection. *IEEE Trans Nucl Sci,* 48, 58–64.

Defrise M, Noo F, Clackdoyle R & Kudo H. 2006. Truncated Hilbert transform and image reconstruction from limited tomographic data. *Inverse Problems,* 22, 1037–1053.

Dey J, Segars WP, Pretorius PH, Walvick RP, Bruyant PP, Dahlberg S & King MA. 2010. Estimation and correction of cardiac respiratory motion in SPECT in the presence of limited-angle effects due to irregular respiration. *Med Phys,* 37, 6453–6465.

Di Bella EVR, Ross SG, Kadrmas DJ, Khare HS, Christian PE, Mcjames S & Gullberg GT. 2001. Compartmental modeling of technetium-99m-labeled teboroxime with dynamic single-photon emission computed tomography: Comparison with static thallium-201 in a canine model. *Invest Radiol,* 36, 178–185.

Du Y, Tsui BM & Frey EC. 2005. Partial volume effect compensation for quantitative brain SPECT imaging. *IEEE Trans Med Imaging*, 24, 969–976.

Eisner R, Churchwell A, Noever T, Nowak D, Cloninger K, Dunn D, Carlson W, Oates J, Jones J, Morris D et al. 1988. Quantitative analysis of the tomographic thallium-201 myocardial bullseye display: Critical role of correcting for patient motion. *J Nucl Med*, 29, 91–97.

Eisner RL, Aaron AM, Worthy MR, Boyers AS, Leon AR, Fajman WA & Patterson RE. 1993. Apparent change in cardiac geometry during single-photon emission tomography thallium-201 acquisition: A complex phenomenon. *Eur J Nucl Med*, 20, 324–329.

Eisner RL, Noever T, Nowak D, Carlson W, Dunn D, Oates J, Cloninger K, Liberman HA. & Patterson RE 1987. Use of cross-correlation function to detect patient motion during SPECT imaging. *J Nucl Med*, 28, 97–101.

Erlandsson K, Buvat I, Pretorius PH, Thomas BA & Hutton BF. 2012. A review of partial volume correction techniques for emission tomography and their applications in neurology, cardiology and oncology. *Phys Med Biol*, 57, R119–R159.

Erlandsson K, Kacperski K, Van Gramberg D & Hutton BF. 2009. Performance evaluation of D-SPECT: A novel SPECT system for nuclear cardiology. *Phys Med Biol*, 54, 2635–2649.

Farncombe TH, Celler AM, Bever CA, Noll D, Maeght J & Harrop R. 2001. The incorporation of organ uptake into dynamic SPECT (dSPECT) image reconstruction. *IEEE Trans Nucl Sci*, 48, 3–9.

Farncombe TH, Noll D, Maeght J & Harrop R. 1999. Dynamic SPECT imaging using a single camera rotation (dSPECT). *IEEE Trans Nucl Sci*, 46, 1055–1061.

Feng B, Gifford HC, Beach RD, Boening G, Gennert MA & King MA. 2006a. Use of three-dimensional Gaussian interpolation in the projector/backprojector pair of iterative reconstruction for compensation of known rigid-body motion in SPECT. *IEEE Trans Med Imaging*, 25, 838–844.

Feng B, Pretorius PH, Farncombe TH, Dahlberg ST, Narayanan MV, Wernick MN, Celler AM, Leppo JA & King MA. 2006b. Simultaneous assessment of cardiac perfusion and function using 5-dimensional imaging with Tc-99m teboroxime. *J Nucl Cardiol*, 13, 354–361.

Fleming JS. 1984. A technique for motion correction in dynamic scintigraphy. *Eur J Nucl Med*, 9, 397–402.

Fleming JS. 1989. A technique for using CT images in attenuation correction and quantification in SPECT. *Nucl Med Commun*, 10, 83–97.

Flotats A & Carrio I. 2004. Cardiac neurotransmission SPECT imaging. *J Nucl Cardiol*, 11, 587–602.

Formiconi AR. 1993. Least-squares algorithm for region-of-interest evaluation in emission tomography. *IEEE Trans Med Imaging*, 12, 90–100.

Frey EC & Tsui BMW. 1997. A new method for modeling the spatially-variant, object-dependent scatter response function in SPECT. *1996 IEEE Nuclear Science Symposium–Conference Record, Vols 1–3*, 1082–1086.

Friedman J, Van Train K, Maddahi J, Rozanski A, Prigent F, Bietendorf J, Waxman A & Berman DS 1989. "Upward creep" of the heart: A frequent source of false-positive reversible defects during thallium-201 stress-redistribution SPECT. *J Nucl Med*, 30, 1718–1722.

Funk T, Kirch DL, Koss JE, Botvinick E & Hasegawa BH. 2006. A novel approach to multipinhole SPECT for myocardial perfusion imaging. *J Nucl Med*, 47, 595–602.

Gagnon D, Tung CH, Zeng L & Hawkins WG. 1999. Design and early testing of a new medium-energy transmission device for attenuation correction in SPECT and PET. *1999 IEEE Nuclear Science Symposium–Conference Record, Vols 1–3*, 1349–1353.

Garcia EV, Faber TL & Esteves FP. 2011. Cardiac dedicated ultrafast SPECT cameras: New designs and clinical implications. *J Nucl Med*, 52, 210–217.

Geckle WJ, Frank TL, Links JM & Becker LC. 1988. Correction for patient and organ movement in SPECT: Application to exercise thallium-201 cardiac imaging. *J Nucl Med*, 29, 441–450.

Germano G, Chua T, Kavanagh PB, Kiat H & Berman DS. 1993. Detection and correction of patient motion in dynamic and static myocardial SPECT using a multidetector camera. *J Nucl Med*, 34, 1349–1355.

Gravier E, Yang Y & Jin M. 2007. Tomographic reconstruction of dynamic cardiac image sequences. *IEEE Trans Image Process*, 16, 932–942.

Gullberg GT, Huesman RH, Dibella EVR & Reutter BW. 2004. Dynamic cardiac single photon emission computed tomography using fast data acquisition systems. In: Zaret, B. L. & Beller, G. A. (eds.) *Clinical Nuclear Cardiology: State of the Art and Future Directions.* Third Edition. Philadelphia, PA: Elsevier Mosby.

Gullberg GT, Huesman RH, Malko JA, Pelc NJ & Budinger TF. 1985. An attenuated projector backprojector for iterative SPECT reconstruction. *Phys Med Biol,* 30, 799–816.

Gullberg GT, Huesman RH, Ross SG, Dibella EVR, Zeng GL, Reutter BW, Christian PE & Foresti SA. 1999. Dynamic cardiac single-photon emission computed tomography. In: Zaret, B. L. & Beller, G. A. (eds.) *Nuclear Cardiology: State of the Art and Future Directions.* Second Edition ed. St. Louis, MO: Mosby, Inc.

Gullberg GT, Morgan HT, Zeng GSL, Christian PE, Di Bella EVR, Tung CH, Maniawski PJ, Hsieh YL & Datz FL. 1998. The design and performance of a simultaneous transmission and emission tomography system. *IEEE Trans Nucl Sci,* 45, 1676–1698.

Gunn RN, Gunn SR, Turkheimer FE, Aston JAD & Cunningham VJ. 2002. Positron emission tomography compartment models: A basis pursuit strategy for kinetic modeling. *J Cereb Blood Flow and Metab,* 22, 1425–1439.

Hebber E, Oldenburg D, Farncombe T & Celler A. 1997. Direct estimation of dynamic parameters in SPECT tomography. *IEEE Trans Nucl Sci,* 44, 2425–2430.

Hicks K, Ganti G, Mullani N & Gould KL. 1989. Automated quantitation of three-dimensional cardiac positron emission tomography for routine clinical use. *J Nucl Med,* 30, 1787–1797.

Hong YT & Fryer TD. 2010. Kinetic modelling using basis functions derived from two-tissue compartment models with a plasma input function: General principle and application to [18F]fluorodeoxyglucose positron emission tomography. *Neuroimage,* 51, 164–172.

Hu J, Boutchko R, Sitek A, Reutter BW, Huesman RH & Gullberg GT. 2008. *Dynamic Molecular Imaging of Cardiac Innervation Using a Dual Head Pinhole SPECT System.* Berkeley: Lawrence Berkeley National Lab-60008.

Huang SC, Hoffman EJ, Phelps ME & Kuhl DE. 1979. Quantitation in positron emission computed-tomography. 2. Effects of inaccurate attenuation correction. *J Comput Assist Tomogr,* 3, 804–814.

Huang SC & Yu DC. 1992. Capability evaluation of a sinogram error-detection and correction method in computed-tomography. *IEEE Trans Nucl Sci,* 39, 1106–1110.

Hudson HM & Larkin RS. 1994. Accelerated image-reconstruction using ordered subsets of projection data. *IEEE Trans Med Imaging,* 13, 601–609.

Hudson HM & Walsh C. 1997. Density deconvolution using spectral mixture models. Proceedings of the Second World Congress of the IASC, Pasadena, CA. 593–599.

Huesman RH. 1984. A new fast algorithm for the evaluation of regions of interest and statistical uncertainty in computed-tomography. *Phys Med Biol,* 29, 543–552.

Huesman RH, Knittel BL, Mazoyer BM, Coxson PG, Salmeron EM, Klein GJ, Reutter BW & Budinger TF. 1995. *Notes on RFIT: A Program for Fitting Compartment Models to Region-of-Interest Dynamic Emission Tomography Data.* Berkeley: Lawrence Berkeley National Laboratory.

Huesman RH, Reutter BW, Zeng GL & Gullberg GT. 1998. Kinetic parameter estimation from SPECT cone-beam projection measurements. *Phys Med Biol,* 43, 973–982.

Humphries T. 2011. *Temporal Regularization and Artifact Correction in Single Slow-Rotation Dynamic SPECT.* PhD Thesis, Simon Fraser University.

Hutton BF, Buvat I & Beekman FJ. 2011. Review and current status of SPECT scatter correction. *Phys Med Biol,* 56, R85–R112.

Hutton BF & Osiecki A. 1998. Correction of partial volume effects in myocardial SPECT. *J Nucl Cardiol,* 5, 402–413.

Iida H, Hayashi T, Eberl S & Saji H. 2003. Quantification in SPECT cardiac imaging. *J Nucl Med,* 44, 40–42.

Jacobson AF, Lombard J, Banerjee G & Camici PG. 2009. 123I-mIBG scintigraphy to predict risk for adverse cardiac outcomes in heart failure patients: Design of two prospective multicenter international trials. *J Nucl Cardiol,* 16, 113–121.

Jaszczak RJ, Greer KL, Floyd CE, Harris CC & Coleman RE. 1984. Improved SPECT quantification using compensation for scattered photons. *J Nucl Med,* 25, 893–900.

Ji SY & Travin MI. 2010. Radionuclide imaging of cardiac autonomic innervation. *J Nucl Cardiol,* 17, 655–666.

Kadrmas DJ, Frey EC & Tsui BMW. 1998. Application of reconstruction-based scatter compensation to thallium-201 SPECT: Implementations for reduced reconstructed image noise. *IEEE Trans Med Imaging,* 17, 325–333.

Kadrmas DJ & Gullberg GT. 2001. 4D maximum *a posteriori* reconstruction in dynamic SPECT using a compartmental model-based prior. *Phys Med Biol,* 46, 1553–1574.

Kamphuis C, Beekman FJ, Van Rijk PP & Viergever MA. 1998. Dual matrix ordered subsets reconstruction for accelerated 3D scatter compensation in single-photon emission tomography. *Eur J Nucl Med,* 25, 8–18.

Kessler RM, Ellis JR & Eden M. 1984. Analysis of emission tomographic scan data—Limitations imposed by resolution and background. *J Comput Assist Tomogr,* 8, 514–522.

Kim AS, Marvin B, Ruddy TD & Wells G. 2010. Patient motion on the GE Discovery CZT camera. *J Nucl Med,* 51 (Supplement 2), 2114.

King MA, Devries DJ, Pan TS, Pretorius PH & Case JA. 1997. An investigation of the filtering of TEW scatter estimates used to compensate for scatter with ordered subset reconstructions. *IEEE Trans Nucl Sci,* 44, 1140–1145.

King MA, Dey J, Johnson K, Dasari P, Mukherjee JM, McNamara JE, Konik A, Lindsay C, Zheng SK & Coughlin D. 2013. Use of MRI to assess the prediction of heart motion with gross body motion in myocardial perfusion imaging by stereotracking of markers on the body surface. *Med Phys,* 40.

King MA, Tsui BMW & Pan TS. 1995. Attenuation compensation for cardiac single-photon emission computed tomographic imaging. 1. Impact of attenuation and methods of estimating attenuation maps. *J Nucl Cardiol,* 2, 513–524.

Kovalski G, Israel O, Keidar Z, Frenkel A, Sachs, J & Azhari H. 2007. Correction of heart motion due to respiration in clinical myocardial perfusion SPECT scans using respiratory gating. *J Nucl Med,* 48, 630–636.

Krogh A. 1919. The supply of oxygen to the tissues and the regulation of the capillary circulation. *J Physiol,* 52, 457–474.

Kudo H, Courdurier M, Noo F & Defrise M. 2008. Tiny a priori knowledge solves the interior problem in computed tomography. *Phys Med Biol,* 53, 2207–2231.

Kuhl DE & Edwards RQ. 1963. Image separation redioisotope scanning. *Radiology,* 80, 653–661.

Lalush DS & Tsui BMW. 2000. Performance of ordered-subset reconstruction algorithms under conditions of extreme attenuation and truncation in myocardial SPECT. *J Nucl Med,* 41, 737–744.

Lange K & Carson R. 1984. EM reconstruction algorithms for emission and transmission tomography. *J Comput Assist Tomogr,* 8, 306–316.

Leppo JA & Meerdink D. 1989. Comparison of the myocardial uptake of a technetium-labeled isonitrile analogue and thallium. *J Circ Res,* 65, 632–639.

Leslie WD, Dupont JO, Mcdonald D & Peterdy AE. 1997. Comparison of motion correction algorithms for cardiac SPECT. *J Nucl Med,* 38, 785–790.

Lim Y, Gottschalk S, Walker R, Schreiner R, Valentino F, Pinkstaff C, Janzo J, Covic J, Perusek A, Anderson J, Kim K, Shand D, Coulman K, King S & Styblo D. 1985. Triangular SPECT system for 3-D total organ volume imaging: Design concept and preliminary results. *IEEE Trans on Nucl Sci,* NS-32, 741–747.

Links JM, Frank TL & Becker LC. 1991. Effect of differential tracer washout during SPECT acquisition. *J Nucl Med,* 32, 2253–2257.

Liu C, Pierce LA, 2ND, Alessio AM & Kinahan PE. 2009. The impact of respiratory motion on tumor quantification and delineation in static PET/CT imaging. *Phys Med Biol,* 54, 7345–7362.

Liu C & Sinusas AJ. 2014. Is assessment of absolute myocardial perfusion with SPECT ready for prime time? *J Nucl Med,* 55, 1573–1575.

Liu H, Chan C, Grobshtein Y, MA T, Liu Y, Wang S, Stacy MR, Sinusas AJ & Liu C. 2015. Anatomical-based partial volume correction for low-dose dedicated cardiac SPECT/CT. *Phys Med Biol,* 60, 6751–6773.

Ljungberg M & Strand SE. 1989. A Monte-Carlo program for the simulation of scintillation camera characteristics. *Comput Methods Programs Biomed,* 29, 257–272.

Lodge MA, Carson RE, Carrasquillo JA, Whatley M, Libutti SK & Bacharach SL. 2000. Parametric images of blood flow in oncology PET studies using [15O]water. *J Nucl Med,* 41, 1784–1792.

Maltz JS. 2000. Region resolvability versus noise level characteristics for joint spatial and kinetic parameter estimation in inconsistent projection dynamic ECT. *IEEE Trans Nucl Sci,* 47, 1143–1148.

Maltz JS. 2001. Optimal time-activity basis selection for exponential spectral analysis: Application to the solution of large dynamic emission tomographic reconstruction problems. *IEEE Trans Nucl Sci,* 48, 1452–1464.

Maltz JS. 2002. Parsimonious basis selection in exponential spectral analysis. *Phys Med Biol,* 47, 2341–2365.

Manglos SH, Gagne GM & Bassano DA. 1993. Quantitative analysis of image truncation in focal-beam CT. *Phys Med Biol,* 38, 1443.

Mao YF & Zeng GSL. 2013. A tailored ML-EM algorithm for reconstruction of truncated projection data using few view angles. *Phys Med Biol,* 58, N157–N169.

Marshall RC, Powers-Risius P, Reutter BW, Taylor SE, Vanbrocklin HF, Huesman RH & Budinger TF. 2001. Kinetic analysis of ^{125}I-iodorotenone as a deposited myocardial flow tracer: Comparison with ^{99m}Tc-sestamibi. *J Nucl Med,* 42, 272–281.

Matsuda H, Ohnishi T, Asada T, LI Z-J, Kanetaka H, Imabayashi E, Tanaka F & Nakano S. 2003. Correction for partial-volume effects on brain perfusion SPECT in healthy men. *J Nucl Med,* 44, 1243–1252.

Matsumoto N, Berman DS, Kavanagh PB, Gerlach J, Hayes SW, Lewin HC, Friedman JD & Germano G. 2001. Quantitative assessment of motion artifacts and validation of a new motion-correction program for myocardial perfusion SPECT. *J Nucl Med,* 42, 687–694.

Matsuo S, Takahashi M, Nakamura Y & Kinoshita M. 1996. Evaluation of cardiac sympathetic innervation with iodine-123-metaiodobenzylguanidine imaging in silent myocardial ischemia. *J Nucl Med,* 37, 712–717.

Matthews J, Bailey D, Price P & Cunningham V. 1997. The direct calculation of parametric images from dynamic PET data using maximum-likelihood iterative reconstruction. *Phys Med Biol,* 42, 1155–1173.

Maublant JC, Marcaggi X, Lusson J-R, Boire J-Y, Cauvin J-C, Jacob P, Veyre A & Cassagnes J. 1992. Comparison between thallium-201 and technetium-99m methoxyisobutyl isonitrile defect size in single-photon emission computed tomography at rest, exercise and redistribution in coronary artery disease. *Am J Cardiol,* 69, 183–187.

Maublant JC, Moins N, Gachon P, Renoux M, Zhang Z & Veyre A. 1993. Uptake of technetium-99m-teboroxime in cultured myocardial cells: Comparison with thallium-201 and technetium-99m-sestamibi. *J Nucl Med,* 34, 255–259.

McCarthy AW & Miller MI. 1991. Maximum-likelihood SPECT in clinical computation times using mesh-connected parallel computers. *IEEE Trans Med Imaging,* 10, 426–436.

McNamara JE, Bruyant P, Johnson K, Feng B, Lehovich A, Gu SX, Gennert MA & King MA. 2008. An assessment of a low-cost visual tracking system (VTS) to detect and compensate for patient motion during SPECT. *IEEE Trans Nucl Sci,* 55, 992–998.

McNamara JE, Pretorius PH, Johnson K, Mukherjee JM, Dey J, Gennert MA & King MA. 2009. A flexible multicamera visual-tracking system for detecting and correcting motion-induced artifacts in cardiac SPECT slices. *Med Phys,* 36, 1913–1923.

Meikle SR, Matthews J, Cunningham VJ, Bailey DL, Livieratos L, Jones T & Price P. 1998. Parametric image reconstruction using spectral analysis of PET projection data. *Phys Med Biol,* 43, 651–666.

Moore SC, Southekal S, Park MA, Mcquaid SJ, Kijewski MF & Muller SP. 2012. Improved regional activity quantitation in nuclear medicine using a new approach to correct for tissue partial volume and spillover effects. *IEEE Trans Med Imaging,* 31, 405–416.

Mukherjee JM, Johnson KL, McNamara JE & King MA. 2010. Quantitative study of rigid-body and respiratory motion of patients undergoing stress and rest cardiac SPECT imaging. *IEEE Trans Nucl Sci,* 57, 1105–1115.

Mukherjee JM, McNamara JE, Johnson KL, Dey J & King MA. 2009. Estimation of rigid-body and respiratory motion of the heart from marker-tracking data for SPECT motion correction. *IEEE Trans Nucl Sci,* 56, 147–155.

Nakata T, Miyamoto K, Doi A, Sasao H, Wakabayashi T, Kobayashi H, Tsuchihashi K & Shimamoto K. 1998. Cardiac death prediction and impaired cardiac sympathetic innervation assessed by MIBG in patients with failing and nonfailing hearts. *J Nucl Cardiol,* 5, 579–590.

Natterer F. 2001. *The Mathematics of Computerized Tomography,* Philadelphia, PA: SIAM.

Nichols TE, Qi J, Asma E & Leahy RM. 2002. Spatiotemporal reconstruction of list-mode PET data. *IEEE Trans Med Imaging*, 21, 396–404.

Noo F, Clackdoyle R & Pack JD. 2004. A two-step Hilbert transform method for 2D image reconstruction. *Phys Med Biol*, 49, 3903–3923.

O'Connor JM, Pretorius PH, Johnson K & King MA. 2013. A method to synchronize signals from multiple patient monitoring devices through a single input channel for inclusion in list-mode acquisitions. *Med Phys*, 40, 122502.

O'Connor MK, Kanal KM, Gebhard MW & Rossman PJ. 1998. Comparison of four motion correction techniques in SPECT imaging of the heart: A cardiac phantom study. *J Nucl Med*, 39, 2027–2034.

Ogawa K, Harata Y, Ichihara T, Kubo A & Hashimoto S. 1991. A practical method for position-dependent Compton-scatter correction in single photon emission CT. *IEEE Trans Med Imaging*, 10, 408–412.

Ogawa K, Ichihara T & Kubo A. 1994. Accurate scatter correction in single photon emission CT. *Ann Nucl Med Sci*, 7, 145–150.

Patton JA, Slomka PJ, Germano G & Berman DS. 2007. Recent technologic advances in nuclear cardiology. *J Nucl Cardiol*, 14, 501–513.

Perrone-Filardi P, Paolillo S, Dellegrottaglie, S Gargiulo P, Savarese, G Marciano, C, Casaretti L, Cecere M, Musella, F Pirozzi, E Parente, A & Cuocolo A. 2011. Assessment of cardiac sympathetic activity by MIBG imaging in patients with heart failure: A clinical appraisal. *Heart*, 97, 1828–1833.

Pretorius PH, King MA, Johnson K, Mukherjee JM, Dey J & Konik A. 2011. Combined respiratory and rigid body motion compensation in cardiac perfusion SPECT using a visual tracking system. *Proceedings of 2011 IEEE Medical Imaging Conference*, 2768–2773.

Pretorius PH, King MA, Pan TS & Hutton BF. 1997. Attenuation correction strategies for multi-energy photon emitters using SPECT. *IEEE Trans Nucl Sci*, 44, 1323–1328.

Raffel DM & Wieland DM. 2010. Development of mIBG as a cardiac innervation imaging agent. *JACC Cardiovasc Imaging*, 3, 111–116.

Reader AJ, Sureau FC, Comtat C, Trebossen R & Buvat I. 2006. Joint estimation of dynamic PET images and temporal basis functions using fully 4D ML-EM. *Phys Med Biol*, 51, 5455–5474.

Renkin EM 1959. Transport of potassium-42 from blood to tissue in isolated mammalian skeletal muscles. *Am J Physiol*, 197, 1205–1210.

Reutter BW, Gullberg GT & Huesman RH. 1998. Kinetic parameter estimation from attenuated SPECT projection measurements. *IEEE Trans Nucl Sci*, 45, 3007–3013.

Reutter BW, Gullberg GT & Huesman RH. 2000. Direct least squares estimation of spatiotemporal distributions from dynamic SPECT projections using a spatial segmentation and temporal B-splines. *IEEE Trans Med Imaging*, 19, 434–450.

Reutter BW, Gullberg GT & Huesman RH. 2002. Effects of temporal modeling on the statistical uncertainty of spatiotemporal distributions estimated directly from dynamic cone-beam SPECT projections. *Phys Med Biol*, 47, 2673–2683.

Reutter BW, Gullberg GT & Huesman RH. 2004. Accuracy and precision of compartmental model parameters obtained from directly estimated dynamic SPECT time-activity curves. *IEEE Trans Nucl Sci*, 51, 170–176.

Reutter BW, Oh S, Gullberg GT & Huesman RH 2005. Improved quantitation of dynamic SPECT via fully 4-D joint estimation of compartmental models and blood input function directly from projections. 2005 IEEE Nuclear Science Symposium and Medical Imaging Conference, Puerto Rico. 2337–2341.

Sabondjian E, Stodilka RZ, Belhocine T, King ME, Wisenberg G & Prato FS. 2009. Small field-of-view cardiac SPECT can be implemented on hybrid SPECT/CT platforms where data acquisition and reconstruction are guided by CT. *Nucl Med Comm*, 30, 718–726. 10.1097/MNM.0b013e32832eabec.

Segars WP & Tsui BMW. 2002. Study of the efficacy of respiratory gating in myocardial SPECT using the new 4D NCAT phantom. *2001 IEEE Nuclear Science Symposium, Conference Records, Vols 1–4*, 1536–1539.

Shepp L & Vardi, Y. 1982. Maximum likelihood reconstruction for emission tomography. *IEEE Trans Med Imaging*, M-1, 113–122.

Shi Y & Karl WC. 2004. Level set methods for dynamic tomography. 2004 IEEE International Symposium on Biomedical Imaging: Macro to Nano. April 15–18, 2004, Washington, DC.

Shrestha U, Alhassen F, Buchko R, Gould RG, Seo Y, Botvinick EH & Gullberg GT. Fully 6D image reconstruction for myocardial perfusion imaging of tracer dynamics, cardiac and respiratory motion. 12th International Meeting on Fully Three-Dimensional Image Reconstruction in Radiology and Nuclear Medicine, 2013, Lake Tahoe, CA. 209–212.

Shrestha U, Alhassen F, Seo Y, Botvinick EH & Gullberg GT. Reconstruction of gated dynamic cardiac SPECT data using spatiotemporal basis functions. 2012 IEEE Nuclear Science Symposium and Medical Imaging Conference, 2012, Anaheim, CA. 2489–2492.

Sitek A, Di Bella EVR, Gullberg GT & Huesman RH. 2002a. Removal of liver activity contamination in teboroxime dynamic cardiac SPECT imaging using factor analysis. *J Nucl Cardiol,* 9, 197–205.

Sitek A, Gullberg GT, Di Bella EVR & Celler A. 2001. Reconstruction of dynamic renal tomographic data acquired by slow rotation. *J Nucl Med,* 42, 1704–1712.

Sitek A, Gullberg GT & Huesman RH 2002b. Correction for ambiguous solutions in factor analysis using a penalized least squares objective. *IEEE Trans Med Imaging,* 21, 216–225.

Slomka PJ, Patton JA, Berman DS & Germano G. 2009. Advances in technical aspects of myocardial perfusion SPECT imaging. *J Nucl Cardiol,* 16, 255–276.

Smith AM, Gullberg GT, Christian PE & Datz L. 1994. Kinetic modeling of teboroxime using dynamic SPECT imaging of a canine model. *J Nucl Med,* 35, 484–495.

Stewart RE, Schwaiger M, Hutchins GD, Chiao P-C, Gallagher KP, Nguyen N, Petry NA & Rogers WL. 1990. Myocardial clearance kinetics of technetium-99m-SQ30217: A marker of regional myocardial blood flow. *J Nucl Med,* 31, 1183–1190.

Strauss HW, Johnson MN, Schoder H & Tamaki N. 2010. Metaiodobenzylguanidine imaging comes of age. A new arrow in the prognostic quiver for heart failure patients. *J Am Coll Cardiol,* 55, 2222–2224.

Tan P, Bailey DL, Meikle SR, Eberl S, Fulton RR & Hutton BF. 1993. A scanning line source for simultaneous emission and transmission measurements in SPECT. *J Nucl Med,* 34, 1752–1760.

Tung CH, Gullberg GT, Zeng GL, Christian PE, Datz FL & Morgan HT. 1992. Nonuniform attenuation correction using simultaneous transmission and emission converging tomography. *IEEE Trans Nucl Sci,* 39, 1134–1143.

Vanzi E, De Cristofaro MT, Ramat S, Sotgia B, Mascalchi M & Formiconi AR. 2007. A direct ROI quantification method for inherent PVE correction: Accuracy assessment in striatal SPECT measurements. *Eur J Nucl Med Mol Imaging,* 34, 1480–1489.

Verhaeghe J, D'Asseler Y, Staelens S, Vandenberghe S & Lemahieu I. 2007. Reconstruction for gated dynamic cardiac PET imaging using a tensor product spline basis. *IEEE Trans Nucl Sci,* 5480–5491.

Verhaeghe J, Van De Ville D, Khalidov I, D'Asseler Y, Lemahieu I & Unser M. 2008. Dynamic PET reconstruction using wavelet regularization with adapted basis functions. *IEEE Trans Med Imaging,* 27, 943–959.

Vunckx K, Atre A, Baete K, Reilhac A, Deroose CM, Van Laere K & Nuyts J. 2012. Evaluation of three MRI-based anatomical priors for quantitative PET brain imaging. *IEEE Trans Med Imaging,* 31, 599–612.

Wakabayashi T, Nakata T, Hashimoto A, Yuda S, Tsuchihashi K, Travin MI & Shimamoto K. 2001. Assessment of underlying etiology and cardiac sympathetic innervation to identify patients at high risk of cardiac death. *J Nucl Med,* 42, 1757–1767.

Watabe H, Jino H, Kawachi N, Teramoto N, Hayashi T, Ohta Y & Iida H. 2005. Parametric imaging of myocardial blood flow with 15O-water and PET using the basis function method. *J Nucl Med,* 46, 1219–1224.

Welch A & Gullberg GT. 1997. Implementation of a model-based nonuniform scatter correction scheme for SPECT. *IEEE Trans Med Imaging,* 16, 717–726.

Wells RG, Timmins R, Klein R, Lockwood J, Marvin B, Dekemp RA, Wei L & Ruddy TD. 2014. Dynamic SPECT measurement of absolute myocardial blood flow in a porcine model. *J Nucl Med,* 55, 1685–1691.

Winant CD, Aparici CM, Zelnik YR, Reutter BW, Sitek A, Bacharach SL & Gullberg GT. 2012. Investigation of dynamic SPECT measurements of the arterial input function in human subjects using simulation, phantom and human studies. *Phys Med Biol,* 57, 375–393.

Xiao J, Verzijlbergen F, Viergever M & Beekman F. 2010. Small field-of-view dedicated cardiac SPECT systems: Impact of projection truncation. *Eur J Nucl Med Molec Imaging,* 37, 528–536.

Yamada T, Shimonagata T, Fukunami M, Kumagai K, Ogita H, Hirata A, Asai M, Makino N, Kioka H, Kusuoka H, Hori M & Hoki N. 2003. Comparison of the prognostic value of cardiac iodine-123 metaiodobenzylguanidine imaging and heart rate variability in patients with chronic heart failure: A prospective study. *J Am Coll Cardiol*, 41, 231–238.

Ye Y, Yu H, Wei Y & Wang G. 2007. A general local reconstruction approach based on a truncated Hilbert transform. *Int J Biomed Imaging*, 2007, 63634.

Yu HY & Wang G. 2009. Compressed sensing based interior tomography. *Phys Med Biol*, 54, 2791–2805.

Zeng GL, Bai CY & Gullberg GT. 1999. A projector/backprojector with slice-to-slice blurring for efficient three-dimensional scatter modeling. *IEEE Trans Med Imaging*, 18, 722–732.

Zeng GL & Gullberg GT. 1990. A study of reconstruction artifacts in cone beam tomography using filtered backprojection and iterative EM algorithms. *IEEE Trans Nucl Sci*, 37, 759–767.

Zeng GL & Gullberg GT. 2010. Spect region of interest reconstruction with truncated transmission and emission data. *Med Phys*, 37, 4627–4633.

Zeng GSL & Gullberg GT. 2012. Null-space function estimation for the interior problem. *Phys Med Biol*, 57, 1873–1887.

Zeng GL, Gullberg GT & Huesman RH. 1995. Using linear time-invariant system theory to estimate kinetic parameters directly from projection measurements. *IEEE Trans Nucl Sci*, 42, 2339–2346.

Zeniya T, Watabe H, Inomata T, Iida H, Sohlberg A & Kudo H. 3D-OSEM reconstruction from truncated data in pinhole SPECT. Nuclear Science Symposium Conference Record, 2007. NSS '07. IEEE, October 26, 2007–November 3, 2007. 4205–4207.

Zeniya T, Watabe H, Sohlberg A, Inomata T, Kudo H & Iida H. Effect of truncation in quantitative cardiac imaging with small field-of-view pinhole SPECT. Nuclear Science Symposium Conference Record, 2006. IEEE, October 29, 2006–November 1, 2006. 3239–3241.

Zhang B & Zeng GL. 2007. Two-dimensional iterative region-of-interest (ROI) reconstruction from truncated projection data. *Med Phys*, 34, 935–944.

Evaluations of cardiovascular diseases with hybrid PET-CT imaging

ANTTI SARASTE, SAMI KAJANDER, AND JUHANI KNUUTI

15.1 INTRODUCTION

The main clinical application of cardiac hybrid imaging so far has been the evaluation of coronary artery disease (CAD). It is well known that the angiographic severity of coronary lesion is a poor predictor of its hemodynamic significance (Gould and Johnson 2010; Tonino et al. 2010; Uren et al. 1994; White et al. 1984). Correspondingly, studies of computed tomographic angiography (CTA) have shown that only 30%–50% of stenoses with more than 50% luminal narrowing are associated with reversible perfusion defects on myocardial perfusion imaging (MPI). On the other hand, many patients with completely normal myocardial perfusion may have nonobstructive CAD (Gaemperli et al. 2007; Hacker et al. 2007; Schuijf et al. 2006).

In the existence of myocardial ischemia, revascularization improves the outcome by reducing the ischemic burden on the left ventricular (LV) myocardium (Hachamovitch et al. 2003; Shaw et al. 2008), while trials have failed to demonstrate superiority of revascularization over modern medical treatment if patients are not stratified by prior ischemia testing (Boden et al. 2007; Frye et al. 2009). In the recent trials, the role of ischemia-guided revascularization in patients with stable CAD has been emphasized (Pijls et al. 2010; Tonino 2009). According to this evidence, testing for myocardial ischemia has been included in the guidelines (Montalescot et al. 2013; Wijns et al. 2010). On the other hand, existence of non-obstructive CAD has been shown to have prognostic information (Min et al. 20011; van Werkhoven et al. 2009). Thus, to offer the most appropriate treatment strategy for stable CAD, both morphology and function are important.

The rapid development in imaging technology and processing software during the last decade has facilitated the fast clinical utilization of cardiac hybrid imaging (Gaemperli 2012). Cardiac hybrid imaging was stimulated by the successful application of hybrid whole-body positron emission tomography (PET)-computed tomography (CT) in oncology. Initially, the CT data were used only for attenuation correction of MPI images. Since then, several technical issues such as spatial shifts and intrinsic mismatch between the electrocardiogram (ECG)-gated CTA and nongated perfusion study have been resolved. In addition, three-dimensional (3-D) volume-rendering techniques have been developed to correlate perfusion information with the coronary anatomy.

This chapter discusses the benefits and future possibilities of hybrid PET-CT imaging in the evaluation of cardiac diseases in the light of available clinical data.

15.2 DETECTION OF CAD

15.2.1 STRENGTHS AND WEAKNESSES OF SEPARATE MODALITIES

15.2.1.1 CORONARY CTA

During the last decade, CTA has developed to a widely used noninvasive anatomical imaging method (Schroeder et al. 2008). Due to major technical advances, it may now be performed quickly and robustly in clinical practice. Although both the indications and technical refinements for CTA are still evolving, the ability to evaluate both the lumen and its surroundings (vessel wall, myocardium) adds an intriguing perspective to previously available diagnostic tools.

The gantry of a CT scanner rotates around the patient to produce a map of x-ray attenuation values as described in the previous chapters of this book. It is possible to determine the spatial relationship of the radiation absorbing structures to create an image of the scanned object. A computed tomogram consists of a matrix of attenuation values presented in shades of grey. In modern scanners, a large volume is set parallel along the longitudinal (Z-) axis of the patient, thus allowing simultaneous collection of data from a large field of view. Fast gantry rotation (with rotation times below 300 msec) and advanced detector technology allow reconstruction of extremely thin (sub-mm) slices and enables one to acquire detailed images with few motion artefacts. Although the 64-detector-row CT with about 4 cm coverage per rotation and imaging time of 6–12 seconds for the coronary artery tree is still the industry standard, more rapid image acquisition is now allowed by the 128-, 256-, and 320-row configurations of which the 320-slice solution offers the advantage of a 16-cm long Z-axis field of view, thus producing images of the entire heart with a single gantry rotation.

An alternative solution to improve the performance of CT is to use a device with dual x-ray sources and detectors, offering superior time resolution and, thus, reduced artefacts in moving targets. In addition, the use of two x-ray sources makes it possible to use two different x-ray energies, resulting in better contrast and presumably enabling more accurate analysis of a coronary artery plaque. For the same purpose, another vendor uses a single x-ray tube system in which the electric current is rapidly changed between two levels to produce different energies. Other advanced versions allowing for more sophisticated analysis of the energy spectra are being developed.

The radiation dose of a cardiac CTA exam may now be contained through the use of modern technology (Knuuti et al. 2014). The mean dose of 64-row cardiac CTA studies was once estimated as 15 mSv for males and 21 mSv for females, whereas the employment of novel devices and protocols has resulted in significant dose reduction. The most efficient of these techniques include fast spiral imaging with a dual-source CT and sequential ("step-and-shoot") imaging with doses near, and in selected cases even below, 1 mSv. Iterative image reconstruction has recently become available by all the major vendors, reducing noise and thus producing good-quality images with low radiation dose. Thus, the modern CT offers the tools for successful imaging of fast-moving small structures such as coronary arteries. Images gated or triggered by ECG for freezing the motion and acquisition are adequate and fast enough for covering the heart during a few seconds.

Prior to the exam, the usual precautions preceding any CT scan with intravenous contrast (allergies, possible renal failure) must be taken into consideration. Administration of beta-blockers to patients with elevated heart rates (HRs) has proved beneficial: the amount of motion artefacts is decreased and, in some cases, use of more potent dose saving protocols is permitted. Administration of oral nitrates before the scan is of use because the compound dilates coronary arteries, improving visibility.

In the scanner room, the patient is placed supine with the arms placed above the head. After initial localizing images, a noncontrast scan may be obtained to evaluate the amount of coronary calcium. Although the presence or absence of calcium deposits does not necessarily implicate whether the patient has significant CAD or not, this low-dose scan may be useful in determining the length of scan of the contrast-enhanced series, as well as in cases in which intravenous contrast is contraindicated. If the patient's HR is very high despite premedication, thus predisposing to motion artefacts, and there are abundant calcifications, it may be advisable to reconsider the possible benefits of CTA compared with other diagnostic methods. In some studies, a set limit for the amount of calcium is suggested before the contrast enhanced scan.

CTA enables direct visualization of coronary stenoses through the use of intravenous iodinated contrast. Usually, 50–100 mL of highly iodinated contrast media (320–400 mg of iodine/L) is administered at 4–6 mL/s followed by a saline flush. A test bolus method may be used to determine the optimal timing of contrast, but automated or semiautomated bolus triggering may produce comparable results with less radiation and contrast media.

If one uses a dose-saving imaging protocol with prospective ECG triggering, data from the late diastolic phase (65%–85% of the RR-interval) are usually collected. However, in patients with a HR above 60–65 per minute, additional systolic images may be of benefit to decrease motion artefacts (of the right coronary artery [RCA] in particular). Usually, this is obtained by widening and repositioning the time window of image acquisition prior to the scan.

Functional images of the heart can be obtained with a retrospectively ECG-gated protocol when data are collected during the whole RR-interval and, hence, all the phases. This enables the evaluations of wall motion, the cardiac valves, and ejection fraction but also causes more radiation to the patient than prospectively gated acquisition does. Retrospective gating is usually mandatory when the patient has a high or irregular rhythm. With extremely high HRs, it may be necessary to use multisegment acquisition modes that collect data of the same location during two or more heartbeats.

Images are evaluated at a workstation that allows efficient handling of large amounts of data and use of modern image processing techniques such as maximal intensity projection, multiplanar reformation, and volume rendering. It is, however, of utmost importance to also evaluate the original transaxial images. Each coronary artery segment is then assessed in terms of possible plaques, stenoses, or anomalies. In routine work, it may be advisable to use a four- to five-point scale in the evaluation of the degree of the luminal narrowing (for example, in terms of diameter stenosis, <30%, 30%–50%, 50%–70%, >70%, occlusion) and at least a rough estimate of plaque characteristics (calcified, mixed, noncalcified).

A large number of single-center studies (Schroeder et al. 2008) and several randomized clinical trials (Budoff et al. 2008, Meijboom et al. 2008, Miller et al. 2008) with 64-detector CT technology have assessed the diagnostic accuracy of coronary CTA in the detection of obstructive CAD as defined by invasive angiography. These studies have shown that the negative predictive value of such scans is excellent, close to 100%, thereby suggesting that CT can reliably rule out the presence of hemodynamically significant CAD. Positive predictive values, however, have been less impressive, usually between 75% and 85%. Most often, this is due to overestimation of detected stenoses. False-positive and false-negative interpretations and decreased accuracy for evaluating the coronary arteries attribute to image artefacts in many cases. The major cause of such diagnostic drawbacks is the presence of dense vascular calcifications causing blooming artefacts. Less frequent causes include motion and insufficient contrast-to-noise ratio due to obesity or poorly chosen imaging parameters.

Current clinical indications recommend that CTA can be used to rule out CAD in symptomatic stable patients with intermediate pretest probability of CAD especially when patients are unable to exercise or ECG is uninterpretable. CTA is also useful in patients with uninterpretable or equivocal stress test. In patients with acute chest pain, intermediate pretest probability of CAD and no ECG or biomarker changes, normal CTA may be used for safely ruling out acute coronary syndromes (Goldstein et al. 2011: Hoffmann et al. 2012; Litt et al. 2012). Another group of patients for whom CTA can provide important information is the evaluation of coronary arteries in patients with new-onset heart failure. On the contrary, patients for whom CTA is not recommended are those with pronounced coronary calcification and patients with high pretest likelihood of

CAD. In addition, CTA should not be used to assess the severity and significance of known CAD or screen asymptomatic subjects for CAD (Montalescot et al. 2013; Schroeder et al. 2008; Wijns et al. 2010).

In conclusion, trials have demonstrated that CTA is an accurate noninvasive imaging technique in the morphological detection of CAD. In most available studies, the negative predictive value has been high, suggesting that the method is reliable for ruling out significant CAD. In particular, this is true in patients with low or medium likelihood of CAD having stable angina. However, a CTA scan may overestimate the severity of atherosclerotic obstructions, thus requiring further testing to guide patient management. This is most evident for patient populations with extensive calcifications and higher likelihood of CAD.

15.2.1.2 PET PERFUSION IMAGING

Myocardial perfusion PET, like single-photon emission computerized tomography (SPECT), is used to evaluate the radiotracer uptake in the myocardium. The uptake kinetics reflects relative regional myocardial blood flow, which can be quantified at rest and during pharmacological stress as described in detail elsewhere in this book. The three main tracers used in the imaging of myocardial perfusion are 15-oxygen-labeled water, 13-nitrogen-labeled ammonia, and generator-produced 82-Rubidium (^{82}Rb). Each of these tracers has specific advantages and limitations that have also been discussed in detail elsewhere in this book. The radiation dose for the patient is relatively small with these PET tracers (typically 0.5–3 mSv), which makes it an attractive combination with coronary CT angiogram.

Compared with MPI with SPECT, PET can measure myocardial radioactivity concentrations with better spatial and contrast resolutions and it has accurate and well-validated attenuation correction. As a result, MPI using PET may be superior to SPECT for the detection of CAD in terms of image quality, certainty of interpretation, and diagnostic accuracy (Montalescot et al. 2013). A large number of studies have reported excellent diagnostic accuracy of PET MPI for the detection of obstructive CAD, defined as ≥50% or ≥70% stenosis on invasive coronary angiography. These studies have been summarized in two recent meta-analyses. One analysis using only ^{82}Rb as a flow tracer (n = 1394 patients) showed pooled sensitivity of 90% and specificity of 88% (Mc Ardle et al. 2012). The other meta-analysis using all available myocardial PET flow tracers (n = 1692 patients) showed pooled sensitivity of 84% and specificity of 81% (Jaarsma et al. 2012). The comparison of perfusion imaging techniques has favored PET over SPECT (Jaarsma et al. 2012; Mc Ardle et al. 2012). Specific patients (e.g., obese) who could benefit from PET instead of SPECT are those prone to severe attenuation artefacts, leading to an equivocal result.

In addition to assessing relative myocardial tracer uptake, PET perfusion imaging enables quantitative measurements of rest and stress myocardial blood flow (MBF) in absolute units that can be used to compute absolute myocardial flow reserve (MFR). This is based on the evaluation of regional tracer kinetics in the blood pool and the myocardium as described earlier in this book (Gould et al. 2013). Quantitative analysis of MBF has potential to further improve diagnostic accuracy of PET in detecting and estimating severity of CAD as well as stratifying patients' risk. A pooled analysis of studies investigating the accuracy of quantitative regional PET MPI in the detection of obstructive CAD in 165 patients showed sensitivity of 93% and specificity of 87% (Saraste et al. 2012). It is particularly helpful for revealing the true extent of CAD in patients with multivessel disease as well as for detection of global reduction of myocardial perfusion due to balanced multivessel disease or microvascular disease (Fiechter et al. 2012; Kajander et al. 2011; Saraste et al. 2012). The latter may be completely missed by conventional, relative analysis of MPI studies.

Clinical experience has demonstrated that in analogy to SPECT, PET MPI provides powerful prognostic information. Large clinical registries have been published recently that include more than 10,000 patients with known or suspected CAD (Dorbala et al. 2013; Fiechter et al. 2012; Williams et al. 2012). The extent and severity of ischemia and scar on PET MPI provided incremental risk estimates of future cardiac and all-cause death for both males and females, compared with traditional risk factors. More importantly, coronary vasodilator dysfunction quantified by PET myocardial flow reserve (MFR) has been shown as a powerful predictor of cardiac mortality in patients with suspected or known CAD between diabetics and nondiabetics (Murthy et al. 2011, 2012; Saraste et al. 2012). In a large cohort of 2783 patients, MFR provided incremental prognostic value over semiquantitative measures of myocardial ischemia and scar as well as other clinical variables for identification of patients at risk of cardiac death (Murthy et al. 2011). The addition of MFR resulted in the correct reclassification of approximately one third of all intermediate-risk patients. Another study demonstrated that diabetic patients without CAD but with impaired flow reserve had event rates comparable to those of patients with prior CAD (Murthy et al. 2012).

PET myocardial blood flow tracers are currently more expensive and less available than SPECT tracers. Cyclotron-produced PET flow tracers such as nitrogen-13 ammonia and oxygen-15 water have excellent extraction rates of tracer uptake in the myocardium but require close proximity of cyclotron due to their short half-lives (10 min and 2 min, respectively). Therefore, generator-produced ^{82}Rb, which has similar biologic activity to thallium-201, is currently the most widely clinically used radiotracer for PET MPI. New PET perfusion tracers labeled with fluorine-18 (^{18}F) that has a half-life of 110 min have been developed for more widespread use in centers without a cyclotron. ^{18}F-Flurpiridaz was recently shown to be safe and good for diagnosis of CAD and is currently in a phase II clinical trial (Berman et al. 2013). In addition to tracer availability, access to scanner time may be also a limiting factor, but the number of PET scanners for clinical work is rapidly increasing. Although the investment costs are quite high with PET, the scans are very rapid, and when the availability of tracers improves, the PET scans can be also financially competitive to other imaging modalities.

A general limitation of perfusion imaging is the lack of information on coronary anatomy, and thus, it is not possible to differentiate between microvascular dysfunction and epicardial stenosis from PET alone. The knowledge on individual coronary anatomy would also help identify perfusion defects for certain coronary lesions. Coronary artery anatomy varies considerably among individuals and may disagree with standardized vascular territories, particularly in the inferior and inferolateral wall, the territories with the largest variability in coronary anatomy (Javadi et al. 2010). Finally, a limitation of MPI is that it reveals only coronary lesions that induce perfusion defects. Although functionally significant CAD is currently the cornerstone for the selection of further medical and invasive therapy, nonobstructive plaques are also common in patients referred for diagnostic evaluation and might have significant implications for patients' prognosis (Min et al. 2011).

15.2.2 HYBRID CARDIAC PET-CT IMAGING: CLINICAL DATA

Hybrid PET-CT imaging enables fusion of coronary CTA and MPI to visualize coronary atherosclerotic lesions and their hemodynamic consequences in a single study (Gaemperli et al. 2012; Danad 2013b). It is now available at selected centers and holds much promise for the detection of obstructive CAD as well as guidance of revascularization through vessel-specific information.

The current evidence of hybrid PET-CT is limited to few rather small single-center studies. The feasibility and clinical robustness of noninvasive hybrid imaging were first documented by Namdar and coworkers in a clinical study involving fusion of ^{13}N-NH$_3$ PET with four-slice CTA in 25 patients with CAD (Namdar 2005). Using invasive coronary angiography combined with PET as the reference standard, the hybrid PET/CTA approach allowed to detect flow-limiting coronary lesions that required a revascularization procedure with a sensitivity, specificity, positive predictive value (PPV), and negative predictive value of 90%, 98%, 82%, and 99%, respectively. Since then, three studies including a total of 260 patients have evaluated the accuracy of PET-CT imaging in the detection of CAD. These studies show pooled sensitivity of 93% and specificity of 91% in the detection of obstructive CAD (Danad 2013b). Comparison of the diagnostic accuracy of hybrid imaging with the single imaging techniques (i.e., either CTA alone or MPI alone) showed superior diagnostic accuracy. Particularly, the PPV was better with hybrid imaging (87%) than with either CTA (65%) or MPI (83%) alone (Danad 2013b). In addition to the detection of stenosis in invasive coronary angiography, hemodynamic significance of the majority of intermediate lesions was confirmed by fractional flow reserve measurement in two studies (Kajander et al. 2010; Danad 2013a). Despite these promising early results, it should be noted that the aforementioned studies included limited number of patients in experienced centers and did not cover the variety of possible hybrid systems. Thus, larger multicenter trials will be needed to confirm these early results.

Combining information on atherosclerotic lesions and perfusion abnormalities has incremental value for estimation of cardiovascular risk as compared with single imaging modalities using hybrid PET-CT

(Maaniitty et al. 2017) and SPECT-CT (Pazhenkottil et al. 2011; van Werkhoven et al. 2009). A limitation of myocardial perfusion SPECT and PET is that they reveal coronary lesions that induce perfusion defects but do not exclude the presence of subclinical nonobstructive coronary atherosclerosis, which may be missed by perfusion imaging.

In a recent study with 864 symptomatic patients with intermediate probability of CAD, PET MPI using $[^{15}O]H_2O$ during adenosine stress was performed when suspected obstructive stenosis was present on CTA. Obstructive CAD was excluded by CTA in about half of the patients, and these had very low annual rates of adverse events during 3.6 years of follow-up. The patients who had suspected obstructive stenosis on CTA underwent PET study, of which about half had normal and half had abnormal perfusion. The annual rate of events was 5 times higher in those with abnormal than normal perfusion. Patients with normal perfusion had an event rate comparable to that of patients without obstructive CAD on CTA. The independent prognostic value of morphological and functional coronary information was also demonstrated in another follow-up study (van Werkhoven et al. 2009). Moreover, a prospective follow-up trial assessed the incremental prognostic value of hybrid imaging over the side-by-side findings from both techniques (Pazhenkottil et al. 2011). A reversible perfusion defect matching with significant stenosis on CTA was associated with a significantly higher event rate (death or myocardial infarction) than either stenosis or perfusion defect alone.

A recent study evaluated the influence of cardiac hybrid 15-oxygen-labeled water PET-CT imaging results on downstream referral for invasive coronary angiography and revascularization rate (Danad et al. 2014). In the presence of equivocal CTA results, negative perfusion study was associated with a revascularization rate of 0%, whereas abnormal perfusion study was associated with a revascularization rate of 59%. In the presence of obstructive CAD by CTA and abnormal perfusion study, revascularization rate was 72%, indicating that hybrid imaging findings had a significant impact on patient management (Danad et al. 2014). Larger prospective studies are needed to assess whether changes in treatment based on hybrid imaging may have a marked impact on the patients' prognosis.

15.2.3 ADDED VALUE OF HYBRID PET-CT: CASE ILLUSTRATIONS

There are several situations in which hybrid imaging can provide clinically beneficial information (Knuuti and Saraste 2013). Those patients with obstructive CAD causing ischemia can be identified accurately, referred for invasive coronary angiography, and considered for revascularization. The use of invasive tests can be potentially focused to those who have the highest probability to benefit from revascularization. In addition to obstructive disease, CTA can efficiently identify subclinical, nonobstructive coronary atherosclerosis in patients with normal myocardial perfusion. Big data show that nonobstructive CAD by CTA has negative impact on prognosis (Min et al. 2011), and these patients can be provided effective secondary prevention. Furthermore, microvascular disease can be uncovered by CTA for some patients (Kajander et al. 2010). A significant fraction of the patients without any coronary atherosclerosis can be guided to less aggressive primary prevention. The following case examples highlight the benefits of hybrid imaging in different clinical scenarios.

In clinical practice, hybrid imaging is often applied in a sequential manner, with additional scans performed only if the results of the initial modality are equivocal. For example, the hemodynamic severity of intermediate stenosis in CTA can be confirmed by immediate perfusion imaging or the culprit lesion responsible for perfusion defect can be localized by CTA. If the sequential approach is used, the order of the scans depends also on the pretest likelihood. It is rational to suggest that patients with low to moderate pretest likelihood of CAD should start with CTA and MPI would be performed only in those patients with obstructive CAD in CTA. This can be justified by a very high negative predictive value of CTA. On the other hand, if the likelihood of CAD is higher, a larger fraction of the patients will have obstructive disease and starting with perfusion imaging would make sense. CTA would be needed only if anatomical information is needed for a positive perfusion result. Naturally, both of these approaches have specific limitations. In the first option, knowledge of coronary function or microvascular disease is missed, whereas in the second option, preclinical atherosclerotic disease is not detected.

Figure 15.1 shows an example of microvascular disease in a 59-year-old man with risk factors of family history and smoking for CAD detected by PET-CT. The patient had atypical chest pain during exercise and

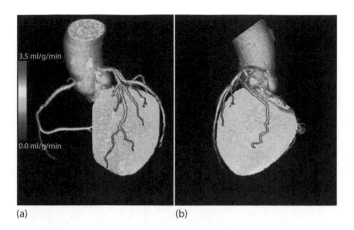

(a)   (b)

Figure 15.1 An example of microvascular disease in a 59-year-old man with risk factors (family history and smoking) for CAD detected by PET-CT. (a) Anterior view and (b) posterior view of the heart.

2-mm horizontal ST depression in lateral chest leads. In the PET-CT study, the coronaries were normal but myocardial perfusion was diffusely reduced (anterior view: Figure 15.1a; posterior view: Figure 15.1b). The absolute MBF during adenosine stress was below 2 mL/g/min (normal value >2.3 mL/g/min) in all myocardial regions as shown in green in Figure 15.1. No epicardial disease was detected in invasive angiography and effective treatment of risk factors was recommended.

Figure 15.2 shows an example of evaluation of hemodynamic significance of CAD by hybrid PET-CTA. The patient was a 63-year-old man with multiple risk factors of CAD (hypercholesterolemia, hypertension, and family history). He was referred for exclusion of CAD due to chest pain on exercise. Exercise ECG showed no ST-segment changes despite mildly reduced functional capacity and chest pain at peak stress. Coronary CTA was performed, and multiplanar reconstructions of the left anterior descending coronary artery (LAD), the left circumflex (LCX), and the RCA are shown in Figure 15.2a. As one can notice, there are multiple, heavily calcified atherosclerotic lesions. There was suspicion of significant luminal obstruction at the site of proximal LAD and middle RCA lesions. In order to assess hemodynamic significance of the lesions, myocardial perfusion was evaluated during adenosine stress by O-15-labeled water PET. Fusion images of 3-D coronary anatomy and quantitative myocardial perfusion map show an area of moderately to severely reduced stress perfusion (1.5–2.0 mL/g/min), as seen in green and blue in the territory of RCA (Figure 15.2c). In contrast, the myocardial perfusion was normal (>2.3 mL/g/min), as seen in red in the areas subtended by the LAD (Figure 15.2b) and LCX (Figure 15.2d). Consequently, percutaneous coronary intervention (PCI) of the RCA was performed due to persisting symptoms despite anti-anginal medication.

Figure 15.3 shows an example of localization of culprit vessel by PET-CT hybrid imaging. A 69-year-old male with Type 2 diabetes, CAD, and stent implanted in the middle RCA some years ago was referred for evaluation of recurrent chest pain. Resting ECG and LV function were normal. Myocardial PET perfusion study with O-15-labeled water and adenosine stress showed abnormally low perfusion in the area of three segments in the posterior wall. Coronary CTA was also performed and showed stenosis in the proximal and distal RCA, whereas in the left coronary artery, atherosclerotic lesions causing mild luminal narrowing (<50%) were seen. Fusion image clarified that the perfusion defect localized within the territory supplied by the RCA (green arrow), whereas the LCX artery is relatively small (red arrow) in this patient.

Figure 15.4 shows an example of balanced three-vessel disease detected by hybrid PET-CTA. The patient was a 52-year-old female with a history of typical angina pectoris. Myocardial perfusion was evaluated with PET and ^{15}O-water during adenosine stress. Images show polar maps of myocardial perfusion with relative normalization to the maximum (Figure 15.4a) or absolute scale from 0 to 3.5 mL/g/min (Figure 15.4b).

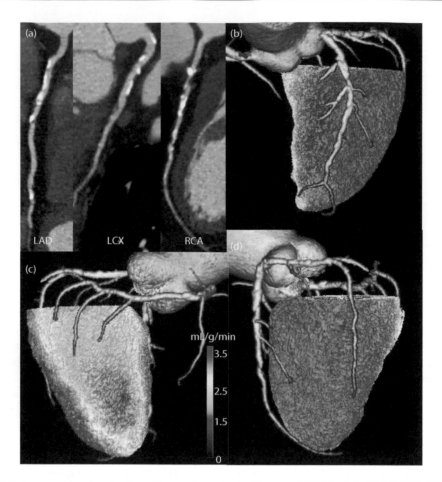

Figure 15.2 An example of evaluation of hemodynamic significance of CAD by hybrid PET-CTA. (a) Coronary CTA and (b, c) hybrid images.

Relative perfusion within the LAD, LCX, and RCA territories was normal (100%, 100%, and 83% of the maximum, respectively), whereas absolute MBF quantification shows severe, balanced reduction of perfusion within all territories (0.75–0.9 mL/g/min; normal value, >2.3 mL/g/min). Coronary CTA revealed the reason for the reduced perfusion. There was a severe stenosis in the left main coronary artery caused by a soft plaque, as shown in Figure 15.4c (arrow), and another calcified lesion causing luminal narrowing in the middle RCA, shown in Figure 15.4d (arrow). Invasive coronary angiography confirmed the findings and the patient was referred for coronary artery bypass surgery.

Figure 15.5 shows an example of a PET-CT study after coronary artery bypass surgery. A 60-year-old man was referred for evaluation of coronary artery bypass grafts due to typical chest pain one year after coronary artery bypass grafting. CTA showed that the graft from the aorta to the LAD, the diagonal branch, and the RCA were open. However, evaluation of the calcified native coronary arteries was challenging and there was a suspicion of occlusion of the distal RCA. Myocardial perfusion was evaluated with PET and ^{15}O-water during adenosine stress. Fusion images of 3-D coronary anatomy and a map of myocardial perfusion show severely reduced (1–1.5 mL/g/min) perfusion in the area subtended by the RCA, confirming RCA occlusion (Figure 15.5a). The perfusion was normal, >2.3 mL/g/min, in the areas subtended by the LAD and LCX (Figure 15.5b). Due to severe symptoms despite optimal medication, the patient was referred for invasive coronary angiography.

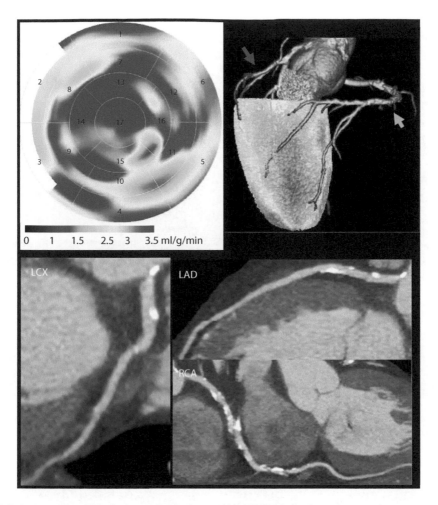

Figure 15.3 An example of localization of culprit vessel by PET-CT hybrid imaging.

15.3 OPEN ISSUES AND FUTURE PERSPECTIVES

As previously discussed, there is a need for confirmation of the diagnostic benefits of hybrid imaging in the evaluation of CAD in large multicenter studies. Furthermore, more information is needed on what kind of patients should undergo integrated examinations for clinical effectiveness and minimization of costs and radiation dose. For example, it can be expected that the proportion of patients needing dual scanning depends on the characteristics of the patient population, especially the pretest likelihood of CAD. Further studies are needed to evaluate whether imaging-guided use of interventions will also influence clinical outcome of patients.

An obvious concern related to hybrid imaging is patient radiation dose that requires careful consideration of the need and benefits of hybrid PET/CT imaging (Knuuti et al. 2014). It is currently assumed that there is a linear relationship between the radiation dose and the risk of cancer, and naturally, all efforts should be made to reduce the exposure to ionizing radiation from imaging. The radiation dose from CTA has been reduced over the last years through improvements in image acquisition protocols, particularly the prospective ECG-triggered sequential scanning. An advantage of short half-life of PET perfusion tracers is that the radiation exposure is low as compared to SPECT perfusion imaging.

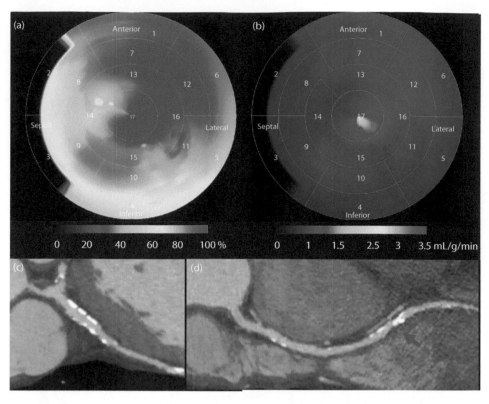

Figure 15.4 An example of balanced three-vessel disease detected by hybrid PET-CTA. (a and b) PET Myocardial perfusion polar maps, (c and d) coronary CTA images.

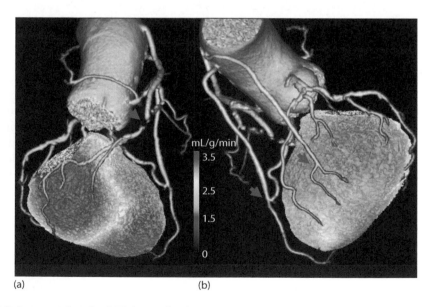

Figure 15.5 An example of a PET-CT study after coronary artery bypass surgery. (a) Posterior view and (b) anterior view.

Besides combination of CTA and perfusion imaging in the detection of CAD, there are other potential approaches where hybrid PET-CT imaging could be applied. Noninvasive assessment of the presence of extensive reversible ischemia and myocardial viability should be considered to guide revascularization of patients with chronic ischemic LV dysfunction (Montalescot et al. 2013). Patients who have viable but ischemic myocardium are at higher risk if they are not revascularized, while the prognosis of patients without dysfunctional and viable myocardium is not improved by revascularization. Several observational studies have shown that the presence of ischemic but viable myocardium is associated with improved outcome after revascularization as compared with pharmacological therapy alone in patients with systolic LV dysfunction due to ischemic heart disease (Allman et al. 2002; Ling et al. 2013). Evaluation of residual glucose metabolism, a hallmark of viable myocardium, by ^{18}F-2-fluoro-2-deoxyglucose (FDG) PET is considered as the most sensitive noninvasive tool to assess myocardial viability (Schinkel et al. 2007). A hybrid PET-CT scanner permits evaluation of viability with FDG-PET immediately in combination with coronary anatomy by CTA and automatic fusion of FDG images with coronary images, but the incremental value of using hybrid PET-CTA imaging in ischemic heart failure remains largely to be studied. It is also worth mentioning that rapid HR and renal dysfunction are common in heart failure and may often preclude the use of contrast-enhanced CTA.

Cardiac molecular imaging has provided techniques and new targeted probes to better understand the pathophysiological mechanism underlying cardiovascular diseases (Nahrendorf et al. 2009; Saraste et al. 2009). Examples of clinical problems that might benefit from molecular imaging include the identification of vulnerable atherosclerotic plaques before rupture and subsequent myocardial infarction, detection of mechanisms that precede LV remodeling and development of heart failure, and the assessment of risk of ventricular arrhythmias by neuronal imaging. Owing to high sensitivity and availability of tracers with low risk of toxicity, PET is the leading imaging technique to proceed with translation of molecular imaging into clinical trials (Knuuti and Bengel 2008). Molecular imaging depends on hybrid imaging approaches, where the nuclear imaging component is used for molecular targeting seen as "hot spot" and the CT or other anatomical imaging modality is used for localization of the molecular signal as well as account for partial volume errors that would cause underestimation of the true regional radiotracer activity.

15.4 CONCLUSIONS

Cardiac PET-CT imaging holds a great promise for the detection of CAD and assessment of the severity of the heart disease. Increasing evidence has shown that the method is very accurate and provides prognostic information and allows guidance of therapy. The technique appears to be mature now for the use in routine clinical cardiology. However, the availability of tracers seems currently to limit the wider use of the approach. The recent development of commercial longer half-life perfusion tracers may make the method widely available. In addition, novel molecular imaging applications may be brought into the clinical work in the near future.

ACKNOWLEDGMENTS

The authors acknowledge financial support from The Academy of Finland Centre of Excellence in Cardiovascular and Metabolic Diseases, Finland, and The Finnish Foundation for Cardiovascular Research, Helsinki, Finland.

REFERENCES

Allman KC, Shaw LJ, Hachamovitch R, Udelson JE. 2002. Myocardial viability testing and impact of revascularization on prognosis in patients with coronary artery disease and left ventricular dysfunction: A meta-analysis. *J Am Coll Cardiol* 39:1151–58.

Berman DS, Maddahi J, Tamarappoo BK et al. 2013. Phase II safety and clinical comparison with single-photon emission computed tomography myocardial perfusion imaging for detection of coronary artery disease: Flurpiridaz f 18 positron emission tomography. *J Am Coll Cardiol* 61:469–77.

Boden WE, O'Rourke RA, Teo KK et al. 2007. Optimal medical therapy with or without PCI for stable coronary disease. *N Engl J Med* 356:1503–16.

Budoff MJ, Dowe D, Jollis JG et al. 2008. Diagnostic performance of 64-multidetector row coronary computed tomographic angiography for evaluation of coronary artery stenosis in individuals without known coronary artery disease: Results from the prospective multicenter ACCURACY (Assessment by Coronary Computed Tomographic Angiography of Individuals Undergoing Invasive Coronary Angiography) trial. *J Am Coll Cardiol* 52:1724–32.

Danad I, Raijmakers PG, Appelman YE et al. 2013a. Hybrid imaging using quantitative h215o PET and CT-based coronary angiography for the detection of coronary artery disease. *J Nucl Med* 54:55–63.

Danad I, Raijmakers PG, Knaapen P. 2013b. Diagnosing coronary artery disease with hybrid PET/CT: It takes two to tango. *J Nucl Cardiol* 20:874–90.

Danad I, Raijmakers PG, Harms HJ et al. 2014. Effect of cardiac hybrid ^{15}O-water PET/CT imaging on downstream referral for invasive coronary angiography and revascularization rate. *Eur Heart J Cardiovasc Imaging* 15:170–9.

Dorbala S, Di Carli MF, Beanlands RS et al. 2013. Prognostic value of stress myocardial perfusion positron emission tomography: Results from a multicenter observational registry. *J Am Coll Cardiol* 61:176–84.

Fiechter M, Ghadri JR, Gebhard C et al. 2012. Diagnostic value of 13n-ammonia myocardial perfusion PET: Added value of myocardial flow reserve. *J Nucl Med* 53:1230–34.

Frye RL, August P, Brooks MM et al. 2009. A randomized trial of therapies for type 2 diabetes and coronary artery disease. *N Engl J Med* 360:2503–15.

Gaemperli O, Schepis T, Koepfli P et al. 2007. Accuracy of 64-slice CT angiography for the detection of functionally relevant coronary stenoses as assessed with myocardial perfusion SPECT. *Eur J Nucl Med Mol Imaging* 34:1162–71.

Gaemperli O, Saraste A, Knuuti J. 2012. Cardiac hybrid imaging. *Eur Heart J Cardiovasc Imaging* 13:51–60.

Goldstein JA, Chinnaiyan KM, Abidov A et al. 2011. The CT-STAT (Coronary Computed Tomographic Angiography for Systematic Triage of Acute Chest Pain Patients to Treatment) trial. *J Am Coll Cardiol* 58:1414–22.

Gould KL, Johnson NP. 2010. Coronary artery disease: Percent stenosis in CAD—A flaw in current practice. *Nat Rev Cardiol* 7:482–4.

Gould KL, Johnson NP, Bateman TM et al. 2013. Anatomic versus physiologic assessment of coronary artery disease. Role of coronary flow reserve, fractional flow reserve, and positron emission tomography imaging in revascularization decision-making. *J Am Coll Cardiol* 62:1639–53.

Hachamovitch R, Hayes SW, Friedman JD et al. 2003. Comparison of the short-term survival benefit associated with revascularization compared with medical therapy in patients with no prior coronary artery disease undergoing stress myocardial perfusion single photon emission computed tomography. *Circulation* 107:2900–7.

Hacker M, Jakobs T, Hack N et al. 2007. Sixty-four slice spiral CT angiography does not predict the functional relevance of coronary artery stenoses in patients with stable angina. *Eur J Nucl Med Mol Imaging* 34:4–10.

Hoffmann U, Truong QA, Schoenfeld DA et al. 2012. Coronary CT angiography versus standard evaluation in acute chest pain. *N Engl J Med* 367:299–308.

Jaarsma C, Leiner T, Bekkers SC et al. 2012. Diagnostic performance of noninvasive myocardial perfusion imaging using single-photon emission computed tomography, cardiac magnetic resonance, and positron emission tomography imaging for the detection of obstructive coronary artery disease: A meta-analysis. *J Am Coll Cardiol* 59:1719–28.

Javadi MS, Lautamaki R, Merrill J et al. 2010. Definition of vascular territories on myocardial perfusion images by integration with true coronary anatomy: A hybrid PET/CT analysis. *J Nucl Med* 51:198–203.

Kajander S, Joutsiniemi E, Saraste M et al. 2010. Cardiac positron emission tomography/computed tomography imaging accurately detects anatomically and functionally significant coronary artery disease. *Circulation* 122:603–13.

Kajander SA, Joutsiniemi E, Saraste M et al. 2011. Clinical value of absolute quantification of myocardial perfusion with (15)o-water in coronary artery disease. *Circ Cardiovasc Imaging* 4:678–84.

Knuuti J, Bengel FM. 2008. Positron emission tomography and molecular imaging. *Heart* 94:360–7.

Knuuti J, Saraste A. 2013. Combined functional and anatomical imaging for the detection and guiding the therapy of coronary artery disease. *Eur Heart J* 34:1954–7.

Knuuti J, Bengel F, Bax JJ et al. 2014. Risks and benefits of cardiac imaging: An analysis of risks related to imaging for coronary artery disease. *Eur Heart J* 35:633–8.

Ling LF, Marwick TH, Flores DR et al. 2013. Identification of therapeutic benefit from revascularization in patients with left ventricular systolic dysfunction: Inducible ischemia versus hibernating myocardium. *Circ Cardiovasc Imaging* 6:363–72.

Litt HI, Gatsonis C, Snyder B et al. 2012. CT angiography for safe discharge of patients with possible acute coronary syndromes. *N Engl J Med* 366:1393–403.

Maaniitty TS, Stenström I, Bax JJ, Uusitalo V, Ukkonen H, Kajander S, Mäki M, Saraste A, Knuuti J. 2017. Prognostic value of coronary CT angiography with selective PET perfusion imaging in coronary artery disease. *JACC Cardiovasc Imaging*. doi: 10.1016/j.jcmg.2016.10.025.

Mc Ardle BA, Dowsley TF, deKemp RA, Wells GA, Beanlands RS. 2012. Does rubidium-82 PET have superior accuracy to SPECT perfusion imaging for the diagnosis of obstructive coronary disease?: A systematic review and meta-analysis. *J Am Coll Cardiol* 60:1828–37.

Meijboom WB, Meijs MF, Schuijf JD et al. 2008. Diagnostic accuracy of 64-slice computed tomography coronary angiography: A prospective, multicenter, multivendor study. *J Am Coll Cardiol* 52:2135–44.

Miller JM, Rochitte CE, Dewey M et al. 2008. Diagnostic performance of coronary angiography by 64-row CT. *N Engl J Med* 359:2324–36.

Min JK, Dunning A, Lin FY et al. 2011. Age- and sex-related differences in all-cause mortality risk based on coronary computed tomography angiography findings results from the International Multicenter CONFIRM (Coronary CT Angiography Evaluation for Clinical Outcomes: An international multicenter registry) of 23,854 patients without known coronary artery disease. *J Am Coll Cardiol* 58:849–60.

Montalescot G, Sechtem U, Achenbach S et al. 2013. Esc guidelines on the management of stable coronary artery disease: The Task Force on the Management of Stable Coronary Artery Disease of the European Society of Cardiology. *Eur Heart J* 34:2949–3003.

Murthy VL, Naya M, Foster CR et al. 2011. Improved cardiac risk assessment with noninvasive measures of coronary flow reserve. *Circulation* 124:2215–24.

Murthy VL, Naya M, Foster CR et al. 2012. Association between coronary vascular dysfunction and cardiac mortality in patients with and without diabetes mellitus. *Circulation* 126:1858–68.

Nahrendorf M, Sosnovik DE, French BA et al. 2009. Multimodality cardiovascular molecular imaging, Part II. *Circ Cardiovasc Imaging* 2:56–70.

Namdar M, Hany TF, Koepfli P et al. 2005. Integrated PET/CT for the assessment of coronary artery disease: A feasibility study. *J Nucl Med* 46:930–5.

Pazhenkottil AP, Nkoulou RN, Ghadri JR et al. 2011. Prognostic value of cardiac hybrid imaging integrating single-photon emission computed tomography with coronary computed tomography angiography. *Eur Heart J* 32:1465–71.

Pijls NH, Fearon WF, Tonino PA et al. 2010. Fractional flow reserve versus angiography for guiding percutaneous coronary intervention in patients with multivessel coronary artery disease: 2-year follow-up of the FAME (Fractional Flow Reserve Versus Angiography for Multivessel Evaluation) study. *J Am Coll Cardiol* 56:177–84.

Saraste A, Nekolla SG, Schwaiger M. 2009. Cardiovascular molecular imaging: An overview. *Cardiovasc Res* 50:1088–94.

Saraste A, Kajander S, Han C, Nesterov SV, Knuuti J. 2012. PET: Is myocardial flow quantification a clinical reality? *J Nucl Cardiol* 19:1044–59.

Schinkel AF, Bax JJ, Poldermans D, Elhendy A, Ferrari R, Rahimtoola SH. 2007. Hibernating myocardium: Diagnosis and patient outcomes. *Curr Probl Cardiol* 32:375–410.

Schroeder S, Achenbach S, Bengel F et al. 2008. Cardiac computed tomography: Indications, applications, limitations, and training requirements: Report of a Writing Group deployed by the Working Group Nuclear Cardiology and Cardiac CT of the European Society of Cardiology and the European Council of Nuclear Cardiology. *Eur Heart J* 29:531–56.

Schuijf JD, Wijns W, Jukema JW et al. 2006. Relationship between noninvasive coronary angiography with multi-slice computed tomography and myocardial perfusion imaging. *J Am Coll Cardiol* 48:2508–14.

Shaw LJ, Berman DS, Maron DJ et al. 2008. Optimal medical therapy with or without percutaneous coronary intervention to reduce ischemic burden: Results from the Clinical Outcomes Utilizing Revascularization and Aggressive Drug Evaluation (COURAGE) trial nuclear substudy. *Circulation* 117:1283–91.

Tonino PA, Fearon WF, De Bruyne B et al. 2010. Angiographic versus functional severity of coronary artery stenoses in the FAME study fractional flow reserve versus angiography in multivessel evaluation. *J Am Coll Cardiol* 55:2816–21.

Uren NG, Melin JA, De Bruyne B et al. 1994. Relation between myocardial blood flow and the severity of coronary-artery stenosis. *N Engl J Med* 330:1782–8.

Tonino PA, De Bruyne B, Pijls NH et al. 2009. Fractional flow reserve versus angiography for guiding percutaneous coronary intervention. *N Engl J Med* 360:213–24.

van Werkhoven JM, Schuijf JD, Gaemperli O et al. 2009. Prognostic value of multislice computed tomography and gated single-photon emission computed tomography in patients with suspected coronary artery disease. *J Am Coll Cardiol* 53:623–32.

White CW, Wright CB, Doty DB et al. 1984. Does visual interpretation of the coronary arteriogram predict the physiologic importance of a coronary stenosis? *N Engl J Med* 310:819–24.

Wijns W, Kolh P, Danchin N et al. 2010. Guidelines on myocardial revascularization: The Task Force on Myocardial Revascularization of the European Society of Cardiology (ESC) and the European Association for Cardio-Thoracic Surgery (EACTS). *Eur Heart J* 31:2501–55.

Williams BA, Dorn JM, LaMonte MJ et al. 2012. Evaluating the prognostic value of positron-emission tomography myocardial perfusion imaging using automated software to calculate perfusion defect size. *Clin Cardiol* 35:E14–21.

Quantitative analyses and case studies of hybrid PET-MRI imaging

LEON J. MENEZES, ELEANOR C. WICKS, AND BRIAN F. HUTTON

16.1 INTRODUCTION

The excellent soft tissue contrast of magnetic resonance imaging (MRI) and the capability of functional imaging at the molecular level by positron emission tomography (PET) should have the potential to create unique additive value in cardiac imaging. As a novel imaging technology, PET/MRI in cardiovascular disease needs to demonstrate its performance and value in clinical practice.

In addition to any clinical value, the multiparametric imaging offered by MRI and PET in a single examination will provide novel possibilities for research. The first direction of research into hybrid PET/MRI has been cross-validation studies. Subsequent studies will address clinical questions.

The main motivation for the development of PET/MRI scanners was to take advantage of the superior soft tissue contrast of MRI and the reduced ionizing radiation exposure in comparison to PET/computed tomography (CT). But there are other advantages specific to cardiac imaging that might be exploited, e.g., the dynamic imaging of the ventricle for improved correction for motion and respiration. However, there are difficulties for quantitative analysis when the generation of attenuation correction (AC) maps of the patient is directly from MRI data.

In this chapter, we provide a review and outlook on future applications of cardiac PET/MRI based on published research and clinical examples.

16.2 WORKFLOW

As discussed in Chapter 4 of this book, AC is fundamental, as is the registration between the AC map and the PET data. The AC maps are estimated using a two-point Dixon sequence (Martinez-Möller et al. 2009). The potential disadvantage of this approach, that bone is excluded, is not so limiting when scanning the thorax, compared to the brain or pelvis. Uncertainty in the attenuation distribution in lungs continues to be a challenge and can lead to unpredictable errors. From PET/CT, motion and, hence, misalignment may lead to artefacts and diagnostic error (Gould et al. 2007). An advantage of PET/MRI over PET/CT is that the Dixon sequences can be repeated without any further radiation. While the acquisition of the image data in PET/MRI is simultaneous (in the case of the Siemens Biograph mMR and GE Signa PET/MR), the time taken for each modality varies considerably. PET acquires a true volume with acquisition times of the order of minutes, depending on the chosen tracer, its half-life, and the injected activity. Therefore, the acquired PET data include the motion that took place during the scan: cardiac and respiratory motion and voluntary and involuntary patient motion.

On the other hand, cardiac MRI sequences are acquired serially with acquisition times of the order of seconds and usually in end-expiration breath-holds. Thus, in cardiac PET/MRI, summed averaged rather than respiratory- or cardiac-gated PET data are compared and fused with end-diastolic MR images from end-expiration. The development and use of dual-gated or self-navigated whole heart cardiac MR sequences (Piccini et al. 2011) may be better suited to PET/MRI scanning than conventional cardiac MR image acquisition (Kramer et al. 2013).

With simultaneous PET/MRI, there is the potential of using the motion signals from one modality to correct both image data sets, e.g., MR-based respiratory motion correction (Chun et al. 2012) and partial volume correction (Erlandsson et al. 2012). However, any extra MRI needed might add to the total length of the scan.

Current MR scanners used in hybrid PET/MRI are based on a 3.0 Tesla field strength, enabling a better signal-to-noise ratio at the expense of more frequent susceptibility effects and radiofrequency inhomogeneities that require fine-tuned acquisition parameters.

16.3 CARDIAC APPLICATIONS IN PET/MRI

16.3.1 MYOCARDIAL PERFUSION

Myocardial perfusion imaging with PET has long been seen as a reference method for validation of other diagnostic approaches (Bengel et al. 2009). A recent meta-analysis of noninvasive perfusion imaging showed that PET achieved the highest diagnostic performance and that first-pass perfusion cardiac MR may provide an alternative without ionizing radiation and a similar diagnostic accuracy as PET (Jaarsma et al. 2012). Three PET perfusion tracers are in clinical use: ^{13}N-NH$_3$, ^{15}O-water, and ^{82}Rubidium (^{82}Rb). However, ^{13}N-NH$_3$ and ^{15}O-water need an on-site cyclotron and the generator for ^{82}Rb is not MR-compatible.

PET/MRI offers a test bed for the validation of new tracers for myocardial perfusion PET or new cardiac MRI perfusion techniques in human subjects under identical conditions. Up until now, such cross-validations took place on standalone machines separated in time and space (Morton et al. 2012). As such, one could perform further validation of ^{18}F-flurpiridaz, a tracer that has certain advantages over the existing PET perfusion tracers against cardiac MR perfusion rather than single-photon emission computed tomography (SPECT) (Berman, Maddahi, and Tamarappoo 2013). More specifically, this new PET tracer has a low positron energy and a long half-life that allows one to transport the radiopharmaceutical to areas distant from a cyclotron, the option of exercise stress testing with tracer injection outside the scanner, and a simplified myocardial blood flow estimation approach. One could also validate blood-oxygen level dependent (BOLD) MRI (Arnold et al. 2012) or arterial spin (ASL) labeling (Zun et al. 2011) against PET perfusion. These latter techniques have potential clinical utility as they do not require the administration of gadolinium-based contrast agents, which are contraindicated in severe renal dysfunction.

Recent research on cardiac perfusion MRI is aimed at increasing the spatial coverage of the heart from at least three slices (basal, mid, and apical, limited by the length of the R–R interval of the electrocardiogram [ECG]) during clinical routine first-pass perfusion cardiac MRI, by using three-dimensional sequences (Manka et al. 2015). The increased diagnostic yield and noninvasive validation would be better performed with simultaneous PET/MRI than against separate SPECT (Jogiya et al. 2014).

One ongoing clinical study of simultaneous ^{3}N-NH$_3$ PET perfusion and first-pass MRI (clinical trial NCT01779869) is addressing the relative contributions of stress MRI and stress PET data and whether the rest PET scan can be omitted in favor of rest first-pass MRI, potentially shortening the scan acquisition time and reducing radiation dose.

16.3.2 Myocardial tissue characterization by PET/MRI

16.3.2.1 VIABILITY

Since the publication of the trial on surgical treatment of ischemic heart failure (STICH) (Velazquez et al. 2011) and the viability substudy (Bonow et al. 2011), the utility of viability testing in patients with left ventricular dysfunction and coronary artery disease prior to revascularization has been questioned. The randomized controlled trial on PET and recovery following revascularization (PARR-2) did not show a significant difference in the composite primary outcome of cardiac death, myocardial infarction (MI), or recurrent hospitalization among those patients randomized to PET-guided management versus standard care without PET (Beanlands et al. 2007), although, in about 25% of patients, the imaging-based recommendations were not followed by the treating physicians.

Fluorodeoxyglucose (^{18}F-FDG) PET under oral glucose loading or hyperinsulinemic-euglycaemic clamping demonstrates viable myocardium, whereas late gadolinium enhancement (LGE) MRI makes use of the property of gadolinium-chelates to accumulate in increased extracellular spaces (such as scarred myocardium). Good agreement between PET and MRI has been demonstrated on separate imaging (Klein et al. 2002), but there are advantages that MRI might have over PET. The quality of ^{18}F-FDG images is relatively poor in diabetics because of decreased ^{18}F-FDG uptake. MRI has a superior spatial resolution of 1–3 mm, which allows differentiation of transmural and subendocardial infarction. Also, MRI could help in classifying areas with reduced or absent ^{18}F-FDG uptake due to myocardial wall thinning, which can be mislabeled as scar by PET alone (a partial volume error). Whether these advantages determine changes in management and lead to improved patient outcomes would require a large multicenter trial. To date, a feasibility study of cardiac PET/MRI including five patients with chronic MI has demonstrated substantial agreement between PET and MRI (Nensa et al. 2013). Figure 16.1 shows an example of viability assessment in PET/MRI.

16.3.2.2 ACUTE MI

Early after acute MI and primary percutaneous coronary intervention (PPCI), cardiac MRI can be performed to characterize the tissue of the reperfused heart, allowing direct visualization and quantification of its constituent components, including the area of "irreversible" myocardial injury or MI delineated by LGE imaging, the area of "reversible" myocardial injury, or the "area-at-risk" delineated by T2-weighted imaging (Francone et al. 2009; Friedrich et al. 2008) and T1-weighted imaging (Ugander et al. 2012), zones of microvascular obstruction (Bekkers et al. 2009), and intramyocardial hemorrhage within the MI (O'Regan et al. 2009). All of these are surrogate predictors of clinical outcome post-PPCI. Nevertheless, how these individual components of the acutely reperfused MI impact on cardiac metabolism in patients with MI is not clear. There are two studies demonstrating that the area of reduced ^{18}F-FDG uptake correlates with the area at risk (Nensa et al. 2015; White, Bulluck, and Frohlich 2015). One study of 28 patients with 6 months follow-up after PPCI for acute MI showed that a fifth of those MI segments that had reduced ^{18}F-FDG uptake but no LGE on MRI had an inferior functional recovery (Rischpler et al. 2015). A case illustration is shown in Figure 16.2.

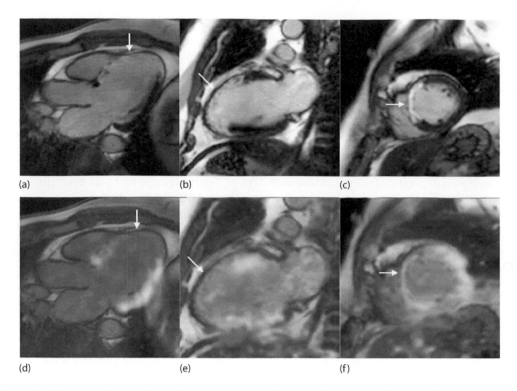

Figure 16.1 Horizontal long-axis (a, d), vertical long-axis (b, e), and short-axis (c, f) views of a 49-year-old man who had had a large anterior MI 1 year earlier, treated with PPCI to the left anterior descending (LAD) artery with a drug-eluting stent. There is transmural LGE of the anterior septal region on MRI (a–c) with corresponding hypometabolism with ^{18}F-FDG PET (d–f) (white arrows).

16.3.2.3 INFLAMMATORY RESPONSE AFTER ACUTE MI

The inflammatory response after acute MI may also be a risk factor for adverse remodeling and the development of heart failure. Monocytes migrating into the heart during the first days after MI have been identified as key players of cardiac inflammation and have been linked to the development of heart failure (Frangogiannis et al. 2002). ^{18}F-FDG PET is frequently used to image inflammation in many clinical situations, as activated inflammatory cells show increased expression of glucose membrane transporter proteins. In patients, the glucose metabolism of cardiac myocytes can be suppressed by a high-fat, carbohydrate-restricted diet and the intravenous (i.v.) administration of unfractionated heparin before ^{18}F-FDG injection (Soussan et al. 2012) (Figure 16.3). Heparin administration induces an activation of lipoprotein lipase, which in turn causes an increase of free fatty acids in the blood. Using these interventions, the human heart switches its energy metabolism from glucose to almost exclusively fatty acids, which thus specifically allows imaging the glucose metabolism in the infarct area. Whether this technique is superior for identifying patients at risk for developing heart failure or can guide future anti-inflammatory therapy after acute MI needs to be further investigated.

16.3.2.4 SYMPATHETIC INNERVATION

[11]C-hydroxyephedrine (HED) is the most commonly used PET tracer used to measure sympathetic integrity. PET provides improved spatial image resolution and tracer uptake quantification compared to ^{123}I-Metaiodobenzylguanidine (MIBG) SPECT (Matsunari et al. 2010). A prospective study of 204 patients with ischemic cardiomyopathy (left ventricular ejection fraction <35%) showed that sympathetic denervation (assessed by ^{11}C-HED PET) is an independent predictor of sudden cardiac death independent of ejection fraction and infarct size (Fallavollita et al. 2014).

Hybrid PET/MRI innervation studies with cardiac MRI left ventricular function, perfusion, and scar assessment could provide a comprehensive prognostic assessment in heart failure patients and may provide

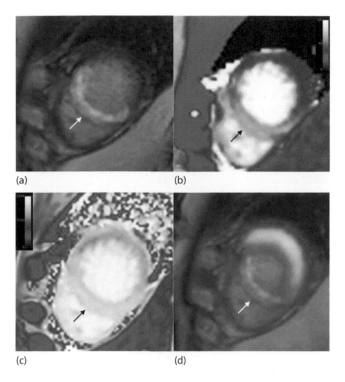

Figure 16.2 Short-axis views of the myocardium of a 47-year-old man, smoker, who presented with chest pain and transient ST elevation inferolaterally. Angiography showed subtotally occluded distal LAD, treated with PPCI to the LAD artery. Cardiac PET/MRI was performed 3 days after the infarction. There is (arrows) transmural LGE of the interventricular septum (a) with matching intracellular and interstitial edema demonstrated together with increased T2 signal and edema on T2 mapping (b), which indicates no salvage. A quantitative T1 map also indicates the area at risk (c). 18F-FDG PET under oral glucose loading shows corresponding hypo metabolism in the infarcted segments (d).

synergistic information for risk stratification and better direct device therapy such as implantable cardioverter defibrillator insertion.

16.3.2.5 SARCOIDOSIS

Cardiac sarcoid is a potentially fatal condition that presents with a wide range of clinical manifestations including conduction abnormalities, tachyarrhythmia, congestive heart failure, cardiomyopathy, and sudden cardiac death (Wicks, Menezes, and Elliott 2015). Cardiac MRI with LGE is a criterion for the diagnosis of cardiac sarcoid. Inflammation imaging with 18F-FDG PET has supplanted 67Gallium imaging. The patient preparation described in imaging the inflammatory response after acute MI earlier (high-fat, low-carbohydrate diet and i.v. heparin) was first used to image cardiac sarcoidosis (Figure 16.3). One series of 51 patients with suspected cardiac sarcoidosis underwent PET/MRI. The combination of uptake of PET and LGE on MRI had greater sensitivity than either modality alone and there was poor topological agreement between the location of 18F-FDG uptake and scar. This may reflect the natural history of the disease with progression from inflammation to scar (Wicks, Menezes, and Pantazis 2014).

16.3.2.6 MYOCARDITIS

Myocarditis is a challenging diagnosis due to the heterogeneity of clinical presentations. The actual incidence of myocarditis is also difficult to determine as endomyocardial biopsy, the diagnostic gold standard, is used infrequently (Caforio et al. 2015). Using the same inflammatory 18F-FDG PET scanning, there have been case reports published on the use of PET/MRI (Nensa et al. 2014). Investigating the diagnostic yield, the ability to direct endomyocardial biopsy, prognosticate patients, and monitor response to therapy in myocarditis using hybrid PET/MRI requires proper study.

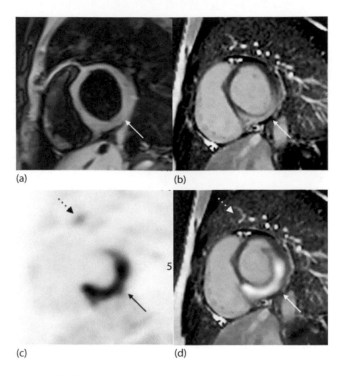

Figure 16.3 Short-axis views of a 29-year-old lady known to have pulmonary sarcoidosis, thought to be quiescent, who presented with 4 weeks history of palpitations. The ECG showed a broad complex tachycardia. There is homogenous STIR signal indicating no edema (white arrow in a). There is mid to epicardial LGE inferolaterally (white arrow in b), which corresponds to focal increased ^{18}F-FDG uptake on PET (following a carbohydrate-restricted diet and then a 12-hour fast) (white arrow in c). The fused images also slow focal uptake in the lung parenchyma (white dotted arrows in d). The patient was treated with immunosuppression with steroids.

16.3.2.7 ANDERSON-FABRY DISEASE

Anderson-Fabry disease is a rare, X-linked, genetic disorder caused by mutations in the gene that encodes the lysosomal enzyme, alpha-galactosidase A. Cardiac involvement is the most common cause of premature mortality. Tissue damage, including fibrosis in the heart and kidney due to progressive accumulation of glycosphingolipids, starts before symptomatic onset of organ failure. In cardiomyocytes, glycosphingolipid storage may directly cause injury and alter the expression of signaling molecules, triggering inflammation, hypertrophy, apoptosis, early interstitial fibrosis, and late cell replacement fibrosis. Thirteen patients with Anderson-Fabry disease have been studied with inflammatory ^{18}F-FDG PET/MRI (Nappi et al. 2015). Of the 13 patients, 6 exhibited focal LGE indicating myocardial fibrosis; 4 of those also had edema on T2-weighted images (short T1 inversion recovery [STIR]), which was associated with focal FDG uptake on the corresponding PET images. Again, larger studies are warranted to investigate the role of inflammatory PET/MRI in disease "staging" and to evaluate response with enzyme replacement therapy.

16.3.3 INTERSTITIAL SPACE

16.3.3.1 T1 MAPPING AND EXTRACELLULAR VOLUME

The intrinsic signal (measured as the MR parameters T1 and T2) of myocardium without a contrast agent can be imaged using rapid sequences providing T1 "maps" (Messroghli et al. 2004; Piechnik et al. 2010). Differences in native T1 can be detected in acute myocardial injury and global interstitial fibrosis. A problem with these techniques is that they measure a composite myocardial signal from both the interstitium and myocytes. Using MR contrast media, however, native and postcontrast T1 maps can be compared, and by calculating the extracellular volume (ECV) fraction the myocardial and interstitial space can be separated (White et al. 2012). Native T1 maps have been used to identify regional edema, fibrosis, amyloidosis (vide infra), Anderson-Fabry disease, and intramyocardial hemorrhage (Moon et al. 2013). Expansion of the ECV occurs principally in diffuse fibrosis, but also with edema or infiltration by amyloid protein.

Possible synergistic uses of T1 mapping and ECV measurement include validation of perfusable tissue index by $H_2^{15}O$. As fibrotic myocardium is unable to exchange water rapidly, a reduction should correlate with fibrosis. Although not validated against biopsy, perfusable tissue index is reduced in patients with advanced dilated cardiomyopathy, correlates with reduced function, and is also reduced in MI (Knaapen et al. 2004, 2005).

Thus, T1-mapping techniques with cardiac MRI to assess diffuse fibrosis, infiltration, and inflammation and PET techniques to assess inflammation and fibrosis could provide new insights into diseases such as hypertrophic and dilated cardiomyopathies, diastolic heart failure, and diabetic cardiomyopathy.

16.3.3.2 AMYLOIDOSIS

Amyloidosis is a clinical disorder that arises from the aggregation of insoluble fibrous deposits of misfolded proteins. Deposition of fibrillary material and the toxic effects of precursor soluble intermediates result in progressive organ dysfunction, which manifest as heart failure with restrictive physiology (Gertz, Dispenzieri, and Sher 2014). The recent development of PET tracers for amyloid plaque identification in Alzheimer's disease has generated potential interest in the evaluation of cardiac amyloidosis, a condition that is imaged using MRI (Fontana et al. 2014) and ^{99m}Tc-3,3-diphosphono-1,2-propanodicarboxylic acid (^{99m}Tc-DPD) SPECT (Hutt et al. 2014). ^{11}C-PIB and ^{18}F-Florbetapir have been used to depict systemic and cardiac amyloid (Antoni et al. 2013; Dorbala et al. 2014; Lee et al. 2015). Validation with T1 mapping and ECV measurements with simultaneous PET/MRI could possibly allow earlier diagnosis and treatment monitoring (Figure 16.4).

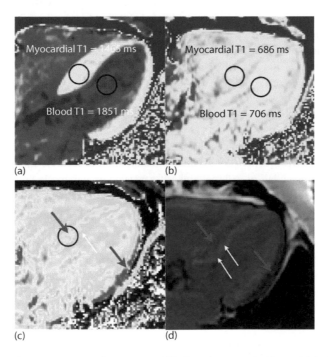

Figure 16.4 Horizontal long-axis views of a 65-year-old lady with systemic AL amyloidosis, showing native T1 mapping (a) with high native T1 values in the myocardium precontrast. All T1 values drop from pre to post contrast (b) (i.e., map becomes all blue) because of the presence of gadolinium throughout. There is a differential ECV/interstitial expansion in the "core" LV septum to endocardium and epicardium (light blue surrounded by dark blue in b). This is shown in the LGE image (white vs red arrows in d) also in the lateral wall, which indicates that there is a nonuniform amyloid distribution. This is confirmed in the ECV map (c) by the blue vs. the epicardium and endocardium in green, which actually merges with the blood pool; i.e., the ECV/interstitial space expansion is as high as blood.

16.3.3.3 ATHEROSCLEROSIS

Several PET tracers have been used to image atherosclerotic plaques. The most frequently used tracer is ^{18}F-FDG. Increased ^{18}F-FDG uptake is related to macrophage infiltration, in culprit carotid plaques in stroke (Menezes et al. 2011). However, ^{18}F-FDG is not an ideal tracer to image the coronaries due to spillover from the myocardium. ^{18}F-fluoride has been shown to identify culprit lesions in recent MI (Joshi et al. 2014), and prognostic studies are underway. ^{18}F-Galacto-RGD PET/CT imaging of αvβ3 expression has been similarly validated in a carotid endarterectomy model (Beer et al. 2014). Atherosclerotic plaques and vessel walls are at the limits of PET spatial resolution. The tracer uptake in the carotids is usually rather low and, as said, susceptible to spillover effects, and obviously cardiac and respiratory motion is also an issue. This might ultimately be addressed using MRI-based partial volume correction and motion correction. As such, hybrid PET/MRI would advance molecular PET imaging of coronaries and vulnerable plaque using novel tracers, correlating with plaque composition by vessel wall MRI and MR angiography.

16.3.3.4 STEM CELL TRACKING

Hybrid PET/MRI could also become the imaging modality of choice for the evaluation and follow-up of new therapeutic techniques aimed at tissue regeneration and myocardial repair. However, preclinical studies have demonstrated that only a small fraction of the transplanted cells are engrafted into the myocardium (Terrovitis et al. 2009). Recent studies have shown that PET imaging can be used for assessment of the kinetics of stem cell therapy in MI (Wu, Abraham, and Kraitchman 2010). Dual labeling of stem cells with iron particles may offer new tools for monitoring the delivery, survival, and migration of cells after cell transplantation (Higuchi et al. 2009). Human endothelial cells labeled with iron oxide nanoparticles were modified using a viral vector so that they expressed the human sodium iodide symporter. After injection of these cells into the heart of nude rats, MRI was able to exactly locate these cells, whereas [124]Iodine PET was ideal for monitoring cell survival (Higuchi et al. 2009). After death of the transplanted cells, the iron oxide particles were taken up by macrophages; consequently, the MR signal did not indicate viable transplanted cells. Therefore, simultaneous PET/MRI might be valuable for monitoring of cell- or gene-based therapy approaches.

16.4 CONCLUSION

Hybrid cardiac PET/MRI may find its niche as a cross-validation platform for new tracers or applications of existing tracers, or new MR sequences, for eventual dissemination to separate standalone MR and PET/CT scanners. However, the potential for external and self-navigated gating, motion correction, partial volume correction, constrained PET reconstruction, and the reduced ionizing radiation compared to PET/CT could find multimodality, multiparametric PET/MRI to have a unique role in rare and possibly common cardiovascular disorders.

With increased scrutiny of both the utilization and the cost of imaging, one of the major hurdles for the clinical application of PET/MRI for cardiovascular applications will be to demonstrate that the combination of these two tests within one imaging session will have incremental value and most importantly will directly impact management decisions and improve patient outcomes. New comparative effectiveness studies will be needed to validate the clinical utility of combined cardiovascular PET/MRI.

ACKNOWLEDGMENTS

This work was supported by the National Institute for Health Research University College London Hospitals Biomedical Research Centre. We gratefully acknowledge the assistance of Celia O'Meara, Dr. Heerajnarain Bulluck, and Dr. Steven K White.

REFERENCES

Antoni, G, M Lubberink, S Estrada et al. 2013. In Vivo Visualization of Amyloid Deposits in the Heart with 11C-PIB and PET. *J Nucl Med* 54 (2): 213–20.

Arnold, JR, TD Karamitsos, P Bhamra-Ariza et al. 2012. Myocardial Oxygenation in Coronary Artery Disease: Insights from Blood Oxygen Level-Dependent Magnetic Resonance Imaging at 3 Tesla. *J Am Coll Cardiol* 59 (22): 1954–64.

Beanlands, RSB, G Nichol, E Huszti et al. 2007. F-18-Fluorodeoxyglucose Positron Emission Tomography Imaging-Assisted Management of Patients with Severe Left Ventricular Dysfunction and Suspected Coronary Disease: A Randomized, Controlled Trial (PARR-2). *J Am Coll Cardiol* 50 (20): 2002–12.

Beer, AJ, J Pelisek, P Heider et al. 2014. PET/CT Imaging of Integrin Avβ3 Expression in Human Carotid Atherosclerosis. *J Am Coll Cardiol* 7 (2): 178–87.

Bekkers, SC, WH Backes, RJ Kim et al. 2009. Detection and Characteristics of Microvascular Obstruction in Reperfused Acute Myocardial Infarction Using an Optimized Protocol for Contrast-Enhanced Cardiovascular Magnetic Resonance Imaging. *Eur Radiol* 19 (12): 2904–12.

Bengel, FM, T Higuchi, MS Javadi et al. 2009. Cardiac Positron Emission Tomography. *J Am Coll Cardiol* 54 (1): 1–15.

Berman, DS, J Maddahi, BK Tamarappoo et al. 2013. Flurpiridaz F 18 PET: Phase II Safety and Clinical Comparison with SPECT Myocardial Perfusion Imaging for Detection of Coronary Artery Disease. *J Am Coll Cardiol* 61 (4): 469–77.

Bonow, RO, G Maurer, KL Lee et al. 2011. Myocardial Viability and Survival in Ischemic Left Ventricular Dysfunction. *Engl Med* 364 (17): 1617–25.

Caforio, ALP, R Marcolongo, C Basso et al. 2015. Clinical Presentation and Diagnosis of Myocarditis. *Heart* 101 (16): 1332–44.

Chun, SY, TG Reese, J Ouyang et al. 2012. MRI-Based Nonrigid Motion Correction in Simultaneous PET/MRI. *J Nucl Med* 53 (8): 1284–91.

Dorbala, S, D Vangala, J Semer et al. 2014. Imaging Cardiac Amyloidosis: A Pilot Study Using 18F-Florbetapir Positron Emission Tomography. *Eur Nucl Med Molec Imaging* 41: 1652.

Erlandsson, K, I Buvat, PH Pretorius et al. 2012. A Review of Partial Volume Correction Techniques for Emission Tomography and Their Applications in Neurology, Cardiology and Oncology. *Phys Med Biol* 57 (21): R119–59.

Fallavollita, JA, BM Heavey, AJ Luisi et al. 2014. Regional Myocardial Sympathetic Denervation Predicts the Risk of Sudden Cardiac Arrest in Ischemic Cardiomyopathy. *J Am Coll Cardiol* 63 (2): 141–49.

Fontana, M, SM Banypersad, TA Treibel et al. 2014. Native T1 Mapping in Transthyretin Amyloidosis. *JACC: Cardiovasc Imaging* 7 (2): 157–65.

Francone, M, C Bucciarelli-Ducci, I Carbone et al. 2009. Impact of Primary Coronary Angioplasty Delay on Myocardial Salvage, Infarct Size, and Microvascular Damage in Patients with ST-Segment Elevation Myocardial Infarction: Insight from Cardiovascular Magnetic Resonance. *J Am Coll Cardiol* 54 (23): 2145–53.

Frangogiannis, NG, S Shimoni, SM Chang et al. 2002. Evidence for an Active Inflammatory Process in the Hibernating Human Myocardium. *Am Pathol* 160 (4): 1425–33.

Friedrich, MG, H Abdel-Aty, A Taylor et al. 2008. The Salvaged Area at Risk in Reperfused Acute Myocardial Infarction as Visualized by Cardiovascular Magnetic Resonance. *J Am Coll Cardiol* 51 (16): 1581–87.

Gertz, MA, A Dispenzieri, and T Sher. 2014. Pathophysiology and Treatment of Cardiac Amyloidosis. *Nature Rev Cardiol* 12 (2): 91–102.

Gould, KL, T Pan, C Loghin et al. 2007. Frequent Diagnostic Errors in Cardiac PET/CT Due to Misregistration of CT Attenuation and Emission PET Images: A Definitive Analysis of Causes, Consequences, and Corrections. *J Nucl Med* 48 (7): 1112–21.

Higuchi, T, M Anton, K Dumler et al. 2009. Combined Reporter Gene PET and Iron Oxide MRI for Monitoring Survival and Localization of Transplanted Cells in the Rat Heart. *J Nucl Med* 50 (7): 1088–94.

Hutt, DF, AM Quigley, J Page et al. 2014. Utility and Limitations of 3,3-Diphosphono-1,2-Propanodicarboxylic Acid Scintigraphy in Systemic Amyloidosis. *Eur Heart J Cardiovasc Imaging* 15 (11): 1289–98.

Jaarsma, C, T Leiner, SC Bekkers et al. 2012. Diagnostic Performance of Noninvasive Myocardial Perfusion Imaging Using Single-Photon Emission Computed Tomography, Cardiac Magnetic Resonance, and Positron Emission Tomography Imaging for the Detection of Obstructive Coronary Artery Disease. *J Am Coll Cardiol* 59 (19). Elsevier Inc.: 1719–28.

Jogiya, R, G Morton, K De Silva et al. 2014. Ischemic Burden by 3-Dimensional Myocardial Perfusion Cardiovascular Magnetic Resonance: Comparison with Myocardial Perfusion Scintigraphy. *Circ Cardiovasc Imaging* 7 (4): 647–54.

Joshi, NV, AT Vesey, MC Williams et al. 2014. 18F-Fluoride Positron Emission Tomography for Identification of Ruptured and High-Risk Coronary Atherosclerotic Plaques: A Prospective Clinical Trial. *Lancet* 383 (9918): 705–13.

Klein, C, SG Nekolla, FM Bengel et al. 2002. Assessment of Myocardial Viability with Contrast-Enhanced Magnetic Resonance Imaging. *Circulation* 105 (2): 162–67.

Knaapen, P, R Boellaard, MJW Götte et al. 2004. Perfusable Tissue Index as a Potential Marker of Fibrosis in Patients with Idiopathic Dilated Cardiomyopathy. *J Nucl Med* 45 (8): 1299–304.

Knaapen, P, WG van Dockum, O Bondarenko et al. 2005. Delayed Contrast Enhancement and Perfusable Tissue Index in Hypertrophic Cardiomyopathy: Comparison between Cardiac MRI and PET. *J Nucl Med* 46 (6): 923–29.

Kramer, CM, J Barkhausen, SD Flamm et al. Society for Cardiovascular Magnetic Resonance Board of Trustees Task Force on Standardized Protocols. 2013. Standardized Cardiovascular Magnetic Resonance (CMR) Protocols 2013 Update. *J Cardiovasc Magn Reson* 15: 91.

Lee, S-P, ES Lee, H Choi et al. 2015. 11C-Pittsburgh B PET Imaging in Cardiac Amyloidosis. *JACC: Cardiovasc Imaging* 8 (1): 50–59.

Manka, R, L Wissmann, R Gebker et al. 2015. Multicenter Evaluation of Dynamic Three-Dimensional Magnetic Resonance Myocardial Perfusion Imaging for the Detection of Coronary Artery Disease Defined by Fractional Flow Reserve. *Circ Cardiovasc Imaging* 8: e003061.

Martinez-Möller, A, M Souvatzoglou, G Delso et al. 2009. Tissue Classification as a Potential Approach for Attenuation Correction in Whole-Body PET/MRI: Evaluation with PET/CT Data. *J Nucl Med* 50 (4): 520–26.

Matsunari, I, H Aoki, Y Nomura et al. 2010. Iodine-123 Metaiodobenzylguanidine Imaging and Carbon-11 Hydroxyephedrine Positron Emission Tomography Compared in Patients with Left Ventricular Dysfunction. *Circ Cardiovasc Imaging* 3 (5): 595–603.

Menezes, LJ, CW Kotze, O Agu et al. 2011. Investigating Vulnerable Atheroma Using Combined (18)F-FDG PET/CT Angiography of Carotid Plaque with Immunohistochemical Validation. *J Nucl Med* 52 (11): 1698–703.

Messroghli, DR, A Radjenovic, S Kozerke et al. 2004. Modified Look-Locker Inversion Recovery (MOLLI) for High-Resolution T1 Mapping of the Heart. *Magn Reson Med* 52 (1): 141–6.

Moon, JC, DR Messroghli, P Kellman et al. 2013. Myocardial T1 Mapping and Extracellular Volume Quantification: A Society for Cardiovascular Magnetic Resonance (SCMR) and CMR Working Group of the European Society of Cardiology Consensus Statement. *J Cardiovasc Magn Reson* 15: 92.

Morton, G, A Chiribiri, M Ishida et al. 2012. Quantification of Absolute Myocardial Perfusion in Patients with Coronary Artery Disease. *J Am Coll Cardiol*, September: 1–10.

Nappi, C, M Altiero, M Imbriaco et al. 2015. First Experience of Simultaneous PET/MRI for the Early Detection of Cardiac Involvement in Patients with Anderson-Fabry Disease. *Eur Nucl Med Molec Imaging* 42: 1025.

Nensa, F, TD Poeppel, K Beiderwellen et al. 2013. Hybrid PET/MR Imaging of the Heart: Feasibility and Initial Results. *Radiology* 268: 366–73.

Nensa, F, TD Poeppel, P Krings, and T Schlosser. 2014. Multiparametric Assessment of Myocarditis Using Simultaneous Positron Emission Tomography/Magnetic Resonance Imaging. *Eur Heart J* 35 (32): 2173–73.

Nensa, F, T Poeppel, E Tezgah et al. 2015. Integrated FDG PET/MR Imaging for the Assessment of Myocardial Salvage in Reperfused Acute Myocardial Infarction. *Radiology* 276 (2): 400–7.

O'Regan, DP, R Ahmed, N Karunanithy et al. 2009. Reperfusion Hemorrhage Following Acute Myocardial Infarction: Assessment with T2* Mapping and Effect on Measuring the Area at Risk. *Radiology* 250 (3): 916–22.

Piccini, D, A Littmann, S Nielles-Vallespin, and MO Zenge. 2011. Spiral Phyllotaxis: The Natural Way to Construct a 3D Radial Trajectory in MRI. *Magn Reson Med* 66 (4): 1049–56.

Piechnik, SK, VM Ferreira, E Dall'Armellina et al. 2010. Shortened Modified Look-Locker Inversion Recovery (ShMOLLI) for Clinical Myocardial T1-Mapping at 1.5 and 3 T within a 9 Heartbeat Breathhold. *J Cardiovasc Magn Reson* 12: 69.

Rischpler, C, N Langwieser, M Souvatzoglou et al. 2015. PET/MRI Early After Myocardial Infarction: Evaluation of Viability with Late Gadolinium Enhancement Transmurality vs. 18F-FDG Uptake. *Eur Heart J Cardiovasc Imaging* 16: 661–9.

Soussan, M, P-Y Brillet, H Nunes et al. 2012. Clinical Value of a High-Fat and Low-Carbohydrate Diet Before FDG-PET/CT for Evaluation of Patients with Suspected Cardiac Sarcoidosis. *J Nucl Cardiol* 20 (1): 120–27.

Terrovitis, J, R Lautamäki, M Bonios et al. 2009. Noninvasive Quantification and Optimization of Acute Cell Retention by In Vivo Positron Emission Tomography after Intramyocardial Cardiac-Derived Stem Cell Delivery. *J Am Coll Cardiol* 54 (17): 1619–26.

Ugander, M, PS Bagi, AJ Oki et al. 2012. Myocardial Edema as Detected by Pre-Contrast T1 and T2 CMR Delineates Area at Risk Associated with Acute Myocardial Infarction. *J Am Coll Cardiol Cardiovasc Imaging* 5 (6): 596–603.

Velazquez, EJ, KL Lee, MA Deja et al. 2011. Coronary-Artery Bypass Surgery in Patients with Left Ventricular Dysfunction. *Engl Med* 364 (17): 1607–16.

White, SK, DM Sado, AS Flett, and JC Moon. 2012. Characterising the Myocardial Interstitial Space: The Clinical Relevance of Non-Invasive Imaging. *Heart* 98 (10): 773–79.

White, SK, H Bulluck, and GM Frohlich. 2015. Hybrid PET/MR Metabolic Imaging of the Reperfused Infarct-New Biology, Future Directions. *J Cardiovasc Magn Reson* 17 (Suppl 1): O41.

Wicks, EC, L Menezes, and A Pantazis. 2014. Novel Hybrid Positron Emission Tomography-Magnetic Resonance (PET-MR) Multi-Modality Inflammatory Imaging Has Improved Diagnostic Accuracy for Detecting Cardiac Sarcoidosis. *Heart* 100 (Suppl 3).

Wicks, EC, LJ Menezes, and PM Elliott. 2015. Improving the Diagnostic Accuracy for Detecting Cardiac Sarcoidosis. *Exp Rev Cardiovasc Ther* 13 (2): 223–36.

Wu, JC, M Abraham, and DL Kraitchman. 2010. Current Perspectives on Imaging Cardiac Stem Cell Therapy. *J Nucl Med* 51 Suppl 1 (May): 128S–136S.

Zun, Z, P Varadarajan, RG Pai, EC Wong, and KS Nayak. 2011. Arterial Spin Labeled CMR Detects Clinically Relevant Increase in Myocardial Blood Flow with Vasodilation. *J Am Coll Cardiol Cardiovasc Imaging* 4 (12): 1253–61.

Merging optical with other imaging approaches

DOUG YEAGER, NICHOLAS DANA, AND STANISLAV EMELIANOV

17.1 INTRODUCTION

Within cardiovascular medicine, the development of improved diagnostic imaging techniques remains an area of active research and development, focusing on a wide range of diseases and pathologies. The successful prevention or management of such conditions necessitates a fundamental understanding of the underlying pathophysiology, early detection of those pathologies, and reliable risk stratification. The application of a wide array of optical imaging modalities provides both clinicians and preclinical researchers with the ability to accomplish these goals by enabling imaging of both biological anatomy and biochemical composition. Optical imaging is particularly well suited for high-sensitivity molecular imaging of pathophysiological processes of the cardiovascular system through the application of fluorescence, bioluminescence, or spectroscopic techniques. *In vivo* applications of these techniques within deep seated tissue (>1 mm), however, suffer from poor spatial resolution and lack of anatomical context for the received signals. These limitations are increasingly being met through the incorporation of additional imaging techniques that are capable of providing images revealing tissue anatomy, such as x-ray, magnetic resonance imaging (MRI), or ultrasound (US). These hybrid optical imaging modalities are increasingly utilized to provide multifaceted characterization of tissue, enabling simultaneous assessment of both morphology and composition.

Clinically, intravascular imaging represents a natural application of hybrid optical imaging techniques. Historically, the diagnosis of coronary atherosclerosis has been limited to morphological measurements, despite increasing evidence that their underlying composition is of great importance in assessing their future risk of inducing acute coronary events. Because the relevant field of view for intravascular imaging is limited to several millimeters, a wide range of optical imaging techniques have been adopted, using catheter-based configurations for improved characterization of atherosclerotic lesions. Hybrid intravascular optical imaging techniques, including systems that are clinically approved and many that are in preclinical development, promise to ultimately provide clinicians with improved platforms for evaluating structural and biochemical characteristics of atherosclerotic plaques to help enable their more accurate diagnosis and subsequently guide more effective treatment.

Beyond intravascular imaging, noninvasive optical imaging techniques provide preclinical researchers with platforms for complimentary assessment of anatomy, molecular composition, and physiological parameters during arterial or cardiac imaging. The use of specially designed hybrid optical platforms has enabled rapid, user-friendly small-animal imaging for *in vivo* evaluation of diseases and monitoring of therapeutic response. Furthermore, the recent expansion of hybrid nonlinear optical microscopy (NLOM) techniques enables label-free, high-resolution imaging for the study of tissues at the biomolecular level. Insights obtained from these hybrid applications are helping guide the development of novel contrast agents, which may, in turn, ultimately lead to next generation clinical diagnoses and therapeutic approaches.

This chapter overviews the current status of hybrid optical imaging techniques utilized within cardiovascular medicine, focusing heavily on emerging multimodality intravascular imaging techniques. Other topics include applications of contrast-enhanced noninvasive imaging techniques, small-animal optical imaging systems, and hybrid NLOM applications for imaging of cardiac and arterial tissues.

17.2 INTRAVASCULAR IMAGING OF CORONARY ATHEROSCLEROSIS

Coronary heart disease remains a leading cause of death throughout industrialized nations, responsible for more than 7 million deaths annually throughout the world (Mendis et al. 2011). This high mortality rate is due, in part, to an inability to appropriately characterize and treat atherosclerotic plaques, the underlying cause of coronary heart disease (Roger et al. 2012). Advances in molecular biology over the past several decades have provided an understanding of coronary pathophysiology involved in atherosclerotic plaque development and destabilization. Plaque progression is a process involving disordered cholesterol metabolism, compromised endothelial function, and loss of vessel patency, with systemic and local inflammatory events occurring throughout its growth. Atherosclerosis develops from endothelial dysfunction, causing monocyte recruitment and eventual progression toward an advanced, thrombotic plaque, as illustrated in Figure 17.1 (Choudhury, Fuster, and Fayad 2004). High levels of low-density lipoprotein or infectious agents and toxins can cause endothelial dysfunction and expression of adhesion molecules and selectins. These molecules then promote accumulation of monocytes, which subsequently form foam cells and an associated amplified inflammatory response with secretion of extracellular matrix proteins (Ross 1995; Mullenix, Andersen, and Starnes 2005; Packard and Libby 2008). Ultimately, production of matrix metalloproteinases can weaken a plaque's fibrous cap, rendering it prone to rupture with subsequent thrombus formation, which is associated with acute complications. Several studies have shown that thin-cap fibroatheromas (TCFA), those most prone to rupture, are typically characterized by the presence of a thin fibrous cap (<65 μm) overlaying a large, lipid-rich pool and accompanied by an increased number of inflammatory macrophages (Friedman and Van den Bovenkamp 1966; Constantinides 1967; Chapman 1968; Falk 1983; Willerson et al. 1984; Davies and Thomas 1985; Muller, Tofler, and Stone 1989; Muller et al. 1994).

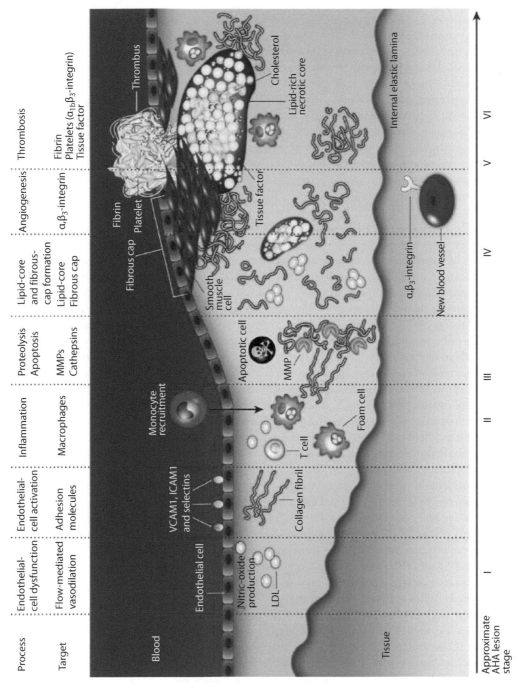

Figure 17.1 Hallmarks of atherosclerotic plaque formation and progression. (From Choudhury RP et al., *Nat Rev Drug Discov*, 3, 913–925, 2004).

Despite this fundamental understanding of the pathophysiology of atheroma progression, its translation into reliable clinical diagnoses remains limited by the capabilities of current imaging modalities. For example, the traditional gold standard for coronary imaging during percutaneous coronary interventions (PCIs), contrast coronary angiography, provides only a two-dimensional projection of vessel patency. As a result, the image provides a primarily morphological characterization of the coronary tree and suffers from inadequate resolution and contrast to detect certain characteristics of plaque stability, such as the thickness of a fibrous cap and underlying composition. Therefore, there remains a need for improved diagnostic imaging modalities that provide interventional cardiologists with greater information regarding the morphology as well as the cellular and molecular composition of atheroma to help guide accurate lesion characterization and to monitor therapy delivery. Numerous intravascular imaging techniques, including those which have already been translated into clinical practice as well as those in preclinical development, are poised to help bridge this gap between diagnostic imaging capabilities and the fundamental understanding of disease pathology (Waxman, Ishibashi, and Muller 2006). Among those are several hybrid imaging techniques that seek to merge modalities with complimentary advantages, including various combinations of intravascular US (IVUS), optical coherence tomography (OCT), coronary angioscopy (CAS), spectroscopic imaging methods, fluorescence imaging techniques, intravascular photoacoustic imaging (IVPA), and intravascular nuclear imaging techniques.

17.2.1 Combined IVUS and OCT

IVUS imaging has provided a tool for arterial assessment within the catheterization laboratory for several decades, particularly within Japan, Europe, and the United States. At the time of intervention, an IVUS catheter is advanced into the coronary arteries and scanned using a high-center-frequency US transducer (20–60 MHz) that is rotated to provide real-time cross-sectional images of arterial morphology to aid in diagnosis and therapy guidance during PCI. Volumetric renderings of coronary artery segments can also be obtained if the catheter is mechanically pulled back during rotation. IVUS is commonly used for the characterization of atherosclerotic lesions of unclear severity based on angiographic assessment to guide selection of the appropriate transcatheter therapy option through its ability to delineate the full vessel wall thickness and echogenicity to provide metrics associated with lesion severity (Fitzgerald and Yock 1993). Despite its ongoing clinical applications, however, IVUS provides limited histopathological information about the imaged artery, and histological studies have reported a generally low sensitivity of the technique's ability to differentiate thrombotic and lipid-rich lesions (Franzen, Sechtem, and Hopp 1998; Nissen and Yock 2001). Efforts to overcome the fundamental limitations of IVUS have led to the investigation of numerous imaging techniques intended to supplement the morphological view provided by IVUS imaging with greater information regarding the compositional or mechanical characteristics of the interrogated arteries. Among these techniques are virtual histology-IVUS (VH-IVUS) and integrated backscatter-IVUS (IB-IVUS), both of which involve analysis of the spectral content of backscattered US signals as a means of predicting tissue composition. Although VH-IVUS and IB-IVUS are clinically available methods intended to improve the ability of IVUS to detect plaque composition, the reliability of these approaches has recently been questioned (Kim et al. 2008; Thim et al. 2010). Finally, IVUS palpography, a technique that involves the assessment of vessel wall deformation at variable intravascular pressures (e.g., systole and diastole) to detect soft versus hard lesions, has also been investigated (Gaglia, Steinberg, and Weissman 2009).

Intravascular OCT has also been introduced as an alternative to IVUS imaging. Intravascular OCT is conceptually analogous to IVUS imaging in that images from both modalities are generated based on signals reflected from tissue structures. However, IVUS signals are received by the direct detection of time of flight of the propagating US pressure transients, whereas in the case of OCT, interferometry techniques are used to detect either time delays or frequency shifts in received optical echoes, which travel five orders of magnitude faster than US waves. As a result, OCT is able to achieve a higher axial resolution than IVUS imaging and can therefore provide greater ability to differentiate vessel wall and atherosclerotic plaque microstructures (Kubo et al. 2007). In addition to general detection of vessel morphology and identification of atherosclerotic plaques, fibrous cap thickness, microcalcifications, neovascularization, thrombus, and macrophage infiltration have all been interrogated by OCT (Tearney et al. 2012). This modality is not without its own limitations, however, as it requires saline flushing to eliminate blood during signal acquisition and is capable of providing only a limited tissue penetration depth of less than 1.25 mm, or even less in lipid-rich regions (Jang et al. 2002).

The combination of these two clinical imaging modalities offers several potential advantages over either IVUS or OCT as a stand-alone solution during PCI. Through combined IVUS/OCT, the advantages of each can be exploited to provide high-resolution interrogation of the vessel microstructure near the luminal surface without sacrificing the ability to image the full vessel wall thickness. Additionally, IVUS is well suited for imaging beyond lipid-rich pools but not beyond calcifications, while the opposite is true for OCT imaging (Raber et al. 2012). Therefore, the possibility of combining the modalities is particularly intriguing as a means of assessing both the fibrous cap thickness and full extent of a plaque burden, including potential positive remodeling, in advanced atherosclerotic lesions. Figure 17.2 demonstrates the improved diagnostic potential of combining the modalities for detecting TCFA plaques, using off-line coregistration of IVUS, or VH-IVUS, and OCT images (Sawada et al. 2008; Gonzalo et al. 2009). The synergistic effect of combined IVUS/OCT is, in part, a consequence of the fact that these vulnerable plaques are diagnosed differently using the two modalities. For OCT, fibrous cap thickness and a diffuse boundary, indicative of a lipid-rich pool, are the primary parameters utilized to differentiate TCFA. For IVUS, the differentiation is based on the extent (>40%) and axial location, relative to the luminal wall, of a plaque. When both modalities are performed, the diagnostic efficacy for differentiating TCFA has been found to increase.

As a result of the demonstrated improved diagnostic capability performing both IVUS and OCT imaging, efforts to develop combined IVUS/OCT imaging catheters are currently under way. While several prototypes have been demonstrated (Li et al. 2010, 2013; Yang et al. 2010; Yin et al. 2010, 2011; Xiang et al. 2014), no fully integrated system or catheter capable of *in vivo* intravascular coronary imaging has been developed to date. The smallest profile of the developed integrated IVUS/OCT catheters is approximately 690 µm, but with an

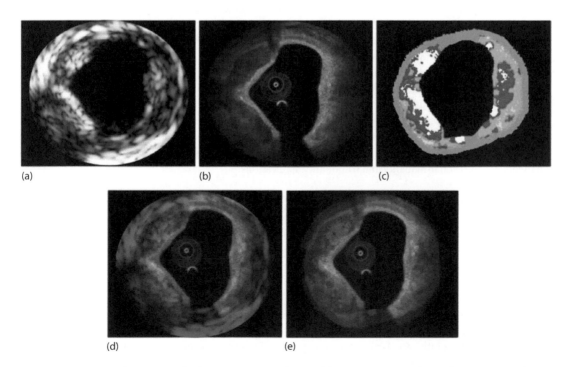

(a) (b) (c)

(d) (e)

Figure 17.2 In panels a–c, matched cross-sections obtained by IVUS (a), OCT (b), and IVUS-VH (c) depict fibrocalcific plaque in the left half of the vessel, and a necrotic core/lipid pool at the opposite side of the calcification. Superimposition of IVUS on top of OCT and IVUS-VH on top of OCT is presented in Panels (d) and (e), respectively. (Adapted from Raber L et al., *EuroIntervention* 8, 98–108, 2012. With permission.)

inflexible distal housing of the catheter (Yin et al. 2011), while other prototypes exhibit shorter inflexible distal ends, but at larger catheter diameters, thus limiting their application to *ex vivo* (Li et al. 2010; Yang et al. 2010; Yin et al. 2010, 2011) and *in vivo* (Xiang et al. 2014) examination of animal models of atherosclerosis and excised human coronary arteries (Li et al. 2013). While further improvements of the integrated systems are still necessary to improve their performance to that comparable to results demonstrated using independent, clinical IVUS and OCT systems, the preliminary results that have been demonstrated using combined IVUS/OCT catheters in recent years are indicative of the promising potential for the hybrid technique to enable improved lesion characterization.

17.2.2 HYBRID CAS TECHNIQUES

CAS represents an alternative approach for intravascular imaging, enabling clear visualization of the luminal arterial wall. Originally introduced as early as the mid-1980s (Spears, Spokojny, and Marais 1985; Sherman et al. 1986), CAS is a clinically approved endoscopic technique that utilizes a fiber bundle to deliver a white light illumination source as well as to collect and return reflected light to a charge-coupled-device, color camera (Suter et al. 2011). The generated CAS images have been demonstrated to enable the differentiation of atheroma composition based, in part, on the color of identified lesions. Fibrous lesions tend to appear as white lesions, whereas lipid-rich plaques tend to exhibit a yellow color upon CAS visualization, likely due to the carotenoid content of intimal cholesterol (Ishibashi et al. 2006). Furthermore, the intensity of the detected yellow color has also been associated with the prevalence of thrombosis on the atheroma, suggesting a potential indicator of lesion vulnerability (Ueda et al. 2004).

Despite its regulatory approval and ability to distinguish different arterial surface compositions, the utilization of CAS remains limited to research applications, and there are currently no clinical indications for the use of CAS over other imaging modalities during coronary intervention. Limitations of this technique include the inability to image beyond the surface of the arterial wall, the requirement of flushing luminal blood during imaging, and catheter diameters that are too large to enable imaging beyond the proximal regions of main coronary arteries. As a research tool, however, CAS continues to be refined and is commonly utilized to help evaluate the efficacy of both pharmacological treatment regimes, such as statin therapies (Takano et al. 2003; Tawakol and Muller 2003), and evaluation of arterial wall response following stent placement (Yamamoto et al. 2009; Yokoyama et al. 2009). Additionally, due to its inability to image beyond the arterial surface, CAS has been utilized in combination with IVUS (Tokuhiro et al. 2000; Koie et al. 2003) and OCT (Kubo et al. 2008), as well as the combination of all three modalities (Kubo et al. 2007), to provide a more comprehensive evaluation of atherosclerotic lesions. Despite these pairings, integrated intravascular probes combining CAS with other intravascular modalities have not yet been developed, and coregistration of CAS images with those obtained from the same arterial cross-section with different techniques remains a challenge for these dual-modality and trimodality applications.

17.2.3 HYBRID INTRAVASCULAR SPECTROSCOPY TECHNIQUES

While both IVUS and OCT imaging provide clinicians with information regarding coronary artery morphology, each modality is limited in its ability to definitively diagnose arterial composition due to potential ambiguities and artifacts in their respective backscattered signals. Intracoronary diffuse reflectance near-infrared spectroscopy (NIRS) represents an additional and clinically approved imaging technique that is capable of detecting lipid-rich plaques. NIRS relies on variations in the molecular overtone and combination vibrations of endogenous molecules, which result in molecular composition-specific optical absorption and scattering within the near-infrared region (NIR) of the electromagnetic spectrum (approximately 1000 nm to 2400 nm). During intravascular NIRS imaging, a broadband NIR source is emitted into the arterial wall and the returned light is received to extract chemical compositional information of the illuminated region. First demonstrated as a tool for assessing atherosclerotic plaques in an animal model of atherosclerosis in the early 1990s (Cassis and Lodder 1993), the capability of NIRS to detect several plaque risk indicators, including lipid pools, fibrous cap, and inflammatory cells was demonstrated nearly a decade later (Moreno et al. 2002), and the modality was first utilized for *in vivo* human imaging for lipid detection soon thereafter (Caplan et al. 2006). Commercialized by InfraReDx (Burlington, Massachusetts) and approved by the Food and Drug Administration in 2008, clinical intravascular NIRS provides a chemogram, representing the likelihood of a lipid-rich plaque within an angular region of an arterial cross-section. A notable advantage of the technique is that this chemogram can be obtained through blood, thus avoiding the need for luminal flushing, which is required for most other optical imaging modalities. However, because NIRS detects remitted light from within highly optically scattering blood and tissue, the received signal is not depth resolved (Brugaletta et al. 2011). Therefore, while the absence or presence of a lipid-rich plaque can be detected using NIRS, it alone cannot provide corresponding morphological context without being coupled with IVUS or OCT.

The current clinical NIRS system couples the technique with IVUS imaging, providing a dual-modality approach to coronary assessment. Both NIRS and IVUS image acquisition are performed simultaneously with a conventional IVUS image display encompassed within a chemogram, revealing the likelihood of lipid-rich plaque within an angular region (Figure 17.3a). In addition to a demonstrated general improvement in the ability to characterize high-risk, lipid-rich plaques (Caplan et al. 2006; Waxman et al. 2009), the hybrid NIRS/IVUS imaging technique is also being used to generate a database from 1000 patients through the Chemometric Observations of Lipid Core Plaque of Interest in Native Coronary Arteries registry, to be used for assessment of the varieties of lipid-rich plaques and their long-term clinical significance and responses to available therapies (Suter et al. 2011). Additionally, a recently launched 9000-patient Lipid-Rich Plaque Study trial is intended to assess whether this hybrid imaging technique can be utilized as a predictor of future coronary events.

Early clinical adoption of NIRS/IVUS has been found through its inclusion as an end point for lipid-lowering clinical trials (Simsek et al. 2012) and is actively being investigated as a tool to help reduce risk of peri-procedural complications during PCI through the Coronary Assessment by Near-infrared of Atherosclerotic Rupture-prone Yellow trial (Goldstein et al. 2009, 2011). As the results of the ongoing clinical trials come out, the true clinical impact of NIRS/IVUS imaging will be further clarified. Interestingly, as NIRS/IVUS continues its expansion within clinical practice, it was not until very recently that a combined NIRS/OCT catheter prototype was reported (Fard et al. 2013). The dual-modality NIRS/OCT catheter utilized a single, dual-clad optical fiber as a conduit for separately carrying the light required for each modality and was demonstrated for imaging *ex vivo* atherosclerotic human coronary artery tissue at a frame rate of 24 Hz (Figure 17.3b and c).

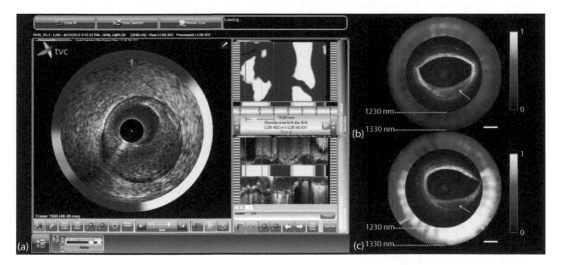

Figure 17.3 Representative hybrid NIRS images. (a) View of the clinically approved InfraReDx TVC NIRS/IVUS imaging system showing both cross-sectional and longitudinal views. (b, c) OCT-NIRS images of cadaver coronary artery *ex vivo*. Yellow color in the ring surrounding the cross-sectional images indicates the presence of lipid within the associated angular direction. ([a] Adapted from Shydo B et al., *J Invasive Cardiol*, 25, 5A–8A, ISNN: 1557–2501, 2013. With permission; [b, c] reprinted from Fard AM, Vacas-Jacques P, Hamidi E, Wang H, Carruth RW, Gardecki JA, and Tearney GJ. 2013. Optical coherence tomography–near infrared spectroscopy system and catheter for intravascular imaging. *Opt Express* 21 (25):30849–30858. With permission of Optical Society of America.)

Relative to the further developed NIRS/IVUS, NIRS/OCT would provide the added benefit of high axial resolution for improved assessment of arterial microstructures, but at the expense of requiring luminal flushing and inability to image the full vessel wall thickness.

Raman spectroscopy, an alternative method for detecting the molecular signatures of biological tissues, has also been investigated as a potential intravascular imaging modality. Rather than molecular absorption spectra, as is the basis for NIRS assessment, Raman spectroscopy relies on frequency shifts of emitted light, which are generated by an energy exchange between photons and molecular components of tissues. The magnitude of this energy exchange, and thus the extent of the frequency shift of received photons, is a function of a given molecule's vibrational and rotational energies. Raman spectroscopy therefore offers the potential for very high molecular specificity. Whereas NIRS is used to differentiate lipid-rich plaques, Raman spectroscopy offers the potential to differentiate specific chemical components such as triglycerides, cholesterol, cholesterol esters, elastin, and collagen (Baraga, Feld, and Rava 1992; Manoharan et al. 1992; Brennan et al. 1997; Romer et al. 1998; van de Poll et al. 2003). Several intravascular Raman catheter prototypes have been developed and applied for arterial assessment including in the presence of luminal blood (Buschman et al. 2000; van de Poll et al. 2003; Motz et al. 2006; Chau et al. 2008). Additionally, it has been shown that Raman spectroscopy enables evaluation of pharmaceutical treatment efficacy through the quantification of plaque composition (van de Poll et al. 2001) as well as detection of vulnerable plaques with a sensitivity and specificity of 79% and 85%, respectively (Motz et al. 2006).

Irrespective of the potential for high intravascular molecular sensitivity, Raman spectroscopy as a stand-alone modality suffers from similar limitations as NIRS, namely, its inability to detect arterial morphology. To this end, an offline approach to couple IVUS imaging with Raman spectroscopy was demonstrated and proposed as a potential future method of providing interventional cardiologists with a means of identifying vulnerable atherosclerotic lesions and monitoring therapy outcomes (Romer et al. 2000). A similar approach combining Raman spectroscopy with IVUS elastography, an IVUS technique that measures mechanical properties of the arterial wall through correlation of US images obtained before and after application of a force, such as pulsatile blood flow, was also proposed (van de Poll et al. 2000). Along the same lines, bench-top designs combining Raman spectroscopy with OCT have also been demonstrated, although the designs have not yet been integrated into a probe design or even utilized for the assessment of arterial tissues (Patil et al. 2008; Evans et al. 2009; Khan et al. 2014). While these proof-of-concept demonstrations of hybrid Raman techniques are still in early stages, they offer the potential for development into an intravascular system analogous to that of NIRS/IVUS.

Despite these encouraging results and demonstrations of hybrid modalities combining Raman spectroscopy with IVUS or OCT, Raman spectroscopy has failed to be translated into the clinic. In comparison to NIRS, Raman spectroscopy typically yields low intensity signals, and the resulting low signal-to-noise ratio is exacerbated by noise introduced within the optical fibers of Raman catheters. While high-wavenumber Raman spectroscopy techniques and new catheter designs have provided hope that the technical challenges preventing commercialization of a hybrid Raman-based approach to plaque assessment will be attained, significant validation is still required (Brennan et al. 2008; Chau et al. 2008).

17.2.4 HYBRID INTRAVASCULAR FLUORESCENCE IMAGING TECHNIQUES

Fluorescence imaging techniques have also been introduced as a means of enhancing the ability to detect atherosclerotic plaque composition. Several distinct fluorescence imaging techniques are underway, including time-resolved fluorescence spectroscopy (TRFS) and near-infrared fluorescence spectroscopy (NIRF). Initial studies were able to differentiate normal arterial tissue from atherosclerotic lesions by assessing the tissue molecular components, including lipid, elastin, and collagen, based on their respective autofluorescence characteristics (Deckelbaum et al. 1987; Richards-Kortum et al. 1989; Baraga et al. 1990). Increased molecular specificity, however, has been achieved through the introduction of TRFS, a technique that relies on the temporal profile of remitted photons following excitation with a pulsed optical source. TFRS has been demonstrated for the *ex vivo* classification of atherosclerotic plaques into three subtypes: intimal thickening, fibrotic and fibrocalcific, and inflamed and necrotic lesions (Marcu et al. 2009). The ability to detect the presence of macrophage foam cells was also demonstrated (Marcu et al. 2005). Further improvements were subsequently made through the introduction of fluorescence lifetime imaging microscopy (FLIM), a TRFS technique that enables the recording of two-dimensional lifetime information. FLIM has been performed utilizing an endoscopic design that permits an improved imaging depth of approximately 4 mm (Elson, Jo, and Marcu 2007; Phipps et al. 2009). The FLIM catheter was utilized for *ex vivo* characterization of 11 human aorta sections at a total of 48 locations, demonstrating that the technique can be used to distinguish clinically relevant plaque features with sensitivities as high as 86% and an all case specificity of 87% (Phipps et al. 2011). Notably, a hybrid imaging probe combining fluorescence spectroscopy with reflectance spectroscopy and Raman spectroscopy has also been described (Šćepanović et al. 2009) and utilized to characterize plaque composition (Scepanovic et al. 2011).

As FLIM and, more generally, TRFS continue development toward clinical translation, they suffer from two primary drawbacks: a limited penetration depth of approximately 250 μm due to the high absorption and scattering of the excitation light and the inability to provide arterial morphology to compliment the compositional assessment (Phipps et al. 2009). As with the previously discussed spectroscopic imaging techniques, combination of TRFS with IVUS or OCT offers the potential to eliminate the later of these limitations. Recently, TRFS has been integrated with IVUS imaging through the development of a multimodality intravascular catheter prototype with an outer diameter of 1.8 mm to enable fluorescence-based assessment of arterial composition under US guidance (Stephens et al. 2009). However, this initial prototype exhibited a fixed fiber orientation and it was necessary to maintain close proximity to the vessel wall to acquire TRFS images. In a modified design, a second-generation TRFS/IVUS catheter with an outer diameter of 2.6 mm incorporated a flushing design to enable intraluminal measurement without totally occluding a vessel or forcing direct contact of the optical fiber while also enabling rotation of both the IVUS transducer and TRFS optical fiber and was demonstrated on *ex vivo* coronary artery specimens as shown in Figure 17.4a and b (Bec et al. 2012). Notably, TRFS/IVUS was subsequently utilized for *in vivo* imaging in a swine model (Bec et al. 2014). A similar multimodality approach combining TRFS with US microscopy and photoacoustic (PA) microscopy was used to characterize *ex vivo* human carotid artery sections to provide complimentary compositional, morphological, and functional characterization of the arterial wall (Sun et al. 2011).

In addition to TRFS/IVUS, preclinical research combining OCT with fluorescence imaging techniques is also under way. An initial demonstration of fluorescence spectroscopy and OCT was utilized to differentiate normal from plaque regions in *ex vivo* postmortem aorta samples using a bench top setup (Barton, Guzman, and Tumlinson 2004). Similarly, a separate bench top prototype combining FLIM with OCT has also been developed and utilized to characterize *ex vivo* human coronary artery specimens (Park et al. 2010).

Several studies have also investigated the potential for combining the anatomical information from IVUS or OCT with NIRF imaging. NIRF is a technique that relies on the labeling of tissues with exogenous contrast agents that fluoresce at longer wavelengths than tissue autofluorescence. Initially introduced for the detection of cathepsin B in atherosclerotic mouse models (Chen et al. 2002), NIRF has since been demonstrated for the detection of numerous molecular targets of relevance to atherosclerosis, including both *in vitro* and *in vivo* studies (Jaffer, Libby, and Weissleder 2009). Notably, a 2.9F, two-dimensional rotational NIRF imaging catheter emitting at 750 nm and capable of high-sensitivity fluorophore detection was developed and demonstrated for the detection of inflammation-regulated cysteine protease activity in atheroma and stent induced injury (Jaffer et al. 2011). In the same study, IVUS pullbacks were also performed at the time of NIRF imaging to enable off-line coregistration of the arterial anatomy with the distribution of detected fluorescence (Figure 17.4c–e). NIRF has also been more directly integrated with OCT imaging through the development of integrated catheters utilizing double-clad optical fiber designs to independently carry the illumination for OCT and NIRF (Yoo et al. 2011; Liang et al. 2012). In one embodiment, a prism at the distal end of the catheter was rotated using a micromotor to scan the optical illumination profiles during imaging of *ex vivo* atherosclerotic tissue samples (Liang et al. 2012). This design, however, causes the outer diameter of the catheter to be over 2 mm, with further miniaturization dependent on the development of smaller profile motor assemblies. A different OCT/NIRF catheter design utilizes a rotational assembly that is comparable in functionality and size to clinical OCT catheters (Yoo et al. 2011).

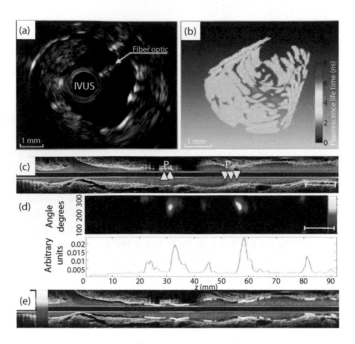

Figure 17.4 Hybrid fluorescence imaging representative images. (a–b) Combined IVUS-FLIM imaging. (a) IVUS cross-section and (b) reconstructed 3-D FLIM images of a stented coronary artery labeled with two coumarin bead inclusions (blue) obtained using a hybrid catheter *ex vivo*. (c–e) Combined IVUS-NIRF imaging. (c) Longitudinal IVUS image of the abdominal aortoiliac arteries. Arrows demarcate IVUS-detectable mildly stenotic plaques (P_1, P_2). (d) Image of NIRF catheter pullback aligned with the angiogram and IVUS image demonstrates NIRF signal in small volume plaques, in >3.0-mm-diameter arteries. Corresponding 1D plot of the angle-averaged 2-D NIRF intensity pullback below. (e) Longitudinal superimposed NIRF and IVUS fusion images (yellow/white = strongest NIRF signal intensity, red/black = lowest NIRF signal intensity). ([a, b] Reprinted from Bec J et al., *J Biomed Opt*, 17, 106012, 2012. With permission; [e] reprinted from *J Am Coll Cardiol*, 57, Jaffer FA, Calfon MA, Rosenthal A, Mallas G, Razansky RN, Mauskapf A, Weissleder R, Libby P, Ntziachristos V, Two-dimensional intravascular nearinfrared fluorescence molecular imaging of inflammation in atherosclerosis and stent-induced vascular injury, 2516–2526, Copyright 2011, with permission from Elsevier.)

Further development and validation of intravascular hybrid imaging probes utilizing the molecular specificity offered by fluorescence techniques and morphological information provided by IVUS or OCT are ongoing. In the case of NIRF, while several fluorophores have been investigated within the preclinical setting, including several that are activated upon interaction with a specific molecular target of interest, most of the fluorophores have not been approved for clinical use and their safe utilization remains to be demonstrated in humans. As these techniques are further refined, their true clinical significance and safety will need to be clarified through the initiation of large-scale clinical trials similar to those being conducted using the clinical NIRS/IVUS imaging system.

17.2.5 COMBINED IVUS AND PA IMAGING

The combination of IVUS with IVPA was recently introduced as an alternative potential hybrid imaging modality for improved characterization of coronary arteries. IVPA is based on the PA effect in which the absorption of nanosecond-pulsed optical illumination results in a rapid thermal expansion of the absorber, with subsequent generation of pressure transients that can be detected by US transducer. The generated pressure amplitude is proportional to the delivered optical fluence, the absorption coefficient of the absorber(s) and a tissue- and temperature-dependent parameter (the Grüneisen parameter), which accounts for the efficiency of conversion from optical to acoustic energy (see Chapter 8 for a more in-depth description of PA imaging fundamentals). Due to the dependence of the PA signal amplitude on the optical absorption properties of the illuminated medium, it is possible to differentiate particular absorbers of interest based on their unique spectral signatures. In this respect, PA imaging is similar to optical spectroscopy techniques such as NIRS. However, PA imaging offers a distinct advantage in comparison to purely optical-based techniques. Because the signal detection is achieved using US transducer, locating the origin of the signal can be achieved based on the US time-of-flight. As a result, PA signals can be readily depth-resolved to reveal the spatial distribution of absorbers. Therefore, IVPA imaging has been introduced as a tool for localizing absorbers within arterial tissues (Sethuraman et al. 2007a; Jansen, van Soest, and van der Steen 2014).

Combined IVUS/IVPA imaging provides a means of supplementing the morphological information provided by IVUS with additional capability for assessing the composition of atherosclerotic lesions based on unique optical absorption properties of specific plaque components or delivered contrast agents (Sethuraman et al. 2007b). Beyond the complimentary nature of the information received from the two modalities, there is the added benefit of shared hardware between IVUS and IVPA. Because the received signals for both modalities are acoustic pressures, a hybrid IVUS/IVPA system can use a single US transducer, analog-to-digital converter, and many of the same signal processing algorithms. This shared instrumentation is an important consideration as IVUS/IVPA looks to become a commercial system in the future as it reduces added cost associated with combining multiple intravascular modalities.

Early demonstrations of the feasibility and potential applications of IVUS/IVPA utilized bench-top experimental setups with external illumination for IVPA signal generation and commercially available IVUS transducers for IVUS transmission as well as receiving of both IVUS and IVPA signals (Sethuraman et al. 2007a, 2007b). Using this setup, IVUS-guided spectroscopic PA (sPA) was performed on *ex vivo* tissue sections from animal models of atherosclerosis by selectively tuning the output wavelength of the excitation light sources within the range of 680 nm to 900 nm in order to differentiate and localize fibrous and lipid components within the arterial wall based on differences in the associated molecular optical absorption spectra (Sethuraman et al. 2008). The ability to localize collagen as well as lipid within atherosclerotic arteries was more thoroughly studied using similar bench top sPA imaging systems using optical excitation within the near infrared wavelength region, particularly near lipid absorption spectral peaks at approximately 1200 nm and 1720 nm (Wang et al. 2010, 2012c, 2012d; Allen et al. 2012).

These early demonstrations of IVPA contrast based on endogenous tissue absorbers have also been expanded to the localization of exogenous absorbers. One clear example is the imaging of metallic stent struts based on the intrinsically high optical absorption of metal relative to endogenous tissue (Su, Wang, and Emelianov 2009). While clinical OCT systems are well suited to image stents and detect apposition, the high contrast of struts in IVUS/IVPA imaging may provide an additional method for guidance of their placement during PCI and may prove to be particularly advantageous for assessment of restenotic arteries in which OCT lacks adequate imaging depth. Additionally, the bench-top IVUS/IVPA system was further utilized to localize macrophages labeled with exogenous contrast agents (Wang et al. 2009). Spherical plasmonic gold nanoparticles, which yield very high PA signal due to the surface plasmon resonance effect, were shown to be phagocytosed by macrophages. Once internalized and aggregated within macrophage endosomes, the nanoparticles exhibited a broadening of their absorption peak which was detected by spectroscopic IVPA (sIVPA) imaging.

As these early applications of PA imaging for detection of atherosclerotic plaque components emerged, custom fabricated integrated IVUS/IVPA catheter prototypes and imaging systems enabling the acquisition of spatially coregistered and temporally consecutive IVUS and IVPA images were simultaneously under development. The initial description of integrated catheters capable of combined IVUS/IVPA imaging included two prototype designs, which were fundamentally different in terms of the method used to direct light onto the arterial wall (Karpiouk, Wang, and Emelianov 2010). In one design, an angled mirror placed distal to the optical fiber was used to direct the PA excitation light onto the arterial wall such that it overlapped with the beam profile of a commercially available 40 MHz center frequency IVUS transducer. In the second design, the mirror is replaced by the utilization of the total internal reflection effect to produce side-fire illumination from an angle-polished optical fiber. Additional integrated IVUS/IVPA catheter designs have subsequently been proposed, including the use of a forward-pulsing hollow cylindrical transducer encompassing an optical fiber with an angled mirror to coaxially redirect both the IVUS and IVPA beams (Wei et al. 2011), as well as an optical fiber with an axicon mirror at the distal tip for PA excitation and microring based acoustic detection, which could allow an all-optical assembly (Hsieh et al. 2012). Despite these novel designs, integrated catheters utilizing side-fire optical fiber (Karpiouk, Wang, and Emelianov 2010) coupled with a conventional IVUS transducer is the design that is currently the most commonly used in ongoing preclinical IVUS/IVPA studies. This is likely due to both the simplicity of fabrication and the risk of damaging a mirror due to the high optical intensities required for IVPA imaging with a single, small-diameter optical fiber delivery. All of these initial prototype integrated IVUS/IVPA catheters were too large and inflexible to be directly applied to *in vivo* coronary artery imaging. However, further miniaturization of the side-fire fiber-based catheter design has been reported, with the smallest integrated IVUS/IVPA catheter reported to date having a diameter of approximately 1.25 mm (Jansen et al. 2011). While these integrated IVUS/IVPA catheters have not yet been demonstrated *in vivo* in human coronary arteries, they have been utilized in the characterization of arterial plaques from both animal models of atherosclerosis and human cadaver artery sections (Figure 17.5).

Using integrated IVUS/IVPA imaging catheters, sIVPA imaging has been demonstrated to further advance imaging of lipid distribution within arterial plaques. Given the association between large lipid pools within atherosclerotic plaques and their susceptibility to rupture, the localization of lipid has been a major focus for the hybrid IVUS/IVPA technique and is a likely first clinical application for IVUS/IVPA imaging. Specifically, two distinct optical absorption peaks of lipid within the NIR region have been utilized for sIVPA imaging of lipid-rich plaques, one located at approximately 1210 nm, as shown in Figure 17.5a and b (Jansen et al. 2011, 2012, 2013; Wang and Emelianov 2011) and the other at 1720 nm, as shown in Figure 17.5c and d (Wang et al. 2012a, 2012b; Jansen et al. 2013). A direct comparison between results obtained from performing sIVPA imaging over the two spectral bands found similar results, with a lower pulse energy required for

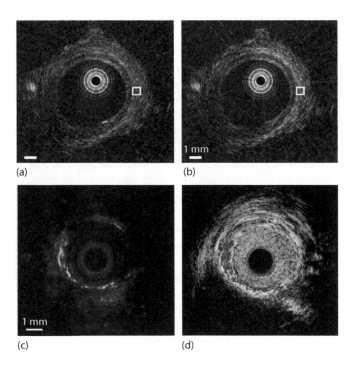

(a)   (b)

(c)   (d)

Figure 17.5 IVUS/IVPA imaging of lipid with atherosclerotic plaques. (a, b) *Ex vivo* IVUS/IVPA images obtained from two distinct locations within a human coronary artery section using sIVPA imaging near 1210 nm. (c) IVPA and (d) IVUS/IVPA cross-sectional images obtained from an atherosclerotic rabbit aorta *in vivo* using 1720 nm excitation. ([a, b] Reprinted from Jansen K, van der Steen AF, van Beusekom HM, Oosterhuis JW, and van Soest G. 2011. Intravascular photoacoustic imaging of human coronary atherosclerosis. *Opt Lett* 36 (5):597–9. With permission of Optical Society of America; [c, d] reprinted from *Ultrasound Med Biol*, 38, Wang B, Karpiouk A, Yeager D, Amirian J, Litovsky S, Smalling R, Emelianov S, *In vivo* intravascular ultrasoundguided photoacoustic imaging of lipid in plaques using an animal model of atherosclerosis, Copyright 2012, with permission from Elsevier.)

detection of lipid at the 1720-nm spectral band but with a reduced achievable imaging depth (Jansen et al. 2013). IVUS/sIVPA has also been expanded to go beyond localizing lipid within plaques to the differentiation of specific cholesterol compositions, much like the aforementioned capability of Raman spectroscopy, as a way of further characterizing lesions and potentially differentiating lipid within plaques from adventitial fat (Jansen et al. 2013).

For the absorption peak located at 1720 nm, the IVPA signal obtained using only a single wavelength was shown to correlate well with regions identified as lipid by both the sIVPA approach and subsequent histological staining (Wang et al. 2012b). In fact, this single wavelength lipid-detection technique was the approach utilized for the first *in vivo* IVUS/IVPA demonstration, which was performed on a rabbit model of atherosclerosis as shown in Figure 17.5c and d (Wang et al. 2012a). While the single wavelength approach offers advantages in terms of the required imaging time and image processing complexity, its sensitivity and specificity should be further quantified on a large sample size to validate the efficacy of the method. Importantly, US/PA imaging for the detection of lipid has also been performed in presence of luminal blood for both of the optical peaks under investigation, a notable advantage over other depth-resolved optical techniques (Wang et al. 2010, 2012b; Allen et al. 2012).

While the localization of endogenous lipid, and potentially collagen, within the coronary arteries is a likely first clinical target for IVUS/IVPA imaging, expansion of the technique to enable greater tissue compositional assessment is also being investigated. The use of plasmonic gold nanoparticles to localize phagocytically active macrophages was expanded through the demonstration that systemically injected, near-infrared absorbing, rod-shaped gold nanoparticles also preferentially localized within atherosclerotic plaques and could be detected using IVUS/IVPA imaging in the presence of luminal blood in *ex vivo* studies on animal models of atherosclerosis (Yeager et al. 2012). The role of nanoparticle contrast agents for IVUS/IVPA was recently further motivated through the demonstration that a continuous wave (CW) laser emitting at the peak absorption wavelength of the nanoparticles could be coupled into the same optical fiber of the integrated catheter to induce selective heating of the tissue immediately surrounding the nanoparticles (Yeager et al. 2013). The same study also showed that the induced temperature rise could be measured by monitoring the intensity of the received IVPA signal during selective CW laser heating, thus providing a potential future platform for selective, controllable photothermal therapy of labeled atherosclerotic plaques as shown in Figure 17.6. The use of metallic nanoparticles for the detection of plaque macrophages represents one of many potential contrast agent types and plaque biomarker targets that may be investigated for IVUS/IVPA imaging in the future. For example, organic dyes and single-walled carbon nanotubes have each been broadly utilized as PA imaging contrast agents, while plaque biomarkers, including the overexpression of cell

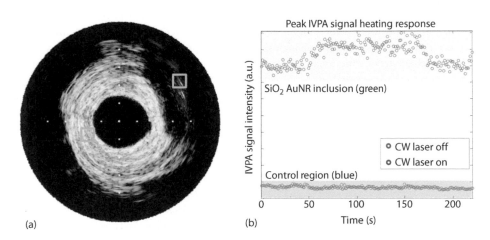

(a) (b)

Figure 17.6 IVUS/IVPA imaging and temperature monitoring using nanoparticle contrast agents. (a) IVUS/IVPA image of a coronary artery labeled with an inclusion of silica-coated gold nanorods *ex vivo*. (b) IVPA signal intensity change during heating of the nanoparticle inclusion (green) and a control region (blue) using a CW laser. (Reprinted from Yeager D et al., *Theranostics*, 4, 36–46, 2013. With permission.)

adhesion molecules, matrix metalloproteinases, and integrins, have all been molecular targets for detection with preclinical PA imaging systems (Luke, Yeager, and Emelianov 2012; Mehrmohammadi et al. 2013). Each of these applications involving exogenous contrast agents could be adapted for IVUS/IVPA imaging applications, although the clinical utilization contrast-enhanced imaging for intravascular characterization of atherosclerotic plaque composition would likely require a lengthy regulatory approval process to ensure the safety and efficacy of specific contrast agents and imaging procedures.

In general, although IVUS/IVPA is still under development as a preclinical system, its ability to not only detect but also localize lipid deposits within the context of the arterial morphology makes it a potentially attractive clinical modality. However, technical hurdles of further miniaturization of the integrated catheters and implementation of laser sources with a significantly higher repetition rate, as well as safety concerns due to the high optical intensities at the distal tip of the catheter must be addressed as the technology progresses towards clinical adoption.

Complimentary to IVPA imaging for the characterization of coronary atherosclerosis, sPA imaging has also been utilized with an *ex vivo* animal tissue model to visualize the optical wavelength-dependent changes that occur when native cardiac tissue is ablated in the course of cardiac arrhythmia treatment (Dana et al. 2014). Cardiac arrhythmia is often treated by the targeted destruction of myocardial tissue with the application of energy via percutaneous catheter access. As hyperthermia denatures proteins and disrupts the native structure of cardiac tissue, changes in the optical absorption of the treated tissue in the NIR regime occur. sPA imaging, being sensitive to absorption as a contrast mechanism, can characterize this tissue and identify hyperthermia lesions down to depths of at least 3 mm in *ex vivo* studies using animal models. Furthermore, it was shown that sPA imaging can differentiate gaps between adjacent lesions on the order of 1 mm. Both of these results are promising to physicians as they demonstrate that sPA imaging can provide physicians feedback on the transmurality and contiguity of lesions, which are paramount to improving cardiac ablation outcomes. While this application is not truly intravascular, particularly at its current stage, a novel combined PA and US imaging catheter designed for combined intracardiac US/PA imaging has been demonstrated and such a catheter could be utilized for ablation monitoring in the future (Nikoozadeh et al. 2012). This represents a first translational step in the development of a clinical platform for PA imaging guidance of cardiac arrhythmia ablation and, potentially, other intracardiac applications.

17.2.6 HYBRID INTRAVASCULAR POSITRON DETECTION AND OCT

Conventional scintigraphic techniques, including positron emission tomography (PET) and single-photon emission computed tomography (SPECT), have a wide array of applications within cardiovascular imaging. These techniques utilize radionuclide contrast agents to detect and monitor physiological processes, thus enabling functional imaging. However, in the case of coronary atherosclerosis, these noninvasive modalities suffer as a result of substantial physiological motion and provide insufficient resolution and sensitivity to effectively characterize functional activity within atherosclerotic lesions. Efforts to circumvent these limitations, however, have resulted in the development and *ex vivo* characterization of several prototype intravascular radiation detection catheters (Lederman et al. 2001; Janecek et al. 2004; Hosokawa et al. 2006; Shikhaliev et al. 2006; Strauss et al. 2006). For such catheters, it is desirable to detect beta radiation, due to its short path length of only a few millimeters, rather than gamma and x-radiation, which have much longer path lengths and could increase undesirable background signal from tissue beyond the coronary adventitia. The distal ends of these intravascular catheters have typically utilized plastic scintillating fibers for beta detection, resulting in improved sensitivity radionuclide detection over noninvasive nuclear imaging techniques.

Intravascular positron detection alone, however, suffers from a lack of anatomical information. Therefore, to compliment the functional information obtained from arterial lesions, a dual positron detection and OCT probe has been developed (Piao et al. 2005). The hybrid probe was demonstrated on *ex vivo* arteries from animal models of atherosclerosis, demonstrating the complimentary functional and microstructure information offered by the two modalities. Despite this initial demonstration of the feasibility to combine these two modalities, intravascular positron detection has not continued to make advances toward a clinical catheter, likely as a result of its limited resolution. Additional challenges include the need for improved catheter design and the optimization of radionuclide delivery and detection protocols, owing to their limited half-life and inherent circulation times.

17.2.7 MERGING INTRAVASCULAR IMAGING WITH CORONARY ANGIOGRAPHY

In addition to the aforementioned combinations of intravascular imaging techniques for complimentary characterization of coronary morphology and composition, significant effort has also focused on the integration of intravascular modalities with noninvasive imaging techniques that enable broader visualization of the coronary tree. For example, methods for integrating IVUS images with both x-ray angiography (Klein et al. 1992; Reiber et al. 2011) and computed tomographic coronary angiography (van der Giessen et al. 2010) have been developed to enable evaluation of the arterial wall morphology within the context of the broader coronary geometry. Despite the existence of well-developed image processing algorithms for combining these invasive and noninvasive modalities, such systems are not used for general clinical applications and remain restricted to research applications (Bourantas et al. 2013). Similarly, the combination of OCT with coronary angiography is also in development, including a hybrid quantitative coronary angiography (QCA) and OCT technique, as demonstrated by Figure 17.7 (Tu et al. 2011). Reconstruction and coregistration of images from the two modalities are accomplished using biplane QCA for vessel reconstruction, followed by anatomical landmarks to overlay the OCT image (Bourantas et al. 2013). While its future application as a clinical

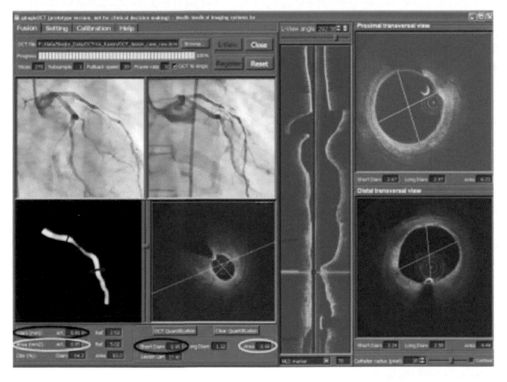

Figure 17.7 X-ray angiography-OCT coregistration and quantification. A sidebranch was manually identified from both x-ray angiography and OCT images to be used as a baseline position to register the two imaging modalities. After the registration, the markers superimposed in the OCT longitudinal view were synchronized with the same markers in the x-ray angiography views. The OCT measurements could be compared with 3-D QCA at the same position. (Reprinted from Tu S et al., *Int J Cardiovasc Imaging*, 27, 197–207, 2011.)

technique remains to be seen, the hybrid OCT/QCA technique is actively being utilized as a research platform to evaluate the accuracy of angiographic measurements (Tsuchida et al. 2007) and to investigate the role of vascular shear and tensile stress on atherosclerotic plaque destabilization (Bourantas et al. 2012).

17.2.8 PROSPECTIVE OUTLOOK FOR HYBRID INTRAVASCULAR OPTICAL IMAGING TECHNIQUES

The wide array of intravascular imaging modalities, and hybrid techniques thereof which are either already approved for clinical use or under active development, are motivated by the need for improved diagnostic capabilities to help identify high-risk atherosclerotic lesions and subsequently guide PCIs. The detection of high-risk atherosclerotic plaques remains an ongoing challenge and one that is inadequately addressed using conventional clinical imaging modalities, as was demonstrated through the results of the Providing Regional Observations to Study Predictors of Events in the Coronary Tree study (Stone et al. 2011; Fleg et al. 2012).

Each of the individual modalities discussed herein offers unique advantages for intravascular imaging despite their individual limitations, as highlighted in Table 17.1. Hybrid imaging techniques seek to intelligently combine the advantages of multiple modalities to yield improved diagnostic potential, with a central focus on identifying atherosclerotic plaque features, both morphological and compositional, which are known to correlate with lesion vulnerability. While IVUS, OCT, and NIRS are currently the only clinically approved intravascular modalities, and IVUS/NIRS is the only approved hybrid intravascular technique, commercialization of one or more of the other emerging techniques is likely in coming years. In order for this to occur, however, several broad challenges must be addressed. For each of the preclinical hybrid intravascular techniques discussed, the combined imaging systems and especially the integrated catheters must be further refined and demonstrated to be safe and effective for *in vivo* diagnoses (Bourantas et al. 2013). If this is achieved, demonstration of the true clinical impact of each particular hybrid technique will be dependent on the results of large-scale clinical trials, such as those currently under way utilizing the NIRS/IVUS hybrid system, and the widespread transition from existing technologies to hybrid systems is likely contingent on results of such trials.

17.3 NONINVASIVE COMBINED US AND PA CAROTID ARTERY IMAGING

Carotid atherosclerosis has been widely studied as a result of both its proximity to the skin surface and, more importantly, its established correlation with cerebrovascular events (Dempsey et al. 2010) and coronary atherosclerosis (Wofford et al. 1991). Measurement of the carotid artery intima-media thickness via B-mode US imaging was introduced as early as the mid-1980s (Pignoli et al. 1986) as a means of assessing atherosclerotic burden and is commonly utilized both for clinical assessment of patient risk and as an end point for clinical trials. Subsequently, variability in discrete echogenic and echolucent lesions identified in carotid artery US images was demonstrated as a method for identifying specific plaque types associated with elevated incidence of embolic brain infarcts (Kakkos et al. 2007). Drawing from these findings, PA imaging was introduced to compliment US assessment of the carotid artery plaques. For example, a novel approach combining US, strain rate, and PA imaging for assessment of diseased carotid arteries was introduced using excised arteries within a pulsatile bench-top setup (Graf et al. 2011, 2012). Using strain rate imaging provides a method of interrogating the vessel wall mechanical properties based on a cross-correlation algorithm measuring displacement generated across the arterial wall during pulsatile flow. Thus, the triple-modality approach utilizes shared instrumentation to provide complimentary information of the carotid artery's structural, compositional, and mechanical characteristics. The feasibility of *in vivo* PA imaging for carotid artery assessment was subsequently demonstrated and shown to be more effective using a curved array transducer for PA detection, although this application did not integrate US imaging (Dima and Ntziachristos 2012). Together, these preliminary studies support the continued investigation into the potential added benefit that PA imaging could provide in the noninvasive evaluation of carotid arteries.

Table 17.1 Advantages and limitation of available and emerging hybrid intravascular imaging techniques

Hybrid intravascular optical imaging modality	Representation of vessel geometry	Visualization of the distribution of plaque	Imaging and quantification of atheroma	Plaque characterization	Identification of features associated with increased vulnerability	Detection of the culprit lesion	Detection of vessel wall inflammation
IVUS/OCT	−	−	+++	+++	+++	+++	+
NIRS/IVUS	−	−	+++	++	++	+	−
NIRS/OCT	−	−	+++	+++	++	+++	+
TRFS/IVUS	−	−	+++	++	++	+	++
NIRF/OCT	−	−	+	++	++	+++	+++
IVUS/IVPA	−	−	+++	++	++	+	++
PET/OCT	−	−	+++	++	++	+++	++
IVUS/Angio	+++	+++	+++	+	+	+	−
OCT/Angio	+++	+	+	++	++	+++	+
NIRS/IVUS/Angio	+++	+++	+++	++	++	+	−

Source: Modified from Bourantas CV, Garcia-Garcia HM, Naka KK, Sakellarios A, Athanasiou L, Fotiadis DI, Michalis LK, Serruys PW, J Am Coll Cardiol, 61, 1369–1378, 2013. With permission.

17.4 CARDIOVASCULAR APPLICATIONS OF HYBRID OPTICAL IMAGING USING MULTIMODAL CONTRAST AGENTS

Hybrid optical imaging techniques combining fluorescence or bioluminescence imaging (BLI) with other molecular and anatomical imaging modalities have been widely investigated in recent years in an attempt to further expand the capabilities for monitoring pathophysiological changes relevant to cardiovascular medicine. Such techniques typically utilize contrast agents designed to provide molecular specificity for at least one imaging modality. Several examples of molecularly targeted, dual- or triple-modality contrast agents have been synthesized and are under investigation for applications in monitoring cardiovascular pathologies. However, due in large part to the use of investigational contrast agents, the majority of such hybrid imaging approaches are limited to preclinical investigations, particularly on small-animal models.

Due to the unique size considerations of small animal models, several imaging modalities using hybrid optical technologies are currently available. Not only does the size of small animal models (on the order of 100 g body weight for murine models) allow for improved optical delivery for deep tissue targets, important for cardiac applications, but also the constrained size allows for sophisticated tomographic approaches, which may be more difficult to implement clinically or on larger animal models. Given this advantage, several commercially available and experimental systems coupling fluorescence, bioluminescence, or PA imaging with an additional imaging modality capable of providing anatomical information have been utilized for preclinical cardiovascular research.

17.4.1 COMBINED X-RAY COMPUTED TOMOGRAPHY AND FLUORESCENCE MOLECULAR TOMOGRAPHY

Fluorescence molecular tomography (FMT) is an optical imaging procedure that enables three-dimensional mapping and quantification of exogenous near-infrared fluorescent molecule distribution through the use of multiple source/detector pairs surrounding the image target. FMT is a powerful tool for *in vivo* small-animal imaging but is incapable of being simply scaled up for optical imaging of large animals or humans due to limitations in imaging depth. Despite its high sensitivity and quantitative capabilities, FMT additionally suffers from a lack of anatomical context. To alleviate this limitation, it has been combined with x-ray computed tomography (CT) imaging using fiducial markers in the animal/sample housing that can be readily identified in both the CT and FMT images (Nahrendorf et al. 2009, 2010; Leuschner et al. 2010; Panizzi et al. 2010; Leuschner and Nahrendorf 2011). This combination has been utilized to enable noninvasive *in vivo* assessment of cardiovascular disease progression, including the monitoring of protease activity in murine atherosclerosis, as shown in Figure 17.8 (Nahrendorf et al. 2009), and cellular and molecular interactions in response to myocardial infarction (Leuschner et al. 2010; Panizzi et al. 2010). Together, these results are indicative of the power of FMT-CT imaging as a platform for investigating cardiovascular pathologies and injury responses at the molecular level in small animals.

17.4.2 COMBINED OPTICAL AND MRI

Recent work has also combined fluorescence tomography and MRI through the use of a dual-contrast magnetofluorescent nanoparticle, CLIO-Cy5.5, which was demonstrated on a mouse myocardial infarct model (Sosnovik et al. 2005, 2007). The study demonstrated that macrophage uptake of CLIO-Cy5.5 48 hours post left coronary artery ligation could be well visualized using both MRI and FMT techniques, with approximately a fourfold increase in signal over the control, which was corroborated by fluorescence microscopy. The hybridization of these two imaging modalities is analogous to the more common fusion of PET/SPECT and x-ray CT imaging, where functional information obtained from PET/SPECT imaging can be merged with anatomical information from x-ray CT. In the case of the optical/MRI hybrid technique, FMT provides high-sensitivity contrast agent detection while MRI is able to provide greater surrounding anatomical context.

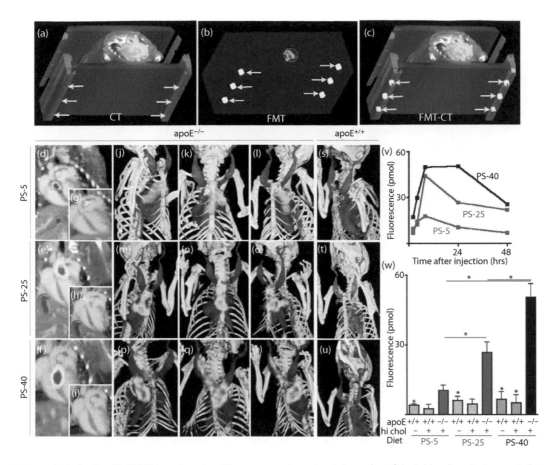

Figure 17.8 *In vivo* FMT-CT imaging. (a–c) Image coregistration is based on fiducial landmarks (arrows) that are incorporated into the animal holder and are identifiable on CT (a) and FMT (b). The software co-aligns these fiducials to create a hybrid data set (c). Fluorescence signal in the aortic root of an apoE$^{-/-}$ mouse is encircled. (d–f) Two-dimensional (2-D) FMT-CT long axis views of apoE$^{-/-}$ mice injected with respective protease sensors. Fluorescence signal is observed in the aortic root and arch, regions with high plaque load and high *ex vivo* fluorescence signal. (g–i) CT-only views of d–f. Arrow heads depict vascular calcification, likely colocalizing with plaques. (j–r) Three-dimensional (3-D) maximum intensity projection of hybrid data sets show skeletal and vascular anatomy and the distribution of fluorescence signal. Most signal is observed in the root and arch; however, (q) and (r) show additional activation of the protease sensor in the carotid artery, also a region predisposed to atheroma build-up in this model. (s–u) FMT-CT after injection of respective sensor into wild type mice. (v) Fluorochrome concentration reflecting protease activity plotted over time. (w) Protease activity 24 hours after injection of sensors. *$p < 0.05$. (Reprinted from Nahrendorf M et al., *Arterioscler Thromb Vasc Biol*, 29, 1444–1451, 2009. With permission.)

Whereas other modalities may use a unified system to register images from the two modalities, image coregistration in FMT/MRI is typically obtained using fiducials as well as the dual-contrast magnetofluorescent probe that colocalizes the fluorescence and MRI signal.

This hybrid contrast agent and multimodal imaging approach has introduced a promising platform to develop the role of FMT and magneto-fluorescent nanoparticles probes for *in vivo* cardiac imaging. Further development and evaluation has led to the simultaneous use of AnxCLIO-Cy5.5 as an apoptosis indicator (Schellenberger et al. 2004) and a novel gadolinium chelate, Gd-DTPA-NBD, as an indicator of necrosis (Sosnovik et al. 2009). The Gd-DTPA-NBD indicator was demonstrated to demarcate lesions of cardiomyocyte necrosis using *ex vivo* delayed enhancement MRI and confirmed with fluorescence reflectance imaging (FRI) imaging in a mouse ligation model, as shown in Figure 17.9. Dual-modality imaging of mice who received the cocktail injection following myocardial infarction identified large regions of apoptotic but viable

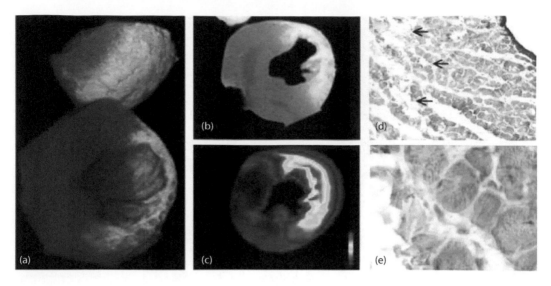

Figure 17.9 Delayed enhancement MRI of Gd-DTPA-NBD in an infarcted mouse heart. The heart was excised 20 minutes after the injection of Gd-DTPA-NBD, bisected and imaged with a T1-weighted 3-D gradient echo sequence. (a) A volume rendered image and (b) a 2-D short axis reconstruction at midventricular level are shown. Accumulation of Gd-DTPA-NBD is seen in the territory of the ligated left coronary artery (anterior wall to inferolateral wall and most of the apex). (c) FRI of Gd-DTPA-NBD shows that the detection of the agent by MRI and by fluorescence imaging correspond very well. (d) Magnification 100× and (e) 400× immunohistochemistry for Gd-DTPA-NBD at the border zone of the infarct shows that the agent accumulates in areas of cardiomyocyte degeneration and expansion of the extracellular space. No accumulation of the agent is seen in areas with intact cardiomyocyte cell membranes. (Reprinted from Sosnovik DE et al., *Circ Cardiovasc Imaging* 2, 6, 460–467, 2009. With permission.)

cardiomyocytes within 4–6 hours following ischemic injury. Such findings have implications for potential clinical treatment strategies, and further development and characterization of AnxCLIO-Cy5.5 contrast agents are ongoing (Chen, Josephson, and Sosnovik 2011).

An additional nanoparticle platform that is designed to mimic endogenous high-density lipoprotein has recently been reported (Cormode et al. 2008; Fay et al. 2013). Consisting of a phospholipid-coated, metallic core structure, the platform has been used to incorporate CT, fluorescence, and MRI contrast agents for multimodal imaging. Because of its size and structural similarity to endogenous high-density lipoprotein, this multifunctional nanoparticle platform offers a clear potential for imaging atherosclerotic plaques (Skajaa et al. 2011). One additional future consideration for this hybrid technique is the potential for both MRI and fluorescence imaging techniques to be performed using intravascular catheters (Ferrari and Wilensky 2007). While the two intravascular techniques have not been combined for hybrid imaging at this stage, their continued development may provide an avenue for the clinical translation of hybrid optical MRI techniques for intravascular applications such as the use of the lipoprotein mimetic contrast agent platform.

In addition to contrast agents combining MRI and fluorescence imaging, BLI combined with a modality providing anatomical/structural imaging such as MRI represents a promising hybrid imaging application to track engrafted cell therapies. Recent work has been done tracking embryonic stem cell treatment response in a murine myocardial infarction model utilizing specialized embryonic stem cells expressing firefly luciferase and tagged with superparamagnetic iron oxide (SPIO) MRI contrast agents (Stone et al. 2011; Zhang et al. 2011b). This methodology represents an intriguing application of dual-contrast imaging probes, which are separable following apoptosis of the engrafted embryonic stem cells by macrophages localizing at the infarct region. Animals that exhibited an improved treatment response, as measured by histology, correlated well to those animals exhibiting SPIO-based contrast on MRI and increased BLI signal, whereas those that exhibited only SPIO-based MRI contrast (in which the embryonic stem cells largely died off and the SPIO contrast probe was phagocytosed by macrophages) showed no significant benefit from treatment. Thus, treatment outcome could be well predicted from combined MRI and BLI imaging using labeled embryonic stem cells.

17.4.3 COMBINED NUCLEAR AND OPTICAL IMAGING

The intelligent design and synthesis of molecularly targeted tracers combining radionuclide- and fluorescence-based contrast for hybrid PET/optical imaging applications have been the focus of significant preclinical research in recent years. Such multimodality contrast agents offer the potential for clinical translation to capitalize on the high-sensitivity molecular information that they provide to help improve diagnoses and guide therapies. Clinically, using a hybrid PET/optical contrast agent and imaging approach, one can envision the initial whole-body PET imaging, followed by the guidance of surgical intervention using fluorescence imaging techniques (Ghosh et al. 2013). For example, in one scenario, dual-modality contrast agents could be utilized to reveal atherosclerotic plaque inflammation to guide the decision of whether or not intervention is appropriate, immediately followed by the use of the fluorescence provided by the same particles to help guide the intervention with a fluorescence imaging intravascular catheter (Majmudar and Nahrendorf 2012). However, the clinical translation of such hybrid PET/optical imaging applications relies on the further characterization of the multimodal contrast agents and their clinical approval.

At this stage, the hybrid approach can also help with the preclinical evaluation of PET imaging agents on small animal models through the reduction in required equipment for evaluation of contrast agent targeting efficacy (Nahrendorf et al. 2010). Given these advantages, it has been shown that FMT measurements using multifunctional contrast agents show high correlation as well as seamless integration and visualization with PET/CT images obtained from both phantoms and small-animal models (Nahrendorf et al. 2010). Given this complimentary contrast between hybrid PET/fluorescence techniques for small-animal imaging, and its potential long-term clinical impact, numerous investigational multifunctional contrast agents have been developed, many with direct applications to the diagnosis and monitoring of cardiovascular pathologies.

Investigational dual-labeled contrast agents combining fluorescence and nuclear imaging have been developed for a wide range of cellular and molecular targets. For example, a dextran-coated magnetofluorescent nanoparticle labeled with the radionuclide ^{64}Cu was utilized for the imaging of macrophages in inflammatory atherosclerosis in mice using PET, MRI, and FMT (Nahrendorf et al. 2008). The study found that elevated signals from the accumulation of contrast agents observed within the aortic root and arch exhibited a target-to-background ratio of 5.1, and accumulated doses within the aortas and carotid arteries were found to be 260% and 392% greater in the atherosclerotic mice than in wild-type controls. A similar platform, comprising the radionuclide ^{18}F-CLIO and a near-infrared fluorochrome, was utilized for the detection of macrophages within atherosclerotic mice aortic aneurysms (Nahrendorf et al. 2011). Signals obtained from the aneurysms were found to be higher than both wild type aorta and atherosclerotic plaques. Dual-modality micelles labeled with the radionuclide ^{111}In and near-infrared fluorescent Cy7-like dye and conjugated with Annexin V, which specifically binds to phosphatidylserine residues that are exposed on the cell surface during apoptosis, have also been demonstrated for the hybrid imaging of apoptosis (Zhang et al. 2011a). Annexin V-labeled, multimodality contrast agents represent a natural extension of a nuclear imaging targeting strategy, which has been widely utilized to investigate apoptosis in cardiac tissue (Narula et al. 2001; Kietselaer et al. 2004, 2007).

Peptides combined with radionuclides and optical dyes or fluorophores have also been investigated as molecularly targeted multifunctional nanoparticles (Azhdarinia et al. 2012). For example, a contrast agent comprising the near-infrared dye IRDye 800CW and the radionuclide ^{111}In was conjugated with a peptide containing the commonly utilized arginine-glysine-aspartic acid (RGD) sequence (Houston et al. 2005). RGD serves as a ligand for $\alpha_v\beta_3$ integrins, which are known to be overexpressed in neovasculature, including many tumors as well as the vasa vasorum of advanced atherosclerotic plaques. A further development of multifunctional $\alpha_v\beta_3$ integrin-targeted contrast agents comprised of a near-infrared fluorescent dye, radionuclide, and a cyclic RGD peptide were synthesized and evaluated *in vivo* using fluorescence imaging, gamma scintigraphy, SPECT, as well as BLI imaging of coinjected luciferin (Edwards et al. 2009). Silica nanoparticle-based, RGD peptide-conjugated contrast agents labeled with Cy5 fluorophores and ^{124}I radionuclides have also been synthesized (Benezra et al. 2011). Additional dual-modality platforms combining *in vivo* PET imaging, near-infrared fluorescence, and targeted peptides have been demonstrated for molecular imaging of matrix metalloproteinase-9 (Azhdarinia et al. 2011; Rodenberg et al. 2011) and interlukin-11 (Wang et al. 2007).

While these agents were not directly applied for cardiovascular imaging, the molecular targets evaluated using the hybrid nuclear/optical imaging are known to be of relevance in the progression of atherosclerosis and other cardiovascular pathologies.

The continued rapid development of multimodality contrast agents, along with the characterization of their functionality and *in vivo* utility, is expected to continue within the preclinical setting. Modifying their surface chemistry to optimize circulation times and redesigning their conjugation to render the attached optical contrast components activatable upon interaction with the targeted biomolecules represent two likely areas of further improvement (Sinusas et al. 2008). Furthermore, the intelligent incorporation of pharmaceutical agents as part of the contrast agent platform offers an additional area of further development (Wickline and Lanza 2003). Such a platform enables multimodality imaging to confirm and quantify therapeutic delivery. In addition to these ongoing preclinical refinements in the design of multimodal, hybrid optical imaging agents, there remains a long clinical approval process, which will need to be first motivated and then successfully completed before they can ever be effectively translated beyond platforms for preclinical research.

17.4.4 PA IMAGING OF CARDIAC TISSUE

The reliable visualization of tissue hemodynamics can provide a wealth of functional information on the target tissue. As previously discussed, PA imaging can utilize near-infrared light to probe the optical absorption of oxy- and deoxy-hemoglobin, the primary chromophores within most tissues, as well as highly absorbing exogenous contrast agents. Through the use of sPA imaging and spectral unmixing techniques, different chromophores can be individually identified to obtain molecular images at depths well beyond the ballistic photon regime.

By using single-wavelength and multispectral PA imaging, it has been shown that PA signal changes resulting from a bolus injection of high-contrast plasmonic gold nanorods (AuNR) can be tracked in near real-time (Taruttis et al. 2010). In particular, it was demonstrated than the single wavelength PA signal originating from the murine jugular vein approximately doubles over the 90 s time frame following AuNR injection. The spectral profile of the absorbers and verification of its correspondence to the expected absorption profile of the injected AuNRs were determined by sPA imaging, allowing the near real-time tracking and identification of changes in exogenous contrast agents at tissue depths of approximately 5 mm. This same imaging system was later used to image macrophage infiltration 48 hours post infarction in another murine model (Taruttis et al. 2013). Additionally, indocyanine green, a near-infrared absorbing organic dye, was conjugated with a dendritic polyglycerol sulfate (dPGS), a targeting agent with demonstrated accumulation in inflamed tissues (Licha et al. 2011). By using sPA imaging in conjunction with PA tomographic approaches, leukocyte uptake of the dPGS-indocyanine green (ICG) probe was demonstrated to be localized in the infarct region at 2 hours post injection.

An additional intriguing application of PA imaging is vibrational PA bond imaging, a term given to the application of sPA at optical wavelengths from approximately 1600–1800 nm for detection of overtones of the molecular bond vibrations for identification of biomolecules of interests. This wavelength range corresponds to a local minimum in the absorption spectrum of water and may represent a second "optical window" in the infrared regime. It was shown, through careful wavelength selection, that fat could be distinguished from proteins using sIVPA imaging in phantom studies (Wang et al. 2012c, 2012d), and the ability to image lipid within this wavelength range using IVPA has been heavily investigated (Wang et al. 2012a, 2012b; Jansen et al. 2013). While not yet adapted to tissue or small animal models, likely due to challenges associated with the high optical attenuation in water at these longer wavelengths, this technique represents a potentially promising new optical regime for hybrid optical imaging to monitor lipid distribution.

17.5 CARDIOVASCULAR APPLICATIONS OF HYBRID NLOM TECHNIQUES

NLOM techniques have recently been applied to help characterize arterial or myocardial tissues using ultra-fast laser excitation to exploit several different nonlinear optical effects, which, in turn, can be utilized to generate high-contrast images of specific molecular structures without the need for immunohistochemical staining procedures and at greater depths than can be achieved using linear optical microscopy techniques. These nonlinear techniques include two-photon-excited fluorescence microscopy (TPEF), sum-frequency generation (SFG), coherent anti-Stokes Raman spectroscopy (CARS), and stimulated Raman scattering (SRS). TPEF relies on the fluorescence from either exogenous or endogenous tissue fluorophores generated upon simultaneous absorption of two photons. SFG refers to a nonlinear scattering process in which two photons interact to produce a single emitted photon at a higher frequency, or shorter optical wavelength, than the two original photons. A special case of SFG, second-harmonic generation (SHG) refers to the process in which the two incident photons are of the same frequency and the resulting photon possessing double their frequency. Both CARS and SRS are label-free nonlinear optical techniques, which enable imaging of tissues based on intrinsic molecular vibrational contrast with significantly higher signal than is achievable using spontaneous Raman spectroscopy.

Different biological molecules within tissues intrinsically produce signals using different nonlinear techniques. For example, fibrillar collagen produces SHG, TPEF can excite autofluorescence of elastin, and CARS can be used to visualize lipid. Therefore, the combination of NLOM techniques provides an avenue for label-free, molecular-specific imaging of disease biomarkers relevant to both atherosclerotic plaques and the more general imaging of myocardial tissue. For example, a combination of SHG to image collagen with TPEF to image elastin was used as a means of imaging the collagen network within the fibrous cap and to differentiate the fibrous cap from underlying layers of atherosclerotic plaques (Lilledahl et al. 2007). The assessment of arterial cells and surrounding extracellular matrix was similarly shown to enable visualization of cell membranes, elastin, and collagen, based on images obtained using CARS, SFG, and TPEF, respectively (Wang, Le, and Cheng 2008). SRS was also combined with SHG microscopy as a hybrid imaging approach to detect cholesterol crystals in atherosclerotic plaques (Suhalim et al. 2012).

These hybrid NLOM techniques are being utilized to study structural and biochemical changes associated with cardiovascular disease progression. For example, features derived from multiple NLOM setups have been used to perform label-free imaging of atherosclerotic arterial tissue, demonstrating their ability to identify atherosclerotic plaque characteristics and classify their progression, as demonstrated by Figure 17.10 (Ko et al. 2010, 2011, 2012; Mostaço-Guidolin et al. 2010, 2011). These techniques have also been employed for the assessment of arterial changes induced following stenting of atherosclerotic pig models (Wang et al. 2011)

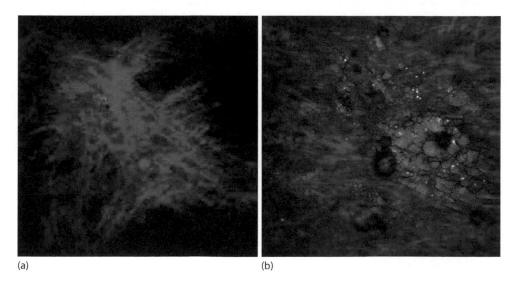

(a) (b)

Figure 17.10 Nonlinear optical images of an advanced plaque obtained at approximate depths of (a) 10 μm and (b) 60 μm from the lumen surface. Blue SHG fibrillary collagen type-I, red/orange CARS lipid-rich structure buried under a fibrous cap. (With kind permission from Springer Science+Business Media: *Biophys Rev*, Nonlinear optical microscopy in decoding arterial disease, 4, 2012, 323–334, Ko A, Risdale A, Mostaco-Guidolin L, Major A, Stolow A, Sowa M.)

as well as the impact of diet on lipid accumulation and macrophage infiltration in atherosclerotic mice models (Lim et al. 2010). Similarly, TPEF and SHG have been utilized to characterize an increase in fibrosis in an animal model of myocardial infarction (Caorsi et al. 2013). Together, these demonstrations of hybrid NLOM techniques for high-contrast imaging of specific biomolecules provide a large opportunity to further characterize the biochemical changes that occur during the progression of cardiovascular pathologies as well as the molecular-specific changes that are induced during their treatment.

17.6 CONCLUSIONS

Cardiovascular imaging has conventionally centered on the morphological assessment of cardiac and vascular tissue. However, improved fundamental understanding of cardiovascular pathologies has elucidated the need to introduce molecular imaging techniques into cardiovascular medicine for improved diagnoses and monitoring of therapeutic response. High-sensitivity molecular imaging techniques, however, tend to provide poor anatomical context to quantify and analyze the location and significance of detected signal. Therefore, hybrid imaging techniques capable of providing the ability to characterize and quantify both morphology and composition are of great interest for cardiovascular imaging applications. A wide array of hybrid optical techniques has been developed and applied for the characterization of heart and vascular pathologies.

Numerous hybrid intravascular imaging modalities are being developed and applied for improved characterization of coronary atherosclerosis. Histopathological studies have shown that thin-cap fibroatheroma, so-called vulnerable plaques, are characterized by a combination of morphological (e.g., fibrous cap thickness <65 um, large plaque volume) and compositional (e.g., lipid pool, macrophage infiltration) characteristics. To this end, and given the limited required penetration depth for coronary artery imaging, hybrid intravascular optical imaging techniques are currently under development to provide a complimentary assessment of coronary anatomy and composition. These hybrid techniques include the clinically approved combined NIRS/IVUS, as well as a multitude of additional pairings, which are currently under development, including combinations of intravascular OCT, Raman spectroscopy, fluorescence imaging techniques, and PA imaging. While each of these hybrid intravascular imaging approaches offers the potential for improved characterization of arterial lesions, the extent of their clinical impact is yet to be determined and will largely remain uncertain until the completion of large-scale clinical trials shed light on their diagnostic advantages.

In addition to intravascular imaging, noninvasive hybrid optical techniques have been widely utilized for preclinical assessment of cardiovascular pathologies, particularly for *in vivo* imaging of small animals. Molecularly targeted hybrid contrast agents have enabled applications combining optical modalities with x-ray CT imaging, MRI, and nuclear imaging as well as contrast-enhanced PA imaging. These intelligently designed probes have enabled complimentary, whole-body anatomical and high-contrast imaging of such targets as apoptotic cells and molecular indicators of myocardial damage or atherosclerotic plaque risk. Lastly, the combination of multiple NLOM techniques enables the label-free imaging of biomolecular structures for unprecedented high-sensitivity and high-resolution tissue analysis.

In summary, hybrid optical techniques offer the potential for high-sensitivity molecular imaging within the context of the underlying anatomy. The continued development of integrated systems promises to help advance preclinical research to better understand cardiovascular pathologies and responses to investigational therapies as well as to potentially improve upon the current gold standard techniques for intravascular imaging and classification of atherosclerotic plaques in the coming years.

ABBREVIATIONS

ADC	Analog-to-digital converter
AuNR	Gold nanorod
BLI	Bioluminescence imaging
CARS	Coherent anti-stokes Raman spectroscopy

CAS	Coronary angioscopy
CTCA	Computed tomographic coronary angiography
CW	Continuous wave
dPGS	Dendritic polyglycerol sulfate
FDA	Food and Drug Administration
FLIM	Fluorescence lifetime imaging
FMT	Fluorescence molecular tomography
FRI	Fluorescence reflectance imaging
IB-IVUS	Integrated backscatter intravascular ultrasound
IVPA	Intravascular photoacoustic
IVUS	Intravascular ultrasound
LDL	Low-density lipoprotein
MNP	Magneto-fluorescent nanoparticles
MRI	Magnetic resonance imaging
NIRF	Near-infrared fluorescence
NIRS	Near-infrared spectroscopy
NLOM	Nonlinear optical microscopy
OCT	Optical coherence tomography
PA	Photoacoustic
PCI	Percutaneous coronary intervention
PET	Positron emission tomography
QCA	Quantitative coronary angiography
RGD	Arginine-glysine-aspartic acid peptide sequence
SFG	Sum-frequency generation
SHG	Second-harmonic generation
sIVPA	Spectroscopic intravascular photoacoustic
SNR	Signal-to-noise ratio
sPA	Spectroscopic photoacoustic
SPECT	Single-photon emission computed tomography
SPIO	Superparamagnetic iron oxide
SRS	Stimulated Raman scattering
TCFA	Thin-cap fibroatheroma
TPEF	Two-photon enhance fluorescence
TRFS	Time-resolved fluorescence spectroscopy
US	Ultrasound
VH-IVUS	Virtual histology intravascular ultrasound

REFERENCES

Allen TJ, A Hall, AP Dhillon, JS Owen, and PC Beard. 2012. Spectroscopic photoacoustic imaging of lipid-rich plaques in the human aorta in the 740 to 1400 nm wavelength range. *J Biomed Opt* 17 (6):061209.

Azhdarinia A, P Ghosh, S Ghosh, N Wilganowski, and EM Sevick-Muraca. 2012. Dual-labeling strategies for nuclear and fluorescence molecular imaging: A review and analysis. *Mol Imaging Biol* 14 (3):261–76.

Azhdarinia A, N Wilganowski, H Robinson, P Ghosh, S Kwon, ZW Lazard, AR Davis, E Olmsted-Davis, and EM Sevick-Muraca. 2011. Characterization of chemical, radiochemical and optical properties of a dual-labeled MMP-9 targeting peptide. *Bioorg Med Chem* 19 (12):3769–76.

Baraga JJ, MS Feld, and RP Rava. 1992. In situ optical histochemistry of human artery using near infrared Fourier transform Raman spectroscopy. *Proc Natl Acad Sci U S A* 89 (8):3473–7.

Baraga JJ, RP Rava, P Taroni, C Kittrell, M Fitzmaurice, and MS Feld. 1990. Laser induced fluorescence spectroscopy of normal and atherosclerotic human aorta using 306–310 nm excitation. *Lasers Surg Med* 10 (3):245–61.

Barton JK, F Guzman, and A Tumlinson. 2004. Dual modality instrument for simultaneous optical coherence tomography imaging and fluorescence spectroscopy. *J Biomed Opt* 9 (3):618–23.

Bec J, DM Ma, DR Yankelevich, J Liu, WT Ferrier, J Southard, and L Marcu. 2014. Multispectral fluorescence lifetime imaging system for intravascular diagnostics with ultrasound guidance: *In vivo* validation in swine arteries. *J Biophotonics* 7 (5):281–5.

Bec J, H Xie, DR Yankelevich, F Zhou, Y Sun, N Ghata, R Aldredge, and L Marcu. 2012. Design, construction, and validation of a rotary multifunctional intravascular diagnostic catheter combining multispectral fluorescence lifetime imaging and intravascular ultrasound. *J Biomed Opt* 17 (10):106012.

Benezra M, O Penate-Medina, PB Zanzonico, D Schaer, H Ow, A Burns, E DeStanchina, V Longo, E Herz, S Iyer, J Wolchok, SM Larson, U Wiesner, and MS Bradbury. 2011. Multimodal silica nanoparticles are effective cancer-targeted probes in a model of human melanoma. *J Clin Invest* 121 (7):2768–80.

Bourantas CV, HM Garcia-Garcia, KK Naka, A Sakellarios, L Athanasiou, DI Fotiadis, LK Michalis, and PW Serruys. 2013. Hybrid intravascular imaging: Current applications and prospective potential in the study of coronary atherosclerosis. *J Am Coll Cardiol* 61 (13):1369–78.

Bourantas CV, MI Papafaklis, KK Naka, VD Tsakanikas, DN Lysitsas, FM Alamgir, DI Fotiadis, and LK Michalis. 2012. Fusion of optical coherence tomography and coronary angiography—*In vivo* assessment of shear stress in plaque rupture. *Int J Cardiol* 155 (2):e24–6.

Brennan III, JF, J Nazemi, J Motz, and S Ramcharitar. 2008. The vPredict™ Optical Catheter System: Intravascular Raman Spectroscopy. *EuroIntervention* 3 (5):635–8.

Brennan JF, TJ Römer, RS Lees, AM Tercyak, JR Kramer, and MS Feld. 1997. Determination of human coronary artery composition by Raman spectroscopy. *Circulation* 96 (1):99–105.

Brugaletta S, HM Garcia-Garcia, PW Serruys, S de Boer, J Ligthart, J Gomez-Lara, K Witberg, R Diletti, J Wykrzykowska, R-J van Geuns, C Schultz, E Regar, HJ Duckers, N van Mieghem, P de Jaegere, SP Madden, JE Muller, AFW van der Steen, WJ van der Giessen, and E Boersma. 2011. NIRS and IVUS for characterization of atherosclerosis in patients undergoing coronary angiography. *JACC Cardiovasc Imaging* 4 (6):647–55.

Buschman HP, ET Marple, ML Wach, B Bennett, TCB Schut, HA Bruining, AV Bruschke, A van der Laarse, and GJ Puppels. 2000. *In vivo* determination of the molecular composition of artery wall by intravascular Raman spectroscopy. *Anal Chem* 72 (16):3771–5.

Caorsi V, C Toepfer, MB Sikkel, AR Lyon, K MacLeod, and MA Ferenczi. 2013. Non-linear optical microscopy sheds light on cardiovascular disease. *PloS One* 8 (2):e56136.

Caplan JD, S Waxman, RW Nesto, and JE Muller. 2006. Near-infrared spectroscopy for the detection of vulnerable coronary artery plaques. *J Am Coll Cardiol* 47 (8 Suppl):C92–6.

Cassis LA, and RA Lodder. 1993. Near-IR imaging of atheromas in living arterial tissue. *Anal Chem* 65 (9):1247–56.

Chapman I. 1968. Relationships of recent coronary artery occlusion and acute myocardial infarction. *J Mt Sinai Hosp N Y* 35 (2):149–54.

Chau AH, JT Motz, JA Gardecki, S Waxman, BE Bouma, and GJ Tearney. 2008. Fingerprint and high-wavenumber Raman spectroscopy in a human-swine coronary xenograft *in vivo*. *J Biomed Opt* 13 (4):040501.

Chen HH, L Josephson, and DE Sosnovik. 2011. Imaging of apoptosis in the heart with nanoparticle technology. *Wiley Interdiscip Rev Nanomed Nanobiotechnol* 3 (1):86–99.

Chen J, CH Tung, U Mahmood, V Ntziachristos, R Gyurko, MC Fishman, PL Huang, and R Weissleder. 2002. *In vivo* imaging of proteolytic activity in atherosclerosis. *Circulation* 105 (23):2766–71.

Choudhury RP, V Fuster, and ZA Fayad. 2004. Molecular, cellular and functional imaging of atherothrombosis. *Nat Rev Drug Discov* 3 (11):913–25.

Constantinides P 1967. Pathogenesis of cerebral artery thrombosis in man. *Arch Pathol* 83 (5):422–8.

Cormode DP, T Skajaa, MM van Schooneveld, R Koole, P Jarzyna, ME Lobatto, C Calcagno, A Barazza, RE Gordon, P Zanzonico, EA Fisher, ZA Fayad, and WJ Mulder. 2008. Nanocrystal core high-density lipoproteins: A multimodality contrast agent platform. *Nano Lett* 8 (11):3715–23.

Dana N, L Di Biase, A Natale, S Emelianov, and R Bouchard. 2014. *In vitro* photoacoustic visualization of myocardial ablation lesions. *Heart Rhythm* 11 (1):150–7.

Davies MJ, and AC Thomas. 1985. Plaque fissuring—The cause of acute myocardial infarction, sudden ischaemic death, and crescendo angina. *Br Heart J* 53 (4):363–73.

Deckelbaum LI, JK Lam, HS Cabin, KS Clubb, and MB Long. 1987. Discrimination of normal and atherosclerotic aorta by laser-induced fluorescence. *Lasers Surg Med* 7 (4):330–5.

Dempsey RJ, R Vemuganti, T Varghese, and BP Hermann. 2010. A review of carotid atherosclerosis and vascular cognitive decline: A new understanding of the keys to symptomology. *Neurosurgery* 67 (2):484–93; discussion 493–4.

Dima A, and V Ntziachristos. 2012. Non-invasive carotid imaging using optoacoustic tomography. *Opt Express* 20 (22):25044–57.

Edwards WB, WJ Akers, Y Ye, PP Cheney, S Bloch, B Xu, R Laforest, and S Achilefu. 2009. Multimodal imaging of integrin receptor-positive tumors by bioluminescence, fluorescence, gamma scintigraphy, and single-photon emission computed tomography using a cyclic RGD peptide labeled with a near-infrared fluorescent dye and a radionuclide. *Mol Imaging* 8 (2):101–10.

Elson DS, JA Jo, and L Marcu. 2007. Miniaturized side-viewing imaging probe for fluorescence lifetime imaging (FLIM): Validation with fluorescence dyes, tissue structural proteins and tissue specimens. *N J Phys* 9:127.

Evans JW, RJ Zawadzki, R Liu, JW Chan, SM Lane, and JS Werner. 2009. Optical coherence tomography and Raman spectroscopy of the *ex-vivo* retina. *J Biophotonics* 2 (6–7):398–406.

Falk E. 1983. Plaque rupture with severe pre-existing stenosis precipitating coronary thrombosis. Characteristics of coronary atherosclerotic plaques underlying fatal occlusive thrombi. *Br Heart J* 50 (2):127–34.

Fard AM, P Vacas-Jacques, E Hamidi, H Wang, RW Carruth, JA Gardecki, and GJ Tearney. 2013. Optical coherence tomography–near infrared spectroscopy system and catheter for intravascular imaging. *Opt Express* 21 (25):30849–58.

Fay F, BL Sanchez-Gaytan, DP Cormode, T Skajaa, EA Fisher, ZA Fayad, and WJ Mulder. 2013. Nanocrystal core lipoprotein biomimetics for imaging of lipoproteins and associated diseases. *Curr Cardiovasc Imaging Rep* 6 (1):45–54.

Ferrari VA, and RL Wilensky. 2007. Intravascular magnetic resonance imaging. *Top Magn Reson Imaging* 18 (5):401–8.

Fitzgerald PJ, and PG Yock. 1993. Mechanisms and outcomes of angioplasty and atherectomy assessed by intravascular ultrasound imaging. *J Clin Ultrasound* 21 (9):579–88.

Fleg JL, GW Stone, ZA Fayad, JF Granada, TS Hatsukami, FD Kolodgie, J Ohayon, R Pettigrew, MS Sabatine, GJ Tearney, S Waxman, MJ Domanski, PR Srinivas, and J Narula. 2012. Detection of high-risk atherosclerotic plaque: Report of the NHLBI Working Group on Current Status and Future Directions. *JACC Cardiovasc Imaging* 5 (9):941–55.

Franzen D, U Sechtem, and HW Hopp. 1998. Comparison of angioscopic, intravascular ultrasonic, and angiographic detection of thrombus in coronary stenosis. *Am J Cardiol* 82 (10):1273–5, a9.

Friedman M, and GJ Van den Bovenkamp. 1966. Role of thrombus in plaque formation in the human diseased coronary artery. *Br J Exp Pathol* 47 (6):550–7.

Gaglia MA, Jr, DH Steinberg, and NJ Weissman. 2009. Intravascular ultrasound: Virtual histology IVUS, integrated backscatter IVUS, and palpography. *Curr Cardiovasc Imaging Rep* 2 (4):268–74.

Ghosh SC, P Ghosh, N Wilganowski, H Robinson, MA Hall, G Dickinson, KL Pinkston, BR Harvey, EM Sevick-Muraca, and A Azhdarinia. 2013. Multimodal chelation platform for near-infrared fluorescence/nuclear imaging. *J Med Chem* 56 (2):406–16.

Goldstein JA, C Grines, T Fischell, R Virmani, D Rizik, J Muller, and SR Dixon. 2009. Coronary embolization following balloon dilation of lipid-core plaques. *JACC Cardiovasc Imaging* 2 (12):1420–4.

Goldstein JA, B Maini, SR Dixon, ES Brilakis, CL Grines, DG Rizik, ER Powers, DH Steinberg, KA Shunk, G Weisz, PR Moreno, A Kini, SK Sharma, MJ Hendricks, ST Sum, SP Madden, JE Muller, GW Stone, and MJ Kern. 2011. Detection of lipid-core plaques by intracoronary near-infrared spectroscopy identifies high risk of periprocedural myocardial infarction. *Circ Cardiovasc Interv* 4 (5):429–37.

Gonzalo N, HM Garcia-Garcia, E Regar, P Barlis, J Wentzel, Y Onuma, J Ligthart, and PW Serruys. 2009. *In vivo* assessment of high-risk coronary plaques at bifurcations with combined intravascular ultrasound and optical coherence tomography. *JACC Cardiovasc Imaging* 2 (4):473–82.

Graf IM, S Kim, B Wang, R Smalling, and S Emelianov. 2012. Noninvasive detection of intimal xanthoma using combined ultrasound, strain rate and photoacoustic imaging. *Ultrasonics* 52 (3):435–41.

Graf IM, J Su, D Yeager, J Amirian, R Smalling, and S Emelianov. 2011. Methodical study on plaque characterization using integrated vascular ultrasound, strain and spectroscopic photoacoustic imaging. Proc. SPIE 7899, Photons Plus Ultrasound: Imaging and Sensing 2011, 789902.

Hosokawa R, N Kambara, M Ohba, T Mukai, M Ogawa, H Motomura, N Kume, H Saji, T Kita, and R Nohara. 2006. A catheter-based intravascular radiation detector of vulnerable plaques. *J Nucl Med* 47 (5):863–7.

Houston JP, S Ke, W Wang, C Li, and EM Sevick-Muraca. 2005. Quality analysis of *in vivo* near-infrared fluorescence and conventional gamma images acquired using a dual-labeled tumor-targeting probe. *J Biomed Opt* 10 (5):054010.

Hsieh BY, SL Chen, T Ling, LJ Guo, and PC Li. 2012. All-optical scanhead for ultrasound and photoacoustic dual-modality imaging. *Opt Express* 20 (2):1588–96.

Ishibashi F, K Aziz, GS Abela, and S Waxman. 2006. Update on coronary angioscopy: Review of a 20-year experience and potential application for detection of vulnerable plaque. *J Interv Cardiol* 19 (1):17–25.

Jaffer FA, MA Calfon, A Rosenthal, G Mallas, RN Razansky, A Mauskapf, R Weissleder, P Libby, and V Ntziachristos. 2011. Two-dimensional intravascular near-infrared fluorescence molecular imaging of inflammation in atherosclerosis and stent-induced vascular injury. *J Am Coll Cardiol* 57 (25):2516–26.

Jaffer FA, P Libby, and R Weissleder. 2009. Optical and multimodality molecular imaging: Insights into atherosclerosis. *Arterioscler Thromb Vasc Biol* 29 (7):1017–24.

Janecek M, BE Patt, JS Iwanczyk, L MacDonald, Y Yamaguchi, H William Strauss, R Tsugita, V Ghazarossian, and EJ Hoffman. 2004. Intravascular probe for detection of vulnerable plaque. *Mol Imaging Biol* 6 (3):131–8.

Jang I-K, BE Bouma, D-H Kang, S-J Park, S-W Park, K-B Seung, K-B Choi, M Shishkov, K Schlendorf, E Pomerantsev, SL Houser, H Thomas Aretz, and GJ Tearney. 2002. Visualization of coronary atherosclerotic plaques in patients using optical coherence tomography: Comparison with intravascular ultrasound. *J Am Coll Cardiol* 39 (4):604–9.

Jansen K, AF van der Steen, HM van Beusekom, JW Oosterhuis, and G van Soest. 2011. Intravascular photoacoustic imaging of human coronary atherosclerosis. *Opt Lett* 36 (5):597–9.

Jansen K, AF van der Steen, HM van Beusekom, G Springeling, W Min, F Mastik, and G van Soest. 2012. Automatic lipid detection in human coronary atherosclerosis using spectroscopic intravascular photoacoustic imaging. Ultrasonics Symposium (IUS), 2012 IEEE International, October 7–10, 2012.

Jansen K, M Wu, AF van der Steen, and G van Soest. 2013. Lipid detection in atherosclerotic human coronaries by spectroscopic intravascular photoacoustic imaging. *Opt Express* 21 (18):21472–84.

Jansen K, G van Soest, and AFW van der Steen. 2014. Intravascular photoacoustic imaging: A new tool for vulnerable plaque identification. *Ultrasound Med Biol* 40 (6):1037–48.

Kakkos SK, JM Stevens, AN Nicolaides, E Kyriacou, CS Pattichis, G Geroulakos, and D Thomas. 2007. Texture Analysis of ultrasonic images of symptomatic carotid plaques can identify those plaques associated with ipsilateral embolic brain infarction. *Eur J Vasc Endovasc Surg* 33 (4):422–9.

Karpiouk AB, B Wang, and SY Emelianov. 2010. Development of a catheter for combined intravascular ultrasound and photoacoustic imaging. *Rev Sci Instrum* 81 (1):014901.

Khan KM, H Krishna, SK Majumder, KD Rao, and PK Gupta. 2014. Depth-sensitive Raman spectroscopy combined with optical coherence tomography for layered tissue analysis. *J Biophotonics* 7 (1–2):77–85.

Kietselaer BL, CP Reutelingsperger, HH Boersma, GA Heidendal, IH Liem, HJ Crijns, J Narula, and L Hofstra. 2007. Noninvasive detection of programmed cell loss with 99mTc-labeled Annexin A5 in heart failure. *J Nucl Med* 48 (4):562–7.

Kietselaer BLJH, CPM Reutelingsperger, GAK Heidendal, MJAP Daemen, WH Mess, L Hofstra, and J Narula. 2004. Noninvasive detection of plaque instability with use of radiolabeled Annexin A5 in patients with carotid-artery atherosclerosis. *N Engl J Med* 350 (14):1472–3.

Kim SW, GS Mintz, YJ Hong, R Pakala, KS Park, AD Pichard, LF Satler, KM Kent, WO Suddath, R Waksman, and NJ Weissman. 2008. The virtual histology intravascular ultrasound appearance of newly placed drug-eluting stents. *Am J Cardiol* 102 (9):1182–6.

Klein HM, RW Gunther, M Verlande, W Schneider, D Vorwerk, J Kelch, and M Hamm. 1992. 3D-surface reconstruction of intravascular ultrasound images using personal computer hardware and a motorized catheter control. *Cardiovasc Intervent Radiol* 15 (2):97–101.

Ko AC, A Ridsdale, MS Smith, LB Mostaco-Guidolin, MD Hewko, AF Pegoraro, EK Kohlenberg, B Schattka, M Shiomi, A Stolow, and MG Sowa. 2010. Multimodal nonlinear optical imaging of atherosclerotic plaque development in myocardial infarction-prone rabbits. *J Biomed Opt* 15 (2):020501.

Ko ACT, LB Mostaço-Guidolin, A Ridsdale, AF Pegoraro, MSD Smith, A Slepkov, MD Hewko, EK Kohlenberg, B Schattka, A Stolow, and MG Sowa. 2011. Using multimodal femtosecond CARS imaging to determine plaque burden in luminal atherosclerosis. Proc. SPIE 7903, Multiphoton Microscopy in the Biomedical Sciences XI, 790318.

Ko ACT, A Ridsdale, LB Mostaço-Guidolin, A Major, A Stolow, and MG Sowa. 2012. Nonlinear optical microscopy in decoding arterial diseases. *Biophys Rev* 4 (4):323–34.

Koie S, H Matsuyama, M Nomura, and H Hishida. 2003. [Intravascular ultrasound and angioscopy]. *Nihon Rinsho* 61 (10):1744–50.

Kubo T, T Imanishi, S Takarada, A Kuroi, S Ueno, T Yamano, T Tanimoto, Y Matsuo, T Masho, H Kitabata, A Tanaka, N Nakamura, M Mizukoshi, Y Tomobuchi, and T Akasaka. 2008. Implication of plaque color classification for assessing plaque vulnerability: A coronary angioscopy and optical coherence tomography investigation. *JACC Cardiovasc Interv* 1 (1):74–80.

Kubo T, T Imanishi, S Takarada, A Kuroi, S Ueno, T Yamano, T Tanimoto, Y Matsuo, T Masho, H Kitabata, K Tsuda, Y Tomobuchi, and T Akasaka. 2007. Assessment of culprit lesion morphology in acute myocardial infarction: Ability of optical coherence tomography compared with intravascular ultrasound and coronary angioscopy. *J Am Coll Cardiol* 50 (10):933–9.

Lederman RJ, RR Raylman, SJ Fisher, PV Kison, H San, EG Nabel, and RL Wahl. 2001. Detection of atherosclerosis using a novel positron-sensitive probe and 18-fluorodeoxyglucose (FDG). *Nucl Med Commun* 22 (7):747–53.

Leuschner F, and M Nahrendorf. 2011. Molecular imaging of coronary atherosclerosis and myocardial infarction: Considerations for the bench and perspectives for the clinic. *Circ Res* 108 (5):593–606.

Leuschner F, P Panizzi, I Chico-Calero, WW Lee, T Ueno, V Cortez-Retamozo, P Waterman, R Gorbatov, B Marinelli, Y Iwamoto, A Chudnovskiy, J-L Figueiredo, DE Sosnovik, MJ Pittet, FK Swirski, R Weissleder, and M Nahrendorf. 2010. Angiotensin-converting enzyme inhibition prevents the release of monocytes from their splenic reservoir in mice with myocardial infarction. *Circ Res* 107 (11):1364–73.

Li BH, ASO Leung, A Soong, CE Munding, H Lee, AS Thind, NR Munce, GA Wright, CH Rowsell, VXD Yang, B H Strauss, F Stuart Foster, and BK Courtney. 2013. Hybrid intravascular ultrasound and optical coherence tomography catheter for imaging of coronary atherosclerosis. *Catheter Cardiovasc Interv* 81 (3):494–507.

Li X, J Yin, C Hu, Q Zhou, K Kirk Shung, and Z Chen. 2010. High-resolution coregistered intravascular imaging with integrated ultrasound and optical coherence tomography probe. *Appl Phys Lett* 97 (13):133702.

Liang S, A Saidi, J Jing, G Liu, J Li, J Zhang, C Sun, J Narula, and Z Chen. 2012. Intravascular atherosclerotic imaging with combined fluorescence and optical coherence tomography probe based on a double-clad fiber combiner. *J Biomed Opt* 17 (7):070501.

Licha K, P Welker, M Weinhart, N Wegner, S Kern, S Reichert, I Gemeinhardt, C Weissbach, B Ebert, R Haag, and M Schirner. 2011. Fluorescence imaging with multifunctional polyglycerol sulfates: Novel polymeric near-IR probes targeting inflammation. *Bioconjug Chem* 22 (12):2453–60.

Lilledahl MB, OA Haugen, C de Lange Davies, and LO Svaasand. 2007. Characterization of vulnerable plaques by multiphoton microscopy. *J Biomed Opt* 12 (4):044005.

Lim RS, A Kratzer, NP Barry, S Miyazaki-Anzai, M Miyazaki, WW Mantulin, M Levi, EO Potma, and BJ Tromberg. 2010. Multimodal CARS microscopy determination of the impact of diet on macrophage infiltration and lipid accumulation on plaque formation in ApoE-deficient mice. *J Lipid Res* 51 (7):1729–37.

Luke GP, D Yeager, and SY Emelianov. 2012. Biomedical applications of photoacoustic imaging with exogenous contrast agents. *Ann Biomed Eng* 40 (2):422–37.

Majmudar MD, and M Nahrendorf. 2012. Cardiovascular molecular imaging: The road ahead. *J Nucl Med* 53 (5):673–6.

Manoharan R, JJ Baraga, MS Feld, and RP Rava. 1992. Quantitative histochemical analysis of human artery using Raman spectroscopy. *J Photochem Photobiol B* 16 (2):211–33.

Marcu L, Q Fang, JA Jo, T Papaioannou, A Dorafshar, T Reil, JH Qiao, JD Baker, JA Freischlag, and MC Fishbein. 2005. *In vivo* detection of macrophages in a rabbit atherosclerotic model by time-resolved laser-induced fluorescence spectroscopy. *Atherosclerosis* 181 (2):295–303.

Marcu L, JA Jo, Q Fang, T Papaioannou, T Reil, JH Qiao, JD Baker, JA Freischlag, and MC Fishbein. 2009. Detection of rupture-prone atherosclerotic plaques by time-resolved laser-induced fluorescence spectroscopy. *Atherosclerosis* 204 (1):156–64.

Mehrmohammadi M, SJ Yoon, D Yeager, and SY Emelianov. 2013. Photoacoustic Imaging for Cancer Detection and Staging. *Curr Mol Imaging* 2 (1):89–105.

Mendis S, P Puska, and B Norrving. 2011. *Global Atlas on Cardiovascular Disease Prevention and Control*. Edited by World Health Organization; World heart Federation; World Stroke Organization. Geneva.

Moreno PR, RA Lodder, KR Purushothaman, WE Charash, WN O'Connor, and JE Muller. 2002. Detection of lipid pool, thin fibrous cap, and inflammatory cells in human aortic atherosclerotic plaques by near-infrared spectroscopy. *Circulation* 105 (8):923–7.

Mostaço-Guidolin LB, AC Ko, DP Popescu, MS Smith, EK Kohlenberg, M Shiomi, A Major, and MG Sowa. 2011. Evaluation of texture parameters for the quantitative description of multimodal nonlinear optical images from atherosclerotic rabbit arteries. *Phys Med Biol* 56 (16):5319–34.

Mostaço-Guidolin LB, MG Sowa, A Ridsdale, AF Pegoraro, MSD Smith, MD Hewko, EK Kohlenberg, B Schattka, M Shiomi, A Stolow, and ACT Ko. 2010. Differentiating atherosclerotic plaque burden in arterial tissues using femtosecond CARS-based multimodal nonlinear optical imaging. *Biomed Optics Express* 1 (1):59–73.

Motz JT, M Fitzmaurice, A Miller, SJ Gandhi, AS Haka, LH Galindo, RR Dasari, JR Kramer, and MS Feld. 2006. *In vivo* Raman spectral pathology of human atherosclerosis and vulnerable plaque. *J Biomed Opt* 11 (2):021003.

Mullenix PS, CA Andersen, and BW Starnes. 2005. Atherosclerosis as inflammation. *Ann Vasc Surg* 19 (1):130–8.

Muller JE, GH Tofler, and PH Stone. 1989. Circadian variation and triggers of onset of acute cardiovascular disease. *Circulation* 79 (4):733–43.

Muller JE, GS Abela, RW Nesto, and GH Tofler. 1994. Triggers, acute risk factors and vulnerable plaques: The lexicon of a new frontier. *J Am Coll Cardiol* 23 (3):809–13.

Nahrendorf M, E Keliher, B Marinelli, F Leuschner, CS Robbins, RE Gerszten, MJ Pittet, FK Swirski, and R Weissleder. 2011. Detection of macrophages in aortic aneurysms by nanoparticle positron emission tomography-computed tomography. *Arterioscler Thromb Vasc Biol* 31 (4):750–7.

Nahrendorf M, P Waterman, G Thurber, K Groves, M Rajopadhye, P Panizzi, B Marinelli, E Aikawa, MJ Pittet, FK Swirski, and R Weissleder. 2009. Hybrid *in vivo* FMT-CT imaging of protease activity in atherosclerosis with customized nanosensors. *Arterioscler Thromb Vasc Biol* 29 (10):1444–51.

Nahrendorf M, H Zhang, S Hembrador, P Panizzi, DE Sosnovik, E Aikawa, P Libby, FK Swirski, and R Weissleder. 2008. Nanoparticle PET-CT imaging of macrophages in inflammatory atherosclerosis. *Circulation* 117 (3):379–87.

Nahrendorf M, E Keliher, B Marinelli, P Waterman, P Fumene Feruglio, L Fexon, M Pivovarov, FK Swirski, MJ Pittet, C Vinegoni, and R Weissleder. 2010. Hybrid PET-optical imaging using targeted probes. *Proc Natl Acad Sci U S A* 107 (17):7910–5.

Narula J, ER Acio, N Narula, LE Samuels, B Fyfe, D Wood, JM Fitzpatrick, PN Raghunath, JE Tomaszewski, C Kelly, N Steinmetz, A Green, JF Tait, J Leppo, FG Blankenberg, D Jain, and HW Strauss. 2001. Annexin-V imaging for noninvasive detection of cardiac allograft rejection. *Nat Med* 7 (12):1347–52.

Nikoozadeh A, CJ Woo, SR Kothapalli, A Moini, SS Sanjani, A Kamaya, O Oralkan, SS Gambhir, and PT Khuri-Yakub. 2012. Photoacoustic imaging using a 9F microlinear CMUT ICE catheter. Ultrasonics Symposium (IUS), 2012 IEEE International, October 7–10, 2012.

Nissen SE, and P Yock. 2001. Intravascular ultrasound: Novel pathophysiological insights and current clinical applications. *Circulation* 103 (4):604–16.

Packard RRS, and P Libby. 2008. Inflammation in atherosclerosis: From vascular biology to biomarker discovery and risk prediction. *Clin Chem* 54 (1):24–38.

Panizzi P, FK Swirski, JL Figueiredo, P Waterman, DE Sosnovik, E Aikawa, P Libby, M Pittet, R Weissleder, and M Nahrendorf. 2010. Impaired infarct healing in atherosclerotic mice with Ly-6C(hi) monocytosis. *J Am Coll Cardiol* 55 (15):1629–38.

Park J, JA Jo, S Shrestha, P Pande, Q Wan, and BE Applegate. 2010. A dual-modality optical coherence tomography and fluorescence lifetime imaging microscopy system for simultaneous morphological and biochemical tissue characterization. *Biomed Opt Express* 1 (1):186–200.

Patil CA, N Bosschaart, MD Keller, TG van Leeuwen, and A Mahadevan-Jansen. 2008. Combined Raman spectroscopy and optical coherence tomography device for tissue characterization. *Opt Lett* 33 (10):1135–7.

Phipps J, Y Sun, R Saroufeem, N Hatami, MC Fishbein, and L Marcu. 2011. Fluorescence lifetime imaging for the characterization of the biochemical composition of atherosclerotic plaques. *J Biomed Opt* 16 (9):096018.

Phipps J, Y Sun, R Saroufeem, N Hatami, and L Marcu. 2009. Fluorescence lifetime imaging microscopy for the characterization of atherosclerotic plaques. *Proc Soc Photo Opt Instrum Eng* 7161:71612g.

Piao D, MM Sadeghi, J Zhang, Y Chen, AJ Sinusas, and Q Zhu. 2005. Hybrid positron detection and optical coherence tomography system: Design, calibration, and experimental validation with rabbit atherosclerotic models. *J Biomed Opt* 10 (4):44010.

Pignoli P, E Tremoli, A Poli, P Oreste, and R Paoletti. 1986. Intimal plus medial thickness of the arterial wall: A direct measurement with ultrasound imaging. *Circulation* 74 (6):1399–406.

Raber L, JH Heo, MD Radu, HM Garcia-Garcia, GG Stefanini, A Moschovitis, J Dijkstra, H Kelbaek, S Windecker, and PW Serruys. 2012. Offline fusion of co-registered intravascular ultrasound and frequency domain optical coherence tomography images for the analysis of human atherosclerotic plaques. *EuroIntervention* 8 (1):98–108.

Reiber JHC, S Tu, JC Tuinenburg, G Koning, JP Janssen, and J Dijkstra. 2011. QCA, IVUS and OCT in interventional cardiology in 2011. *Cardiovasc Diagn Ther* 1 (1):57–70.

Richards-Kortum R, RP Rava, M Fitzmaurice, LL Tong, NB Ratliff, JR Kramer, and M Feld. 1989. A one-layer model of laser-induced fluorescence for diagnosis of disease in human tissue: Applications to atherosclerosis. *IEEE Trans Biomed Eng* 36 (12):1222–2.

Rodenberg E, A Azhdarinia, ZW Lazard, M Hall, SK Kwon, N Wilganowski, EA Salisbury, M Merched-Sauvage, EA Olmsted-Davis, EM Sevick-Muraca, and AR Davis. 2011. Matrix metalloproteinase-9 is a diagnostic marker of heterotopic ossification in a murine model. *Tissue Eng Part A* 17 (19–20):2487–96.

Roger VL, AS Go, DM Lloyd-Jones, EJ Benjamin, JD Berry, WB Borden, DM Bravata, S Dai, ES Ford, CS Fox, HJ Fullerton, C Gillespie, SM Hailpern, JA Heit, VJ Howard, BM Kissela, SJ Kittner, DT Lackland, JH Lichtman, LD Lisabeth, DM Makuc, GM Marcus, A Marelli, DB Matchar, CS Moy, D Mozaffarian, ME Mussolino, G Nichol, NP Paynter, EZ Soliman, PD Sorlie, N Sotoodehnia, TN Turan, SS Virani, ND Wong, D Woo, and MB Turner. 2012. Heart disease and stroke statistics—2012 update: A report from the American Heart Association. *Circulation* 125 (1):e2–220.

Romer TJ, JF Brennan, 3rd, M Fitzmaurice, ML Feldstein, G Deinum, JL Myles, JR Kramer, RS Lees, and MS Feld. 1998. Histopathology of human coronary atherosclerosis by quantifying its chemical composition with Raman spectroscopy. *Circulation* 97 (9):878–85.

Romer TJ, JF Brennan, 3rd, GJ Puppels, AH Zwinderman, SG van Duinen, A van der Laarse, AF van der Steen, NA Bom, and AV Bruschke. 2000. Intravascular ultrasound combined with Raman spectroscopy to localize and quantify cholesterol and calcium salts in atherosclerotic coronary arteries. *Arterioscler Thromb Vasc Biol* 20 (2):478–83.

Ross R. 1995. Cell biology of atherosclerosis. *Ann Rev Physiol* 57 (1):791–804.

Sawada T, J Shite, HM Garcia-Garcia, T Shinke, S Watanabe, H Otake, D Matsumoto, Y Tanino, D Ogasawara, H Kawamori, H Kato, N Miyoshi, M Yokoyama, PW Serruys, and K Hirata. 2008. Feasibility of combined use of intravascular ultrasound radiofrequency data analysis and optical coherence tomography for detecting thin-cap fibroatheroma. *Eur Heart J* 29 (9):1136–46.

Scepanovic OR, M Fitzmaurice, A Miller, CR Kong, Z Volynskaya, RR Dasari, JR Kramer, and MS Feld. 2011. Multimodal spectroscopy detects features of vulnerable atherosclerotic plaque. *J Biomed Opt* 16 (1):011009.

Šćepanović OR, Z Volynskaya, C-R Kong, LH Galindo, RR Dasari, and MS Feld. 2009. A multimodal spectroscopy system for real-time disease diagnosis. *Rev Sci Instrum* 80 (4):043103.

Schellenberger EA, D Sosnovik, R Weissleder, and L Josephson. 2004. Magneto/optical Annexin V, a multimodal protein. *Bioconjug Chem* 15 (5):1062–7.

Sethuraman S, SR Aglyamov, JH Amirian, RW Smalling, and SY Emelianov. 2007a. Intravascular photoacoustic imaging using an IVUS imaging catheter. *IEEE Trans Ultrason Ferroelectr Freq Control* 54:978–86.

Sethuraman S, JH Amirian, SH Litovsky, RW Smalling, and SY Emelianov. 2008. Spectroscopic intravascular photoacoustic imaging to differentiate atherosclerotic plaques. *Opt Express* 16 (5):3362–7.

Sethuraman S, JH Amirian, SH Litovsky, RW Smalling, and SY Emelianov. 2007b. *Ex vivo* Characterization of atherosclerosis using intravascular photoacoustic imaging. *Opt Express* 15 (25):16657–66.

Sherman CT, F Litvack, W Grundfest, M Lee, A Hickey, A Chaux, R Kass, C Blanche, J Matloff, L Morgenstern, W Ganz, HJC Swan, and J Forrester. 1986. Coronary angioscopy in patients with unstable angina pectoris. *N Engl J Med* 315 (15):913–9.

Shikhaliev PM, T Xu, JL Ducote, B Easwaramoorthy, J Mukherjee, and S Molloi. 2006. Positron autoradiography for intravascular imaging: Feasibility evaluation. *Phys Med Biol* 51 (4):963–79.

Simsek C, HM Garcia-Garcia, RJ van Geuns, M Magro, C Girasis, N van Mieghem, M Lenzen, S de Boer, E Regar, W van der Giessen, J Raichlen, HJ Duckers, F Zijlstra, T van der Steen, E Boersma, and PW Serruys. 2012. The ability of high dose rosuvastatin to improve plaque composition in non-intervened coronary arteries: Rationale and design of the Integrated Biomarker and Imaging Study-3 (IBIS-3). *EuroIntervention* 8 (2):235–41.

Sinusas AJ, F Bengel, M Nahrendorf, FH Epstein, JC Wu, FS Villanueva, ZA Fayad, and RJ Gropler. 2008. Multimodality cardiovascular molecular imaging, part I. *Circ Cardiovasc Imaging* 1 (3):244–56.

Skajaa T, DP Cormode, PA Jarzyna, A Delshad, C Blachford, A Barazza, EA Fisher, RE Gordon, ZA Fayad, and WJ Mulder. 2011. The biological properties of iron oxide core high-density lipoprotein in experimental atherosclerosis. *Biomaterials* 32 (1):206–13.

Sosnovik DE, E Garanger, E Aikawa, M Nahrendorf, JL Figuiredo, G Dai, F Reynolds, A Rosenzweig, R Weissleder, and L Josephson. 2009. Molecular MRI of cardiomyocyte apoptosis with simultaneous delayed-enhancement MRI distinguishes apoptotic and necrotic myocytes *in vivo*: Potential for midmyocardial salvage in acute ischemia. *Circ Cardiovasc Imaging* 2 (6):460–7.

Sosnovik DE, M Nahrendorf, N Deliolanis, M Novikov, E Aikawa, L Josephson, A Rosenzweig, R Weissleder, and V Ntziachristos. 2007. Fluorescence tomography and magnetic resonance imaging of myocardial macrophage infiltration in infarcted myocardium *in vivo*. *Circulation* 115 (11):1384–91.

Sosnovik DE, EA Schellenberger, M Nahrendorf, MS Novikov, T Matsui, G Dai, F Reynolds, L Grazette, A Rosenzweig, R Weissleder, and L Josephson. 2005. Magnetic resonance imaging of cardiomyocyte apoptosis with a novel magneto-optical nanoparticle. *Magn Reson Med* 54 (3):718–24.

Spears JR, AM Spokojny, and HJ Marais. 1985. Coronary angioscopy during cardiac catheterization. *J Am Coll Cardiol* 6 (1):93–7.

Stephens DN, J Park, Y Sun, T Papaioannou, and L Marcu. 2009. Intraluminal fluorescence spectroscopy catheter with ultrasound guidance. *J Biomed Opt* 14 (3):030505.

Stone GW, A Maehara, AJ Lansky, B de Bruyne, E Cristea, GS Mintz, R Mehran, J McPherson, N Farhat, SP Marso, H Parise, B Templin, R White, Z Zhang, and PW Serruys. 2011. A prospective natural-history study of coronary atherosclerosis. *N Engl J Med* 364 (3):226–35.

Strauss HW, C Mari, BE Patt, and V Ghazarossian. 2006. Intravascular radiation detectors for the detection of vulnerable atheroma. *J Am Coll Cardiol* 47 (8, Supplement):C97–100.

Su JL, B Wang, and SY Emelianov. 2009. Photoacoustic imaging of coronary artery stents. *Opt Express* 17 (22):19894–901.

Suhalim JL, CY Chung, MB Lilledahl, RS Lim, M Levi, BJ Tromberg, and EO Potma. 2012. Characterization of cholesterol crystals in atherosclerotic plaques using stimulated Raman scattering and second-harmonic generation microscopy. *Biophys J* 102 (8):1988–95.

Sun Y, AJ Chaudhari, M Lam, H Xie, DR Yankelevich, J Phipps, J Liu, MC Fishbein, JM Cannata, KK Shung, and L Marcu. 2011. Multimodal characterization of compositional, structural and functional features of human atherosclerotic plaques. *Biomed Opt Express* 2 (8):2288–98.

Suter MJ, SK Nadkarni, G Weisz, A Tanaka, FA Jaffer, BE Bouma, and GJ Tearney. 2011. Intravascular optical imaging technology for investigating the coronary artery. *JACC Cardiovasc Imaging* 4 (9):1022–39.

Takano M, K Mizuno, S Yokoyama, K Seimiya, F Ishibashi, K Okamatsu, and R Uemura. 2003. Changes in coronary plaque color and morphology by lipid-lowering therapy with atorvastatin: Serial evaluation by coronary angioscopy. *J Am Coll Cardiol* 42 (4):680–6.

Taruttis A, E Herzog, D Razansky, and V Ntziachristos. 2010. Real-time imaging of cardiovascular dynamics and circulating gold nanorods with multispectral optoacoustic tomography. *Opt Express* 18 (19):19592–602.

Taruttis A, M Wildgruber, K Kosanke, N Beziere, K Licha, R Haag, M Aichler, A Walch, E Rummeny, and V Ntziachristos. 2013. Multispectral optoacoustic tomography of myocardial infarction. *Photoacoustics* 1 (1):3–8.

Tawakol A, and J Muller. 2003. Through the looking glass: An angioscopic view of the effect of statin therapy on coronary artery plaques. *J Am Coll Cardiol* 42 (4):687–9.

Tearney GJ, E Regar, T Akasaka, T Adriaenssens, P Barlis, HG Bezerra, B Bouma, N Bruining, J-M Cho, S Chowdhary, MA Costa, R de Silva, J Dijkstra, C Di Mario, D Dudeck, E Falk, MD Feldman, P Fitzgerald, H Garcia, N Gonzalo, JF Granada, G Guagliumi, NR Holm, Y Honda, F Ikeno, M Kawasaki, J Kochman, L Koltowski, T Kubo, T Kume, H Kyono, C Chi Simon Lam, G Lamouche, DP Lee, MB Leon, A Maehara, O Manfrini, GS Mintz, K Mizuno, M-A Morel, S Nadkarni, H Okura, H Otake, A Pietrasik, F Prati, L Räber, MD Radu, J Rieber, M Riga, A Rollins, M Rosenberg, V Sirbu, PWJC Serruys, K Shimada, T Shinke, J Shite, E Siegel, S Sonada, M Suter, S Takarada, A Tanaka, M Terashima, T Troels, S Uemura, GJ Ughi, HMM van Beusekom, AFW van der Steen, G-A van Es, G van Soest, R Virmani, S Waxman, NJ Weissman, and G Weisz. 2012. Consensus standards for acquisition, measurement, and reporting of intravascular optical coherence tomography studies: A report from the International Working Group for Intravascular Optical Coherence Tomography Standardization and Validation. *J Am Coll Cardiol* 59 (12):1058–72.

Thim T, MK Hagensen, D Wallace-Bradley, JF Granada, GL Kaluza, L Drouet, WP Paaske, HE Botker, and E Falk. 2010. Unreliable assessment of necrotic core by virtual histology intravascular ultrasound in porcine coronary artery disease. *Circ Cardiovasc Imaging* 3 (4):384–91.

Tokuhiro K, Y Uchida, K Kawamura, H Sakuragawa, H Masuhara, H Oosawa, and N Koyama. 2000. Evaluation of annuloaortic ectasia by angioscopy and IVUS "report of 2 cases". *Diagn Ther Endosc* 7 (1):35–45.

Tsuchida K, WJ van der Giessen, M Patterson, S Tanimoto, HM Garcia-Garcia, E Regar, JM Ligthart, AM Maugenest, G Maatrijk, JJ Wentzel, and PW Serruys. 2007. *In vivo* validation of a novel three-dimensional quantitative coronary angiography system (CardiOp-B): Comparison with a conventional two-dimensional system (CAAS II) and with special reference to optical coherence tomography. *EuroIntervention* 3 (1):100–8.

Tu S, NR Holm, G Koning, Z Huang, and JH Reiber. 2011. Fusion of 3D QCA and IVUS/OCT. *Int J Cardiovasc Imaging* 27 (2):197–207.

Ueda Y, T Ohtani, M Shimizu, A Hirayama, and K Kodama. 2004. Assessment of plaque vulnerability by angioscopic classification of plaque color. *Am Heart J* 148 (2):333–5.

van de Poll SWE, CL De Korte, AFW Van der Steen, GJ Puppels, and A Van der Laarse. 2000. Coronary atherosclerotic plaque characterization using IVUS elastography and Raman spectroscopy. Ultrasonics Symposium, 2000 IEEE, October 2000.

van de Poll SWE, K Kastelijn, TC Bakker Schut, C Strijder, G Pasterkamp, GJ Puppels, and A van der Laarse. 2003. On-line detection of cholesterol and calcification by catheter based Raman spectroscopy in human atherosclerotic plaque *ex vivo*. *Heart* 89 (9):1078–82.

van De Poll SW, TJ Romer, OL Volger, DJ Delsing, TC Bakker Schut, HM Princen, LM Havekes, JW Jukema, A van Der Laarse, and GJ Puppels. 2001. Raman spectroscopic evaluation of the effects of diet and lipid-lowering therapy on atherosclerotic plaque development in mice. *Arterioscler Thromb Vasc Biol* 21 (10):1630–5.

van der Giessen AG, M Schaap, FJ Gijsen, HC Groen, T van Walsum, NR Mollet, J Dijkstra, FN van de Vosse, WJ Niessen, PJ de Feyter, AF van der Steen, and JJ Wentzel. 2010. 3D fusion of intravascular ultrasound and coronary computed tomography for *in-vivo* wall shear stress analysis: A feasibility study. *Int J Cardiovasc Imaging* 26 (7):781–96.

Wang B, and S Emelianov. 2011. Thermal intravascular photoacoustic imaging. *Biomedic Opt Express* 2 (11):3072–8.

Wang B, A Karpiouk, D Yeager, J Amirian, S Litovsky, R Smalling, and S Emelianov. 2012a. *In vivo* intravascular ultrasound-guided photoacoustic imaging of lipid in plaques using an animal model of atherosclerosis. *Ultrasound Med Biol* 38 (12):2098–103.

Wang B, A Karpiouk, D Yeager, J Amirian, S Litovsky, R Smalling, and S Emelianov. 2012b. Intravascular photoacoustic imaging of lipid in atherosclerotic plaques in the presence of luminal blood. *Opt Lett* 37 (7):1244–6.

Wang W, S Ke, S Kwon, S Yallampalli, AG Cameron, KE Adams, ME Mawad, and EM Sevick-Muraca. 2007. A new optical and nuclear dual-labeled imaging agent targeting interleukin 11 receptor alpha-chain. *Bioconjug Chem* 18 (2):397–402.

Wang HW, TT Le, and JX Cheng. 2008. Label-free Imaging of Arterial Cells and Extracellular Matrix Using a Multimodal CARS Microscope. *Opt Commun* 281 (7):1813–22.

Wang HW, V Simianu, MJ Locker, JX Cheng, and M Sturek. 2011. Stent-induced coronary artery stenosis characterized by multimodal nonlinear optical microscopy. *J Biomed Opt* 16 (2):021110.

Wang B, JL Su, J Amirian, SH Litovsky, R Smalling, and S Emelianov. 2010. Detection of lipid in atherosclerotic vessels using ultrasound-guided spectroscopic intravascular photoacoustic imaging. *Opt Express* 18 (5):4889–97.

Wang P, HW Wang, M Sturek, and JX Cheng. 2012c. Bond-selective imaging of deep tissue through the optical window between 1600 and 1850 nm. *J Biophoton* 5 (1):25–32.

Wang P, P Wang, H Wang, and J Cheng. 2012d. Mapping lipid and collagen by multispectral photoacoustic imaging of chemical bond vibration. *J Biomed Opt* 17 (9):96010–1.

Wang B, E Yantsen, T Larson, AB Karpiouk, S Sethuraman, JL Su, K Sokolov, and SY Emelianov. 2009. Plasmonic intravascular photoacoustic imaging for detection of macrophages in atherosclerotic plaques. *Nano Lett* 9 (6):2212–7.

Waxman S, SR Dixon, P L'Allier, JW Moses, JL Petersen, D Cutlip, JC Tardif, RW Nesto, JE Muller, MJ Hendricks, ST Sum, CM Gardner, JA Goldstein, GW Stone, and MW Krucoff. 2009. *In vivo* validation of a catheter-based near-infrared spectroscopy system for detection of lipid core coronary plaques: Initial results of the SPECTACL study. *JACC Cardiovasc Imaging* 2 (7):858–68.

Waxman S, F Ishibashi, and JE Muller. 2006. Detection and treatment of vulnerable plaques and vulnerable patients: Novel approaches to prevention of coronary events. *Circulation* 114 (22):2390–411.

Wei W, X Li, Q Zhou, KK Shung, and Z Chen. 2011. Integrated ultrasound and photoacoustic probe for co-registered intravascular imaging. *J Biomed Opt* 16 (10):106001.

Wickline SA, and GM Lanza. 2003. Nanotechnology for molecular imaging and targeted therapy. *Circulation* 107 (8):1092–5.

Willerson JT, WB Campbell, MD Winniford, J Schmitz, P Apprill, BG Firth, J Ashton, T Smitherman, L Bush, and LM Buja. 1984. Conversion from chronic to acute coronary artery disease: Speculation regarding mechanisms. *Am J Cardiol* 54 (10):1349–54.

Wofford JL, FR Kahl, GR Howard, WM McKinney, JF Toole, and JR Crouse. 1991. Relation of extent of extracranial carotid artery atherosclerosis as measured by B-mode ultrasound to the extent of coronary atherosclerosis. *Arterioscler Thromb Vasc Biol* 11 (6):1786–94.

Xiang L, L Jiawen, JJ, M Teng, L Shanshan, Z Jun, D Mohar, A Raney, S Mahon, M Brenner, P Patel, KK Shung, Z Qifa, and C Zhongping. 2014. Integrated IVUS-OCT imaging for atherosclerotic plaque characterization. *IEEE J Sel Top Quantum Electron* 20 (2):7100108.

Yamamoto M, K Okamatsu, S Inami, M Takano, S Yokoyama, T Ohba, C Ibuki, N Hata, Y Seino, and K Mizuno. 2009. Relationship between neointimal coverage of sirolimus-eluting stents and lesion characteristics: A study with serial coronary angioscopy. *Am Heart J* 158 (1):99–104.

Yang HC, J Yin, C Hu, J Cannata, Q Zhou, J Zhang, Z Chen, and KK Shung. 2010. A dual-modality probe utilizing intravascular ultrasound and optical coherence tomography for intravascular imaging applications. *IEEE Trans Ultrason Ferroelectr Freq Control* 57 (12):2839–43.

Yeager D, YS Chen, S Litovsky, and S Emelianov. 2013. Intravascular photoacoustics for image-guidance and temperature monitoring during plasmonic photothermal therapy of atherosclerotic plaques: A feasibility study. *Theranostics* 4 (1):36–46.

Yeager D, A Karpiouk, B Wang, J Amirian, K Sokolov, R Smalling, and S Emelianov. 2012. Intravascular photoacoustic imaging of exogenously labeled atherosclerotic plaque through luminal blood. *J. Biomed. Optics* 17 (10):106016.

Yin J, X Li, J Jing, J Li, D Mukai, S Mahon, A Edris, K Hoang, KK Shung, M Brenner, J Narula, Q Zhou, and Z Chen. 2011. Novel combined miniature optical coherence tomography ultrasound probe for *in vivo* intravascular imaging. *J Biomed Opt* 16 (6):060505.

Yin J, HC Yang, X Li, J Zhang, Q Zhou, C Hu, KK Shung, and Z Chen. 2010. Integrated intravascular optical coherence tomography ultrasound imaging system. *J Biomed Opt* 15 (1):010512.

Yokoyama S, M Takano, M Yamamoto, S Inami, S Sakai, K Okamatsu, S Okuni, K Seimiya, D Murakami, T Ohba, R Uemura, Y Seino, N Hata, and K Mizuno. 2009. Extended follow-up by serial angioscopic observation for bare-metal stents in native coronary arteries: From healing response to atherosclerotic transformation of neointima. *Circ Cardiovasc Interv* 2 (3):205–12.

Yoo H, JW Kim, M Shishkov, E Namati, T Morse, R Shubochkin, JR McCarthy, V Ntziachristos, BE Bouma, FA Jaffer, and GJ Tearney. 2011. Intra-arterial catheter for simultaneous microstructural and molecular imaging *in vivo*. *Nat Med* 17 (12):1680–4.

Zhang R, W Lu, X Wen, M Huang, M Zhou, D Liang, and C Li. 2011a. Annexin A5-conjugated polymeric micelles for dual SPECT and optical detection of apoptosis. *J Nucl Med* 52 (6):958–64.

Zhang H, H Qiao, A Bakken, F Gao, B Huang, YY Liu, W El-Deiry, VA Ferrari, and R Zhou. 2011b. Utility of dual-modality bioluminescence and MRI in monitoring stem cell survival and impact on post myocardial infarct remodeling. *Acad Radiol* 18 (1):3–12.

FUTURE CHALLENGES OF HYBRID IMAGING TECHNIQUES

Hybrid instrumentation versus image fusion: Path to multibrid visualization

ERNEST V. GARCIA AND MARINA PICCINELLI

18.1 INTRODUCTION

With the advent of the acceptance of the results of the Fractional Flow Reserve Versus Angiography for Multivessel Evaluation trial (Pijls et al. 2010) that among 50%–90% coronary lesion anatomic data alone are inaccurate in determining their physiologic relevance, the need for integrated assessment of cardiac anatomy and physiology has been elucidated. This realization has promoted the use of multimodality imaging and the query as to how best to integrate these results.

Much had been written as to the bright future of hybrid scanning (Beyer et al. 2011a,b,c). Hybrid scanning generally implies an imaging system with hardware that acquires images with two types of information, such as physiologic and anatomic. Image fusion generally implies software methods that register the images with one type of information onto images with another type of information. Importantly, in cardiac image fusion of two modalities, software methods are always required to account for heart motion and patient breathing, which are usually inevitably different in the two imaging modalities by which images are sequentially acquired.

Criticisms of a multimodality cardiac imaging approach usually start with concerns over radiation exposure to the patient and increased healthcare costs (Beyer et al. 2011b,c; Di Carli and Hachamovitch 2007). Radiation concerns are discussed in Chapter 19 of this book. In general, the rationale for a multimodality approach is *not* that each patient should be submitted to every possible cardiac imaging modality for improved diagnostic accuracy, but rather that all cardiac imaging modalities performed on a patient should be integrated in one display, interpreted by one cardiac imaging expert, resulting in one clinical report. Clearly, the challenges to the routine use of this approach are multifactorial, with such issues as technical, scientific, clinical, quantitative, and socioeconomic (political). The remainder of this chapter will address these challenges.

18.2 MULTIBRID VISUALIZATION

As well discussed in this book, the topic of integration is generally confounded as to whether multimodality fused imaging relates to how the data are acquired or how they are visualized. More precisely, the term hybrid, as in hybrid images, implies a new type of image generated from the information of two other types of images and thus relates to the visualization of the image rather than how the information is acquired. As we continue to realize the advantages of this colocalization of multispectral images into one display, the integration of more than two types of clinical information is desired, further emphasizing the emergence of a multibrid (Sasaki 2008) cardiac image (or n-brid, where n is the number of fused modalities).

18.3 BENEFITS AND CHALLENGES OF HYBRID INSTRUMENTATION/HARDWARE TECHNIQUES

18.3.1 BENEFITS

Although the term hybrid has different meanings in the imaging field, it means the combination of two varieties of imaging hardware that are physically coupled in the same imaging equipment. As discussed in this book, typical examples are positron emission tomography (PET)/computed tomography (CT), single-photon emission CT (SPECT)/CT, and PET/cardiovascular magnetic resonance (CMR). The benefits of the hybrid scanning approach are many (Gaemperli et al. 2011; Kaufmann 2009; Kaufmann and Di Carli 2009; Saraste and Knuuti 2012). The two types of information provided by the hybrid approach are often complementary, such as anatomic and physiologic. But beyond this attribute, the advantage of the hybrid approach is that the two types of information are acquired either simultaneously, such as in the PET/CMR approach, or, more commonly, sequentially in a short time window, such as in PET/CT and SPECT/CT. Moreover, the patient is not moved to a different imaging table between imaging modalities and general shape and file format compatibility between modalities are usually realized. This compatibility facilitates the software fusion algorithms necessary to register the three-dimensional (3-D) cardiac distribution from one modality onto the complementary modality considering the need for correcting for heart motion and patient breathing (Slomka and Baum 2009; Slomka et al. 2008).

The advantage of the hybrid approach is optimized in simultaneous imaging, as in PET/CMR, since both types of information are obtained when the heart is at the same exact physiological state (Saraste and Knuuti 2012). Another pairing of modalities that benefits the hybrid approach is when a less expensive and fast acquisition modality is combined to a more expensive and slower modality, the most appropriate example being a SPECT/CT cardiac scanner, where the CT scanner has been simplified to perform only transmission scans for attenuation correction and gated CT scans for measuring calcium score. This assumes that the CT simplification has resulted in a significant reduction of hardware cost.

18.3.2 Challenges

Clearly, one of the challenges for this hardware coupling hybrid approach would be to couple more than two imaging scanners into one device, such as PET/CT/CMR or any other combination. Although this multiple (>two) hardware scanning approach has been investigated in a couple of settings, the technical challenges and cost that have to be overcome make it a prohibitive option.

Another challenge for the hybrid scanning approach is whether the approach is *cost-effective* (Beanlands et al. 2007; Flotats et al. 2011; Fraser et al. 2006). Two issues are relevant to the question of cost-effectiveness. The first one is related to updating the equipment. Take, for example, PET/CT. Even if the original purchase of the equipment has the latest PET and latest CT scanners that have been coupled, as hardware innovations in either modality are commercially released, the imaging laboratory administrators of the hybrid equipment are challenged with whether to do an expensive fork-lift upgrade of both scanners or wait for the time when there are new versions of both the PET and CT scanners.

The second issue related to the reduced cost-effectiveness of the hybrid equipment is that most hybrid scanning systems, like PET/CT or SPECT/CT, do not perform simultaneous imaging of both modalities. Thus, while the emission scan is being performed, the CT scanner is idle and not producing clinical revenue.

Yet another challenge for the hybrid scanning approach is the training necessary for licensing and certification in two modalities by the personnel (physicians and technologists) in one imaging laboratory (Beanlands et al. 2007; Flotats et al. 2011; Fraser et al. 2006). This is further complicated by the political turf battles in many countries as to which department is allowed to perform a specific imaging procedure and the training necessary for interpreting the hybrid results.

18.4 BENEFITS OF SOFTWARE IMAGE FUSION TECHNIQUES

By software image fusion here, we mean the coregistration of the 3-D distribution of one modality onto one or more other modalities or the same modality at a different state using software algorithms. As explained previously, even in hybrid cardiac scanning techniques, additional software fusion techniques are required since the multimodality acquisitions are usually acquired sequentially and in different total acquisition times, necessitating corrections for heart movement and patient breathing (Slomka et al. 2008; Slomka and Baum 2009). Thus, coregistration improvements in software cardiac image fusions automatically also translate to improvements in hybrid scanning techniques. Additionally, since software cardiac software fusion techniques (Faber et al. 2004; Peifer et al. 1990, 1992) predate cardiac hybrid scanning techniques (Groves et al. 2009; Kaufmann 2009; Namdar et al. 2005), software techniques "paved the way" (Slomka and Baum 2009) for the clinical application of hybrid scanners. Moreover, the benefits of software fusion techniques overcome most of the challenges of hybrid hardware techniques since the software approach is not limited to two modalities and is cost-effective since there is no high cost of hybrid scanners or idle scanners, and there is even less concern as to cross-training, licensing, or turf battles since each modality may be acquired in the politically correct imaging laboratory for a specific institution. Nevertheless, this does not overcome the issue of the training necessary to interpret the fused information. In Table 18.1, a concise list of the pros

Table 18.1 Pros/cons of hybrid scanning vs. software fusion

	Hybrid scanning	Software fusion
Cost effectiveness/upgradability	−	++
Acquisition time	+	++
Radiation concerns	−	+
Image portability	++	−
Processing time	+	−
Quantitative synergism	−	+++
Multibrid visualization	−	+++

and cons of the hybrid scanning versus the software fusion approaches is presented. The challenges of hybrid image fusion in general are covered in the next section.

18.5 FUTURE CHALLENGES

The future challenges of image fusion are multifactorial and include issues that are technical, synergistic, scientific, clinical, and socioeconomic. Issues related to the scientific and clinical challenges of how best to fuse the various cardiac anatomic and functional information are well covered in other chapters in this book.

18.5.1 TECHNICAL CHALLENGES

There are plenty of technical challenges (Slomka and Baum 2009; Slomka et al. 2008), summarized in Table 18.2. Perhaps the greatest challenge to the routine clinical application of multimodality cardiac image fusion is that it should be totally automatic. Total automation would promote reproducibility of results between institutions, but more importantly, it would eliminate the potentially time-consuming process of manually aligning two totally different modalities. An important step in total automation of software fusion is the automatic segmentation of the pertinent cardiac anatomic structures (Faber et al. 2011; Woo et al. 2009). For example, in fusing cardiac CT (and computed tomographic angiography [CTA]) information to PET or SPECT emission information, it would be facilitated by the automatic identification of the endocardium and epicardium of the left and right ventricles (Heimann and Meinzer 2009; Paragios 2002; Petitjean and Dacher 2011; Tsai et al. 2003). Since the CT coronary angiographic information is inherently registered with the biventricular CT anatomic information, by coregistering the emission myocardium to the CT myocardium, the CT coronary angiogram is automatically registered (Piccinelli et al. 2014).

Another technical challenge of software image fusion is the degree of deformation (Slomka and Baum 2009), i.e., nonlinear, nonrigid coregistration that should take place for the two cardiac modalities to perfectly align spatially. This is particularly true of cardiac soft tissue deformation between two scans due to respiration, cardiac motion, different physiological effects, and treatment *effects* (Beyer et al. 2011b; Slomka and Baum 2009).

A technical challenge that has received little attention is the need for interactive 3-D cardiac region of interest (ROI) tools that either generate volumetric measurements or cut through overlapping fused structures for cross-sectional visualization of important hidden information. For example, Figure 18.1 shows a 3-D volume rendered CTA image set fused to perfusion SPECT information where the coronary artery cross-sectional information shows a distal patent stent demonstrating that the distal perfusion defect is due to proximal vessel calcification and not a blocked stent.

Finally, a technical challenge that is slowly being overcome is the portability of images from the different modalities into a multimodality workstation with the software to perform the image fusion (Gaemperli and Kaufmann 2008). This issue is being helped by the progress in multimodality PACS (Picture Archiving and

Table 18.2 Technical challenges of software cardiovascular image fusion

- Automation
- Cardiac segmentation
- Cardiac motion and respiratory motion
- Nonlinear coregistration/deformation
- Interactive 3D ROI tools
- Image portability
- Quantitative synergism

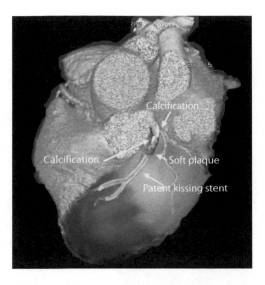

Figure 18.1 Volume-rendered display with cross-sectional LAD visualization. The CTCA is rendered to show the epicardial surface, and quantitated perfusion information is overlaid in color. In this patient, the black region corresponds to a perfusion abnormality in the anterior apical region. Coronary artery data from CTCA are visualized in a multiplanar reformat; that is, the original artery cross-section from the image data is overlaid on the volume rendering. Note that both calcified and soft plaque can be identified and the patency of the stent can be evaluated. The abnormality seen in the LAD can be corresponded with the perfusion abnormality. CTCA: computerized tomographic coronary angiography; LAD: left anterior descending coronary artery.

Communication Systems) infrastructure and better defined DICOM (Digital Imaging and Communications in Medicine) standards.

18.5.2 SYNERGISTIC CHALLENGES

It has been pointed out that the goal of cardiac imaging fusion should be to generate hybrid or multibrid images where the multimodality information is synergistic to the point that this fused image is greater than the sum of its parts (Bax et al. 2007; Gaemperli and Kaufmann 2008). One of the most direct ways to accomplish this goal is to use quantitative information from one modality to improve the quantitative information of the other modality and to use both quantitative and complementary types of information to improve the overall diagnostic information content. Experts continue to emphasize that the future of cardiovascular multimodality imaging is directly dependent on our ability to improve our quantification of flow, function, and anatomic variables (Bax et al. 2007; Beller 2010).

Importantly, distinction first has to be made as to the differences between qualitative image fusion and quantitative image fusion. In qualitative image fusion, the two modality image sets are scaled and aligned in the same angular projection and then software is used so that one image is "painted" onto the other image set for a diagnostician to visually interpret the two types of information (Gaemperli et al. 2011). Although these are usually aesthetically, nice-looking images, in this qualitative fusion approach, there is no possibility to measure any integrated quantitative parameters or to improve the scientific content of one modality from the other modality. In quantitative image fusion, the software has access to quantitative measurements from one modality, which may be used to improve the image or complement measurements from the other modality.

An example of an application of quantitative image fusion is in the integrated measure of myocardium at risk by incorporating the intersection of quantitative physiologic mass at risk from a SPECT or PET myocardial perfusion imaging study with the quantitative anatomic mass at risk from the same patient's corresponding coronary tree (Faber et al. 2004) (Figure 18.2). Two other examples of improved quantification through synergistic image fusion are (1) the use of the better defined valve plane from a cardiac CT study to define the valve plane of the emission study (Slomka et al. 2009a) and (2) the use of the better defined LV myocardium

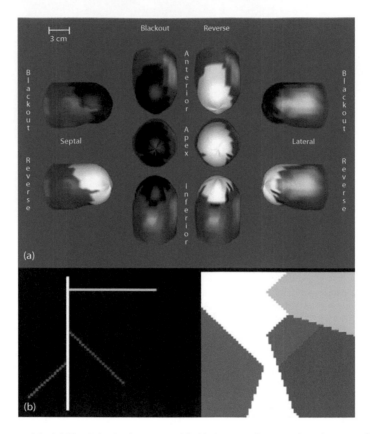

Figure 18.2 Mass at risk. (a) Physiological mass at risk. Various surface-rendered views of the left ventricle (LV) with perfusion color coded onto the surface. Each shows abnormal perfusion during stress (blacked-out regions) and areas that normalize at after rest (white-out regions). White-out regions identify vascular territories that are at risk for myocardial infarction. (b) Anatomical mass at risk. Anatomical mass at risk is related to the territories of the left ventricle supplied by each coronary artery. These two pictures (left and right) should be superimposed but are separated for the sake of clarity. The left picture shows a coronary artery as it would look if the LV were flattened into a 2-D plane. The regions shown on the right picture are those areas that are closer to the artery of the same color than to any other artery. Thus, the green area in the right picture is supplied by the green artery in the left picture. The red area in the right picture corresponds to the arterial bed of that portion of the red artery, which is distal to an occlusion. Thus, this is the area at risk for myocardial infarction, as determined by this model of coronary arteries and their territories plus their alignment with the LV surface. The mass of this area is then computed to obtain the anatomical mass-at-risk.

from the CT scan to correct the emission tracer concentration including for partial volume effect (Li et al. 2013; Suh et al. 2012).

18.5.3 SOCIOECONOMIC CHALLENGES

The technological improvements in the cardiac imaging field of the last decades have allowed to study any given patient with a wide range of methodologies that vary from the technical point of view as well as the clinical and the financial ones. This is further complicated by the interpretation of one composite cardiac image that has been fused with information from multiple modalities. A joint European multiorganization task force (Fraser et al. 2006) has identified that these challenges are related to how clinicians, hospital managers, and insurers adapt to these changes; how the different approaches should be reconciled to the benefit of the patient; and what are the implications for the education and training of future cardiologists, radiologists, and other specialists in cardiovascular imaging. This task force recommends that these challenges may be overcome through a joint cardiac imaging service (Figure 18.3). This joint service would facilitate the clinical

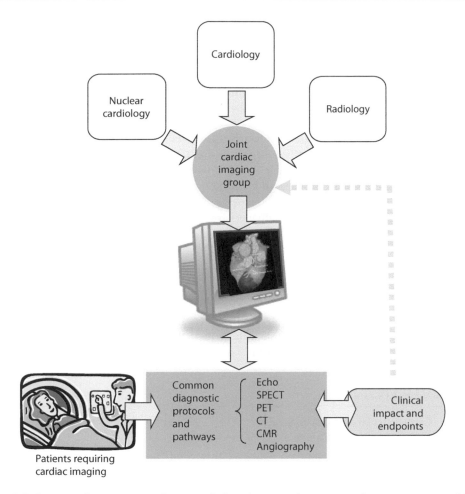

Figure 18.3 Suggested organization of joint, multidisciplinary cardiac imaging diagnostic services. Note the multibrid visualization as the focus of the diagnostic process. (Adapted from Fraser AG et al., *Eur J Echocardiography*, 7, 268–273, 2006.)

diagnostic services where experts in all methods collaborate (Gaemperli et al. 2011). This approach would emphasize an organ-based approach where the diagnostician emphasizes the integration of results into clinical decision making and on the impact on clinical outcomes (Gaemperli et al. 2011).

18.6 CONCLUSION

Multimodality cardiovascular image fusion, whether obtained from hybrid scanning or software fusion alone, has been shown to enhance the process of diagnosing a patient with myocardial flow limiting disease (Faber et al. 2004; Gaemperli et al. 2007; Groves et al. 2009; Jaarsma et al. 2012; Namdar et al. 2005; Rispler et al. 2007; Santana et al. 2009; Slomka et al. 2009). We do agree with those who have determined that integration of multiple types of cardiac information, specifically physiological and anatomical data, is most efficiently achieved by their software fusion acquired on separate scanners (Gaemperli and Kaufmann 2008).

Moreover, software fusion is preferably performed quantitatively over qualitative approaches. Quantitative fusion allows important quantitative cardiac information to be preserved and utilized, and just as important, it allows the synergistic use of mutually complementary quantitative information to improve image and diagnostic accuracy. Although there are a number of technical challenges that still need to be overcome for ideal, clinically useful software fusion, these challenges will be met as research funding continues to fuel the necessary work to perform these developments.

Finally, although the hybrid technology has to date emphasized the acquisition of images from two cardiac imaging modalities to create a third, fused, new image, the focus should change to the visualization process so that more than two types of modalities performed on a patient should be integrated in one multibrid display, interpreted by one cardiac imaging expert, resulting in one clinical report. This multibrid approach will use complementary imaging information to facilitate the detection and quantification of the burden of calcified and noncalcified plaques, the quantification of vascular reactivity and endothelial health, the identification of flow-limiting coronary stenosis, and potentially, the identification of high-risk plaques by using a fusion of morphology and biology with molecularly targeted PET imaging (Di Carli and Hachamovitch 2007; Gaemperli and Kaufmann 2008). Moreover, this multibrid visualization will transcend the diagnostic process, making its way into the operating room of the future for the advancement of minimally invasive cardiac surgery (Kpodonu 2010). Clearly, these clinical advances will be accompanied by socioeconomic challenges that will have to be overcome for the benefit of the patient.

CONFLICTS OF INTEREST

Dr. Ernest Garcia receives royalties from the sale of the Emory Cardiac Toolbox and has an equity position with Syntermed, Inc., which markets Heartfusion. The terms of these arrangements have been reviewed and approved by Emory University in accordance with its conflict of interest policies.

ABBREVIATIONS

CMR	Cardiovascular magnetic resonance
DICOM	Digital Imaging and Communications in Medicine
FAME	Fractional Flow Reserve Versus Angiography for Multivessel Evaluation
MPI	Myocardial perfusion imaging
PACS	Picture archiving and communication systems
ROI	Regions of interest

REFERENCES

Bax JJ, Beanlands RS, Klocke FJ et al. 2007. Diagnostic and clinical perspectives of fusion imaging in cardiology: Is the total greater than the sum of its parts? *Heart* 93:17–22.

Beanlands RS, Chow BJ, Dick A et al. 2007. CCS/CAR/CANM/CNCS/CanSCMR joint position statement on advanced noninvasive cardiac imaging using positron emission tomography, magnetic resonance imaging and multidetector computed tomographic angiography in the diagnosis and evaluation of ischemic heart disease—Executive summary. *Can J Cardiol* 23:107–19.

Beller GA. 2010. Recent advances and future trends in multimodality cardiac imaging. *Heart Lung Circ* 19:193–209.

Beyer T, Freudenberg LS, Townsend DW et al. 2011a. The future of hybrid imaging—Part 1: Hybrid imaging technologies and SPECT/CT. *Insights Imaging* 2:161–69.

Beyer T, Townsend DW, Czernin J et al. 2011b. The future of hybrid imaging—Part 2: PET/CT. *Insights Imaging* 2:225–34.

Beyer T, Freudenberg LS, Czernin J et al. 2011c. The future of hybrid imaging—Part 3: ET/MR, small-animal imaging and beyond. *Insights Imaging* 2:235–46.

Di Carli MF, Hachamovitch R. 2007. New technology for noninvasive evaluation of coronary artery disease. *Circulation* 115:1464–80.

Faber TL, Santana CA, Garcia EV et al. 2004. Three-dimensional fusion of coronary arteries with myocardial perfusion distributions: Clinical validation. *J Nucl Med* 45:745–53.

Faber TL, Santana CA, Piccinelli M et al. 2011. Automatic alignment of myocardial perfusion images with contrast-enhanced cardiac computed tomography. *IEEE Trans Nucl Science* 58:2296–302.

Flotats A, Knuuti J, Gutberlet M et al. 2011. Hybrid cardiac imaging: SPECT/CT and PET/CT. A joint position statement by the European Association of Nuclear Medicine (EANM), the European Society of Cardiac Radiology (ESCR) and the European Council of Nuclear Cardiology (ECNC). *Eur J Nucl Med Mol Imaging* 38:201–12.

Fraser AG, Buser PT, Bax JJ et al. 2006. The future of cardiovascular imaging and non-invasive diagnosis: A joint statement from the European Association of Echocardiography, the Working Groups on Cardiovascular Magnetic Resonance, Computers in Cardiology, and Nuclear Cardiology of the European Society of Cardiology, the European Association of Nuclear Medicine and the Association for European Paediatric Cardiology. *Eur J Echocardiogr* 7:268–73.

Gaemperli O, Bengel FM, Kaufmann P. 2011. Cardiac hybrid imaging. *Eur Heart J* 32:2100–8.

Gaemperli O, Kaufmann PA. 2008. Hybrid cardiac imaging: More than the sum of its parts? *J Nucl Cardiol* 15:123–6.

Gaemperli O, Schepis T, Valenta I et al. 2007. Cardiac image fusion from stand-alone SPECT and CT: Clinical experience. *J Nucl Med* 48:696–703.

Groves AM, Speechly-Dick ME, Kayani I et al. 2009. First experience of combined cardiac PET/64-detector CT angiography with invasive angiographic validation. *Eur J Nucl Med Mol Imaging* 36:2027–33.

Heimann T, Meinzer HP. 2009. Statistical shape models for 3D medical segmentation: A review. *Med Image Anal* 13:543–63.

Jaarsma C, Leiner T, Bekkers SC et al. 2012. Diagnostic performance of noninvasive myocardial perfusion imaging using single-photon emission computed tomography, cardiac magnetic resonance, and positron emission tomography imaging for the detection of obstructive coronary artery disease: A meta-analysis. *J Am Coll Cardiol* 59:1719–28.

Kaufmann PA. 2009. Cardiac hybrid imaging: State-of-the-art. *Ann Nucl Med* 23:325–31.

Kaufmann PA, Di Carli MF. 2009. Hybrid SPECT/CT and PET/CT imaging: The next step in noninvasive cardiac imaging. *Semin Nucl Med* 39:341–47.

Kpodonu J. 2010. Hybrid cardiovascular suite: The operating room of the future. *J Card Surg* 25:704–9.

Li S, Sinusas AJ, Dobrucki LW et al. 2013. New approach to quantification of molecularly targeted radiotracer uptake from hybrid cardiac SPECT/CT: Methodology and validation. *J Nucl Med* 54:2175–81.

Namdar M, Hany TF, Koepfli P et al. 2005. Integrated PET/CT for the assessment of coronary artery disease: A feasibility study. *J Nucl Med* 46:930–35.

Paragios N. 2002. A variational approach for the segmentation of the left ventricle in cardiac image analysis. *Int J Comp Vision* 50:345–62.

Peifer JW, Garcia EV, Cooke CD et al. 1992. 3-D Registration and visualization of reconstructed coronary arterial trees on myocardial perfusion distributions. *Visualization in Biomedical Computing*, Richard A. Robb, Editor, Proc. SPIE 1808, pp. 225–34.

Peifer J, Ezquerra NF, Cooke CD et al. 1990. Visualization of multimodality cardiac imagery. *IEEE Trans Biomed Eng* 37:744–56.

Piccinelli M, Faber T, Arepalli CD et al. 2014. Automatic detection of left and right ventricles from CTA enables efficient alignment of anatomy with myocardial perfusion data. *J Nucl Cardiol* 21:96–108.

Pijls NH, Fearon WF, Tonino PA et al. 2010. Fractional flow reserve versus angiography for guiding percutaneous coronary intervention in patients with multivessel coronary artery disease: 2-year follow-up of the FAME (Fractional Flow Reserve Versus Angiography for Multivessel Evaluation) study. *J Am Coll Cardiol* 56:177–84.

Petitjean C, Dacher JN. 2011. A review of segmentation methods in short axis cardiac MR images. *Med Image Anal* 15:169–84.

Rispler S, Keidar Z, Ghersin E et al. 2007. Integrated single-photon emission computed tomography and computed tomography coronary angiography for the assessment of hemodynamically significant coronary artery lesions. *J Am Coll Cardiol* 49:1059–67.

Santana CA, Garcia EV, Faber TL et al. 2009. Diagnostic performance of fusion of myocardial perfusion imaging (MPI) and computed tomography coronary angiography. *J Nucl Cardiol* 16:201–11.

Saraste A, Knuuti J. 2012. Cardiac PET, CT, and MR: What are the advantages of hybrid imaging? *Curr Cardiol Rep* 14:24–31.

Sasaki M. 2008. Multi-brid inflation and non-Gaussianity. *Prog Theor Phys* 120:159–74.

Slomka PJ, Berman DS, Germano G. 2008. Application and software techniques for integrated cardiac multi-modality imaging. *Expert Rev Cardiovasc Ther* 6:27–41.

Slomka PJ, Baum RP. 2009. Multimodality image registration with software: State-of-the-art. *Eur J Nucl Med Mol Imaging* 36:S44–55.

Slomka PJ, Cheng VY, Dey D et al. 2009. Quantitative analysis of myocardial perfusion SPECT anatomically guided by coregistered 64-slice coronary CT angiography. *J Nucl Med* 50:1621–30.

Suh JW, Kwon OK, Scheinost D et al. 2012. CT-PET weighted image fusion for separately scanned whole body rat. *Med Phys* 39:533.

Tsai A, Yezzi A, Wells W et al. 2003. A shape-based approach to the segmentation of medical imagery using level sets. *IEEE Trans Med Imaging* 22:137–54.

Woo J, Slomka PJ, Dey D et al. 2009. Geometric feature-based multimodal image registration of contrast-enhanced cardiac CT with gated myocardial perfusion SPECT. *Med Phys* 32:5467–79.

Concerns with radiation safety

MATHEW MERCURI AND ANDREW J. EINSTEIN

19.1 INTRODUCTION

The use of medical imaging has greatly improved our ability to accurately diagnose disease and manage patient care. However, as in all aspects of health care, it is important that the benefits of a procedure outweigh the risks involved. Whereas the benefits of medical imaging, while often difficult to quantify, are widely appreciated, optimizing this benefit-to-risk ratio requires that we know something about the risks involved. Most radiologic and all nuclear medicine procedures expose patients and operators to ionizing radiation. Radiation exposure can result in detrimental health effects, including skin injury, genetic defects, and cancer. Thus, it is important that physicians both avoid inappropriate procedures and limit exposure to levels that are "as low as reasonably achievable" (ALARA principle) (International Commission on Radiological Protection [ICRP] 103, 2007).

Hybrid imaging utilizes both radiologic and nuclear medicine modalities. The combined use of these modalities may result in enhanced imaging capabilities. However, from the perspective of radiation protection, hybrid imaging poses some unique issues for patients and operators. For example, the use of two or more radiologic or nuclear medicine modalities will expose the patient to radiation from each modality, which may result in a relatively higher dose of radiation than what is expected from a single modality. This may also result in greater potential for radiation-related effects. Furthermore, radiologic and nuclear medicine modalities differ in how they produce (and thus expose patients to) radiation, creating challenges in monitoring and quantifying the total exposure and subsequent dose from hybrid procedures that use both.

This chapter will outline some special considerations for radiation safety when using hybrid imaging. We will begin with an outline of the common terms used in radiation monitoring and the types of injury and other negative health effects that can result from radiation exposure. Next, we will review the history of ionizing radiation use in medicine and the evidence for radiation-induced negative health effects based on epidemiologic and laboratory studies. Special attention will be given to clinical cardiac populations. We will then review the different imaging modalities that use ionizing radiation. This will include an examination of how radiation is produced, how exposure/dose is measured, and the range of expected exposures/doses for each modality. Herein, we will present some measurement issues and special considerations for radiation safety that arise from the integration of radiologic with nuclear medicine imaging modalities. The chapter concludes with suggestions on how to minimize radiation dose to patients undergoing hybrid imaging, ways of communicating risks to patients, and considerations for following patients who have or are expected to receive a high dose. Throughout this chapter, we identify gaps in the current literature and present considerations for future study.

19.2 TERMINOLOGY

19.2.1 IONIZING VS. NON-IONIZING RADIATION

Many devices used in medicine utilize electromagnetic radiation in order to image anatomical features or physiologic processes for the purpose of diagnosis or to guide an intervention. In many cases, this radiation may be of sufficient kinetic energy to liberate an electron from an atom or molecule and thus form ions when interacting with matter. This is known as ionizing radiation. Ionizing radiation is required to produce the tissue reactions and stochastic effects as described in Section 19.2.4 (Radiation-induced negative health effects). Radiologic modalities, such as computerized tomography (CT) and fluoroscopy, emit ionizing radiation in the form of x-rays. Nuclear medicine modalities, such as single-photon emission CT (SPECT) and positron emission tomography (PET), require the use of radiopharmaceuticals that emit ionizing radiation in the form of both gamma and x-rays. Some imaging modalities are less harmful to biological tissue on account that they do not emit energy sufficient to strip electrons. For example, magnetic resonance imaging (MRI) uses radio waves, which are of insufficient energy to be ionizing. Likewise, ultrasound devices use high-frequency sound (or pressure) waves to produce images. Sound waves differ from electromagnetic radiation in how they interact with matter, including biological tissue.

19.2.2 DOSE METRICS

Ionizing radiation produced by medical imaging modalities may be in the form of either x-rays or gamma-rays. Biological tissue that is exposed to such radiation may absorb some of the energy. The *absorbed dose* (D) is defined as the amount of energy per tissue unit mass, and is measured in Grays (Gy), a special term for Joules per kg used in the context of radiation protection. Different types of ionizing radiation vary in their potential to damage biological tissue. The *equivalent dose* (H) accounts for such differences by applying a "radiation weighting" factor (w_r) to the absorbed dose. Thus, the equivalent dose is the product of the radiation weighting factor and the absorbed dose ($H = w_r \times D$). For x-rays, this weighting factor is 1. Biological tissues differ in their sensitivity to radiation. The *effective dose* (ED) accounts for such differences and can be estimated from the weighted sum of the equivalent doses to each tissue. The equivalent doses are weighted according to tissue weighting factors (w_t). The sum of w_t over all body tissues is 1. Both the equivalent and effective doses are measured in Sieverts (Sv), another special term, used for a weighted Gray. The *collective ED* denotes the sum of all EDs to all individuals in a specified population over a specified period of time. The total dose to an individual from repeated exposures is commonly referred to as the *cumulative ED*.

19.2.3 Exposure metrics

The dose metrics described previously cannot be directly measured for an individual undergoing a procedure. Thus, it is unlikely that such information will be available to healthcare providers for monitoring and tracking purposes. However, additional, more easily measured metrics have been developed that can provide an indication of exposure. These can be used as quality metrics, such as in setting diagnostic reference levels (often referred to as DRLs), for the purpose of minimizing patient risk.

Exposure from radiologic imaging modalities can be measured using *air kerma* (AK). While AK typically denotes the amount of kinetic energy of the liberated charged particles per unit mass (in this case air), the exposure monitoring equipment actually measures the amount of energy transferred to the charged particles. This AK is measured at some defined distance from the radiation emitting source and is measured in Grays. A related measure is the *kerma area product* (KAP), which takes into account the area of the x-ray field. The KAP is measured as the product of Grays and the area of exposure (e.g., $Gy \times cm^2$). Due to the three-dimensional (3-D) nature of the examination, the estimation of radiation exposure from CT requires a consideration of volume. The *CT dose index* ($CTDI_{vol}$) indicates the radiation dose from a single CT slice, based on a phantom model. In order to account for the radiation from the entire scan, the $CTDI_{vol}$ is multiplied by the irradiated length (i.e., length of the body portion scanned) to yield the *dose length product* (DLP), measured in Grays per unit length (e.g., $Gy \times cm$). While this metric intends to estimate dose, the individual patient may differ in size and composition from the phantom, and thus, DLP may indeed be more an indication of radiation exposure than of dose. However, these issues may be somewhat mitigated through the use of the size-specific dose estimate, a variation of the $CTDI_{vol}$ that reflects anthropometric parameters of the patient (Boone et al. 2011). Nuclear medicine examinations expose the patient to radiation via an injected radiopharmaceutical rather than an external ionizing radiation field. As the radiopharmaceutical decays, ionizing radiation is emitted. *Administered activity* indicates the number of decays per second of the radiopharmaceutical and is measured in units of Becquerels (Bq). Radiation exposure from nuclear medicine is typically expressed using administered activity.

Radiation exposure metrics can be used to estimate the ED from a procedure. With knowledge of spatial parameters of the examination and patient-specific anthropometric data, exposure metrics from radiologic examinations (specifically, fluoroscopy-based procedures), such as KAP, can be used to estimate organ and whole-body EDs (National Council on Radiation Protection and Measurements [NCRP] 168, 2010). ED from CT can be estimated by multiplying the DLP by a conversion factor (Huda et al. 2008). The value of the conversion factor is dependent on what part of the body is scanned. Estimates of ED from nuclear medicine procedures can be calculated using models that account for administered activity, the radionuclide used, and standard patient biokinetic characteristics and habitus (ICRP 53, 1988; 80, 1998; 106, 2008). While ED was developed as a population-level metric and thus should not be considered an accurate depiction of radiation dose in an individual patient, converting exposure metrics to ED estimates facilitates the direct comparison of patient doses (and thus, relative potential risk) between different imaging modalities. Furthermore, this allows for some estimation of cumulative dose for patients who undergo one or more procedures using different modalities, including hybrid imaging techniques such as PET/CT or SPECT/CT.

19.2.4 Radiation-induced negative health effects

Exposure to ionizing radiation can cause detrimental health effects. Such effects include, but are not limited to, skin injury, cataracts, genetic disorders, and cancer (e.g., Dunlap 1942; Ron 1998; Koenig et al. 2001; Pearce et al. 2012). Radiation-induced negative health effects can be divided into two general categories: tissue reactions and stochastic effects.

19.2.4.1 TISSUE REACTIONS

In the past, some radiation-related effects were considered *deterministic* in nature, in that the severity of damage to biological tissue increased relative to the radiation dose, with defined dose thresholds below which injury could not occur. These effects include skin depilation, erythema, and necrosis; cataracts; and lung fibrosis. Visible effects of radiation occur within a few hours to a few weeks from the time of exposure to ionizing radiation, with possible additional late effects many years after exposure. The mechanism for resulting injury was understood to be mainly from DNA or chromosomal damage leading to cellular death (Stone et al. 2003). However, we now know that cellular death cannot explain all the effects observed in irradiated tissue. This is especially true for late (often more severe) effects. Observed tissue injury may be the result of multiple processes, including cellular death (clonogenic cell death and/or apoptosis), and inflammatory responses that alter cell function and molecular cell signaling (Denham et al. 2001; ICRP 118, 2012). Thus, resulting injuries are no longer considered strictly "predetermined" from dose, but rather modifiable (ICRP 118, 2012). A more suitable term for this class of radiation-related effects is *tissue reactions*. The threshold for a given tissue reaction is typically defined as the dose above which there is a 1% incidence of that injury (ICRP 118, 2012). For example, skin erythema can occur above a peak skin dose of 2 Gy. More serious effects, such as skin necrosis or desquamation, can occur at much higher peak skin doses (>15 Gy) (see Koenig et al. 2001; Balter et al. 2010).

19.2.4.2 STOCHASTIC EFFECTS

Stochastic effects are those where the probability of the effect occurring is a function of the radiation dose but the severity of that effect is not. Stochastic effects include genetic disorders and cancer (e.g., solid tumor, leukemia). The mechanism for stochastic effects is DNA alterations in living cells. Unlike tissue reactions, stochastic effects can occur at any dose (i.e., there is no threshold). However, there is controversy regarding the validity of this assertion, which we will cover in more detail in Section 19.4.2 (Evidence for stochastic effects).

19.3 BRIEF HISTORY OF RADIATION USE IN MEDICINE

The end of the nineteenth century brought advancements in physics that made modern medical imaging possible. First was the discovery of x-rays by Roentgen in 1895, which led to the first x-ray photograph, a precursor to the radiogram. Roentgen's application of x-ray photography was almost immediately put to medical use. This technique, facilitated by the widespread availability of the cathode-ray tube, provided the basis for all radiologic imaging modalities. Next was the discovery of radioactivity in 1896, credited to both Henri Becquerel and Marie and Pierre Curie. Radioactivity would provide the basis for nuclear medicine.

Over the next three decades, medical applications of ionizing radiation were therapeutic or for simple 2-D imaging of bone. In 1924, de Hevesy and colleagues introduced a radiotracer method to study bone metabolism in animals. This technique allowed for additional diagnostic methods, such as investigation of *in vivo* circulation or blood flow. Imaging the distribution of an injected or ingested radiopharmaceutical was made possible with the development of scintillation detectors (e.g., gamma camera) in the 1950s. Again, these images were restricted to a single plane. By the 1960s, computer technology and software development had advanced to where radiopharmaceutical emissions data could be reconstructed to produce a topographical "mapping" of the imaged portion of the body. As described in the previous chapters of this book, this led to the development of SPECT, PET, and the x-ray-based CT. Thus, by the early 1970s, medical imaging could be used to acquire 3-D representations of the human body. More recent technological innovations have produced hybrid imaging techniques as focused in this book. These combine functional information from nuclear medicine with anatomical information from radiologic modalities. Hybrid techniques using ionizing radiation include SPECT/CT and PET/CT as introduced respectively in Chapters 1 and 2 of this book. More recent hybrid imaging capitalizes on the 3-D capabilities of MRI technology, in the form of PET/MRI as described in Chapter 4 of this book.

The use of medical imaging has increased steadily since its introduction to medicine over a century ago. However, the number of procedures per capita has increased dramatically in recent decades (United Nations Scientific Committee on the Effects of Atomic Radiation [UNSCEAR] 2008; Smith-Bindman et al. 2012).

This is likely due to both technological innovation and its application for increasingly complex procedures. Commensurate with increasing use is a dramatic rise in the collective dose. The NCRP estimates that during the period 1980 to 2006, the U.S. population's annual per capita radiation ED from (non-therapeutic) medical sources increased sixfold, such that the proportion of the total dose attributed to medical imaging sources is now approximately equal to that from all other sources combined (NCRP 160, 2009). Worldwide, the estimated contribution of medical sources to the collective dose doubled over the same period (UNSCEAR 1988, annex C; 2008, annex A). Much of this increase is attributed to CT, where although individual examination doses have decreased, the rate of use has increased (NCRP 160, 2009; Smith-Bindman et al. 2012).

As the use of medical imaging had increased over the past century, regulation and safety standards have evolved to meet our understanding of the negative effects of ionizing radiation. Early on, radiation safety was focused at limiting doses to levels that would not produce visible effects, such as skin injury or hair loss (Jones 2005). While there were restrictions for medical use, much of the safety initiatives were aimed at minimizing occupational exposure (Jones 2005). In the late 1920s, organizations whose purpose was to provide scientific support and provide recommendations regarding protection began to form. Such organizations include precursors to the ICRP, and the NCRP. By the mid-twentieth century, the recommendations of these organizations began to include concern for those effects that are more difficult to both observe and directly attribute to radiation exposure, such as cancer or genetic damage. Recognizing the need to balance the substantial benefits of medical imaging with the potential negative effects of ionizing radiation, modern radiation protection in medicine is primarily focused on a program of procedural justification, and dose optimization (ICRP 26, 1977; 103, 2007).

19.4 EVIDENCE FOR RADIATION-INDUCED EFFECTS

19.4.1 EVIDENCE FOR TISSUE REACTIONS

Immediately after the discovery of x-rays, evidence of tissue reactions directly attributable to radiation was reported in the scientific literature (Codman 1896; Daniel 1896; Miller 1995). These were for the most part skin injury and hair loss. *In vitro* and *in vivo* laboratory studies in animals demonstrate the damaging effects of ionizing radiation on biological tissue (Biological Effects of Ionizing Radiation [BEIR] VII 2006). Evidence of radiation-related tissue reactions in clinical populations is based primarily on case studies of patients undergoing fluoroscopy-guided procedures, such as cardiac catheterization or electrophysiology studies (U.S. Food and Drug Administration 1994; Koenig et al. 2001). Cohort studies in cardiac populations have thus far yielded few tissue reactions (Vano et al. 2001, 2005; Padovani et al. 2005). This is expected, however, as fluoroscopy-guided cardiac procedures typically administer a peak skin dose far below the threshold to induce skin injury.

19.4.2 EVIDENCE FOR STOCHASTIC EFFECTS

Reports of ionizing radiation-induced stochastic effects among laboratory workers and radium dial painters soon followed the emergence of tissue reactions (Miller 1995; Fry 1998). Radiation-induced leukemia was first reported in 20 radiologists and radioisotope workers in the 1940s (Dunlap 1942). Traditionally, evidence for stochastic effects has come from cohort studies of atomic bomb survivors, patients undergoing radiation therapy, and nuclear industry workers (Cardis et al. 1995; Ron 1998, 2003; Little 2001a,b; Little et al. 2009), and more recently from studies of pediatric populations undergoing CT procedures (Pearce et al. 2012; Mathews et al. 2013). The pediatric CT population aside, the radiation dose for individuals in these populations is typically higher than that seen in clinical populations undergoing radiologic or nuclear-medicine-based examinations or procedures. The magnitude of risk of stochastic effects resulting from radiation doses in the range typical of medical imaging is less certain compared to that in the relatively higher dose populations described earlier. Brenner and colleagues (2003) argued that there is good evidence of stochastic risk with acute doses of >50 mSv and protracted doses of >100 mSv and reasonable evidence at very low acute doses (i.e., >5 mSv). Models have been developed to project risk in doses <50 mSv. The most common is the "linear non-threshold" (LNT) model, which assumes that there is potential for stochastic effects at any dose and that risk for such increases linearly with dose (NCRP 136, 2001; Upton 2003).

The LNT model has been challenged for low doses on the basis that it does not account for natural cellular level protective mechanisms (e.g., DNA repair, cell elimination) or adaptive responses (e.g., hormesis/radiation homeostasis), whereby an initial radiation exposure prepares a cell to withstand subsequent larger exposures, thus reducing the magnitude of biological effects (Olivieri et al. 1984). On the contrary, cellular interactions, whereby irradiated cells signal nearby unirradiated cells to undergo mutogenesis (i.e., bystander effects) may result in damage in excess of LNT model predictions (Zhou et al. 2000; Brenner et al. 2001; Hall 2003). Thus, other models, including linear quadratic, threshold, downward curving, and hormetic have been proposed. Despite these concerns, the NCRP (136, 2001), ICRP (99, 2005), and the UNSCEAR (2010), based on reviews and syntheses of the existing evidence, regard the LNT model as that best representing the current evidence and advocate its use for purposes of protection and regulation. Furthermore, a comprehensive review of the evidence presented in the 2006 BEIR report concluded that, for purposes of radiation protection, the LNT model is the one that best fits the available data (BEIR 2006).

There is a scarcity of evidence of stochastic effects in clinical populations undergoing medical imaging. This is likely due to the substantial sample size and long-term follow-up requirements that studies need to overcome the number of potential confounders and the multiyear latency period between dose and observed onset of disease, respectively (Land 1980). A number of studies, specifically focused at children, have been proposed or are currently in progress to overcome these obstacles (Einstein 2012). Recently completed studies of children undergoing CT demonstrate an increased risk of solid tumor formation (most notably brain cancer) and leukemia (Pearce et al. 2012; Mathews et al. 2013). The findings of these studies provide support for the LNT model.

19.5 IMAGING MODALITIES THAT USE IONIZING RADIATION

19.5.1 RADIOLOGIC PROCEDURES

Radiologic imaging modalities utilize x-rays emitted from a source external to the patient. These x-rays are generated when an electron emitted from a cathode inside a vacuum tube removes an electron from a target atom (located on an anode). An electron from a higher shell fills the vacancy on the target atom, emitting an x-ray in the process. The stream of x-rays emitted from the vacuum tube is of sufficient energy to be considered ionizing radiation. When the x-ray beam comes into contact with biological tissue, it may be absorbed, scattered, or pass through without incident. As mentioned previously, absorbed x-rays can cause damage to biological tissue.

Radiologic procedures include conventional radiography (e.g., orthopedic and chest x-rays, mammography), fluoroscopy, and CT. These procedures collectively account for the majority of imaging examinations. As a result of their frequency of use and the magnitude of the acute dose from some modalities, this class of procedures accounts for the majority of the population collective radiation dose attributable to non-therapeutic medical sources. The biggest culprit is CT, accounting for roughly half of this collective dose (NCRP 160, 2009). The ED from a CT examination can vary dramatically depending on the procedure parameters, and the body area imaged. Current dose estimates for cardiac CT indicate that EDs of <1 mSv can be achieved with modern equipment and best practices (Achenbach et al. 2010; Leipsic et al. 2010; Schuhbaeck et al. 2013). This is lower than fluoroscopy-driven cardiac diagnostics (e.g., cardiac catheterization) and even (in the best case scenario) in the range of a chest x-ray (average ED approximately 0.1 mSv) (Mettler et al. 2009; Schuhbaeck et al. 2013).

CT is the most commonly used radiologic modality in hybrid imaging. Thus, consideration of radiation dose from CT is noteworthy in hybrid imaging. Furthermore, the use of CT in conjunction with nuclear medicine modalities (e.g., PET, SPECT) is becoming more common in clinical practice. In fact, PET/CT systems are increasingly replacing standalone PET, as such systems are no longer available from many imaging device manufacturers (Hricak et al. 2010).

19.5.2 NUCLEAR MEDICINE PROCEDURES

Imaging modalities used in nuclear medicine rely on the radioactive decay of inhaled, injected, or ingested radiopharmaceutical to produce images of physiological processes and function. For example, in PET, when the radiopharmaceutical decays, it releases positrons. These positrons collide with electrons. This collision results in an annihilation of both particles and the release of gamma rays. The gamma rays are "collected" by detectors to produce images. Thus, radiologic and nuclear medicine imaging modalities differ with respect to radiation in two important ways: (1) the source of the radiation (external to the patient vs. internal) and (2) the type of radiation (x-ray vs. both gamma ray and x-ray).

Like x-rays, gamma rays are of sufficient energy to be considered ionizing radiation. However, gamma rays differ from x-rays. First, gamma rays are formed inside the nucleus of an atom, whereas x-rays emanate from an atomic orbital. Second, the gamma rays used in medical imaging typically have a shorter wavelength and, thus, higher energy. This allows gamma rays used in medical imaging to more easily penetrate matter. For purposes of radiation protection, gamma rays and x-rays are considered to be reasonably similar in their potential to damage biological tissue (i.e., "relative biological effectiveness" [RBE]) (ICRP 60, 1991; 92, 2003). Thus, all gamma rays and x-rays are assigned the same radiation weighting factor when calculating an equivalent dose (ICRP 103, 2007). This is notwithstanding some laboratory evidence, which suggests that low-energy x-rays have a higher RBE compared to high-energy x-rays and gamma rays, due to their ability to produce lower energy secondary electrons (Hill 2004; Hunter and Muirhead 2009).

Nuclear medicine includes procedures using PET or SPECT systems. Radionuclide myocardial perfusion imaging (MPI) is a common nuclear medicine diagnostic procedure in cardiology. The estimated dose from typical nuclear cardiology procedures range from 2 to >20 mSv, depending on the type of procedure and the radioisotope and protocol used (Einstein et al. 2007; Mettler et al. 2008). For example, a cardiac stress test using 3.5 mCi (129.5 MBq) thallium-201 has an estimated ED of 22 mSv, compared to one using 27.5 mCi (1017.5 MBq) ^{99m}Tc-sestamibi where the ED is estimated at 8 mSv (Einstein et al. 2007). Actual patient doses from nuclear medicine procedures vary from estimated doses as a function of individual patient biokinetics and habitus (Mercuri et al. 2012). While nuclear-medicine-based imaging is used less often than radiologic procedures, due to its relatively higher average acute dose, nuclear medicine accounts for a quarter of the collective dose to the U.S. population from medical sources (NCRP 160, 2009).

19.5.3 SPECIAL CONSIDERATIONS FOR HYBRID IMAGING

Although various hybrid imaging techniques exist, those that combine PET or SPECT with CT are of most concern from the perspective of radiation protection. The combination of these modalities brings about multiple challenges for radiation dose recording and monitoring and subsequent protection strategies. The first challenge is related to qualitative differences between radiologic (i.e., CT) and nuclear medicine modalities (i.e., PET or SPECT) in how radiation is produced and measured. In radiologic modalities, which use radiation fields that are generated externally to the patient, the magnitude of exposure (and subsequent dose) is contingent on the intensity of the radiation field, distance from the x-ray tube, field size, and duration exposure. Exposure is quantified by the amount of energy transferred from the radiation field to the tissue (or air) in question and is estimated using kerma-based metrics. Contrast this with nuclear medicine modalities, where radiation exposure to the patient is due to the radioactive decay of pharmaceutical agents. In this case, exposure is quantified using the administered activity relative to the concentration of the radiopharmaceutical. Unfortunately, these different exposure metrics are incompatible on account that they are measured in different units (i.e., Grays vs. Becquerels), and thus, the individual radiation burden from each modality used in the hybrid procedure must be reported separately.

One solution is to convert the different exposure metrics into a common measure. As mentioned previously, ED may serve as a means to quantify the burden of radiation risk to the patient from qualitatively different modalities. Again, it should be emphasized that ED was not developed for use at the individual level; however, it may be adequate for the purpose of estimating relative dose until which time a more valid and reliable measure is developed. However, estimates of ED from individual hybrid imaging procedures are more prone to error than are individual radiologic or nuclear medicine procedures. Whereas radiologic procedure doses can vary relative to differences in patient body size and tissue distribution relative to the radiation source, and nuclear medicine procedures relative to patient biokinetics and habitus, dose estimates from individual hybrid imaging procedures are affected by all of these factors.

Table 19.1 Typical EDs from nonmedical exposures and medical imaging sources

Source	Typical ED (mSv)
Nonmedical exposures	
Backscatter for airport screening	0.0008
One way flight, Helsinki to New York	0.05
Miner or nuclear industry worker (annual)	2
Background radiation to public (annual, worldwide)	2.4
Average annual limit, radiation workers	20
Lifetime occupational limit (Germany)	400
Noncardiac medical imaging	
Chest x-ray, posteroanterior	0.02
Chest x-ray, posteroanterior and lateral	0.1
Mammogram	0.7
Head CT	2
Abdominal CT	10
Nuclear cardiology	
Lowdose ^{99m}Tc stress-only (450 MBq)	3
One day rest/stress or stress/rest ^{99m}Tc (450/1350 MBq)	13
Two day ^{99m}Tc (750/750 MBq)	11
^{201}Tl rest-redistribution (92 MBq)	11
Dual isotope (U.S. protocol) (120 MBq Tl/1110 MBq ^{99m}Tc)	22
^{18}F Fluorodeoxglucose (275 MBq)	5
^{82}Rb rest/stress (1665/1665 MBq)	2
^{13}N-ammonia rest/stress (555/555 MBq)	2
^{15}O-water rest/stress (500/500 MBq)	1
^{153}Gd Line Source attenuation correction	0.001
Cardiac CT	
Attenuation correction	
CT attenuation correction scan	< 1
Calcium scoring	
Electron beam CT	1
Multidetector-row CT	3
Coronary CT angiography	
Prospectively triggered, 100 kVp	2
Prospectively triggered, 120 kVp	3
Retrospectively gated ESTCM, 120 kVp	14
Retrospectively gated, 120 kVp	20
Cardiac catheterization	
Diagnostic catheterization	7
Percutaneous coronary intervention	20

Source: Adapted from Einstein AJ, Knuuti J, *Eur Heart J*, 33, 573–578, 2012.

The second challenge is that radiation doses from individual CT, PET, and SPECT procedures may be relatively high among all medical imaging procedures, and thus, when combined, the result has the potential for a relatively high dose of radiation to the patient, compared to other imaging modalities. For example, in one report, ED from undergoing a combination myocardial perfusion and coronary angiography study with SPECT/CT using a dual isotope protocol was estimated as high as 41.5 mSv (Rispler et al. 2007). This likely represents the worst case scenario, as protocol optimization in each modality, facilitated by advances in technology since the time of that study, can greatly reduce the dose. Typical EDs from individual cardiac imaging exams, the components of hybrid imaging studies, are presented in Table 19.1, where they are compared to doses from other medical and nonmedical radiation sources. It is evident here that use of optimized technique, e.g., performing MPI using a low-dose ^{99m}Tc-based protocol rather than a dual isotope protocol, can result in a considerable reduction in dose to the patient.

In the published literature on hybrid cardiac imaging, a variety of protocols have been used, with a corresponding variety of radiation doses; these are summarized in Table 19.2. This range is well illustrated in a study by Husmann et al. (2009), who performed stress/rest MPI with ^{99m}Tc-tetrofosmin and CT angiography repeated twice, using both prospective triggering and retrospective gating. ED from the MPI portion of the study averaged about 10 mSv; ED from CT angiography averaged 2.2 mSv for prospective triggering but 19.7 mSv for retrospective gating. Thus, the total dose for the hybrid study was 11.8 mSv using the prospective triggering CT protocol and 30.4 mSv for the retrospective protocol. When considering just the stress portion of the MPI portion, i.e., stress-only imaging, the total dose for the study was 5.4 mSv and 24.1 mSv, respectively.

Even where it is possible to integrate the radiation burden from each modality used in hybrid imaging, a further challenge is how to prospectively record and monitor doses. Exposure information and subsequent dose estimates can be incorporated into imaging devices. Many imaging devices used in modern radiologic procedures offer a means to both estimate and record radiation exposure. This information, used in conjunction with procedure and patient parameters required to estimate dose, could be transferred to the Digital Imaging and Communications in Medicine (DICOM) header as part of a standard dose report. However,

Table 19.2 Reported EDs for cardiac hybrid imaging procedures

Reference	Imaging device	Study type	Radiopharmaceutical	Estimated ED (mSv)
Rispler et al. 2007	SPECT/CT	R/S MPI w. CTAC + rE-g CTA + CS	^{201}Tl (R), ^{99m}Tc-sestamibi (S)	41.5
Preuss et al. 2008	SPECT/CT	R/S MPI w. CTAC	^{99m}Tc-tetrofosmin or ^{99m}Tc-sestamibi	9.8
Husmann et al. 2009	SPECT/CT	S/R MPI w. CTAC + pE-t CTA	^{99m}Tc-tetrofosmin	11.8
Husmann et al. 2009	SPECT/CT	S/R MPI w. CTAC + rE-g CTA	^{99m}Tc-tetrofosmin	30.4
Husmann et al. 2009	SPECT/CT	S MPI w. CTAC + pE-t CTA	^{99m}Tc-tetrofosmin	5.4
Husmann et al. 2009	SPECT/CT	S MPI w. CTAC + rE-g CTA	^{99m}Tc-tetrofosmin	24.1
Pazhenkotti et al. 2010[a]	SPECT/CT	pE-t CTA + S/R MPI w. CTAC	^{99m}Tc-tetrofosmin	4.6–8.6
Pazhenkottil et al. 2011	SPECT/CT	S/R MPI w. CTAC + pE-t CTA	^{99m}Tc-tetrofosmin	12
Herzog et al. 2011[b]	SPECT/CT	S MPI w. CTAC + pE-t CTA	^{99m}Tc-tetrofosmin	4.7
Flotats et al. 2011[c]	PET/CT	R/S MPI w. CTAC + pE-t CTA + CS	^{82}Rb	13–15
Kajander et al. 2009	PET/CT	R/S MPI w. CTAC + pE-t CTA + CS	^{15}O-water	9.5
Kajander et al. 2009	PET/CT	R/S MPI w. CTAC + rE-g CTA + CS	^{15}O-water	22
Gould et al. 2008[d]	PET/CT	R/S MPI w. CTAC + CTA	^{82}Rb	7

Note: R: rest; S: stress; CTA: Coronary computed tomography angiography; CTAC: CT attenuation correction; pE-t: Prospective ECG-TRIGGERING; rE-g: retrospective ECG-gating; CS: calcium score.

[a] MPI was not performed if CTA showed normal result.
[b] ^{99m}Tc administered activity adapted to BMI was as follows: BMI < 25 kg/m^2: 150 MBq, BMI ≥ 25 kg/m^2: 200 MBq.
[c] Estimate based on data from Javadi et al. (2008).
[d] CTAC for whole exam was based on the post stress test only.

critical information necessary to estimate dose from nuclear medicine procedures is not currently part of the system described previously (Mercuri et al. 2012). Therefore, a process must be developed such that this information (e.g., administered activity, radiopharmaceutical administered, etc.) can be entered into the system and then integrated with dose information from the CT component to provide a total dose estimate for a given procedure. This will require the joint effort of medical societies to develop consensus on what should be recorded and how dose should be estimated and manufacturers to develop a means to capture and transmit this information to the DICOM header (Mercuri et al. 2012).

19.6 MINIMIZING RISK

The best way to minimize risk from radiation exposure is to ensure that only those patients who can benefit from undergoing an ionizing radiation-based imaging examination do so (i.e., justification) and that the minimal amount of radiation needed to acquire the clinically necessary information is used (i.e., optimization). Although optimization strategies can significantly reduce exposure and dose, judicious use of appropriateness criteria may have the most impact. Physicians should limit referring patients for hybrid imaging to when the diagnostic utility is expected to be high or where indicated by evidence-based clinical practice guidelines. Referring physicians should work closely with radiologists and imaging cardiologists to ensure that procedures are appropriate. Wherever possible, one could consider using hybrid devices where one component modality does not use ionizing radiation. For example, PET/MRI may be a useful alternative to PET/CT in the event the physicians seek to minimize radiation exposure to the patient. Furthermore, physicians should ensure that each component of the hybrid imaging protocol is essential to yield the necessary clinical information. For example, when performing a myocardial perfusion examination with SPECT/CT or PET/CT, the CT component may be useful for attenuation correction and anatomical reference but may also offer opportunity for investigation of coronary calcium, obviating the need for an additional CT scan for calcium scoring (Einstein et al. 2010). When using hybrid imaging technology, routine acquisition of both functional and anatomical information may be unwarranted unless there is strong clinical justification for both. Thus, hybrid imaging technology offers greater opportunity for "test layering," which in some cases may be unnecessary.

Once the appropriate examination is selected, dose optimization can be achieved through the use of best practices for a given procedure as described in the next two paragraphs. However, with hybrid imaging, dose optimization strategies must account for each component modality—how dose is optimized differs between radiologic and nuclear medicine modalities. Again, it is important that clinicians find the appropriate balance between dose reduction and image quality, so that the utility of the investigation is maintained.

Radiation exposure from CT is primarily determined by scan parameters. Reducing x-ray tube current and potential can greatly reduce radiation dose. These parameters should be chosen in consideration of patient morphology and habitus, as well as the image reconstruction algorithm used. In general, iterative reconstruction methods enable the use of lower tube current and/or potential (Nelson et al. 2011). When performing coronary CT angiography studies, the use of ECG-controlled tube current modulation reduces the exposure to the patient. Newer scan modes such as axial step-and-shoot, volume, and high-pitch helical imaging reduce dose even more (Trattner et al. 2014).

Protocol selection can greatly affect the magnitude of exposure from nuclear medicine procedures. When performing MPI in patients with low pretest probability of coronary artery disease, one could consider the use of "stress-only" or "stress-first" protocols in patients (Chang et al. 2010). Radiopharmaceutical selection can influence the magnitude of dose. For example, dual-isotope protocols show higher doses despite lower exposure rates, due to longer half-life of the radiopharmaceutical. Radiation dose reduction can be achieved with the use of ^{99m}Tc agents when performing SPECT, wherever possible (Einstein et al. 2007). Recent studies suggest that the use of a Rubidium-82 or a N-13 Ammonia PET protocol has the potential to reduce dose further compared to standard SPECT techniques (Einstein et al. 2007; Dilsizian et al. 2009; Hunter et al. 2010; Senthamizhchelvan et al. 2010; Small et al. 2013).

19.7 CONCLUSION

Hybrid imaging technology can enhance a clinician's ability to acquire information necessary to improve diagnosis and optimize patient outcomes. However, this technology presents challenges for reducing risk from radiation exposure. Risk of radiation-related effects could be reduced through a combination of appropriate use, including avoidance of unneeded layered testing, and dose optimization. Careful attention to these strategies will ensure a more favorable benefit-to-risk ratio for patients undergoing hybrid medical imaging procedures.

REFERENCES

Achenbach S, Marwan M, Ropers D et al. 2010. Coronary computed tomography angiography with a consistent dose below 1 mSv using prospectively electrocardiogram-triggered high-pitch spiral acquisition. *Eur Heart J* 31:340–6.

Balter S, Hopewell JW, Miller DL, Wagner LK, Zelefsky MJ. 2010. Fluoroscopically guided interventional procedures: A review of radiation effects on patients' skin and hair. *Radiology* 254(2):326–41.

Boone JM, Strauss KJ, Cody DD, McCollough CH, McNitt-Ray MF, Toth TL, Goske MJ, Frush DP. 2011. *Size-Specific Dose Estimates (SSDE) in Pediatric and Adult Body CT Examinations*. Report of AAPM Task Group 204. College Park, MD: American Association of Physicists in Medicine.

Brenner DJ, Little JB, Sachs RK 2001. The Bystander Effect in Radiation Oncogenesis: II. A Quantitative Model. *Radiat Res* 155:402–8.

Brenner DJ, Doll R, Goodhead DT et al. 2003. Cancer risks attributable to low doses of ionizing radiation: Assessing what we really know. *Proc Natl Acad Sci USA* 100(24):13761–6.

Cardis E, Gilbert ES, Carpenter L et al. 1995. Effects of low doses and low dose rates of external ionizing radiation: Cancer mortality among nuclear industry workers in three countries. *Radiat Res* 142:117–32.

Chang SM, Nabi F, Xu J, Raza U, Mahmarian JJ. 2010. Normal stress-only versus stress/rest myocardial perfusion imaging: Similar patient mortality with reduced radiation exposure. *J Am Coll Cardiol* 55(3):221–30.

Codman EA. 1896. The cause of burns from x-rays (letter). *Boston Med Surg* 135:610–1.

Committee to Assess Health Risks from Exposure to Low Levels of Ionizing Radiation, Nuclear Radiation Studies Board, Division on Earth Life Studies, National Research Council of the National Academies. 2006. *Health Risks from Exposure to Low Levels of Ionizing Radiation: BEIR VII Phase 2*. Washington, DC: National Academies Press; 2006.

Daniel J. 1896. The x-rays. *Science* 3(67):562–3.

Denham JW, Hauer-Jensen M, Peters LJ. 2001. Is it time for a new formalism to categorize normal tissue radiation injury? *Int Radiat Oncol Biol Phys* 50(5):1105–6.

Dilsizian V, Bacharach SL, Beanlands RS et al. 2009. ASNC imaging guidelines for nuclear cardiology procedures: PET myocardial perfusion and metabolism clinical imaging. *J Nucl Cardiol* 16:651–80.

Dunlap E. 1942. Effects of radiation on blood and hemopoietic tissues, including spleen, thymus and lymph nodes. *Arch Pathol* 34:562–608.

Einstein AJ. 2012. Effects of radiation exposure from cardiac imaging: How good are the data? *J Am Coll Cardiol* 59(6):553–65.

Einstein AJ, Moser KW, Thompson RC, Cerqueira MD, Henzlova MJ. 2007. Radiation dose to patients from cardiac diagnostic imaging. *Circulation* 116(11):1290–305.

Einstein AJ, Johnson LL, Bokhari S, Son J, Thompson RC, Bateman TM, Hayes SW, Berman DS. 2010. Agreement of visual estimation of coronary artery calcium from low-dose CT attenuation correction scans in hybrid PET/CT and SPECT/CT with standard Agatston score. *J Am Coll Cardiol* 56(23):1914–21.

Einstein AJ, Knuuti J. 2012. Cardiac imaging: Does radiation matter? *Eur Heart J* 33:573–8.

Flotats A, Knuuti J, Gutberlet M, Marcassa C, Bengel FM, Kaufmann PA, Rees MR, Hesse B. 2011. Hybrid cardiac imaging: SPECT/CT and PET/CT. A joint position statement by the European Association of Nuclear Medicine (EANM), the European Society of Cardiac Radiology (ESCR) and the European Council of Nuclear Cardiology (ECNC). *Eur Nucl Med Molec Imaging* 38:201–12.

Fry SA. 1998. Studies of U.S. radium dial workers: An epidemiological classic. *Radiat Res* 150(Suppl.):S21–S29.

Gould KL, Pan T, Loghin C, Johnson NP, Sdringola S. 2008. Reducing radiation dose in rest-stress cardiac PET/CT by single poststress cine CT for attenuation correction: Quantitative validation. *J Nucl Med* 49:738–45.

Hall EJ. 2003. The bystander effect. *Health Phys* 85(1):31–5.

Herzog BA, Husmann L, Buechel RR et al. 2011. Rapid cardiac hybrid imaging with minimized radiation dose for accurate non-invasive assessment of ischemic coronary artery disease. *Int Cardiol* 153(1):10–13.

Hill MA. 2004. The variation in biological effectiveness of x-rays and gamma rays with energy. *Radiat Prot Dosimetry* 112(4):471–81.

Hricak H, Choi BI, Scott AM, Sugimura K, Muellner A, von Schulthess GK, Reiser MF, Graham MM. Dunnick NR, Larson SM 2010. Global trends in hybrid imaging. *Radiology* 257(2):498–506.

Huda W, Ogden KM, Khorasani MR. 2008. Converting dose–length product to effective dose at CT. *Radiology* 248(3):995–1003.

Hunter N, and Muirhead CR. 2009. Review of relative biological effectiveness dependence on linear energy transfer for low-LET radiations. *J Radiol Prot* 29(1):5–21.

Hunter C, Ziadi M, Etele J et al. 2010. New effective dose estimates for rubidium-81 based on dynamic PET/CT imaging in humans. *J Nucl Med* 51(supplement 2):1429.

Husmann L, Herzog BA, Gaemperli O et al. 2009. Diagnostic accuracy of computed tomography coronary angiography and evaluation of stress-only single-photon emission computed tomography/computed tomography hybrid imaging: Comparison of prospective electrocardiogram-triggering vs. retrospective gating. *Eur Heart J* 30:600–7.

International Commission on Radiological Protection (ICRP). 1977. Recommendations of the International Commission on Radiological Protection. ICRP Publication 26. *Ann ICRP* 1(3).

International Commission on Radiological Protection (ICRP). 1988. Radiation dose to patients from radiopharmaceuticals. ICRP Publication 53. *Ann ICRP* 18(1–4):1–388.

International Commission on Radiological Protection (ICRP). 1991. Recommendation of the International Commission on Radiological Protection. ICRP Publication 60. *Ann ICRP* 21(1–3).

International Commission on Radiological Protection (ICRP). 1998. Radiation dose to patients from radiopharmaceuticals (addendum 2 to ICRP Publication 53): ICRP Publication 80. *Ann ICRP* 28(3):1–123.

International Commission on Radiological Protection (ICRP). 2003. Relative biological effectiveness, radiation weighting and quality factor. ICRP Publication 92. *Ann ICRP* 33(4).

International Commission on Radiological Protection (ICRP). 2005. Low-dose extrapolation of radiation-related cancer risk. ICRP Publication 99. *Ann ICRP* 2005;35(4).

International Commission on Radiological Protection (ICRP). 2007. The 2007 recommendations of the International Commission on Radiological Protection. ICRP Publication 103. *Ann ICRP* 37(2–4):1–332.

International Commission on Radiological Protection (ICRP). 2008. Radiation dose to patients from radiopharmaceuticals (a third addendum to ICRP Publication 53). ICRP Publication 106. *Ann ICRP* 38(1–2):1–198.

International Commission on Radiological Protection (ICRP). 2012. ICRP statement on tissue reactions and early and late effects of radiation in normal tissues and organs—Threshold doses for tissue reactions in a radiation protection context. ICRP Publication 118. *Ann ICRP* 41(1/2):1–322.

Javadi M, Mahesh M, McBride G et al. 2008. Lowering radiation dose for integrated assessment of coronary morphology and physiology: First experience with step-and-shoot CT angiography in a rubidium 82 PET-CT protocol. *J Nucl Cardiol* 15:783–90.

Jones C. 2005. A review of the history of US radiation protection regulations, recommendations, and standards. *Health Phys* 88(2):105–24.

Kajander S, Ukkonen H, Sipila H, Teras M, Knuuti J. 2009. Low radiation dose imaging of myocardial perfusion and coronary angiography with a hybrid PET/CT scanner. *Clin Physiol Funct Imaging* 29:81–8.

Koenig TR, Wolff D, Mettler FA, Wagner LK. 2001. Skin injuries from fluoroscopically guided procedures: Part 1, characteristics of radiation injury. *AJR Am Roentgenol* 177:3–11.

Land CE. 1980. Estimating cancer risks from low doses of ionizing radiation. *Science* 209:1197–203.

Leipsic J, LaBounty TM, Heilbron B et al. 2010. Estimated radiation dose reduction using adaptive statistical iterative reconstruction in coronary CT angiography: The ERASIR Study. *AJR Am Roentgenol* 195: 655–60.

Little MP. 2001a. Comparison of the risks of cancer incidence and mortality following radiation therapy for benign and malignant disease with the cancer risks observed in the Japanese A-bomb survivors. *Int Radiat Biol* 77:431–64.

Little MP. 2001b. Cancer after exposure to radiation in the course of treatment for benign and malignant disease. *Lancet Oncol* 2:212–20.

Little MP, Wakeford R, Tawn J, Bouffler SD, de Gonzalez AB. 2009. Risks associated with low doses and low dose rates of ionizing radiation: Why linearity may be (almost) the best we can do. *Radiology* 251:6–12.

Mathews JD, Forsythe AV, Brady Z et al. 2013. Cancer risk in 680 000 people exposed to computed tomography scans in childhood or adholescene: Data linkage study of 11 million Australians. *Br Med J* 346:f2360.

Mercuri M, Rehani MM, Einstein AJ. 2012. Tracking patient radiation exposure: Challenges to integrating nuclear medicine with other modalities. *J Nucl Cardiol* 19:895–900.

Mettler FA, Huda W, Yoshizumi TT, Mahesh M. 2008. Effective doses in radiology and diagnostic nuclear medicine: A catalog. *Radiology* 248:254–63.

Mettler FA, Bhargavan M, Faulkner K et al. 2009. Radiologic and nuclear medicine studies in the United States and worldwide: Frequency, radiation dose, and comparison with other radiation sources 1950–2007. *Radiology* 253:520–31.

Miller RW. 1995. Delayed effects of external radiation exposure: A brief history. *Radiat Res* 144(2):160–9.

National Council on Radiation Protection and Measurements (NCRP 2001). *Evaluation of the Linear-Nonthreshold Dose-Response Model for Ionizing Radiation. Report No. 136.* Bethesda, MD: NCRP, 2001.

National Council on Radiation Protection and Measurements (NCRP 160). *Ionizing Radiation Exposure of the Population of the United States: 2006. NCRP Report No. 160.* Bethesda, MD: National Council on Radiation Protection and Measurements, 2009.

National Council on Radiation Protection and Measurements (NCRP 2010). *Radiation Dose Management for Fluoroscopically-Guided Interventional Medical Procedures. NCRP Report No. 168.* Bethesda, MD: NCRP, 2010.

Nelson RC, Feuerlein S, Boll DT. 2011. New iterative reconstruction techniques for cardiovascular computed tomography: How do they work, and what are the advantages and disadvantages? *J Cardiovasc Comput Tomogr* 5(5):286–92.

Olivieri G, Bodycote J, Wolff S. 1984. Adaptive responses of human lymphocytes to low concentrations of radioactive thymidine. *Science* 223(4636):594–7.

Padovani R, Bernardi G, Quai E et al. 2005. Retrospective evaluation of occurrence of skin injuries in interventional cardiac procedures. *Radiat Prot Dosimetry* 117(1–3):247–250.

Pazhenkottil AP, Herzog BA, Husmann L et al. 2010. Non-invasive assessment of coronary angiography and SPECT: A novel dose-saving fast-track algorithm. *Eur Nucl Med Molec Imaging* 37(3):522–7.

Pazhenkottil AP, Nkoulou RN, Ghadri JR et al. 2011. Impact of cardiac hybrid single-photon emission computed tomography/computed tomography imaging on choice of treatment strategy in coronary artery disease. *Eur Heart J* 32(22):2824–9.

Pearce MS, Salotti JA, Little MP et al. 2012. Radiation exposure from CT scans in childhood and subsequent risk of leukaemia and brain tumours: A retrospective cohort study. *Lancet* 380:499–505.

Preuss R, Weise R, Lindner O, Fricke E, Fricke H, Burchert W. 2008. Optimization of protocol for low dose CT-derived attenuation correction in myocardial perfusion SPECT imaging. *Eur Nucl Med Molec Imaging* 35(6):1133–41.

Rispler S, Keidar Z, Ghersin E et al. 2007. Integrated single-photon emission computed tomography and computed tomography coronary angiography for the assessment of hemodynamically significant coronary artery lesions. *J Am Coll Cardiol* 49(10):1059–67.

Ron E. 1998. Ionizing radiation and cancer risk: Evidence from epidemiology. *Radiation Research* 150(5):S30–S41.

Ron E. 2003. Cancer risks from medical radiation. *Health Phys* 85(1):47–59.

Schuhbaeck A, Achenbach S, Layritz C et al. 2013. Image quality of ultra-low radiation exposure coronary CT angiography with an effective dose <0.1 mSv using high-pitch spiral acquisition and raw data-based iterative reconstruction. *Eur Radiol* 23:597–606.

Senthamizhchelvan S, Bravo PE, Esaias C et al. 2010. Human biodistribution and radiation dosimetry of 82Rb. *J Nucl Med* 51:1592–99.

Small GR, Wells G, Schindler T, Chow BJ, Ruddy TD. 2013. Advances in cardiac SPECT and PET imaging: Overcoming the challenges to reduceradiation exposure and improve accuracy. *Can Cardiol* 29:275–84.

Smith-Bindman R, Miglioretti D, Johnson E, Lee C, Feigelson H, Flynn M, Greenlee R, Kruger R, Hornbrook M, Roblin D. 2012. Use of diagnostic imaging studies and associated radiation exposure for patients enrolled in large integrated health care systems, 1996–2010. *J Am Med Assoc* 307(22):2400–9.

Stone HB, Coleman CN, Anscher MS, McBride WH. 2003. Effects of radiation on normal tissue: Consequences and mechanisms. *Lancet Oncol* 4:529–36.

Trattner S, Pearson GDN, Chin C, Cody DD, Gupta R, Hess CP, Kalra MK, Kofler JM, Krishnam MS, Einstein AJ. 2014. Standardization and optimization of ct protocols to achieve low dose. *J Am Coll Radiol* 11(3):271–8.

United Nations Scientific Committee on the Effects of Atomic Radiation. *Sources, Effects and Risks of Ionizing Radiation. Exposures from Medical Uses of Radiation, Annex c. 1988 Report to the General Assembly with Annexes.* New York: United Nations, 1988.

United Nations Scientific Committee on the Effects of Atomic Radiation. *Sources and Effects of Ionizing Radiation. Medical Radiation Exposures, Annex A. 2008 Report to the General Assembly with Annexes.* New York: United Nations, 2008.

UNSCEAR 2010 Report. Summary of low-dose radiation effects on health. Report of the United Nations Scientific Committee on the Effects of Atomic Radiation 2010. Available at: www.unscear.org/docs /reports/2010/UNSCEAR_2010_Report_M.pdf. Accessed November 13, 2013.

Upton AC. 2003. The state of the art in the 1990's: NCRP Report no. 136 on the scientific basis for linearity in the dose-response relationship for ionizing radiation. *Health Phys* 85(1):15–22.

U.S. Food and Drug Administration. *Public Health Advisory: Avoidance of Serious X-ray-Induced Skin Injuries to Patients during Fluoroscopy-Guided Procedures.* Rockville, MD: Center for Devices and Radiological Health, 1994.

Vano E, Goicolea J, Galvan C, Gonzalez L, Meiggs L, Ten JI, Macaya C. 2001. Skin radiation injuries in patients following repeated coronary angioplasty procedures. *Br J Radiol* 74:1023–31.

Vano E, Gonzalez L, Guibelalde E, Aviles P, Fernandez M, Prieto C, Galvan C. 2005. Evaluation of risk of deterministic effects in fluoroscopically guided procedures. *Radiation Protection Dosimetry* 117(1–3):190–4.

Zhou H, Randers-Pehrson G, Waldren CA, Vannais D, Hall EJ, Hei TK. 2000. Induction of a bystander mutagenic effect of alpha particles in mammalian cells. *Proc Natl Acade Sci USA* 97(5):2099–104.

Future directions for the development and application of hybrid cardiovascular imaging

ALBERT J. SINUSAS

20.1 INTRODUCTION

This chapter will summarize the overall goals for developing hybrid imaging systems, multimodality imaging probes, and integrated hybrid imaging approaches for the evaluation of cardiovascular disease. There may be advantages and challenges associated with any new hybrid technology, which will require careful evaluation and review. A systematic approach for evaluation of new hybrid technologies will be discussed relevant to assessing the advantage and impact of the new approaches for research and clinical purposes, with the ultimate goal of improvement of health care outcomes. The future implementation and acceptance of any new technology require education of all relevant parties, along with culture change, and organizational and institutional commitment. These advancements in multimodality imaging technology should lead to improved image quantification and diagnostic accuracy, thereby promoting "precision medicine," and a more personalized approach to healthcare delivery.

20.2 HYBRID IMAGING—TECHNOLOGICAL ADVANTAGE

Multimodality hybrid imaging aims to take advantage of the different physical properties of image formation for each imaging modality to create an image set with complementary (or even synergistic) information. The aim is to produce a superior image set for either research purposes, clinical applications, or a combination of both. For research applications, hybrid imaging methodology might simply improve quantification of a targeted molecular probe or imaging approach or provide anatomical localization, as would be the case for hybrid single-photon emission computed tomography/computed tomography (CT), positron emission tomography (PET)/CT, or PET/magnetic resonance (MR) imaging. Alternatively, the hybrid imaging technology may provide new insight into a molecular pathway or pathophysiological process or provide a link between molecular processes and associated anatomical or physiological events. Clinical application of hybrid imaging technology may result in improved diagnostic accuracy and prognostication or facilitate monitoring of disease process or therapeutic intervention. For example, improved performance for risk estimation by using simultaneous PET and cardiovascular magnetic resonance (CMR) imaging with incorporation of cardiac or respiratory motion correction (Petibon et al. 2017) or improved diagnostic categorization of patients according to differential patterns of fused PET-CMR images as in the case of evaluation of cardiac sarcoidosis or amyloidosis (Quail and Sinusas 2017).

Thus, the success of any multimodality imaging approach is dependent on acquiring information that is more valuable than that from individual components, or even the composite of the components. Ideally, one should obtain unique information that can be derived only using a truly integrated hybrid technology, as opposed to acquiring images on two separate imaging systems and fusion of the datasets following acquisition.

20.3 HYBRID IMAGING—TECHNOLOGICAL CHALLENGES

In contrast, if the information from different imaging modalities provides the same information simply in different image spaces, redundancy is introduced. Combining technologies in a single hybrid device could also potentially reduce the efficacy of any given technology. In the example of PET-CMR imaging, the addition of an MR magnet in the PET imaging space may limit subject eligibility for PET scanning, increase operational costs, prohibit the use of generator produced PET radioisotopes within the imaging room, and also complicate attenuation correction of PET scans (Quail and Sinusas 2017). As such, the creation of a hybrid device should provide a true advantage over fusion of images acquired on separate devices.

20.4 HYBRID IMAGING PROBES

The application of hybrid imaging will require the parallel development of hybrid multimodality imaging probes. As outlined in Chapters 10 and 11 of this book and prior review articles (Stendahl and Sinusas 2015a, 2015b), the development of these multimodality probes offers advantages for validation of new molecularly targeted probes and hybrid imaging technologies and may also provide complementary and unique information. Since biological processes are complex and involve the interaction of multiple signaling pathways and physiological changes, the ability to track different processes in parallel offers a unique advantage. Taking advantage of the individual strength of different modality probes, multimodality imaging probes provide a powerful mechanism to enhance the assessment of critical pathophysiological processes and therapies. The visualization and colocalization of multiple molecular targets or the evaluation of molecular and physiological biomarkers with structural changes will be critical in understanding these complex disease processes at the molecular level, could lead to new therapeutics, and will enable the evaluation of novel therapeutics. These hybrid probes, in combination with hybrid imaging technology, have facilitated the emergence of theranostics, which are characterized by the integration of diagnostic probes with targeted therapeutics.

20.5 IMPROVED IMAGE QUANTIFICATION

Hybrid imaging systems often bring together complimentary technologies that allow for improved image quantification, with the corrections for image resolution, cardiac and respiratory motion, scatter between organs, tissue attenuation, and partial volume errors (Slomka 2017). Many hybrid imaging systems bring together components that have complementary strengths, for example, high resolution versus high sensitivity. The improvement in image quantification may be through anatomical colocalization with the correction for partial volume errors or the use of kinetic modeling. More specifically, the kinetic modeling of a tracer may be improved with better identification and characterization of specific tissue compartments.

20.6 SWOT ANALYSIS OF NEW HYBRID TECHNOLOGY

When considering implementation of a new hybrid imaging technology, one needs to take into account a structured *strengths, weaknesses, opportunities*, and *threats* (SWOT) analysis, which involves evaluation of the relevant internal *strengths* and *weaknesses* of any given technology, as well as external *opportunity* and *threats* associated with competitive technologies, or the use of the technologies separately (Ciarmiello and Hinna 2016). This analysis needs to involve all the stakeholders: promoters of technology, researchers, clinicians, and decision makers. Only recently has structured SWOT analysis been introduced in healthcare systems in evaluation of new technologies (van Wijngaarden, Scholten, and van Wijk 2012; Kashyap et al. 2013).

A scoring system can be used to assign importance of factors that might present new opportunities or threats based on the likelihood of impact on current practices within an organization. Similarly, strengths and weaknesses can be assessed using a scoring system that allows the factors to be identified according to their significance (major, minor, neutral) and level of importance (high, medium, low) (Ciarmiello and Hinna 2016). Again, any careful structured quantitative analysis of the relative advantages and disadvantages of a hybrid technology must involve all the stakeholders.

20.7 CULTURE LAG

Introduction of any new hybrid systems may lead to anxiety, confusion, and the inefficient deployment of the new resources. Culture lag is considered an important aspect of social change and evolves, accumulating as a result of invention, discovery, and dispersion (Brinkman and Brinkman 1997). Any delay in developing the appropriate knowledge and skills needed to optimally implement a technology may impact on the efficient use of new hybrid imaging resources within the research or healthcare environment.

The implementation of any new hybrid technology may involve the emergence of new professional identity and intercollegiate interactions. The advancement of a new technology may necessitate the establishment of new professional training pathways and creation of professional society guidelines for training and appropriate utilization, or even creation of new professional organizations or societies. This might lead to changes in certification policy and professional licensing, which may result in an occupational shift and domain ownership. New technology can also require and result in changes in state or federal policy, as the shift in technology may have relevance to regulatory or licensing authorities (Griffiths 2016).

20.8 CHANGE IN PRACTICE

Introduction of new hybrid equipment will foster new imaging techniques and approaches that will lead to new clinical partnerships and clinical pathways and may eventually result in a change in clinical practice (see Figure 20.1). This transition may require restructuring of the environment and creation of new operational communities. To ensure appropriate adaptation and utilization of a new hybrid technology, new quality control measures and appropriate use criteria will need to be established. Also, any new hybrid technology will need to be evaluated in the context of existing technologies in balance with the needs of the patients and support of local and federal regulatory agencies. Without financial reimbursement for use of an emerging hybrid technology, the technology is not likely to gain widespread utilization and will remain a research instrument at select institutions.

For any changes in practice to occur, there needs to be education and training of imaging technologists and physicians, and this needs to be followed by outreach and education of the clinical community. The outreach must be directed to clinical subspecialty organizations involved in establishing clinical guidelines

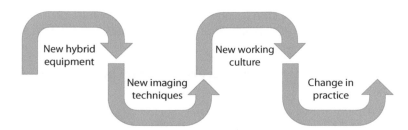

Figure 20.1 Steps in the transition of hybrid technology to clinical practice. The introduction of new hybrid imaging technology requires: optimization of application and approach, retraining of community, change in work culture, and redesign of clinical work flow and practice.

and appropriate use criteria and followed by outreach to patient advocacy groups. This final stage of outreach must incorporate with patient education.

20.9 FUTURE DIRECTIONS

The future of hybrid cardiovascular imaging will likely involve the delivery of novel theranostics and the integration of multimodality probes and hybrid imaging technology with image-guided therapeutic interventions. One future goal would be to bring molecular and/or physiological images and coregistered anatomical images generated on a hybrid imaging system into the anatomic space of a cardiovascular interventional suite or a hybrid operating room in order to direct therapies using image guidance. The approach might involve the intra-arterial or intramyocardial delivery of imageable therapeutics (theranostics) under direct image guidance. This could involve the use of external hybrid imaging systems or multi-modality hybrid imaging catheters, or both types of hybrid systems. This image-guided approach would allow for tracking of the initial delivery and retention of therapies whether they are biochemical, cellular, genetic, or polymer based. New therapies may involve the local delivery of therapeutic agents via engineered polymers that are molecularly targeted to the area and slowly release the therapy based on the local environment. These polymers can be bioresponsive so that the degradation of the polymer and slow and sustained release of the therapy is based on the local environmental conditions (i.e., pH or enzyme activity) (Purcell et al. 2014).

The advancements in multimodality cardiovascular imaging technology and probes outlined in this textbook should lead to improved image quantification and diagnostic accuracy, thereby promoting a more precise and personalized approach to healthcare delivery.

REFERENCES

Brinkman R, and J Brinkman. 1997. "Cultural lag: Conception and theory." *Int J Soc Econ* 24 (6):609–627.
Ciarmiello A, and L Hinna. 2016. "SWOT analysis and stakeholder engagement for comparative evaluation of hybrid molecular imaging modalities." In *PET-CT and PET-MRI in Neurology*, edited by Mansi L, Ciarmiello A, 271–282. Switzerland: Springer International.
Griffiths M. 2016. "The impact of new hybrid imaging technology on the nuclear medicine workforce: Opportunities and challenges." International Conference on Nuclear Medicine and Radiation Therapy, Cologne, Germany, July 14–15, 2016.
Kashyap R, M Dondi, D Paez, and G Mariani. 2013. "Hybrid imaging worldwide challenges and opportunities for the developing world: A report of the technical meeting organized by IAEA." *Semin Nucl Med* 43 (3):208–223.
Petibon Y, NJ Guehl, TG Reese, B Ebrahimi, MD Normandin, TM Shoup, NM Alpert, G El Fakhri, and J Ouyang. 2017. "Impact of motion and partial volume effects correction on PET myocardial perfusion imaging using simultaneous PET-MR." *Phys Med Biol* 62 (2):326–343. doi: 10.1088/1361-6560/aa5087.

Purcell BP, D Lobb, MB Charati, SM Dorsey, RJ Wade, KN Zellars, H Doviak, S Pettaway, CB Logdon, JA Shuman, PD Freels, JH Gorman, 3rd, RC Gorman, FG Spinale, and JA Burdick. 2014. "Injectable and bioresponsive hydrogels for on-demand matrix metalloproteinase inhibition." *Nat Mater* 13 (6):653–661. doi: 10.1038/nmat3922.

Quail MA, and AJ Sinusas. 2017. "PET-CMR in heart failure—Synergistic or redundant imaging?" *Heart Fail Rev*. doi: 10.1007/s10741-017-9607-6.

Slomka P. 2017. "Hybrid quantitative imaging: Will it enter clinical practice?" *J Nucl Cardiol* 1–3. doi: 10.1007 /s12350-017-0868-1.

Stendahl JC, and AJ Sinusas. 2015a. "Nanoparticles for cardiovascular imaging and therapeutic delivery, Part 1: Compositions and features." *J Nucl Med* 56 (10):1469–1475. doi: 10.2967/jnumed.115.160994.

Stendahl JC, and AJ Sinusas. 2015b. "Nanoparticles for cardiovascular imaging and therapeutic delivery, Part 2: Radiolabeled probes." *J Nucl Med* 56 (11):1637–1641. doi: 10.2967/jnumed.115.164145.

van Wijngaarden JD, GR Scholten, and KP van Wijk. 2012. "Strategic analysis for health care organizations: The suitability of the SWOT-analysis." *Int J Health Plann Manage* 27 (1):34–49.

Index

T - #0470 - 071024 - C476 - 254/178/21 - PB - 9780367781743 - Gloss Lamination